T0180601

Advances in Intelligent Systems and Computing

Volume 1040

The series "Advances in Intelligent Systems and Computing" contains publications on theory, applications, and design methods of Intelligent Systems and Intelligent Computing. Virtually all disciplines such as engineering, natural sciences, computer and information science, ICT, economics, business, e-commerce, environment, healthcare, life science are covered. The list of topics spans all the areas of modern intelligent systems and computing such as: computational intelligence, soft computing including neural networks, fuzzy systems, evolutionary computing and the fusion of these paradigms, social intelligence, ambient intelligence, computational neuroscience, artificial life, virtual worlds and society, cognitive science and systems, Perception and Vision, DNA and immune based systems, self-organizing and adaptive systems, e-Learning and teaching, human-centered and human-centric computing, recommender systems, intelligent control, robotics and mechatronics including human-machine teaming, knowledge-based paradigms, learning paradigms, machine ethics, intelligent data analysis, knowledge management, intelligent agents, intelligent decision making and support, intelligent network security, trust management, interactive entertainment, Web intelligence and multimedia.

The publications within "Advances in Intelligent Systems and Computing" are primarily proceedings of important conferences, symposia and congresses. They cover significant recent developments in the field, both of a foundational and applicable character. An important characteristic feature of the series is the short publication time and world-wide distribution. This permits a rapid and broad dissemination of research results.

** Indexing: The books of this series are submitted to ISI Proceedings, EI-Compendex, DBLP, SCOPUS, Google Scholar and Springerlink **

More information about this series at http://www.springer.com/series/11156

Pradeep Kumar Mallick · Valentina Emilia Balas ·
Akash Kumar Bhoi · Gyoo-Soo Chae
Editors

Cognitive Informatics and Soft Computing

Proceeding of CISC 2019

 Springer

Editors
Pradeep Kumar Mallick
School of Computer Engineering
Kalinga Institute of Industrial Technology
(KIIT) Deemed to be University
Bhubaneswar, Odisha, India

Akash Kumar Bhoi
Department of Electrical and Electronics
Engineering, Sikkim Manipal Institute
of Technology
Sikkim Manipal University
Rangpo, India

Valentina Emilia Balas
Faculty of Engineering
Aurel Vlaicu University of Arad
Arad, Romania

Gyoo-Soo Chae
Division of Information
and Communication
Baekseok University
Cheonan-si, Ch'ungch'ong-namdo
Korea (Republic of)

ISSN 2194-5357   ISSN 2194-5365 (electronic)
Advances in Intelligent Systems and Computing
ISBN 978-981-15-1450-0   ISBN 978-981-15-1451-7 (eBook)
https://doi.org/10.1007/978-981-15-1451-7

This Springer imprint is published by the registered company Springer Nature Singapore Pte Ltd.
The registered company address is: 152 Beach Road, #21-01/04 Gateway East, Singapore 189721, Singapore

Committee

Co-conveners

Dr. P. Gopi, Associate Professor, Department of EEE, AITSR

Dr. Sandeep Kumar Satapathy, Department of CSE, VBIT, Hyderabad

P. Bhaskara Prasad, Associate Professor, Department of EEE, AITSR

S. Muqthiar Ali, Associate Professor, Department of EEE, AITSR

Dr. Akash Kumar Bhoi, Assistant Professor, Department of EEE, SMIT

Reviewer Board:

Dr. Debahuti Mishra, ITER, SOA University, Odisha

Dr. Sachidananda Dehury, FM University, Odisha

Dr. Brojo Kishore Mishra, CVRCE, Odisha

Dr. Sandip Vijay, ICFAI University, Dehradun

Dr. Shruti Mishra, Department of CSE, VBIT, Hyderabad

Dr. Sashikala Mishra, IIIT Pune

Dr. Ebrahim Aghajari, Islamic Azad University of Ahvaz, IRAN

Dr. Sudhakar Mande, DBIT, Mumbai, India

International Committee

Dr. Atilla ELÇİ, Aksaray University, Turkey

Dr. Hongyan Yu, Shanghai Maritime University, Shanghai

Dr. Benson Edwin Raj, Fujairah Women's College, Fujairah, UAE

Dr. Mohd. Hussain, Islamic University of Madinah, Saudi Arabia

Dr. Vahid Esmaeelzadeh, Iran University of Science and Technology, Narmak, Tehran, Iran

Dr. Avinash Konkani, University of Virginia Health System, Virginia, USA

Dr. Yu-Min Wang, National Chi Nan University, Taiwan

Dr. Ganesh R. Naik, University of Technology, Sydney, Australia

Dr. Steve S. H. Ling, University of Technology, Sydney, Australia

Dr. Hak-Keung Lam, King's College London, UK

Dr. Frank H. F. Leung, Hong Kong Polytechnic University, Hong Kong

Dr. Yiguang Liu, Sichuan University, China

Dr. Jasni Mohamad Zain, Professor, UMP, Malaysia

Dr. D. N. Subbareddy, Professor, Korea

Dr. Mohd Al Azawi, HOD, Department of CSE, OCMT, Oman

Dr. Mastan Mohamad, HOD, University of Oman, Oman

National Committee

Dr. Kishore Sarawadekar, IIT-BHU, Varanasi, India

Dr. T. Kishore Kumar, NIT Warangal, Warangal, AP, India

Dr. Anil Kumar Vuppala, IIIT Hyderabad, India

Dr. Ganapati Panda, IIT Bhubaneshwar, Odisha

Dr. Preetisudha Meher, NIT Arunachal Pradesh

Dr. C. Subramani, IIT Roorkee

Dr. R. Arthi, Department of ECE, VBIT, Hyderabad

Dr. Brahma Reddy, Department of ECE, VBIT, Hyderabad
Dr. Sachidananda Dehury, FM University, Odisha
Dr. Brojo Kishore Mishra, CVRCE, Odisha
Dr. Inderpreet Kaur, Chandigarh University
Dr. R. Gunasundari, PEC, Puducherry, India
Dr. Ragesh G., SCE, Kuttukulam Hills, Kottayam, India
Dr. Debashish Jena, NITK, India
Dr. N. P. Padhy, IIT Roorkee
Dr. Sashikala Mishra, IIIT Pune
Dr. Subhendu Pani, OEC, Odisha
Prof. Aksash Kumar Bhoi, SMIT, Sikkim
Dr. J. Arputha Vijaya Selvi, KCE, Tamil Nadu
Dr. Punal M. Arabi, ACSCE, Bangalore
Dr. Mihir Narayan Mohanty, ITER, SOA University
Dr. K. Krishna Mohan, Professor, IIT Hyderabad
Dr. G. N. Srinivas, Professor, JNTUCE, Hyderabad
Dr. M. Padmavathamma, Professor, SVUCCMIS, SVU, Tirupati
Dr. S. Basava Raju, Regional Director, VTU, Karnataka
Dr. R. V. Raj Kumar, Professor, IIT Kharagpur
Dr. Allam Appa Rao, Chairman, NTTTR, Chennai
Dr. V. V. Kamakshi Prasad, COE, JNTUCE, Hyderabad
Dr. M. Surya Kalavathi, Professor, JNTUCE, Hyderabad
Dr. A. Govardhan, Rector, JNTUH, Hyderabad
Dr. K. Siva Kumar, Professor, IIT Hyderabad
Dr. S. Soumitra Kumar Sen, IIT Kharagpur
Dr. N. V. Ramana, VTU, Karnataka
Dr. D. V. L. N. Somayajulu, Professor, NIT Warangal
Dr. Atul Negi, Professor, HCU, Hyderabad
Dr. P. Sateesh Kumar, Associate Professor, IIT Roorkee
Dr. C. Sashidhar, CE, JNTUA
Dr. M. Sashi, NIT Warangal

Organizing Committee

Mr. B. Murali Mohan, Associate Professor, Department of EEE, AITSR
Mr. P. Suresh Babu, Associate Professor, Department of EEE, AITSR
Mr. K. Harinath Reddy, Associate Professor, Department of EEE, AITSR
Mrs. S. Sarada, Associate Professor, Department of EEE, AITSR
Mr. C. Ganesh, Associate Professor, Department of EEE, AITSR
Mr. R. Madhan Mohan, Associate Professor, Department of EEE, AITSR
Mr. M. Pala Prasad Reddy, Associate Professor, Department of EEE, AITSR
Mr. L. Baya Reddy, Associate Professor, Department of EEE, AITSR
Mr. M. Ramesh, Associate Professor, Department of EEE, AITSR
Mr. N. Sreeramula Reddy, Associate Professor, Department of EEE, AITSR
Mr. D. Sai Krishna Kanth, Associate Professor, Department of EEE, AITSR
Mr. S. Srikanta Deekshit, Associate Professor, Department of EEE, AITSR

Mr. P. Ravindra Prasad, Associate Professor, Department of EEE, AITSR
Mr. M. Sai Sandeep, Associate Professor, Department of EEE, AITSR
Mr. B. Madhusudhan Reddy, Associate Professor, Department of EEE, AITSR
Mr. P. Pamuletaiah, Associate Professor, Department of EEE, AITSR
Mr. S. Sagar Reddy, Associate Professor, Department of EEE, AITSR
Mr. A. Bhaskar, Associate Professor, Department of EEE, AITSR
Mr. P. Shahir Ali Khan, Associate Professor, Department of EEE, AITSR
Ms. K. Reddy Prasanna, Associate Professor, Department of EEE, AITSR
Mr. M. G. Mahesh, Associate Professor, Department of EEE, AITSR

Tecnical Committee

Dr. N. Mallikarjuna Rao, Director, IQAC, AITSR
Dr. B. Abdul Rahim, Dean, Professional Bodies, AITSR
Dr. M. C. Raju, Dean, Department of R&D, AITSR
Prof. M. Subba Rao, Dean, Department of Student Welfare, AITSR
Dr. CH. Nagaraju, HOD, Department of ECE, AITSR
Dr. M. Rudra Kumar, HOD, Department of CSE, AITSR
Dr. A. Hemantha kumar, HOD, Department of ME, AITSR
Dr. Y. Sriramulu, HOD, Department of CE, AITSR
Dr. K. Prasanna, HOD, Department of IT, AITSR
Dr. B. B. N. Prasad, HOD, Department of H&S, AITSR
Dr. C. Madan Kumar Reddy, HOD, Department of MCA, AITSR
Dr. P. Subramanyam, HOD, Department of MBA, AITSR

Preface

International Conference on *Cognitive Informatics and Soft Computing* (CISC-2019) was held at Annamacharya Institute of Technology & Sciences, Rajampet, from 9–10 April 2019. The outcomes of CISC-2019 are achieved with the book volume *Cognitive Informatics and Soft Computing*, which covers the fields *like* Cognitive Informatics, Computational Intelligence, Advanced Computing and Hybrid Intelligent Models and Applications. All the selected papers which were presented during the conference have been screened through a double-blind peer review with the support of review board members along with the national and international committee members. We would like to acknowledge the governing members and leaders of the Annamacharya Institute of Technology & Sciences for providing the infrastructure and venue to organize high-quality conference. Moreover, we would like to extend our sincere gratitude to the reviewers, technical committee members and professionals from the national and international forums for extending their great support during the conference.

Bhubaneswar, India	Dr. Pradeep Kumar Mallick
Arad, Romania	Dr. Valentina Emilia Balas
Rangpo, India	Dr. Akash Kumar Bhoi
Cheonan-si, Korea (Republic of)	Dr. Gyoo-Soo Chae

Contents

About the Editors

Pradeep Kumar Mallick is currently working as Associate Professor in the School of Computer Engineering, Kalinga Institute of Industrial Technology (KIIT) Deemed to be University, Odisha, India. He has completed his Post-Doctoral Fellow (PDF) in Kongju National University, South Korea; Ph.D. from Siksha 'O' Anusandhan University; M.Tech. (CSE) from Biju Patnaik University of Technology (BPUT); and M.C.A. from Fakir Mohan University, Balasore, India. Besides academics, he is also involved in various administrative activities, Member of Board of Studies, Member of Doctoral Research Evaluation Committee, Admission Committee, etc. His areas of research include Algorithm Design and Analysis, and Data Mining, Image Processing, Soft Computing, and Machine Learning. He has published several book chapters and papers in national and international journals and conference proceedings.

Valentina Emilia Balas is currently an Associate Professor at the Department of Automatics and Applied Software, "Aurel Vlaicu" University of Arad (Romania). She holds a Ph.D. in Applied Electronics and Telecommunications from the Polytechnic University of Timisoara (Romania). The author of more than 160 research papers in peer-reviewed journals and international conference proceedings, her research interests are in Intelligent Systems, Fuzzy Control, Soft Computing, Smart Sensors, Information Fusion, Modeling, and Simulation. She is the Editor-in-Chief of the International Journal of Advanced Intelligence Paradigms (IJAIP), serves on the Editorial Boards of several national and international journals, and is an evaluator expert for national and international projects.

Akash Kumar Bhoi completed his B.Tech. (Biomedical Engineering) at the Trident Academy of Technology (TAT), Bhubaneswar, India, and his M.Tech. (Biomedical Instrumentation) at Karunya University, Coimbatore, India, in 2009 and 2011, respectively. He has completed his Ph.D. (in Biomedical Signal Processing) at Sikkim Manipal University, Sikkim, India, he is currently serving as an Assistant Professor at the Department of Electrical and Electronics Engineering (EEE) and as a Faculty Associate in the R&D Section of Sikkim Manipal Institute of Technology (SMIT), Sikkim Manipal University. He has published several book chapters and papers in national and international journals and conference proceedings.

Gyoo-Soo Chae completed his B.Sc. (Electronics) and M.Sc. (Electrical Engineering) at Kyungpook National University, South Korea. He subsequently completed his Ph.D. at Virginia Polytechnic Institute and State University, VA, USA. Currently, he is a Professor at Baekseok University, South Korea. His research interests are in the areas of Microwave Theory, Antenna Design, and Measurement. He has filed nine patents with the Korean Patent Office, and has published numerous papers in international journals and conference proceedings.

Design of Power Efficient and High-Performance Architecture to Spectrum Sensing Applications Using Cyclostationary Feature Detection

Kadavergu Aishwarya and T. Jagannadha Swamy

Abstract Cognitive radio spectrum sensing is one of the novel techniques in wireless communications. In this process, a wide variety of techniques are available for detecting the spectrum availability to send the secondary signal frequency signals in the absences of the other primary signal frequencies. In this cognitive radio spectrum sensing, speed of operation of the network is one of the important factors for efficient data handling and transmission process. Cyclostationary feature detection is one of the efficient methods for Cognitive Radio spectrum sensing applications. The speed and power of the cyclostationary feature detection-based spectrum sensing architecture in cognitive radio networks can be improved by implementing the advanced multiplication techniques like Vedic multipliers for test statistic computing module deployed in the architecture. To detect the presence the signal over the provided Frequency band, continuous sensing of spectrum is required. This involves numerous multiplications. The proposed model with help of Vedic multipliers reduces the power consumption as well as increases the performance of the architecture. The simulation results are equated with Booth multiplier. It shows better results when it is implemented in the test statistic module. The complete design is implemented in Verilog and tested using Xilinx ISE and Xilinx Vivado tool.

Keywords Cognitive radio · Spectrum sensing · Cyclostatioary detection · Test statistic module · Booth multiplication technique · Vedic multiplier

1 Overview

As technology is growing day to day, the need for the advancement in the communication system also increasing gradually. Cognitive radio is one of the best-used technology that enhances the system capacity by allocating the free licensed spectrum

K. Aishwarya (✉) · T. Jagannadha Swamy
ECE Department, GRIET, Hyderabad, India
e-mail: aishwaryaeng123@gmail.com

T. Jagannadha Swamy
e-mail: jagan.tata@griet.ac.in

© Springer Nature Singapore Pte Ltd. 2020
P. K. Mallick et al. (eds.), *Cognitive Informatics and Soft Computing*,
Advances in Intelligent Systems and Computing 1040,
https://doi.org/10.1007/978-981-15-1451-7_1

bands of primary users to the secondary by using the spectrum sensing technique [1–3]. Sensing techniques implemented with better performance will help in the speed of operation. There are several spectrum sensing techniques. These detection techniques have their own advantages and disadvantages. Matched filter detection technique involves precise information of the target user and it consumes huge power and has high complex architecture that leads to huge cost. On the other hand, energy detection is well known for its effortlessness and ease of hardware implementation. But it is not supportive to use under low SNR conditions [4].

The main focus of this paper is on the performance of test statistic module deployed in the cyclostationary based spectrum sensing CR network. Cyclostationary sensing detection is a technique for sensing primary user broadcasts by manipulating the cyclostationary topographies of the received signal frequencies. It is best for low SNR calculation [5, 6]. In this sensing process, huge computations of multiplications are required to perform the test statistics. With the help of CORDIC, BOOTH technique, the filtering operation takes more time. So, it takes more power consumption to perform entire process. In this process, to reduce the power consumption we deployed Vedic multiplier in the design in test statistic module. After the simulations and with the obtained results, the proposed Vedic multiplier shows good results and used to improve the speed of operation and also reduces the power handling efficiency and also improves the efficiency of the entire module for spectrum sensing applications.

The remaining paper is presented as follows. In Sect. 2, discussed the system design and the importance of test statistic module in the design. In Sect. 3, the advantage of Vedic multiplier over Booth multiplier discussed. In Sect. 4 simulation results of Vedic multiplier and comparison results followed by conclusion and future scope in Sect. 5.

2 System Model

The system Model comprises of selective sampler for reconfigurability and memory efficient. The OFDM symbols in wireless communication networks involve wide range of subcarriers in the multiples of powers of two. The Memory requirement to store such subcarriers will increase automatically. The selective sampling Technique with its control unit implemented using Finite State Machine will transform the received OFDM symbols into 64 subcarriers. Thus, this is one of the techniques to reduce memory consumption [7, 8]. The communication system works based on frequency modulations. The frequency shifting is one of the important features involved in sensing techniques. This can be realized by FFT and shift filters. But they require enormous computation resources. Implementing CORDIC module in place of FFT improvises the hardware. CORDIC involves shift and add operations. And, to obtain higher clock frequencies it can be easily pipelined [8].

A signal X(n) with its time changing prospect of its autocorrelation $E[X(n)X^*(n - T)]$ is cyclic, then it is said to be cyclostationary [9]. In communication system, we mainly prefer OFDM symbols to avoid noise scattering of the original signal. Occurrence of periodic frequency samples can be examined by detecting frequency area models of autocorrelation signals [9]. The evaluation of autocorrelation can be calculated as Eq. (1).

$$R(T) = \frac{1}{N} \sum_{n=0}^{N-1} \left(x(n)x^*(n-T) \right) \tag{1}$$

As autocorrelation of OFDM signals is cyclic for interruption T, Its Fourier series extension is given as Eq. (2).

$$R(T, \alpha) = \frac{1}{N} \sum_{n=0}^{N-1} \left(x(n)x^*(n-T) \right) e^{\frac{(-j2\pi\alpha n)}{N}} \tag{2}$$

Neymann Pearson Suggestion is helpful to test the occurrence of cyclostationary signal and the test statistic is calculated as Eq. (3).

$$T_S = N \times P \times \psi^{-1} \times P^T \tag{3}$$

where P^T is the transpose of $A = [$ real $\{ \mathcal{R}(T, \alpha)\}$ img $\{ \mathcal{R}(T, \alpha)\}]$, that can be recalled as $A = [X_t(\alpha)Y_t(\alpha)]$. In the same way, ψ^{-1} is the inverse of covariance matrix ψ and it is stated as Eq. (4).

$$\psi^{-1} = \frac{1}{PS - RQ} \begin{bmatrix} P & -Q \\ -R & S \end{bmatrix} \tag{4}$$

where $P = \frac{1}{N} \sum_{n=0}^{N-1} x_t^2(n, \alpha)$;

$$Q = R = \frac{1}{N} \sum_{n=0}^{N-1} x_t(n, \alpha) * y_t(n, \alpha) \text{ and } S = \frac{1}{N} \sum_{n=0}^{N-1} y_t^2(n, \alpha)$$

$$T_\delta = N \left[\frac{X_t(\alpha)^2 * S + y_t(\alpha)^2 * P - 2 * X_t(\alpha) * y_t(\alpha) * Q}{P * S - Q^2} \right] \tag{5}$$

Sequentially, the value of the threshold T_S is computed under null hypothesis and it is given by $T_S = F_{x_2^2}^{-1}(1 - P_{FA})$ where P_{FA} is the prospect of False Alarm. This threshold value is pre-calculated and is stored in a control unit. Finally, the output of this recognition process is found by calculating the rate of T_δ as stated in Eq. (5) and linking with the pre-calculated threshold value T_S [10].

Figure 1 describes the system model for cyclostationary based Cognitive Radio Architecture. The autocorrelator module is helpful in detecting the cyclostationary signal with its periodicity. The selective sampling technique is helpful in reducing memory requirements. CORDIC module is used in place of twiddle factor calculation in FFT to reduce hardware [9, 10]. Later, the data samples from autocorrelator and the

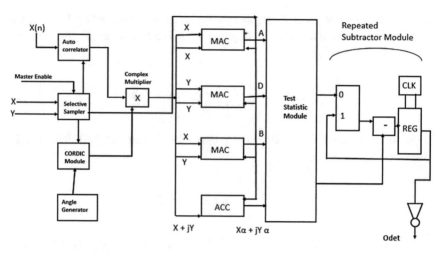

Fig. 1 System model for CS-based architecture

CORDIC module are passed through complex multipliers simultaneously to achieve required frequency shift. The output from complex multiplier is fed through MAC blocks to calculate the matrix elements required to calculate test statistic value. The repeated subtractor module subtracts denominator from the numerator in multiples of threshold value which is pre-calculated. The final Most Significant Bit (MSB) generated is considered as the final detector output. Based on that value we consider if output is present or not.

3 Vedic Multiplier Implementation

From Fig. 1 the Test Statistic Module requires numerous Multipliers to calculate test statistic value continuously which should be compared against pre-calculated threshold value. So far, the previous section gave a brief description about memory efficient by implementing the selective sampling technique and hardware efficient with the help of CORDIC module. Now, shall see on performance and power improvement using Vedic multiplier architecture. The advantage of Vedic over booth multiplication technique is discussed in this section.

The term performance indicates speed, this can be achieved by implementing Vedic multiplier in place of Booth Multiplier. Multipliers play a vital role in low power and high-speed applications. The fastness of multiplication operation depends on the number of partial products involved in it. Lesser the partial products with high bit higher will be the speed of operation [11–13]. Booth Multiplication Technique is advantageous over signed number multiplication but confined to only a smaller number of bits.

Algorithm for Booth technique:

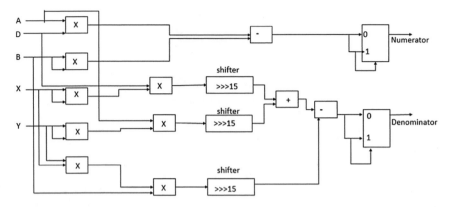

Fig. 2 Test statistic computing module

Let us assume Multiplicand be M and Multiplier be Q. A be a register which is initialized to Zero.

1. If $Q_0 \, Q_{-1}$ is same, i.e., 00 or 11 then perform right arithmetic shift by 1 bit.
2. If $Q_0 \, Q_{-1} = 10$ then perform $A = A - M$, and then perform arithmetic right shift.
3. If $Q_0 \, Q_{-1} = 01$ then perform $A = A + M$, and then perform arithmetic right shift.

Booth multiplier is not an efficient architecture to implement in the design, because of high bit of continuous multiplications are required here in the design.

Figure 2 represents the test statistic computing module. The term cyclostationary means continuous sensing of spectrum which continuously needs to undergo multiplication process to calculate test statistic value which should be compared against pre-calculated threshold value. Thus, deploying Vedic multiplier in test statistic computing module will improve performance and power [13].

Vedic multiplier is derived from the concept of Vedic mathematics. Vedic mathematics is one of the vast subjects that involve sutras to simplify the difficulties in complex operations. It is generally part of four Vedas called books of wisdom. They contain several modern mathematical terms with advanced techniques to simplify the calculations corresponding to time. The whole concepts of mathematics in those books are constructed as 16 sutras nothing but called formulae and include 16 Upasutras called sub formulae from Atharva Veda. The sutras so far defined in those Vedas can be directly applied to the mathematical applications for easy simplification. Vedic multiplication is one among the technique used in our present design to make the design power and delay efficient. Vedic mathematics is based on natural principles.

The Vedic multiplier that we are deploying the design is based on an algorithm Urdhva Tiryakbhyam (Vertical and Crosswise) of ancient Indian Vedic Mathematics. It exactly resources "Vertically and crosswise". It is constructed on a new idea over which the group of all partial products can be done with the simultaneous addition of these partial products. The parallelism in producing partial products and their addition is found using Urdhava Triyakbhyam (Fig. 3).

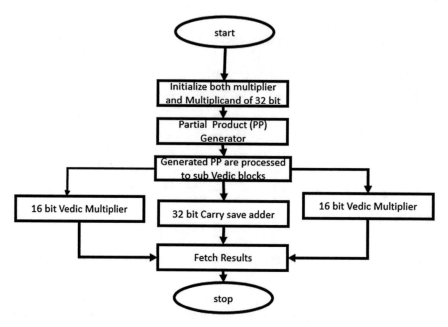

Fig. 3 Flowchart for an vedic multiplication

Generally, this Vedic block is built from a very basic level. Let's see for one such basic step. It is 2 × 2 Vedic multiplier, from which 4 × 4 is derived and so on in terms of multiples of 2. In the similar way to build 32-bit multiplier we define 16 bit from which 32 bit is derived. The fastness of the multiplier can further be increased by choosing the latest adder known as carry-save adder. The architecture is described in Fig. 4.

Vedic Multipliers are built from the basic architectures. The Crisscross method used in Vedic Architectures will reduce the delay and also the signal power consumption. For the system involved with larger bit will have an advantage when it is built using Vedic. The test statistic module is one such system involved in CR networks will operate faster with the help of fastest multiplier like VEDIC architecture [13].

4 Simulation Results

The complete design is implemented in Verilog code using Xilinx Vivado. A prototype of the design is built using Xilinx Vivado. The simulations are observed Using ISIM simulator. The output simulations for the complete design with Vedic Multiplier architecture is shown in Fig. 5. The complete design with normal conventional booth multiplication technique and the Vedic multiplication technique are implemented using Verilog code and the results are compared and tabulated as Table 1.

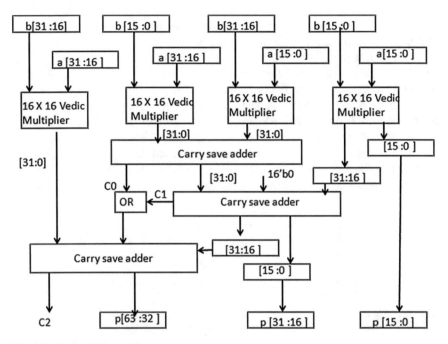

Fig. 4 Vedic multiplier architecture

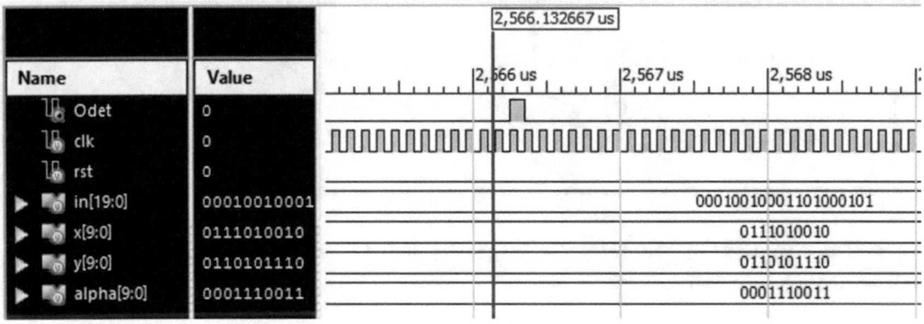

Fig. 5 Complete design simulation with Vedic multiplier

Table 1 Comparision between Booth and Vedic multiplier	Item/technique	With booth (%)	With Vedic (%)
	No. of registers	0.149	0.132
	No. of LUTS	13.44	16.32
	No. of bonded IOB's	25.23	25.23
	No. of buffers	3.125	3.125
	Delay (ns)	27.532	23.943
	Power (watt)	171.039	118.027
	CPU bust time	35.97	33.24

The results describe that the performance can be increased compared to the booth multiplication technique and the hardware required is also less compared to the conventional Booth technique. The power estimations among the two techniques are also found and described as follows in Figs. 9 and 10 show the hardware utilization between two techniques and Power Estimation in Fig. 7 (Figs. 6 and 8).

The RTL schematic for complete architecture is shown in Fig. 6. The RTL schematic for the Test Statistic Module using Vedic multiplier is shown in Fig. 8.

The Vedic Architecture is implemented in the test statistic module and the results are analyzed and related with the conventional Booth multiplier technique. The Vedic Multiplier architecture is advantageous in this case as this test statistic module involves in frequent multiplications, Vedic technique is helpful compared to Booth.

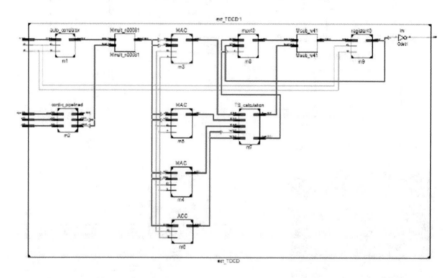

Fig. 6 The complete design output, odet indicates signal detection for the given input X, Y along with orientation of the signal alpha

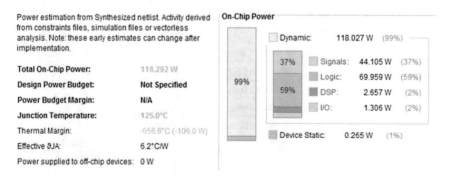

Fig. 7 Power estimation using Vedic implementation

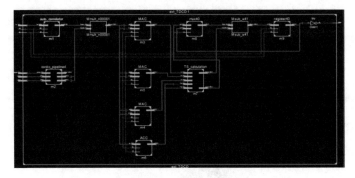

Fig. 8 Its the vedic results ananlysed using Xilinx vivado

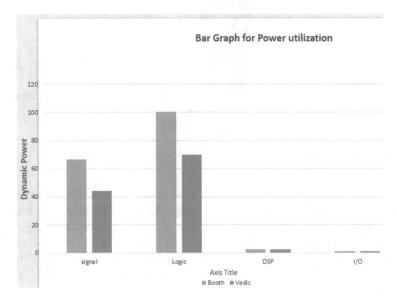

Fig. 9 Bar graph representing power utilization

The Architecture can further be implemented on FPGA and later a prototype can be built which is helpful to build a path between communication and VLSI.

5 Conclusion and Future Scope

Test statistic computing module is an important component in the Cognitive radio Architecture. Improving its performance will increase the speed of operation since it involves the continuous sensing of the spectrum. Implementing Faster multiplication Module in it will be advantageous. With the Vedic multiplier implementation,

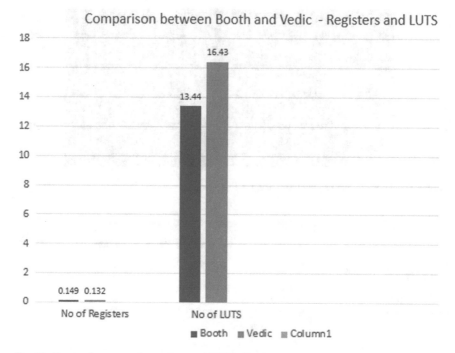

Fig. 10 Bar graph representing register and LUT utilization

the hardware resources got reduced by 11.40%, power consumption got reduced by 30.99% and the performance got increased by 13.035%. The results are analyzed using Xilinx ISE and Xilinx Vivado. In Future we can Build it on a prototype over Chip module that reduces the Hardware Resources. Further research includes implementing the complete design on real-time equipment, later building a prototype of the design that to be implemented in the real-time hardware systems. Thus, it paves a path between communication and VLSI.

References

1. Ghasemi, A., Sousa, E.S.: Spectrum sensing in cognitive radio networks requirements, challenges and design trade-offs. IEEE **46**(4), 32–39 (2008)
2. Yucek, T., Arslan, H.: A survey of spectrum sensing algorithms for cognitive radio applications. IEEE Commun. Surv. Tutorials **11** (2009)
3. Gardner, W.A., Franks, L.: Characterization of cyclostationary random signal processes. IEEE Trans. Inf. Theory **21**(1), 4–14 (1975)
4. Razhavi, B.: Cognitive radio design challenges and techniques. IEEE J. Solid-State Circ. (JSSC) **45**(8), 1542–1553 (2010)
5. Vijay, G., Bdira, E.B.A., Ibnkahla, M.: Cognition in wireless sensor networks: a perspective. IEEE Sensors **11**(3), 582–592 (2011)
6. Joshi, G.P., Nam, S.Y., Kim, S.W.: Cognitive radio wireless sensor networks: applications, challenges and research trends. Sens. **13**(9), 11196–11228 (2013)

7. Murthy, M.S., Shrestha, R.: VLSI architecture for cyclostationary feature detection-based spectrum sensing for cognitive-radio wireless networks and its ASIC implementation. In: IEEE Computer Society Annual Symposium on VLSI, pp. 69–74 (2016)

8. Murthy, M.S., Shrestha, R.: Reconfigurable and memory-efficient cyclostationary spectrum sensor for cognitive-radio wireless networks. IEEE (2017)

9. Chaudhari, S., Koivunen, V., Poor, V.: Autocorrelation-based decentralized sequential detection of OFDM signals in cognitive radios. IEEE Trans. Signal Process. 57(7), 2690–2700 (2009)

10. Dandawate, A.V., Giannakis, G.B.: Statistical tests for presence of cyclostationarity. IEEE Trans. Signal Process. 42(9), 2355–2369 (1994)

11. Hanumantharaju, M.C., Jayalaxmi, H., Renuka, R.K.: A high speed block convolution using ancient indian vedic mathematics. In: International Conference on Computational Intelligence and Multimedia Applications (2007)

12. Saha, P.K., Banerjee, A., Dandapat, A.: High speed low power complex multiplier design using parallel adders and subtractors. Int. J. Electron. Electr. Eng. (IJEEE) 7(II), 38–46 (2009)

13. Tiwari, H.D., Gankhuyag, G., Kim, C.M., Cho, Y.B.: Multiplier design based on ancient Indian vedic mathematics. In: International SoC Design Conference (2008)

Detection of Epileptic Seizure Based on ReliefF Algorithm and Multi-support Vector Machine

Hirald Dwaraka Praveena, C. Subhas and K. Rama Naidu

Abstract In recent decades, epileptic seizure classification is the most challenging aspect in the field of health monitoring systems. So, a new system was developed in this research study for improving the accuracy of epileptic seizure classification. Here, epileptic seizure classification was done by using Bonn University Electroencephalogram (EEG) dataset and Bern-Barcelona EEG dataset. After signal collection, a combination of decomposition and transformation techniques (Hilbert Vibration Decomposition (HVD) and Dual-Tree Complex Wavelet Transform (DTCWT) was utilized for determining the subtle changes in frequency. Then, semantic feature extraction (permutation entropy, spectral entropy, Tsallis entropy, and hjorth parameters (mobility and complexity) were utilized to extract the features from collected signals. After feature extraction, reliefF algorithm was used for eliminating the irrelevant feature vectors or selecting the optimal feature subsets. A Multi-binary classifier: Multi-Support Vector Machine (M-SVM) was helpful in classifying the EEG signals such as ictal, normal, interictal, non-focal, and focal. This research work includes several benefits; assists physicians during surgery, earlier detection of epileptic seizure diseases, and cost-efficient related to the existing systems. The experimental outcome showed that the proposed system effectively distinguishes the EEG classes by means of Negative Predictive Value (NPV), Positive Predictive Value (PPV), f-score and accuracy.

Keywords Dual-Tree complex wavelet transform · Electroencephalogram · Hilbert vibration decomposition · Multi-support vector machine and ReliefF algorithm

H. D. Praveena (✉)
Department of ECE, JNTUA, Ananthapuramu 515002, India
e-mail: hdpraveena@gmail.com

C. Subhas
Department of ECE, JNTUA College of Engineering, Kalikiri 517234, India
e-mail: schennapalli@gmail.com

K. Rama Naidu
Department of ECE, JNTUA College of Engineering, Ananthapuramu 515002, India
e-mail: kramanaidu@gmail.com

© Springer Nature Singapore Pte Ltd. 2020
P. K. Mallick et al. (eds.), *Cognitive Informatics and Soft Computing*,
Advances in Intelligent Systems and Computing 1040,
https://doi.org/10.1007/978-981-15-1451-7_2

13

1 Introduction

Epileptic seizure is a neurological disorder that affects all aged people worldwide. In India, around five million people are affected by epileptic seizure [1–3]. People with epileptic seizure are unaware of the present situation that leads to accidents or may get injured and also it affects memory impairment, difficulties in moving the body and even death [4, 5]. Currently, the detection of epileptic seizure is done on the basis of direct observation of EEG signal by the experts [6]. The manual observation consumes more time because the EEG dataset comprises of large size of data due to an increasing number of seizure population [7]. The visual examination of EEG signal is time-intensive, so the computerized epileptic seizure detection systems are developed for determining the epileptic source or for making a diagnosis about the epilepsy type [8, 9]. Presently, the computerized epileptic seizure detection has made an impact on the research community because of its functional applications [10]. The key contribution of this research is to accomplish epileptic seizure classification by developing an effective system.

In this research paper, the EEG signals were collected from the dataset: Bonn University EEG and Bern-Barcelona EEG dataset. The unwanted artifacts or machinery noises in the collected EEG signals were eliminated using low pass type 2 Chebyshev filter. The main advantage is, no ripple in the pass-band. After signal pre-processing, a combination of decomposition and transformation were carried-out using HVD and DTCWT, which helps to evaluate the subtle changes in frequency. Features were extracted from pre-processed signals using semantic feature extraction. The semantic feature extraction includes five features; permutation entropy, spectral entropy, Tsallis entropy, and Hjorth parameters (mobility and complexity) in order to obtain feature subsets from the set of data inputs by the rejecting redundant and irrelevant features. ReliefF algorithm was utilized to select the optimal feature subsets. The Output of reliefF algorithm specifies, which features of the EEG signals were vital in describing the dataset signals. These optimal feature values were applied to M-SVM classifier as the input to classify EEG classes such as, ictal, normal, interictal, non-focal and focal.

This research work is structured as follows. In Sect. 2, a broad survey of recent papers in epileptic seizure classification was presented. An effective supervised system is developed (semantic feature extraction with reliefF algorithm and M-SVM) for epileptic seizure classification in Sect. 3. In Sect. 4, comparative and quantitative examination of existing and proposed systems were presented. The inference of the results were discussed in fifth section.

2 Literature Survey

Researchers developed numerous new methodologies in epileptic convulsion detection. The evaluations of a few essential contributions to the existing literature papers are presented here.

Al-Sharhan and Bimba [11] developed a new approach for tackling the multiple EEG channels. Usually, the optimization approaches were very effective in reducing the overall dimensionality of data and also helps in preserving the class-dependent features. In this research paper, an effective multi-parent cross-over Genetic Algorithm (GA) was developed to optimize the features, which were utilized in the epileptic seizures classification. The developed optimization approach encodes the spatial and temporal filter estimates and then classification was carried-out using SVM classifier. The developed system was tested on the online available datasets; UCI center for intelligent systems and machine learning. The experimental phase confirmed that the developed system out-performed the existing systems by means of sensitivity, accuracy, and specificity. Additionally, the developed system needed a superior pre-processing approach to further enhance the classification accuracy.

Kalbkhani and Shayesteh [12] presented a new approach for epileptic seizure classification obtained from the stock-well transform. The amplitudes of stock-well transform in 5 sub-bands were computed, namely, beta, theta, delta, gamma, and alpha for classifying the EEG signals as ictal, healthy, and interictal. After computing the amplitudes, kernel principal component analysis was used for extracting the useful features from the feature values. At last, classifier named nearest neighbour was applied to classify the classes as ictal, healthy and interictal. Here, the developed approach was only suitable for uniform pattern data not for non-uniform pattern data.

Misiūnas et al. [13] presented a new system for automatic EEG classification on the basis of different epilepsy types; I and II groups. The First group comprises of patients with structural focal epilepsy and group II patients do not have the causal lesion and benign focal. The developed system consists of three phases; peak detection, identification of EEG peak parameters and EEG signals classification using artificial neural network. The developed system was verified on a real-time dataset. A major problem in the developed system was high computational time related to the other conventional approaches.

Wang et al. [14] developed an effective system for epileptic convulsion detection. In this literature paper, the classification of epileptic seizure was evaluated using a simulated dataset. After signal collection, notch filter, weighted wavelet threshold and bandpass filtering approaches were utilized for removing the unwanted artifacts from the collected EEG signals. The pre-processed EEG signals were given as the input for empirical mode decomposition approach for decomposing the pre-processed EEG signals into several segments. Finally, Teager energy operator, curve length and PSD methods were utilized for extracting the features from decomposed signals. The drawback of the research work was the developed system did not concentrate on the signal classification.

To overcome the above-mentioned problems, a new system (ReliefF algorithm and M-SVM) was implemented for improving the performance of epileptic seizure detection.

3   Proposed System

Epileptic seizure is the second most common brain disorder, so the automatic detection of this disease significantly enhances the patient's quality of life. In this research, the proposed system comprises of seven major phases: signal collection, pre-processing, decomposition, transformation, extraction the features, selecting the features and classification process. The working flow of proposed system denoted in Fig. 1. The detailed explanation about this proposed system is described below.

3.1   Data Collection and Pre-processing of Raw EEG Signal Using Chebyshev Type 2 Low Pass Filter

In this research study, for experimental investigations, Bonn University EEG and Bern-Barcelona EEG datasets are used. **Bern-Barcelona EEG dataset** contains five epilepsy patients with 7500 signal sets that are separated into two categories: non-focal and focal signals [15]. Here, non-focal signals are obtained from the normal brain areas and the focal signals are obtained from the ictal visual inspection-brain areas. **Bonn University EEG dataset** has five subsets (F, S, Z, N and O) with 500 signals [16]. These EEG signals are recorded by twelve-bit analog to digital converter and one twenty-eight channel amplifier. The sub-set Z and sub-set O are obtained from healthy volunteers and S, N, and F subsets are obtained from intracranial and ictal electrodes for observing intracranial and ictal epileptic activity.

After collecting the EEG signals, Chebyshev type 2 low pass filter is used to remove the unwanted artifacts. It is also called as digital or analog filter, which has more pass-band ripple and steeper roll-off than the Butterworth filters. The developed filter effectively reduces the error between the range of filters and idealized filter characteristics. The general formula of low pass Chebyshev type 2 filter is given in the Eq. (1).

$$\text{Chebyshev type 2 low pass filter} = \frac{1}{\sqrt{1 + \rho^2 T_n^2 \left(\frac{w}{w_0}\right)}} \tag{1}$$

where ρ is denoted as ripple factor, cut-off frequency as w_0 (60 Hz), and T_n as Chebyshev polynomial of sixth order. Figure 2 denotes the sample input signal and pre-processed EEG signal.

Fig. 1 Work flow of
proposed system

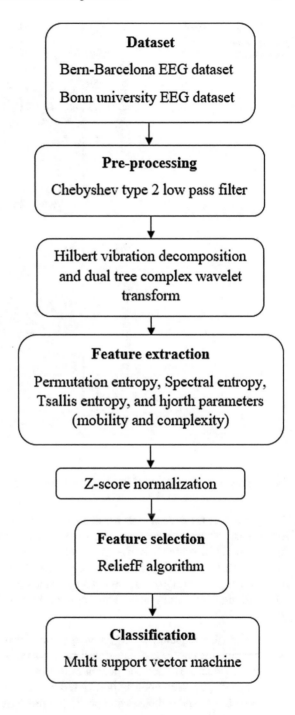

Fig. 2 a Sample input
signal. **b** Sample
pre-processed signal

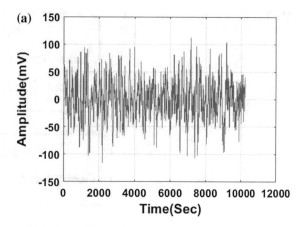

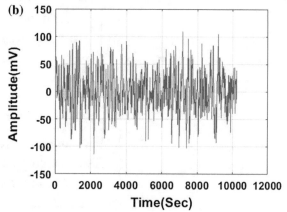

3.2 Hilbert Vibration Decomposition

The pre-processed EEG signals are decomposed using the HVD method that works
on the basis of three assumptions; every harmonics envelope is different, the original
EEG signal is the summation of quasi harmonics and pre-processed signals are longer
compared to slowest component in the original EEG signal. Usually, the components
of the pre-processed signals are in contact with the super harmonics, those amplitudes
are dissimilar from each other. The steps involved in HVD approach is described
below

- Calculate the frequency of largest energy component.
- Determine the envelope of the largest component.
- Finally, deduct the largest component value from the original EEG signal. Then, the
  subsequent largest component is achieved by the repetition of the three steps mentioned above. The sample decomposed pre-processed EEG signal is graphically
  denoted in Fig. 3.

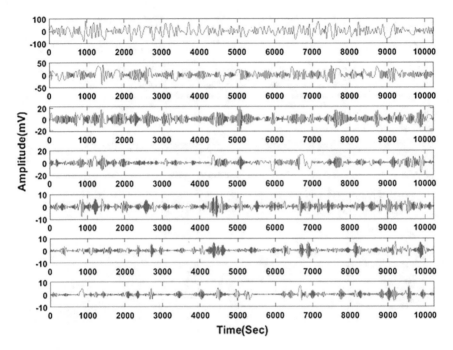

Fig. 3 Decomposed EEG signal

3.3 Dual-Tree Complex Wavelet Transform

The respective decomposed EEG signals are utilized for transformation by applying DTCWT. In signal processing, DTCWT provides more frequency and time resolution in EEG signals, because of its frequency and time localization ability. Additionally, DTCWT describes the characteristics of decomposed EEG signals that help to decreases the feature degradation. Here, the low and high-frequency information is extracted by applying long and short time windows. The decomposed EEG signals have 60 Hz frequency, which is split into four sub-bands with the range of 0–8 Hz, 8–15 Hz, 15–30 Hz and 30–60 Hz, respectively. The DTCWT is very appropriate to investigate the non-stationary signals. Here, symlets 7 is used as the wavelet and scaling functions of DTCWT. The general Eq. of DTCWT $d(t)$ is specified in the Eq. (2).

$$d(t) = \sum_{-\infty}^{\infty} s(n)\emptyset(t - n) + \sum_{m=0}^{\infty} \sum_{n=-\infty}^{\infty} w(m, n)2^{\frac{m}{2}} \psi\left(2^m t - n\right) \qquad (2)$$

where $\psi(t)$ is denoted as the real-valued band-pass wavelet, $\phi(t)$ is represented as the low pass scaling function, $s(n)$ as scaling coefficients and $w(m, n)$ is denotes as the wavelet coefficients. Figure 4 denotes the sub-bands of decomposed signals.

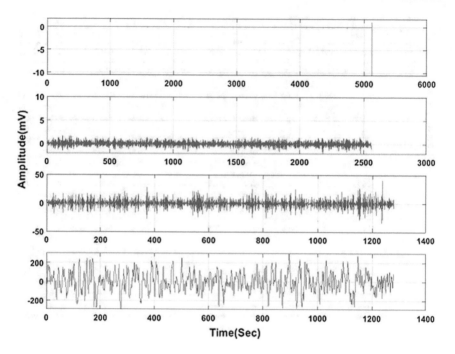

Fig. 4 Sub-bands of decomposed EEG signal

3.4 Semantic Feature Extraction with Z-Score Normalization

After transformation, semantic feature extraction is performed on the transformed EEG signal. Usually, feature extraction is described as the action of mapping EEG data from data space to the feature space and also feature extraction converts the large data into a comprised data that helps to reduce the system complexity. In this research study, semantic feature extraction is performed using Permutation entropy, Spectral entropy, Tsallis entropy, and Hjorth parameters (mobility and complexity). The main advantage of using several features helps to recognize the data occlusion and clutter. After extracting the features from transformed EEG signal, Z-score normalization is used to convert the extracted features to unit variance and zero mean. In this research paper, the length of extracted features is fifty-five. The standard deviation and mean of the feature vector are determined by using the Eq. (3).

$$y_{\text{new}} = \frac{y_{\text{old}} - \text{mean}}{\text{standard deviation}} \qquad (3)$$

where y_{old} is denoted as original value, and $y_{\text{new}} y_{\text{new}}$ is represented as new value. The normalized feature vectors are given as the input for reliefF feature selection to obtain the optimal feature vectors. The detailed explanation about the reliefF feature selection is described below.

Datasets	Number of features	Selected features
Bern-Barcelona EEG dataset	55	5
Bonn University EEG dataset	55	5

Table 1 Features selected after reliefF algorithm

3.5 Feature Selection Using ReliefF Algorithm

The reliefF algorithm was utilized to select the optimal features for performing better classification after extracting the features from EEG signals. While dealing with incomplete and huge data, reliefF algorithm is the best choice for feature selection. At first, reliefF algorithm randomly selects the instances i_s and then search for k nearest neighbour. In case, if the k nearest neighbour search occurred in the dissimilar classes s is called as nearest miss N and for similar classes is named as nearest hit h. Then, the reliefF algorithm updates the quality estimation $Q[a]$ for all attributes a on the basis of i_s, N and h values. If the instances i_s and h have same values, then the attribute a is categorized into two instances with the same class that is desirable to reduce the quality estimation $Q[a]$. In contrast, if the instances i_s and h have dissimilar values, then the attribute a is categorized into two instances with the dissimilar class that is desirable to improve the quality estimation $Q[a]$.

The entire procedure is repeated for n times, where n is denoted as a user-defined parameter. In this research study, the user-defined parameter is set as twenty. In reliefF algorithm, the quality estimation $Q[a]$ is updated by utilizing the Eq. (4).

$$Q[a] = Q[a] - \frac{\sum_{k-1}^{k} d_h(k)}{n, k} + \sum_{c-1}^{c-1} P_c \frac{\sum_{k-1}^{k} d_N(k)}{n, k} \tag{4}$$

where P_c is denoted as prior class, and d is represented as distance between the selected instances i_s. Table 1 describes the features that exists in the datasets and the features selected after applying reliefF algorithm. The selected features are used for classification by employing M-SVM classifier.

3.6 Classification Using Multi-Support Vector Machine

Usually, regular SVM is a two-class classification methodology. Hence, it is essential to concentrate on the multi-binary classification issues for extending the normal SVM classifier to multi-class SVM classifier. In conventional SVM classification approach, the multi-class classification is converted into nth two-class and ith two-class issues, where class i is different from the remaining classes. The two important methodologies in SVM classifiers are One-Against-All (1-a-a) and One-Against-One

(1-a-1). In this scenario, 1-a-a approach gives solution to create a binary classifier for every class that helps to separate the objects in the same class. In nth class, 1-a-a approach generates nth binary classifiers and the ith classifier is trained with the data samples in ith class with the positive labels and the residual data samples are trained with negative labels. The result of nth class in 1-a-a approach relates with the 1-a-1 approach for obtaining the highest output value. In addition, the 1-a-1 approach is the resultant of previous researches on two-class classifier.

The M-SVM classifier generates all probable two-class classifiers from the training sets of nth classes, and it trains only two out of nth classes that result in $n \times (n-1)/2$ classifiers. In M-SVM, decision function is an active way to moderate the multi-class problems that is constructed by assuming all the nth classes. The M-SVM classification technique is an extension of SVM, which is mathematically represented in the Eqs. (5), 6), and (7).

$$\min \Phi(w, \xi) = {}^{1}\!/2 \sum_{m=1}^{k} (w_m \cdot w_m) + \sum_{i=1}^{l} \sum_{m \neq yi} \xi_i^m \tag{5}$$

Subjected to,

$$\left(W_{yi}.xi\right) + b_{yi} \geq \left(W_{yi}.xi\right) + b_m + 2 - \xi_i^m \tag{6}$$

$$\xi_i^m \geq 0, i = 1, 2, \ldots, l, m, yi \in \{1, 2, \ldots k\}, m \neq yi \tag{7}$$

where ξ_i^m is indicated as the slack variables, l is considered as the training data point, c is represented as the user-defined positive constant, k is stated as the number of classes, and yi is represented as the class of training data values xi At last, the decision function is expressed in the Eq. (8).

$$f(x) = \arg \max[(w_i.x) + b_i], i = 1, 2, \ldots k \tag{8}$$

4   Experimental Result and Analysis

In this segment, the proposed system was simulated using MATLAB (version 2018a) with 3.0 GHZ-Intel i5 processor, 1 TB hard disc, and 8 GB RAM. The proposed system performance was related to other existing methodologies on the reputed databases: Bern-Barcelona dataset and Bonn University dataset of EEG signals for estimating the proposed system efficiency. The proposed system is validated by means of PPV, NPV, accuracy and f-score.

4.1 Performance Metric

Performance metric is defined as the regular measurement of outcomes and results that develops a reliable information about the effectiveness and efficiency of proposed system. Also, it is the procedure of reporting, collecting and analysing information about the performance of a group or individual. The mathematical Eq. of accuracy, PPV, NPV and f-score are represented in the Eqs. (9) (10) (11), and (12), respectively.

$$Accuracy = \frac{TP + TN}{FN + TP + TN + FP} \times 100 \tag{9}$$

$$PPV = \frac{TP}{FP + TP} \times 100 \tag{10}$$

$$NVP = \frac{TN}{FN + TN} \times 100 \tag{11}$$

$$F\text{-score} = \frac{2TP}{(2TP + FP + FN)} \times 100 \tag{12}$$

where, FP is represented by False Positive, TP is indicated by True Positive, FN is considered as False Negative, and TN specified as True Negative.

4.2 Quantitative Analysis Using Bern-Barcelona EEG Dataset

In this sub-section, Bern-Barcelona EEG database is used to assess the proposed system performance for classifying non-focal and focal EEG signals. Here, the performance evaluation is validated for 50 non-focal and 50 focal EEG signals with 80% training and 20% testing of signals. Table 2 represents the performance evaluation of the specified system for non-focal and focal EEG signal classification by means of PPV, NPV, accuracy and f-score. In addition, Table 2 evaluates the proposed system performance with reliefF and without reliefF algorithm. The PPV, NPV, f-score, and accuracy of the proposed system (without reliefF feature selection) is 74.30, 78.11, 77.81, and 78.90%. Similarly, the PPV, NPV, f-score, and accuracy of the proposed system (with reliefF feature selection) is 98.78, 98.30, 98.69, and 98.50%. Table 2 clearly illustrates that the M-SVM classifier improves the accuracy in non-focal and

Table 2 Performance evaluation of proposed system using Bern-Barcelona EEG dataset

Feature selection	PPV (%)	NPV (%)	F-score (%)	Accuracy (%)
Without reliefF algorithm	74.30	78.11	77.81	78.90
With reliefF algorithm	98.78	98.30	98.69	98.50

Fig. 5 Graphical comparison of proposed system using Bern-Barcelona EEG dataset

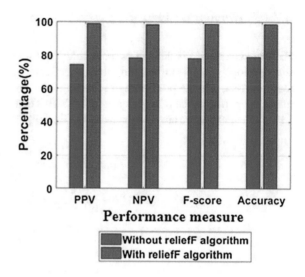

focal EEG signal classification up to 20–24% in Bern-Barcelona EEG dataset, while using reliefF feature selection. The graphical representation of PPV, NPV, *f*-score and accuracy of this proposed system is represented in Fig. 5.

4.3 Quantitative Analysis Using Bonn University EEG Dataset

Here, the performance evaluation of proposed system is described for 50 ictal signal, 50 normal signal, and 50 interictal EEG signals with 80% training and 20% testing of signals. In Table 3, the proposed system achieved 96.27% of PPV, 98.71% of NPV, 94.17% of f-score and 95.73% of accuracy using reliefF algorithm, which shows 1.2–2% of improvement in ictal, normal, and interictal EEG signal classification compared to without using reliefF algorithm. The graphical comparison of PPV, NPV, accuracy and f-score of proposed system for ictal, normal, and interictal EEG signal classification is denoted in Fig. 6.

Table 3 Performance evaluation of proposed system using Bonn University EEG dataset

Feature selection	PPV (%)	NPV (%)	*F*-score (%)	Accuracy (%)
Without reliefF algorithm	94.33	94.49	93	93.87
With reliefF algorithm	96.27	98.71	94.17	95.73

Fig. 6 Graphical comparison of proposed system using Bonn University EEG dataset

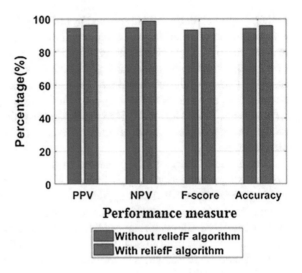

4.4 Comparative Analysis

Comparative analysis of proposed method and existing method is represented in Table 4. Sriraam and Raghu [17] implemented an effective method for detection of epileptic seizure. Initially, the EEG signals were taken from the database: Bern Barcelona EEG dataset. In this research, totally twenty-six features are considered from dissimilar domains like frequency, time, information theory, etc. Among these

Table 4 Comparison of proposed and existing methods

References	Dataset	Features considered	Classification methodology	Accuracy (%)
Sriraam and Raghu [17]	Bern Barcelona EEG dataset	Twenty-six features	SVM	92.15
Ahammad et al. [18]	Bonn University EEG database	Statistical and wavelet features	Linear classifier	84.2
Nicolaou and Georgiou [19]	Bonn University EEG database	Permutation entropy feature	SVM	84.1775
Das et al. [20]	Bonn University EEG database	DTCWT	ANN and SVM	83.5
Proposed work	Bern-Barcelona EEG dataset	Permutation entropy, spectral entropy, Tsallis entropy, and Hjorth parameters (mobility and complexity)	M-SVM	98.50
	Bonn University EEG dataset			95.73

features, optimal features were selected by utilizing Wilcoxon rank-sum test. Finally, SVM classifier was used for classification. The developed system achieved 92.15% of accuracy in EEG signal classification. Additionally, Ahammad et al. [18] developed a new unsupervised epileptic seizure system using statistical and wavelet features. The obtained feature values were applied to linear classifier to classify the EEG classes: normal, ictal and interictal EEG signals. This research work was experimented on a dataset (i.e. Bonn University EEG database) for validating the performance accurately. The developed system almost achieved 84.2% of accuracy in EEG signal classification.

Nicolaou and Georgiou [19] presented an effective epileptic seizure identification system using permutation entropy feature. After extracting the feature information from raw EEG data, SVM was applied to classify the three classes: ictal, interictal and normal EEG signals. In this research study, Bonn University EEG database was utilized to analyze the performance of developed system. Here, the developed system achieved 84.1775% of accuracy in EEG signal classification. In addition, Das et al. [20] collected the EEG signals from Bonn University EEG database. After collecting the EEG signals, DTCWT was used to improve the classification performance. At last, the classification was done using SVM and artificial neural network that achieved 83.5% of classification accuracy. Though, the proposed system achieved 98.50% of classification accuracy in Bern-Barcelona EEG dataset and 95.73% of classification accuracy in Bonn University EEG database.

The EEG signals comprise of numerous features and high data-space volume, which leads to "curse of dimensionality" problem. So, feature selection is essential for optimizing the features, which is fit for better classification. The effectiveness of feature selection is denoted in Tables 2 and 3.

5 Conclusion

The main outcome of this research is to afford an effective supervised system for biomarking of (ictal, normal, and interictal) and (focal and non-focal) EEG signals. The proposed system helps physicians in treating more subjects, minimizing the errors caused by manual classification, and clinical decision making. Here, a combination of signal transformation and decomposition methods are utilized to avoid signal overlapping and also for providing higher flexibility in choosing a particular EEG signal. Then, permutation entropy, spectral entropy, Tsallis entropy, and Hjorth parameters (mobility and complexity) are used for extracting the features from pre-processed signal. After extracting the features, reliefF algorithm is used to select the optimal features. The reliefF algorithm is an efficient feature selection approach because it determines the feature weights by resolving the convex optimization issue. These optimal features are used to classify the EEG classes by applying M-SVM classification method. Related to other existing systems in epileptic seizure detection, the proposed system achieved superior performance by means of accuracy, which showed 3–14% improvement in classification accuracy. In future work, a new system will be designed for analyzing the stages of an epileptic seizure or other diseases like Alzheimer's.

References

1. Hussein, R., Palangi, H., Ward, R.K., Wang, Z.J.: Optimized deep neural network architecture for robust detection of epileptic seizures using EEG signals. Clin. Neurophysiol. **130**, 23–37 (2018)
2. Hussein, R., Elgendi, M., Wang, Z.J., Ward, R.K.: Robust detection of epileptic seizures based on L1-penalized robust regression of EEG signals. Expert Syst. Appl. **104**, 153–167 (2018)
3. Karthick, P.A., Tanaka, H., Khoo, H.M., Gotman, J.: Prediction of secondary generalization from a focal onset seizure in intracerebral EEG. Clin. Neurophysiol. **129**, 1030–1040 (2018)
4. Kevric, J., Subasi, A.: Comparison of signal decomposition methods in classification of EEG signals for motor-imagery BCI system. Biomed. Signal Process. **31**, 398–406 (2017)
5. Mohammadi, M.R., Khaleghi, A., Nasrabadi, A.M., Rafieivand, S., Begol, M., Zarafshan, H.: EEG classification of ADHD and normal children using non-linear features and neural network. Biomed. Eng. Lett. **6**, 66–73 (2016)
6. Arunkumar, N., Ramkumar, K., Venkatraman, V., Abdulhay, E., Fernandes, S.L., Kadry, S., Segal, S.: Classification of focal and non-focal EEG using entropies. Pattern Recogn. Lett. **94**, 112–117 (2017)
7. Raghu, S., Sriraam, N.: Classification of focal and non-focal EEG signals using neighborhood component analysis and machine learning algorithms. Expert Syst. Appl. **113**, 18–32 (2018)
8. Sharma, M., Dhere, A., Pachori, R.B., Acharya, U.R.: An automatic detection of focal EEG signals using new class of time-frequency localized orthogonal wavelet filter banks. Knowl.-Based Syst. **118**, 217–227 (2017)
9. Tjepkema-Cloostermans, M.C., de Carvalho, R.C., van Putten, M.J.: Deeplearning for detection of focal epileptiform discharges from scalp EEG recordings. Clin. Neurophysiol. **129**, 2191–2196 (2018)
10. Chu, H., Chung, C.K., Jeong, W., Cho, K.H.: Predicting epileptic seizures from scalp EEG based on attractor state analysis. Comput. Meth Prog. Bio. **143**, 75–87 (2017)
11. Al-Sharhan, S., Bimba, A.: Adaptive multi-parent crossover GA for feature optimization in epileptic seizure identification. Appl. Soft Comput. **75**, 575–587 (2018)
12. Kalbkhani, H., Shayesteh, M.G.: Stockwell transform for epileptic seizure detection from EEG signals. Biomed. Signal Process. **38**, 108–118 (2017)
13. Misiūnas, A.V.M., Meškauskas, T., Samaitienė, R.: Algorithm for automatic EEG classification according to the epilepsy type: Benign focal childhood epilepsy and structural focal epilepsy. Biomed Signal Proces. **48**, 118–127 (2019)
14. Wang, C., Yi, H., Wang, W., Valliappan, P.: Lesion localization algorithm of high-frequency epileptic signal based on Teager energy operator. Biomed Signal Proces. **47**, 262–275 (2019)
15. Andrzejak, R.G., Schindler, K., Rummel, C.: Nonrandomness, nonlinear dependence, and nonstationarity of electroencephalographic recordings from epilepsy patients. Phys. Rev. E **86**, 046206 (2012)
16. Andrzejak, R.G., Lehnertz, K., Rieke, C., Mormann, F., David, P., Elger, C.E.: Indications of nonlinear deterministic and finite dimensional structures in time series of brain electrical activity: dependence on recording region and brain state. Phys. Rev. E **64**, 061907 (2001)
17. Sriraam, N., Raghu, S.: Classification of focal and non-focal epileptic seizures using multi-features and SVM classifier. J. Med. Syst. **41**, 160 (2017)
18. Ahammad, N., Fathima, T., Joseph, P.: Detection of epileptic seizure event and onset using EEG. Biomed. Res. Int. (2014)

19. Nicolaou, N., Georgiou, J.: Detection of epileptic electroencephalogram based on permutation entropy and support vector machines. Expert Syst. Appl. **39**, 202–209 (2012)
20. Das, A.B., Bhuiyan, M.I.H., Alam, S.S.: A statistical method for automatic detection of seizure and epilepsy in the dual tree complex wavelet transform domain. In: Proceedings of International Conference on IEEE Informatics on Electronics & Vision (ICIEV), pp. 1–6 (2014)

Blockchain-Based Shared Security Architecture

Shaji N. Raj and Elizabeth Sherly

Abstract A prototype model named Shared Security Architecture is introduced, that enables to link relational database management system, distributed DBMS and blockchain technology. Blockchain technology is commonly used in cryptocurrencies but the innovative use of this technology can boost cybersecurity and protect databases more effectively. Three different models have been explored with a combination of traditional data storage (RDBMS), distributed data storage and blockchain technology to construct new security for the data storage system. These models allow direct access to frequently used data, which requires read-write property and restricted access to data which requires write property with immutability without paying much on trust and security.

Keywords Blockchain · Security · Centralized system · Hashing

1 Introduction

A blockchain is a continuously growing list of records, secured using cryptography [1]. As of now, centralized databases play a major role in many database applications, while blockchain is more secure than other databases in some way, a trade-off between blockchain and conventional databases is a good choice. Other than trust, secure and robustness, blockchain can do nothing more than a centralized database [2]. The central administration in centralized databases restricts updation and can perform operations only in a central location, which has advantages over blockchain confidentiality. Centralized database is both read-write controlled, while blockchain is write-controlled only. Blockchains are also considered slower in performance than regular databases. Furthermore, as long as the blocks are stored in the memory and

S. N. Raj (✉)
SAS SNDP Yogam College, Mahatma Gandhi University, Kottayam, India
e-mail: ammashajinraj@gmail.com

E. Sherly
IIITM-K, Technopark, Kazhakkoottam, Kerala, India
e-mail: shely@iitmk.ac.in

© Springer Nature Singapore Pte Ltd. 2020
P. K. Mallick et al. (eds.), *Cognitive Informatics and Soft Computing*,
Advances in Intelligent Systems and Computing 1040,
https://doi.org/10.1007/978-981-15-1451-7_3

hard disk of computers or nodes, it can also be vulnerable. Considering the above scenario, this paper concentrates on a trade-off between relational databases and blockchain databases, distributed databases and blockchain and also a combination of these three databases.

The paper is structured as follows: After introducing the background information on Blockchain and related concepts, the literature review follows. Section 3 presents the proposed methodology and the new algorithm. Section 4 presents the results and evaluations. The paper concludes in Sect. 5, with future work.

2 Background Literature

Researchers developed various techniques for privacy using the blockchain technique. David Shrier, Weige Wu, Alex Pentland, of Connection Science & Engineering, Massachusetts Institute of Technology published a paper entitled "Blockchain & Infrastructure" describing the need for blockchain-based identity authentication. Using blockchain, a global identity was created for each user [3]. Loi et al. discovered that among the 19,366 existing Ethereum contracts, 8833 are vulnerable [4]. The researches show the new blockchain technology had a number of security and privacy issues. In June 2016, the smart contract DAO (Decentralized Autonomous Organization) was attacked [5] and the attacker exploited a recursive calling vulnerability, resulting in DAO losing about 60 million dollars. The hackers attacked the bitcoin in March 2014, causing the collapse of MtGox, with a value of 450 million dollars [4]. The blockchain vulnerability risk of 51% involves private key security, criminal activity, double spending transaction privacy leakage, criminal smart contracts, vulnerabilities in a smart contract, under-optimized smart contract, underpriced operations, etc. [5].

There are a number of security models for databases available, but 100% protection is still a daunting problem for researchers. The proposed Shared Security Architecture (SSA) for different databases is novel, maintaining centralized and decentralized databases together. Therefore, SSA benefits both the advantages, wherein blockchains are suited for the record of certain functions, while a centralized database is entirely appropriate for other functions.

3 Shared Security Architecture (SSA) for Different Databases

The blockchain technologies are widely used in cryptocurrencies but due to its decentralization and immutability properties, it is now used in many applications of information technology. Since many of the existing applications are in centralized databases, and considering blockchain's trust, robustness and distributed over existing databases, a fusion of both provides a better scenario. However, researchers and technologists identified this area as potential research, but not much work has been carried out and it is in a very nasal stage. SSA focuses on creating new scenarios

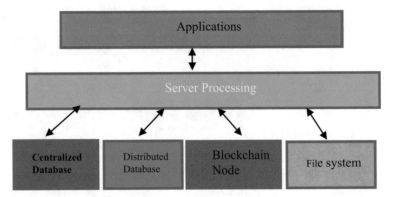

Fig. 1 Shared security architecture model

which combine centralized, distributed and blockchain technology. They are defined as three trade-offs, given as follows:

Trade-off 1: Centralized data storage with blockchain technology.
Trade-off 2: Distributed data storage with blockchain technology.
Trade-off 3: Combination of centralized and distributed data storage with blockchain technology.

Figure 1 illustrates the proposed scenario named Shared Security Architecture (SSA) combining centralized (SQL), distributed (Mongo DB) and blockchain technology to ensure the security of data and prevents the different types of attacks. Normally digital data is stored in a central server managed by big data companies and in any single point of failure, the data will be damaged.

3.1 Alikeness Blockchain Algorithm (ABA)

An Alikeness Blockchain Algorithm is proposed for SSA, in which three trade-offs have been implemented with certain customization for each one.

In this method, permanent inputs are hashed and stored in a database and updatable data are stored in an encrypted form and the hash value of the data is stored in the blockchain. The flow of Alikeness blockchain Algorithm (ABA) is depicted in Fig. 2.

3.2 Centralized Data Storage with Blockchain Technology

One of the challenges facing now by developers is the choice between a centralized database and blockchain technology. The centralized database still has some advantages over blockchain technology, especially in its confidentiality and performance. Blockchain is distributed, shared across boundaries of trust in different nodes, which is viewable to anyone. Therefore, the confidentiality of any data stored is essentially zero, though it has its own proof of validity and its own proof of authorization. Since,

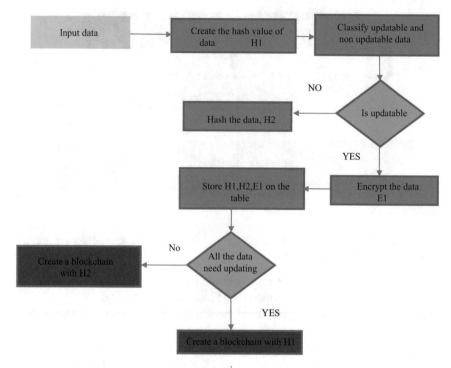

Fig. 2 Block diagram of Alikeness blockchain algorithm

in many cases, data needs confidentiality, the public blockchain may not be very useful in many business applications. In such cases, a trade-off between centralized data and blockchain can be adopted without storing a full record of transactions in a blockchain.

Algorithm: Centralized data storage with blockchain technology

if(all writers are trusted)
{
Use Centralized database system   }
Else if(need Public verification)

{ Use centralized system with Public blockchain   }
Else
{
Use a centralized system with Private blockchain   } }

For centralized database, Alikeness Blockchain Algorithm (ABA) is tuned and modified accordingly. The proposed model has three phases named.

1. Hashing phase, 2. Linking Phase, 3. Blockchain phase.

Hashing phase: In the hashing phase, the hash value of the entire data is taken and then finds the updatable and non-updatable data. Then all the permanent input data are hashed together and encrypted and stored in a database.

Linking phase: The second phase named linking phase is an efficient phase, which is used for creating a blockchain. The blockchain contains a hash of the previous block, a hash of the current data and hash of non-updatable data.

Blockchain Phase: In this phase, a blockchain is created and the hashed data are put into the blockchain. This block will hold the hash of the previous block. In this blockchain, the first record is generated and its hash value is calculated. On the next record, the hash value calculation includes the hash of the previous record, which is immutable. If a hacker tries to modify data, the administrator or system will verify the whole data in the blockchain, compare them to the data in the database and then exclude any that don't match up. In this system login credentials are also stored in the form of the hash value. When an intruder enters any of the SQL keywords, the algorithm takes it as user input and creates the hash value of the data and the hashed data is compared with data in the block.

3.3   Distributed Data Storage with Blockchain Technology

The modified version of ABA for distributed data storage with blockchain is described below. The proposed model can be used in two different ways according to the application. They are,

(A)   Private distributed blockchain database system.
(B)   Private distributed system with public blockchain.

These systems have three phases namely, storage, registration and verification.

1. Storage: A secured system is an environment in which all the data in the system are not accessed by unauthorized users. In the first case (A), a blockchain is created based on a database system, that has properties of distributed and blockchain, named as distributed blockchain database system. In the second case (B), the data is stored in a distributed database then hashed signature is placed on the blockchain. This blockchain is used for verification in the verification phase. Thus private distributed database allows high security rather than a completely 'distributed database'.

2. Registration: In this phase, the hash value is registered into the blockchain. Once the blockchain is created, a signature is returned that cryptographically links the fingerprint of the data to an entry in the blockchain.

3. Verification: The identity of the data can be verified in this phase. Using the digital signature, at any point in the future, the properties of the component can be verified. The cryptographic proof is the signature, the verifier regenerates the signature and confirms it and it matches the correct entry in the blockchain.

4   Evaluation

Blockchain technology is one of the promising technologies to protect data from hackers. The tampering of data can be prevented by the distributed nature of blockchain. In turn, data that is stored on a blockchain can easily remain authentic, since the security of every transaction is recorded. When an attacker attempts to change the data present in the block named Block 4, correspondingly, the hash of the block also changes. But, Block 5 still contains the old hash of the Block 4; thus the hacker cannot modify the block. The advantage of the proposed technique is that no declarative specification of input is required. Also in real-time monitoring techniques, this algorithm prevents all types of attacks.

For experiment purpose, an application is created in blockchain-based on PHP, MongoDB and MySQL to store students data for analysis. The security analysis based on different scenarios is conducted for injection attack, broken authentication and data exposure and access control. We compared the centralized database system with a centralized blockchain system, distributed with a distributed blockchain system and blockchain system. The table given below shows the percentage of the success rate of each scenario in different situation conducted with hundreds of samples. As compared to a centralized system, the blockchain-based centralized system is 6.5% more secure in injection, 13.5% in data exposure and 10% more in access control. As compared to a distributed system, blockchain-based distributed system is 6, 7.5 and 5% more secure in injection, data exposure and access control, respectively. The experiment results showed centralized, distributed with blockchain model having higher security. The advantage of these systems is more resilient against many security threats. In our experiment, a blockchain-based data storage is more secure than any other system. The integration of blockchain technology with existing technology is more secure. Table 1 shows different operations, such as injection, data exposure and access control performance in different scenarios.

Table 1 Performance evaluation of different databases

Operations	Centralized	Centralized with blockchain	Distributed	Distributed with blockchain
Injection	92	98.5	93	99
Data exposure	85	98.5	91	98.5
Access control	89	99	92	97

5 Conclusion

This paper proposes three different types of models with blockchain technology to provide high security for a web application. Considering the high significance of centralized databases in many existing applications, and also the security of blockchain technology, we propose a trade-off between databases, combining centralized, distributed and blockchain databases. While keeping the confidentiality and integrity of centralized databases, blockchain's trust and robustness benefited to it. Since performance is an issue in blockchain, these trade-offs enable to provide a better catch and as per the requirement of the application, it is left to the choice of the user for selection of scenario. The Alikeness Blockchain Algorithm proposed is tuned for different trade-offs and experimentation is conducted with 145 data manually created. In security analysis of our experiment, it is found that centralized with blockchain and distributed with blockchain show better results for injection attacks, data exposure and access control.

References

1. Bitcoin: En.wikipedia.org (2018). [Online]. Available https://en.wikipedia.org/wiki/Bitcoin. Accessed 1 Mar 2018
2. Greenspan, G.: Blockchains vs centralized databases | MultiChain, Multichain.com (2018). [Online]. Available https://www.multichain.com/blog/2016/03/blockchains-vs-centralized-databases/. Accessed 3 Mar 2018
3. Getsmarter.com, 2018. [Online]. Available https://www.getsmarter.com/career-advice/wp-content/uploads/2017/07/mit_ blockchain_transactions_report.pdf. Accessed 5 Jan 2018
4. The Daily Beast: The Daily Beast (2018). [Online]. Available http://www.thedailybeast.com. Accessed 5 Feb 2018
5. E. Foundation: Critical updateRe: DAOVulnerability, Blog.ethereum.org (2018). [Online]. Available https://blog.ethereum.org/2016/06/17/critical-update-re-dao-vulnerability/. Accessed 7 Jan 2018

Assessment of Table Pruning and Semantic Interpretation for Sentiment Analysis Using BRAE Algorithm

G. V. Shilpa and D. R. Shashi Kumar

Abstract We propose bilingually compelled recursive auto-encoders (BRAE) to learn semantic expression embedding (smaller vector portrayals for phrases), which can recognize the expressions with various semantic implications. The BRAE is prepared in a way that limits the semantic separation of interpretation counterparts. Also, it augments the semantic separation of non-translation combinations at the same time. The model identifies how to insert each expression semantically in two dialects and also identifies how to change semantic inserting space in one dialect to the other. We assess our proposed strategy on two end-to-end SMT assignments (express table pruning and interpreting with phrasal semantic likenesses) which need to quantify semantic likeness between a source expression and its interpretation. The detailed tests demonstrate that the BRAE is strikingly compelling in these two assignments.

Keywords BRAE algorithm · Deep neural networks · Recursive auto-encoder · SMT · Semantic analysis · Deep neural networks (DNN)

1 Introduction

Because of the groundbreaking limit of highlight learning also, portrayal, deep (multi-layer) neural networks (DNN) have made an incredible progress in discourse and image processing. As of now, measurable machine interpretation (SMT) network has seen a solid enthusiasm for adjusting also, applying DNN to numerous assignments, for example, word arrangement, interpretation certainty estimation, express reordering forecast, interpretation displaying, and dialect demonstrating. The majority of

G. V. Shilpa (✉)
Department of Computer Science and Engineering, Vemana Institute of Technology, Bangalore, India
e-mail: shilpa.gv@vemanait.edu.in; gvshilpa03@gmail.com

D. R. Shashi Kumar
Department of Computer Science and Engineering, Cambridge Institute of Technology, Bangalore, India
e-mail: shashikumar.cse@citech.edu.in

© Springer Nature Singapore Pte Ltd. 2020
P. K. Mallick et al. (eds.), *Cognitive Informatics and Soft Computing*, Advances in Intelligent Systems and Computing 1040, https://doi.org/10.1007/978-981-15-1451-7_4

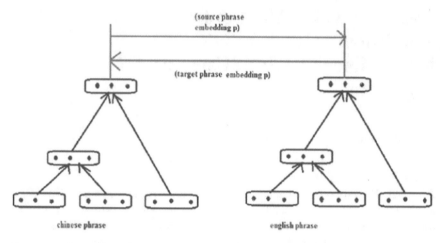

Fig. 1 Example of BRAE model

these works endeavor to make strides a few parts in SMT in light of word implanting, which changes over a word into a thick, low-dimensional, genuine esteemed vector portrayal. In this way, keeping in mind the end goal to effectively apply DNN to display the entire interpretation process, for example, demonstrating the interpreting procedure, learning conservative vector portrayals for the essential phrasal interpretation units is the basic and central work.

In this paper, we explore the articulation which addresses progression of words with a veritable regarded vector. Thus, accepting the articulation is a noteworthy bit of its internal words, we propose BRAE to learn semantic articulation embeddings. We utilize a case to clarify our model. As shown in Fig. 1, the Chinese expression on the left and the English expression on the privilege are interpretations with each other. With the educated model, we can unequivocally quantify the semantic similarity between a source state and an elucidation confident.

The preliminaries show that up to 72% of the articulation table can be discarded without enormous decreasing on the translation quality, and in deciphering with phrasal semantic similarities up to 1.7 BLEU score change over the condition of the craftsmanship standard can be cultivated. In many real-time situations, it is expected that the speaker's emotions need to be identified. In this context, BRAE helps in this aspect by exploiting all possible emotions exhibited by the public.

2 Literature Survey

As of now, state installing has drawn increasingly and more consideration. There are three primary viewpoints taking care of this undertaking in monolingual dialects. One strategy considers the expressions as pack of words what's more, utilizes a convolution model to change the word embeddings to state embeddings. Many researchers have found that the identification of sentiment/emotions in text or speech data is essential. Many computational linguistic researches have proved their applications in many real-time situations. Many published articles have shown that various areas such as forensic science, author identification, market analysis, prediction, and many other fields require sentiment analysis.

This sort of methodologies does not take the word arranged into account and loses much data. This sort of semi-directed expression inserting is truth be told performing phrase grouping with deference to the expression mark. Several researchers have explored many possibilities of identifying human emotions by applying various sentiment analysis techniques. These researchers found that the importance of identification of emotions is very important in several computational linguistic tasks. The semi-directed expression is a common phenomenon in natural language text inputs. Analysis of them using suitable computational methods shows that phrase identification and disambiguation is mandatory.

3 BRAE Model

This segment presents the bilingually compelled recursive auto-encoders (BRAE) that is roused by two perceptions. To start with, the recursive auto-encoder gives a sensible structure system to insert each expression. As shown in Fig. 2, x_1 and x_2 integrate to get y_1, x_3 to get y_2 and x_4 to get y_3. Further, y_1 and y_2 and in turn combining with y_3 generates the complete semantics of the text entered.

As these translation equivalents share the same semantic meanings, we utilize top notch state interpretation matches as preparing corpus in this work. In like manner, we propose the bilingually compelled recursive auto-encoders (BRAE), whose fundamental objective is to limit the semantic separation between the expressions and their interpretations.

$$y_1 = f(W(1)[x_1; x_2] + b)$$
$$y_2 = f(W(1)[y_1; x_3] + b)$$
$$y_3 = f(W(1)[y_2; x_4] + b)$$
$$xi = Lei \in Rn \tag{1}$$

Figure 2 outlines an occurrence of a RAE connected to a parallel tree, in which a standard auto-encoder (in box) is re-utilized at every hub. The standard auto-encoder goes for taking in a theoretical portrayal of its information. For two kids $c_1 = x_1$ and $c_2 = x_2$, the auto-encoder registers the parent vector y_1

$$p = f(W(1)[c_1; c_2] + b(1)) \tag{2}$$

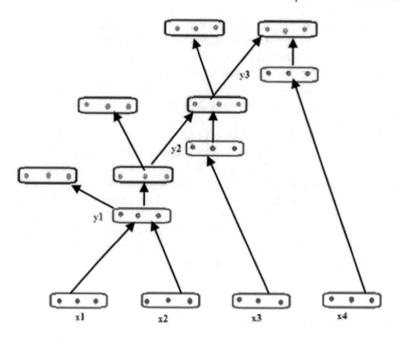

Fig. 2 RAE for four word phrase

where we increase the parameter framework.

$W(1) \in R_n \times 2n$, by the connection of two children

$$[c_1; \; c_2] \in R_2 n \times 1$$

In the wake of including an inclination term b(1), we apply a component astute initiation capacity such as $f = \tan h \, (\cdot)$, which is utilized as a part of our tests.

$$[c_{01}; \; c_{02}] = f(2)(W(2)p + b(2)) \tag{3}$$

where c_{01} and c_{02} are recreated children, $W(2)$ and $b(2)$ are parameter grid and predisposition term for recreation individually, and $f(2) = \tan h \, (\cdot)$. To get the ideal theoretical portrayal of the information sources, the standard auto-encoder attempts to limit the recreation blunders between the data sources also, the remade ones amid preparing:

$$Erec([c_1; \; c_2]) = 1/2\|[c_1; \; c_2] - [c_{01}; \; c_{02}]\|2 \tag{4}$$

Given $y_1 = p$, we can utilize Eq. 2 again to register y_2 by setting the kids to be $[c_1; c_2] = [y_1; x_3]$. A similar auto-encoder is re-utilized until the vector of the entire expression is created. For unsupervised expression inserting, the main objective is to limit the total of recreation blunders at every hub in the ideal double tree:

$$\text{RAE}\,\theta(x) = \arg\,\min \in y A(x) \Sigma\, s\, 2y\, \text{Erec}([c_1;\ c_2]s) \tag{5}$$

where x is the rundown of vectors of an expression, and $A(x)$ means all the conceivable paired trees that can be worked from inputs x.

3.1 Decoding with Phrasal Similarities

Other than using the semantic similarities to prune the articulation table, we also use them as two edifying features like the articulation translation probability to coordinate understanding speculations assurance sincerely busy deciphering. Further, we identify the similar phrases in the entered text input and decode them suitably. This will be performed by editing the features involved in them such as happy, sad, excited, depressed, and indifferent. As illustrated in the above RAE equation, we propose vectors of an expression which generates all possible conceivable paired trees.

3.2 Applications of BRAE

Other than using the semantic similarities to prune the articulation table, we furthermore use them as two edifying features like the articulation understanding probability to coordinate elucidation hypotheses assurance sincerely busy deciphering. Semantic similarities will definitely create problems in analyzing the text documents. Due to the existence of polysemy words in any language, we come across semantic similarities. Hence, it is essential to identify and disambiguate polysemy word/s encountered in an input text. Some researchers have found the ways to eliminate ambiguities in semantics by removing polysemy words with equivalent synonym words. It is evident from published articles that this work has substantial strength due to its complexities. Further, these techniques are applied and successfully obtained the results on Indian regional languages.

3.3 Extension of BRAE

The expressions having a similar significance are translation equivalents in various dialects, yet they summarize in one dialect. These expressions show varieties in evaluating the semantics of the entered text. First of all, converting the input text into an equivalent vector expression is a challenging task. This needs to be done by applying suitable computational linguistic approaches. This is due to the complexities involved in each regional language in this world. In this way, our model can be effortlessly adjusted to learn semantic state embeddings utilizing summarizes. The BRAE pseudocode has been extended for substantial study in obtaining translation equivalents in many Indian regional languages. The handwritten rules are extended to map to many real-time situational texts in natural languages. But still, it is implemented with limitations, as new words will be always appended for languages.

4 Conclusion

The paper explored the bilingually compelled recursive auto-encoders in learning state embeddings, which can perceive phrases with different semantic ramifications. With the objective to restrict the semantic partition between translation partners and intensify the semantic separate between non-understanding consolidations at the same time, the academic model can semantically embed any articulation in two lingos and can change the semantic space in one vernacular to the following. The complexity of language processing increases as it involves polysemy words in them. As mentioned earlier, it is very essential to identify and disambiguate polysemy words in the entered text input. Situation becomes much complex when the input text has many polysemy words in a sentence. Two SMT errands are incorporated which are start to finish to test the power of the proposed presentation at taking in the semantic articulation embeddings. The test comes about exhibiting that the BRAE is astoundingly reasonable in state table pruning and unraveling with phrasal semantic similarities.

References

1. Zhang, J., Liu, S., Li, M., Zhou, M., Zong, C.: Bilingually-constrained Phrase Embeddings for Machine Translation. National Laboratory of Pattern Recognition, CASIA, Beijing, P.R. China
2. Auli, M., Galley, M., Quirk, C., Zweig, G.: Joint language and translation modeling with recurrent neural networks. In: Proceedings of the 2013 Conference on Empirical Methods in Natural Language Processing, pp. 1044–1054 (2013)
3. Li, P., Liu, Y., Sun, M.: Recursive autoencoders for ITG-based translation. In: Proceedings of the Conference on Empirical Methods in Natural Language Processing (2013)
4. Liu, L., Watanabe, T., Sumita, E., Zhao, T.: Additive neural networks for statistical machine translation. In: 51st Annual Meeting of the Association for Computational Linguistics, pp. 791–801 (2013)
5. Duh, K., Neubig, G., Sudoh, K., Tsukada, H.: Adaptation data selection using neural language models: experiments in machine translation. In: 51st Annual Meeting of the Association for Computational Linguistics, pp. 678–683 (2013)
6. Gao, J., He, X., Yih, W.-T., Deng, L.: Learning semantic representations for the phrase translation model. arXiv preprint arXiv:1312.0482 (2013)
7. Vaswani, A., Zhao, Y., Fossum, V., Chiang, D.: Decoding with largescale neural language models improves translation. In: Proceedings of the 2013 Conference on Empirical Methods in Natural Language Processing, pp. 1387–1392 (2013)

8. Yang, N., Liu, S., Li, M., Zhou, M., Yu, N.: Word alignment modeling with context dependent deep neural network. In: 51st Annual Meeting of the Association for Computational Linguistics (2013)
9. Zou, W.Y., Socher, R., Cer, D., Manning, C.D.: Bilingual word embeddings for phrase-based machine translation. In: Proceedings of the 2013 Conference on Empirical Methods in Natural Language Processing, pp. 1393–1398 (2013)
10. Zens, R., Stanton, D., Xu, P.: A systematic comparison of phrase table pruning techniques. In: Proceedings of the 2012 Joint Conference on Empirical Methods in Natural Language Processing and Computational Natural Language Learning, pp. 972–983 (2012)

Feature Extraction and Classification Between Control and Parkinson's Using EMG Signal

Roselene Subba and Akash Kumar Bhoi

Abstract The main objective of the proposed work is to classify and differentiate Parkinson's disease from other neuromuscular disease with the help of Electromyogram (EMG) signal. An Electromyogram signal detects the electric potential generated by muscle cells when these muscle cells contract or relax. However, these electromyography signal itself failed to differentiate between these neuromuscular diseases as their symptoms are almost the same. Therefore, certain features were examined and studied *like* average distance peaks, discrete wavelet functions, entropy, band power, peak-magnitude-to-RMS ratio, mean complex cepstrum and maximum value of single-sided amplitude, etc. These features were extracted, and with these features, we can differentiate between these neuromuscular diseases including Parkinson's disease. Two classifiers were used for detection and classification of Parkinson's, they were Decision Tree and Naive Bayes. However, the accuracy in Decision Tree was found out to be 88.38%, while the accuracy in Naive Bayes was found out to be 54.07%.

Keywords Electromyogram · Parkinson's disease · Feature extraction · Decision Tree · Naive Bayes Classifier

1 Introduction

Saara M. Rissanen et al. have presented a method where acceleration signal features were extracted using 12 different EMG and which is intended to generate feature vectors with higher dimensions [1]. Histogram and crossing rate (CR) analysis-based approach of sEMG could be providing novel morphological information in the Parkinson's disease (PD) [2].

R. Subba · A. K. Bhoi (✉)
Department of Electrical & Electronics Engineering, Sikkim Manipal Institute of Technology,
Sikkim Manipal University, Gangtok, Sikkim, India
e-mail: akash730@gmail.com; akash.b@smit.smu.edu.in

R. Subba
e-mail: subbaroselene06@gmail.com

© Springer Nature Singapore Pte Ltd. 2020   45
P. K. Mallick et al. (eds.), *Cognitive Informatics and Soft Computing*,
Advances in Intelligent Systems and Computing 1040,
https://doi.org/10.1007/978-981-15-1451-7_5

Kugler, P. et al. have studied about the surface electromyography (sEMG) during dynamic movements which could also be one of the possible modality movements [3]. For better understanding between control and Parkinson's disease, pattern recognition method could be used as an significant tool [4]. Jeon, H. et al. have proposed machine-learning algorithms to predict unified Parkinson's disease rating scale (UPDRS) [5].

The tremor analysis in PD can also be done using muscle activation which is clinically measured from sEMG, and results have been shown that sEMG analysis proven to be a discriminate marker for healthy and PD subjects [6–9]. Chowdhury, R et al. in their review article targeted two prominent areas, i.e., (i) preprocessing of EMG signal, where different areas were focused *like* inherent noise in the electrode, movement artifact, electromagnetic noise, cross talk, internal noise, inherent instability of the signal, Electrocardiographic (ECG) artifacts (ii) processing and classification of EMG signals, where various techniques were discussed *like* wavelet analysis, higher-order statistics (HOS), empirical mode decomposition (EMD), independent component analysis (ICA). Further statistical features were also considered for performing classification among different hand motions [10]. This idea further extended, and specific classification approach between healthy and PD is adopted in the proposed work. In our earlier work, frequency-domain power was calculated and considered as one of the significant features for classification among control and PD [11]. Here, several other factors have been taken into account for classification work.

2 Methodology

This work also uses the same gait dynamics in neurodegenerative disease data [12] base as implemented in the earlier study [11]. The proposed work follows the steps *such as* EMG signal segmentation, preprocessing [11], feature extraction and classification as depicted in Fig. 1. Moreover, certain features like average distance peak, discrete wavelet function, entropy, band power, peak-magnitude-to-RMS ratio, mean complex cepstrum (Fig. 2) are extracted from 10 s segmented healthy and PD subjects and further processed for classification.

These features were derived from each subjects of healthy and PD for classification and detection of Parkinson's disease.

Average Distance Peak

In order to find the average distance peak of the input EMG signal, the following parameters needed to be derived

- Peak
- Difference between the peaks
- Average distance of the peaks.

Fig. 1 Proposed
methodology

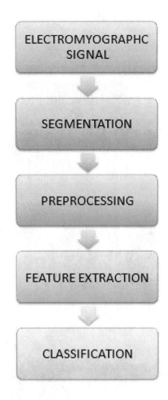

Automatic multiscale-based peak detection (AMPD) could be adopted which calculates and finds out the peak points of a matrix operation and also determines the local maxima. This could be viewed in the following stages (Fig. 3).

Wavelet Decomposition

The EMG signal that we have needs to be compressed in order to extract its features and to study it in detail. In our case, we have compressed the signal using the wavelet transform and discrete cosine transform as it can investigate the signal in short duration of signals.

DWT is given by

$$W(j, k) = W_\emptyset(j_0, k) + W_\emptyset(j, k) \tag{1}$$

where $W_\emptyset$ is a scaling function,

$$W_\emptyset(j_0, k) = \frac{1}{\sqrt{M}} \sum_n s(n) W_\emptyset(n) \tag{2}$$

$$W_\emptyset(j, k) = \frac{1}{\sqrt{M}} \sum_n s(n) W_\emptyset(n) \tag{3}$$

Fig. 2 Features extracted for classification between control and PD

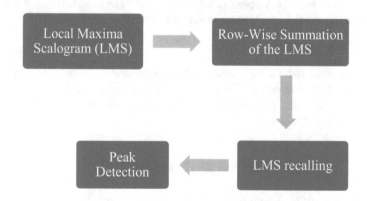

Fig. 3 Steps involved in the automatic multiscale-based peak detection (AMPD)

$J \& K$ are scale and location index. Similarly, inverse discrete wavelet transform (IDWT) reconstructs the wavelet coefficients into original signal,

$$s(n) = \frac{1}{\sqrt{M}} \left[\sum_K W_\emptyset(j,k)\emptyset_{j_0,k}(n) + \sum_{j=j_0}^{\infty} \sum_k W_\emptyset(j,k)\emptyset_{j,k}(n) \right] \quad (4)$$

Entropy

The uncertainty of EMG signal could be measured using entropy. It could also measure the chaos level in a system, and the nonlinear time series EMG signal could be quantified with its degree of complexity [13]. Let E be a set of finite discrete random variables $E = \{e_1, e_2, \ldots, e_m\}$, $e_i \in R^d$, Shannon entropy, $S(E)$ is defined as [14]

$$S(E) = -c \sum_{i=0}^{m} p(e_i) \ln p(e_i) \quad (5)$$

where $c =$ positive constant, $p(e_i) =$ probability of $e_i \in E$ and satisfying;

$$\sum_{i=0}^{m} p(e_i) = 1 \quad (6)$$

Band Power

$$\text{Power of a signal} = \frac{\text{sum of the absolute square of its time}-\text{domain samples}}{\text{signal length}} \quad (7)$$

The band power is nothing but signal power where power of a signal is equal to the total sum of absolute square of its time domain samples divided by its signal length. The band powers are the sum of the powers in a certain frequency band by the total sum of the powers in all the other frequency bands.

Peak-Magnitude-to-RMS Ratio

Peak-magnitude-to-RMS ratio provides the largest absolute value in the input EMG signal to the root-mean-square value of the input EMG signal. The peak-magnitude-to-RMS ratio is derived using

$$\frac{\|X\|_{\infty}}{\sqrt{\frac{1}{N} \sum_{n=1}^{N} |X_n|^2}} \quad (8)$$

Complex Cepstrum

Complex cepstrum could also be considered as one of the EMG features. It may be derived using following expression for a sequence e which is calculated by taking Fourier transform of e by computing complex natural logarithm. The resulting inverse Fourier transforms sequence could be

$$\hat{e} = \frac{1}{2\pi} \int_{-\pi}^{\pi} \log[E(e^{j\omega}d\omega)] \qquad (9)$$

The **maximum value of single-sided amplitude** is also considered as one of the decisive features of EMG signal.

3 Classification

The extracted features from healthy and PD subjects as mentioned are fed to the two classifiers, i.e., Decision Tree (DT) [11] Naive Bayes Classifier (NBC) [15]. Figures 4 and 5 show the classification results using six different features. However, for classification approach, wavelet decomposition features are not taken into consideration due to excessive usage in earlier studies.

Different notations for the features used in Figs. 4 and 5 are, e.g., band power: BP, Shannon entropy: SE, mean complex cepstrum: MCC, peak-magnitude-to-RMS

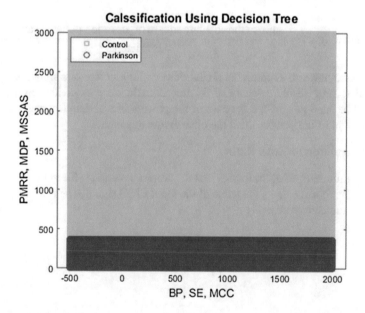

Fig. 4 The classification between Parkinson's and control signal using Decision Tree

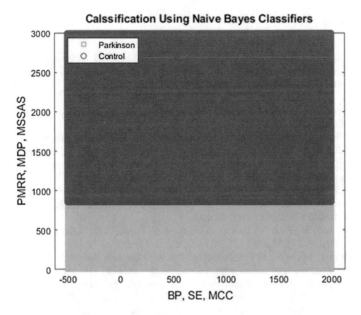

Fig. 5 The classification between Parkinson's and control signal using Naive Bayes Classifier

ratio: PMRR, mean difference between peaks: MDP, maximum value of single-sided amplitude spectrum: MSSAS.

The error using Decision Tree is found to be 11.62% (accuracy: 88.38%) and error using Naïve Bayes Classifier: 45.93% (accuracy: 54.07%) between healthy and PD subjects by considering six sets of feature values.

4   Conclusion

The visual representation of EMG signals for healthy and PD is moreover similar, and it is not possible to make any clinical decision via visual analysis. The EMG signals used in this work have been collected from the foot using force resistive sensors for gait analysis purpose. However, these EMG signals have been studied by extracting suitable features by considering time-frequency domain scenario and fed to the classifiers. The DT performance is found to be more prominent in comparison with NBC. This analysis provides an significant path to explore further finding by exploring such EMG signals for control and PD classification work.

References

1. Rissanen, S.M., Kankaanpää, M., Meigal, A., Tarvainen, M.P., Nuutinen, J., Tarkka, I.M., Airaksinen, O., Karjalainen, P.A.: Surface EMG and acceleration signals in Parkinson's disease: feature extraction and cluster analysis. Med. Biol. Eng. Comput. **46**(9), 849–858 (2008)
2. Rissanen, S., Kankaanpää, M., Tarvainen, M.P., Nuutinen, J., Tarkka, I.M., Airaksinen, O., Karjalainen, P.A.: Analysis of surface EMG signal morphology in Parkinson's disease. Physiol. Meas. **28**(12), 1507 (2007)
3. Kugler, P., Jaremenko, C., Schlachetzki, J., Winkler, J., Klucken, J., Eskofier, B.: Automatic recognition of Parkinson's disease using surface electromyography during standardized gait tests. In: 2013 35th Annual International Conference of the IEEE Engineering in Medicine and Biology Society (EMBC), pp. 5781–5784. IEEE (2013, July)
4. Putri, F.T., Caesarendra, W., Ariyanto, M., Pasmanasari, E.D.: Electromyography gait test for Parkinson disease recognition using artificial neural network classification in Indonesia. Majalah Ilmiah Momentum **12**(2) (2016)
5. Jeon, H., Lee, W., Park, H., Lee, H., Kim, S., Kim, H., Park, K.: Automatic classification of tremor severity in Parkinson's disease using a wearable device. Sensors **17**(9), 2067 (2017)
6. Muthuraman, M., et al.: A new diagnostic test to distinguish tremulous Parkinson's disease from advanced essential tremor. Mov. Disord. **26**(8), 1548–1552 (2011)
7. Kugler, P., et al.: Automated classification of Parkinson's disease and essential tremor by combining electromyography and accelerometer signals. Basal Ganglia **3**(1), 61 (2013)
8. Rissanen, S.M., et al.: Discrimination of EMG and acceleration measurements between patients with Parkinson's disease and healthy persons. In: 2010 Annual International Conference of the IEEE, Engineering in Medicine and Biology Society EMBC, pp. 4878–4881 (2010)
9. Askari, S., et al.: An EMG-based system for continuous monitoring of clinical efficacy of Parkinson's disease treatments. In: 2010 Annual International Conference of the IEEE, Engineering in Medicine and Biology Society EMBC, pp. 98–101 (2010)
10. Chowdhury, R., Reaz, M., Ali, M., Bakar, A., Chellappan, K., Chang, T: Surface electromyography signal processing and classification techniques. Sensors **13**(9), 12431–12466 (2013)
11. Bhoi, A.K.: Classification and clustering of Parkinson's and healthy control gait dynamics using LDA and K-means. Int. J. Bioautomation **21**(1) (2017)
12. Hausdorff, J.M., Lertratanakul, A., Cudkowicz, M.E., Peterson, A.L., Kaliton, D., Goldberger, A.L.: Dynamic markers of altered gait rhythm in amyotrophic lateral sclerosis. J. Appl. Physiol. **88**, 2045–2053 (2000)
13. Phung, D.Q., Tran, D., Ma, W., Nguyen, P., Pham, T.: Using Shannon entropy as EEG signal feature for fast person identification. In: ESANN, vol. 4, issue No. 1, pp. 413–418 (2014, April)
14. Kannathal, N., Choo, M.L., Acharya, U.R., Sadasivan, P.K.: Entropies for detection of epilepsy in eeg. Comput. Methods Prog. Biomed. **80**(3), 187–194 (2005)
15. Machado, J., Balbinot, A.: Executed movement using EEG signals through a Naive Bayes classifier. Micromachines **5**(4), 1082–1105 (2014)

Utilization of Data Analytics-Based Approaches for Hassle-Free Prediction Parkinson Disease

S. Jeba Priya, G. Naveen Sundar and D. Narmadha

Abstract Individuals with Parkinson's disease don't have a sufficient substance called dopamine since a few nerves in the brain lose their functionality. Individuals with Parkinson's disease are in deceptive and damaging condition . Diagnosing this disease on the basis of the motor and cognitive shortage is extremely critical. Machine learning approaches are utilized to settle on prescient choices via preparing the machines to learn with the trained information. It assumes a fundamental role in foreseeing Parkinson's disease in its beginning periods. In this paper, our primary goal is to build up an advanced algorithm to accomplish good classification accuracy utilizing data mining techniques. In this procedure, we distinguish some current algorithms (e.g., Naïve Bayes, decision tree, discriminant, and random forest) and its execution is broken down. Result acquired through these grouping algorithms is moderately prescient. During the time spent in the computation of these algorithms, Naïve Bayes can construct the framework with the high precision rate of 94.11%.

Keywords Parkinson's disease · Naïve Bayes · Decision tree · Random forest · Classification · Regression

1 Introduction

Parkinson's disease (PD) is a dynamic degenerative disease. More than 6.3 million individuals were influenced because of this illness. This is considered as the second biggest occurring neurodegenerative disease [1]. Data mining is the computational procedure of finding covered up and conceivably valuable data. In the distinct view,

S. J. Priya (✉) · G. N. Sundar · D. Narmadha
CSE Department, KITS, Coimbatore, India
e-mail: jebapriya@karunya.edu

G. N. Sundar
e-mail: naveensundar@karunya.edu

D. Narmadha
e-mail: narmadha@karunya.edu

© Springer Nature Singapore Pte Ltd. 2020
P. K. Mallick et al. (eds.), *Cognitive Informatics and Soft Computing*,
Advances in Intelligent Systems and Computing 1040,
https://doi.org/10.1007/978-981-15-1451-7_6

designs are human decipherable. The illustrative techniques for data mining are clustering, association rule discovery, and sequential pattern discovery. For the prognosticative strategy, it will discover the estimation of attributes utilizing estimations of different attributes and the strategies which correspond to the predictive techniques are classification, regression, and deviation detection. The Parkinson's disease extent is expanded to 20% in 2020. In past days, the senior people were struck by this disease. Nowadays, this disease influences the youngsters also. This disease strikes the different parts of the brain. The neural cells get influenced and wind up in death [2]. The different examinations with respect to the Parkinson's disease demonstrate that the patients have the issue of interpreting the feelings from speech [3], detachment in facial expressions [4], likewise affecting the color vision and difference sensitivity of the human eye [5]. While the patient got influenced by this PD, there is no fix, there are no drugs to moderating the movement of the ailment, and there are no techniques to anticipate it. The right diagnosis of PD is just 75%. In the event that the patients get affected for more than 5–10 years without appropriate diagnosis, 70% of the neurons of the brain will be influenced [5]. Human–computer interaction assumes a fundamental role in foreseeing the PD patients. By utilizing HCI innovation, interaction can be made among humans and the PCs. Diagnostics markers can be utilized to gauge physiological, psychological, and mental conditions of the people and the outcomes will be estimated and put away. By utilizing this reading, the expectation accuracy of the PD patients will be determined [6].

In this paper, we are concentrating on data mining grouping algorithms that are talented in foreseeing certain parameter for the given information. Here, we are using three classification algorithms to examine and investigate the dataset identified with Parkinson's disease. The paper is sorted out so that the related works and the current strategies used to anticipate Parkinson's disease utilizing data mining are examined in Sect. 2. Section 2.1 gives the overview of the executed classification algorithms. The discussions and results will be pictured in the Sect. 4.

2 Related Work

There are many data mining algorithms are existing to calculate the accuracy of PD. There are several factors to predict the Parkinson's disease. Some of the factors to identify such as voice data; i.e., voice recognition is also one of the solution [7]. By using brain imaging such as FP-CIT-SPECT and based on frequency also PD can be predicted. Motor learning in PD describes the ability to incorporate learning skills. It provides controversial results. Two stages of motor learning are classified into explicit and implicit learning. Explicit is the actual performance, verbal or visual information. Implicit learning is outside conscious attempt to learn it. The different phases of motor learning are cognitive, associative, and autonomous phase. Cognitive phase labels the operation that has to be performing through interpretation of verbal instructions. Associative phase refines the movement through practice. Autonomous phase is the final phase, and it becomes more automatic progression to this level of learning allowing to perform the skills anywhere [8].

By using pen-and-tablet device, the spiral and handwriting template has been drawn and the severity of the persistence of Parkinson's disease. Spiral drawing has been recorded by the volunteers, and the velocity and the pressure will be calculated. The main objective is to differentiate PD-affected and normal people. For this method, a dataset has 44 age matched subjects and it is subdivided in two groups, namely 20 healthy patient data and 24 affected patients' information. Then, they applied a broad machine learning technique to build a method for classifying the unknown subjects based on the line-drawing approach or spiral approach. This method was able to achieve a prediction accuracy of 91% [9].

In this method, data preprocessing steps were involved and then clustering using EM has been done to cluster the data which is already preprocessed. PCA is used to reduce the dimensionality problem of the attribute. Noise in the data can be reduced by using classification and regression methods and designing the algorithm for decision rules related to the data. Fuzzy rule-based system, input fuzzification, generating membership function, extracting fuzzy rules, and output defuzzification were utilized as an analysis method for disease prediction, and they achieved good accuracy percentage [10].

In hybrid intelligent system, the main aim is to optimize the computation time and to increase the prediction accuracy of Parkinson's disease. UPDRS—unified Parkinson's disease rating scale—is used to assess the stage of PD. Discovering the relationship between speed signal property and UPDRS score is an important task in PD diagnosis. Incremental machine learning technique is used in the UPDRS attribute for PD prediction. Incremental support vector machine is used to predict the total_UPDRS and motor_UPDRS. Supervised learning models will analyze the data used for classification and regression. The hybrid intelligent method is effective in predicting UPDRS, and it leads to increase in the accuracy performance [11] (Fig. 1).

2.1 Proposed Method

One of the principle issues of Parkinson's is the degeneration of dopamine and, when that happens, the speed and the compression would be unruly, which would really result in undesirable moment. This work of dopamine and the dopamine transporter is observed by an imaging called FP-CIT SPECT. The primary sign of FP-CIT SPECT is separating mellow or unverifiable Parkinson's patients. Parkinson's disease (PD) is a typical disorder, and the diagnosis of Parkinson's malady is clinical and depends on the nearness of characteristic motor side effects. The precision of the clinical finding of PD is as yet constrained (Fig. 2).

Dataset which is used in this analysis is taken from the Kaggle Web site. It consists of 21 voice parameters, and it has thousands of patients' information. UPDRS range is found to be dependent variable using regression. Since UPDRS range is unified Parkinson's disease rating scale to detect the progression patient had PD or

S. J. Priya et al.

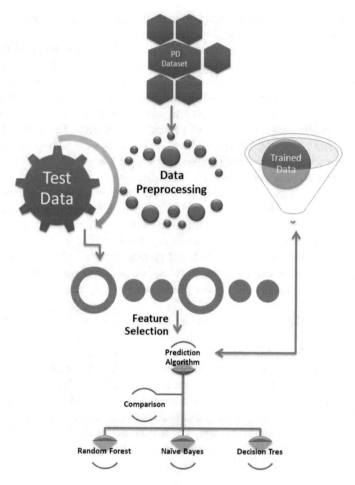

Fig. 1 Architecture diagram of proposed diagram

Fig. 2 Comparisons of
accuracy between classifiers

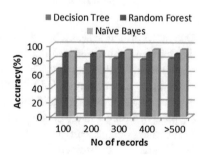

Fig. 3 Comparisons of error
rate between classifiers

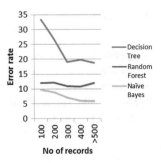

not. Basically data goes through a series of steps during preprocessing. Data cleaning, data integration, data transformation, and data reduction are the various stages.
Data cleaning can be done by filling the missing values or removing the noisy data.
Different representations of the same data are put together, and the conflicts can be
rectified known as data integration. Normalization, aggregation, and generalization
for the data will be processed, and this method is called as data transformation. Data
warehouse can be used to reduce the data known as data reduction. Generally, data
preprocessing can be done by can be done normalization and standardization. Here,
in this dataset, both normalization and standardization have been applied.

After preprocessing stage, the dataset is isolated into two gatherings to be test
data and training data. In the subsequent stage, feature selection is being carried out.
Training data is about 70%, and the remaining 30% amounts to test data. Feature
selection is otherwise called as variable selection or attributes selection. In this proposed algorithm, we have used the Pearson correlation coefficient method in which
P value will be determined, i.e., the linear correlation between two attributes. By
utilizing this feature selection method, only important dependent attributes can be
inferred (Fig. 3).

Further, some classifier has been used for the classification process. Decision tree,
random forest, and Naïve Bayes are the three different types of classifier used for the
comparison purpose. Decision tree is totally depending on the output decision. The
attributes age, jitter, and UPDRS are considered. In age attribute, the subtree will be
formed as (10–30), (30–50), and (50–70). The classification proceeds for the further
attributes, and it finishes until decision cannot be taken. The accuracy achieved by this
method is 81.63. In Naïve Bayes classifier method, the attributes similarity will be
calculated and it will be combined. While combining few attributes, the probability
value may vary. So in this regard, Bayes theorem can be used. In step 2, the frequency
of the variant is calculated and a tabulation is done. Then, likelihood tabulation is
done for any of the variant, and the probabilities are calculated and finally imported
in the Bayes theorem, from which we will get the P value ($P < 0.5$-PD not present, $P
> 0.5$-PD present). Accuracy attained by Naïve Bayes algorithm is 94.11%. Random
forest is one of the simplest algorithms which can be used both for classification
and regression. Random forest algorithm has been used to develop multiple CART
model with various samples and different initial variables. The accuracy obtained by
this algorithm is 89.23%.

3 Conclusion

In this research paper, a review has been made on the various machine learning models used for Parkinson's disease prediction. After analyzing the pitfalls of the existing model, a new model has been proposed which includes a combination of machine learning model for accurate disease prediction. The proposed work has attained an increase in the prediction accuracy of Parkinson's disease.

References

1. Parkinson's Australia: Parkinson's—description, incidence and theories of causation. http://www.parkinsons.org.au/information_sheets
2. Blonder, L.X., Gur, R.E., Gura, R.C.: The effects of right and left hemiparkinsonism on prosody. Brain Lang. **36**, 193–207 (1989)
3. Ariatti, A., Benuzzi, F., Nichelli, P.: Recognition of emotions from visual and prosodic cues in Parkinson's disease. Neurol. Sci. **29**, 219–227 (2008)
4. Dara, C., Monetta, L., Pell, M.D.: Vocal emotion processing in Parkinson's disease: reduced sensitivity to negative emotions. Brain Res. **1188**, 100–111 (2008)
5. Fearnley, J.M., Lees, A.J.: Ageing and Parkinson's disease: substantia nigra regional selectivity. Brain. **114**(5), 2283–2301 (1991)
6. Kalman, Y.M.: HCI markers: a conceptual framework for using human-computer interaction data to detect disease processes. In: The 6th mediterranean conference on information systems (MCIS), Limassol, Cyprus (2011)
7. Smith, M.E., Ramig, L.O., Dromey, C., Perez, K.S., Samandari, R.: Intensive voice treatment in Parkinson disease: laryngostroboscopic findings. J. Voice **9**(4), 453–459 (1995)
8. Marinelli, L., Quartarone, A., Hallet, M., Ghilardi, M.F.: The many facts of motor learning and their relevance for Parkinson's diseases. J. Clin. Nerophysiology **128**(7), 1127–1141 (2017)
9. Kotsavasiloglou, C., Kostikis, N., Hristu-Varsakelis, D., Arnaoutoglou, M.: Machine learning-based classification of simple drawing movements in Parkinson's disease. J. Biomed. Signal Process. Control **31**, 174–180 (2017)
10. Nilashi, M., bin Ibrahim, O., et al.: An analytical method for diseases prediction using machine learning techniques. J. Comput. Chem. Eng. **106**(2), 212–223 (2017)
11. Nilashi, M., et al.: A hybrid intelligent system for the prediction of Parkinson's disease progression using machine learning techniques. J. Biocybern. Biomed. Eng. **38**(1), 1–15 (2018)

A System for the Study of Emotions with EEG Signals Using Machine Learning and Deep Learning

Vasupalli Jaswanth and J. Naren

Abstract Human life deals with a lot of emotions. Analyzing the emotions using EEG signals plays a pivotal role in determining the internal/inner state of a particular human. EEG deals with the spontaneous electrical activity of neurons as recorded from multiple electrodes placed in the interior region of the brain. Initially, EEG signals are captured and preprocessed for the removal of noise signals. Selection of appropriate classification techniques in emotion analysis is an important task. The classifiers like k-nearest neighbor (k-NN), SVM, LDA were evaluated. Performances of the classifiers in analyzing a wide range of emotions (arousal and valence emotions) were examined. The results examined demonstrated that emotion analysis using EEG signals is highly advantageous and efficient than the existing traditional recognition systems.

Keywords EEG · Machine learning · Feature extraction · Classification

1 Introduction

The advancement of technology in recent times has led to solutions in a wide variety of problems. Despite all advancements, the mental state of a person has not been properly accessed. Emotions play a vital role in defining a person's character. So the motive is to scrutinize different emotions of a person and determine an individual's mental health. Depression, Social Anxiety Disorder, Stress are the most serious problems which invade an individual mental and physical health and are prevailing in recent times. Depression is associated with loss of interest, guilt feeling, poor concentration, low self-esteem and loneliness. Stress is kind of psychological pain experienced when people feel that they are strained. Social Anxiety Disorder is the fear of interacting with people due to the feeling of being negatively scrutinized and

V. Jaswanth · J. Naren (✉)
SASTRA Deemed University, Tirumalaisamudram, Thanjavur, Tamil Nadu, India
e-mail: naren@cse.sastra.edu

V. Jaswanth
e-mail: 121015124@sastra.ac.in

© Springer Nature Singapore Pte Ltd. 2020
P. K. Mallick et al. (eds.), *Cognitive Informatics and Soft Computing*,
Advances in Intelligent Systems and Computing 1040,
https://doi.org/10.1007/978-981-15-1451-7_7

judged by people. Diagnosis of the above-mentioned problems would help a lot of younger generations to lead a happy and peaceful life.

Brain is the center for all emotions. The four lobes of the brain are significant in inducing emotions in which the frontal lobe that controls important cognitive skills is responsible for effecting the mental state. Analyzing the emotions using EEG has not been properly invaded. The main motive is to detect human's mental state using EEG signals. Electroencephalograph (EEG) provides an easy approach to determine the responses of a person to external stimuli. EEG signal is a simple recording of the electric potentials developed by the neuronal activity in the brain. Electrodes attached to scalp analyze the electrical impulses in the brain and sends signals to the computer for recording the results. EEG analyzing techniques help us to familiarize with the diagnosis of depression, SAD and stress which is an extension to the paper.

Series of tasks involved in the paper are summarized as follows: A DEAP dataset has been taken and preprocessed to remove the artifacts. The preprocessing techniques that have been used are Discrete Wavelet Transforms, Laplacian Transforms, etc. The data that has been preprocessed is fed into machine learning classifiers. Then emotions (arousal and valence emotions) are distinguished using classifiers, accuracy of each classifier has been recorded which is used for the further extension of research.

Machine learning classifiers play a crucial role in analyzing emotions. Feature selection and extraction are the main challenging tasks involved. To recognize the emotions in real-time, Short Time Fourier Transform (STFT) is used [1]. The features are identified and extracted from the used database. The next step is a feature selection technique that has been used where the most required optimum features have been identified. The optimized features are further fed into three conventional classifiers KNN, SVM and logistic regression and performance and accuracy have been evaluated and recorded.

2   Literature Survey

Isabelle et al. presented a database consisting of EEG data for the analyses of emotions. A reconstruction of various brain regions was made for improving the classification results [2].

Wang et al. elucidated on how emotions can be analyzed for inferring the personality of participants from EEG signals with the help of SVM classifier. Five dimensions of personality traits were predicted with a performance of round 80% [3].

Zhang et al. explained how feature extraction and analysis with an interbrain EEG for emotion tagging by video watching. The performance measure between interbrain features and amplitude feature were examined. Inter brain EEG gave promising results [3].

Andrey et al. studied the influence of depression through EEG by facial expressions [4].

Mandeep et al. designed an emotion classifier through evoked EEG in four quadrants which developed a real-time emotion recognition system through BCI gave a decent accuracy [5].

3 Architecture

See Fig. 1.

4 Architectural Framework

The framework gives us sequence of steps underwent in the analysis of emotions using EEG. The above-mentioned framework has been arrived based on previous work that has been done by other people related to emotion analysis. Elaboration of each step by correlating with a few papers related to emotion analysis has been done below.

4.1 EEG Dataset

Dataset plays a major part in the analysis of emotion. A dataset can be a live dataset or can be a publicly available one. Yong-Jin Liu et al. used a live dataset where selected participants were made to watch few emotion-inducing scenes. The EEG signals can be captured using Emotive epoch, which is an EEG signal detecting device. This

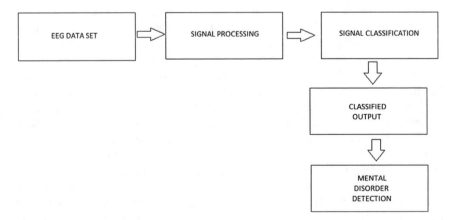

Fig. 1 Architecture for emotion analysis using EEG

Emotive epoch captures the patient's brain state and displays it to us along with the EEG, signal quality, etc. [1].

Wei-Long Zheng et al. used a dataset called DEAP and SEED to determine the performance of machine learning algorithms used. The DEAP dataset used for the proposed study consists of EEG signals of 32 people. These signals were captured while they were watching 40 excerpts of music videos, each of one minute long. The EEG signals were examined at 128 Hz and recorded using 32 active electrodes. The SEED dataset was also used for the same study. The distinction between SEED and DEAP is that SEED comprises of 3 sessions where the gap between each session is slightly longer than that of the DEAPldataset [6]. D. Jude Hemanth et al. used a DEAP data set for the work done. As already mentioned the database consists of emotions induced by musical videos which have a predefined arousal and valence level. Mohammad Soleymani et al. used a dataset which is a subset of mahnob hci database. The database contains recordings based on 2 experiments that had emotional responses to short videos used. 20 video clips were used out of which 14 were clips from movies which were chosen based on a previous basic study. The video clips were kept along to provide time for inducing emotions and conveying the content. The work done was to make a comparative study between emotion analysis using EEG and image processing. So in order to capture the facial expressions, a front video was used which recorded at 60 frames per second. From the study a model is designed to analyze emotions using a SEED dataset for maintaining uniqueness since it has not been used in such studies more often.

4.2 Feature Extraction and Selection

Feature extraction is a process of taking a raw dataset and extracting the variables (features) that can be used for a particular study. Feature selection is used to select a group of optimal features from the features extracted, for a more accurate and relevant approach. However, Feature Selection is not more used since the features obtained from feature extraction are optimal ones. In accordance with the previous studies, there are few feature extraction methods which are frequently used.

An elaboration of a few previous works on feature extraction methods is as follows: Liu et al. [1] used a short-time Fourier transform (STFT) for feature extraction which implements sliding window technique. TF analysis describes the EEG signals by visualizing them in 2 dimensions. STFT basically involves the decomposition of long EEG signals into smaller components and each component is evaluated individually, their Fourier transform is calculated for time t. In a nut shell, STFT analyses each frame of the signal which moves with time. The frame can be assumed to be stationary and its Fourier transform can be evaluated only if the window that moves with time is narrower [1]. The relation between variation in frequency and time can be evaluated by moving the window along the time axis. In the above study 105 EEG features were detected [1]. Then the extracted features are fed into a linear Discriminant analysis which is a feature selection algorithm. The above algorithm reduces dimensions by removing redundant and dependent features by transforming features from a higher dimensional space to space with lower dimensions by maximizing fisher separation criterion. The model designed for the paper also uses short time Fourier transforms as the feature extraction method.

Zheng et al. [6] has concluded that the below combinations are perfect for emotion analysis: power spectral density (PSD), differential asymmetry (DA), rational asymmetry (RASM), asymmetry (ASM), and differential caudality (DCAU) features from the EEG. The differential entropy features are defined using Gaussian distribution formula. The ASM is the combinations of DASM and RASM features. The differences between DE features are stated as DCAU features [6].

Pliho et al. [7] have chosen wavelet transform for the extraction of features. Wavelet transform decomposes the original signal into a series of component signals and maps the one-dimensional signal into a two-dimensional signal. The decomposition of signals via WT generates a group of wavelet coefficients that describes the energy as equation consisting of time and frequency [7]. The wavelet is chosen due to its near optimized time-frequency localization property and then the features are extracted [8]. For more details refer [7].

Since there are limited number of data points in the dataset, feature selection is necessary. The feature selection method which is planned to use in the model is relief algorithm [9] that is an extension of relief algorithm. The algorithm works by selecting a particular instance and find its k nearest hits and misses. The hits are considered to be belonging to the same feature and misses are instances belonging to different features. Then the weight vector is calculated from the obtained results [9]. Then the highest vector features are selected so that the features to be used are reduced and not more than data points of the dataset which has been used.

4.3 Classification

Classification is the process of segregating or classifying the given data/features into their respective classes (here, the classes are referred to by emotions like happy, sad, angry, fear) using machine learning algorithms. An important part of classification is determining and recording the accuracy of algorithms implemented for future use.

Liu et al. [1] used SVM classifier. SVM uses a kernel which is actually a hyperplane that separates or classifies the data into their respective classes. SVM algorithm is so effective that it actually projects the data onto the dimension that is feasible for classification and a kernel can be chosen accordingly to classify testing data. LIBSVM toolbox is used to implement SVM. The proposed work classifies 8 discrete emotions (3 positive, 4 negative and neutrality emotions). The original SVM only solve 2 class problem. For multiclass classification, LIBSVM uses $k(k-1)/2$ classifiers where k is the number of classes. The classification part included three levels. In each level, the emotions are further classified based on valence and arousal.

Zheng et al. [6] used couple of algorithms which include KNN, logistic regression, SVM and discriminative graph regularized extreme learning machine learning

(GELM). KNN works by taking a particular data point and finding all the data that is near to it by any approach (proposed work used Euclidian distance approach) and classifies the data to class which of maximum for its k nearest neighbors. Here k represents the number of nearest neighbors. The k is made to five and cross-validation technique has been used in the proposed paper. For SVM, LIBSVM tool with a linear kernel is used along with cross-validation technique. The cross-validation technique involves all parts of the data to be used for both training and testing. The entire data is divided into an appropriate number of parts where each part is used for training while the other parts can be used for testing so that all the parts of data have been sufficiently trained to classify new data. GELM is a form of neural network [10]. Peng et al. proposed GELM since it neutralizes the differences between samples from same classes and causes them to share the same properties [11]. GELM shows better results in comparison with other algorithms for facial [11] and emotion recognition.

Soleymani et al. [12] used SVM, Discriminant Analysis (DA), decision tree, KNN for classification purpose. Discriminant analysis is an algorithm in which the entire data is been categorized using clusters so that it can be classified into classes. Linear Discriminant Analysis (LDA), Quadratic Discriminant Analysis (QDA) has been implemented by the paper referenced above. Decision tree is an effective algorithm since it uses various rules called splits. The rules used can be of any number. The more the rules, the more complex the tree is. Each tree has a node which is either a child node or a node that makes decision and also has branches. Each branch has a subtree. The classifier goes from leaf node to decision node which develops some decisions to classify the data accurately. 10 splits were used and evaluated using cross-validation technique by Soleymani et al. [12]. KNN showed the highest accuracy among all the above-mentioned classifiers used and DA classifiers showed the least accuracy [12]. The extension of the paper is planned to implement SVM, KNN, decision tree algorithms for classification because of the higher efficiency and accuracy in accordance with previous works done.

5 Conclusion

Emotion Analysis, a broad research area that relates to people directly has been elaborately studied. An overview of the previous work done in emotion analysis has been elaborated and a few extraction methods, classifiers to determine the accuracy of each classifier have been implemented. The classifiers are selected based on the accuracy of previous work. The paper could be further extended to determine the emotion of a person using the above-mentioned classifiers.

References

1. Liu, Y.J., Member, S., Yu, M., Zhao, G., Song, J., Ge, Y.: Real - time movie - induced discrete emotion recognition from EEG signals. IEEE Trans. Affect. Comput. **3045**(c), 1–14 (2017). https://doi.org/10.1109/TAFFC.2017.2660485
2. Becker, H., Fleureau, J., Guillotel, P., Wendling, F., Merlet, I., Albera, L., Member, S.: Emotion recognition based on high-resolution EEG recordings and reconstructed brain sources. IEEE Trans. Affect. Comput. **3045**(1949), 1–14 (2017). https://doi.org/10.1109/TAFFC.2017. 2768030
3. Ding, Y., Hu, X., Xia, Z., Liu, Y., Member, S., Zhang, D.: Inter-brain EEG feature extraction and analysis for continuous implicit emotion tagging during video watching. IEEE Trans.Affect. Comput. (c), 1 (2018). https://doi.org/10.1109/TAFFC.2018.2849758
4. Bocharov, A.V., Knyazev, G.G., Savostyanov, A.N.: Depression and implicit emotion processing: an EEG study (Dépression et traitement des émotions implicites: une étude). Clin. Neurophysiol. (Neurophysiol. Clini.) **47**(3) (2017)
5. Singh, M. I., Singh: Development of a real time emotion classifier based on evoked EEG, 7. M. Biocybern. Biomed. Eng. **37** (2017)
6. Zheng, W., Zhu, J., Lu, B., Member, S.: Identifying stable patterns over time for emotion recognition from EEG. IEEE Trans.Affect. Comput. 1–15 (2018)
7. Piho, L., Tjahjadi, T., Member, S.: A mutual information based adaptive windowing of informative EEG for emotion recognition. IEEE Trans. Affect. Comput. (2018). https://doi.org/10. 1109/TAFFC.2018.2840973
8. Murugappan, M., Ramachandran, N., Sazali, Y.: Classification of human emotion from EEG using discrete wavelet transform. J. Biomed. Sci. Eng. 390–396 (2010). https://doi.org/10. 4236/jbise.2010.34054
9. Sikonja, M.R.: Theoretical and empirical analysis of relief and relief. Mach. Learn. J. **2003**, 23–69 (2003)
10. Huang, G.B., Zhu, Q.Y., Siew, C.K.: Extreme learning machine: theory and applications. Neurocomputing **70**, 489–501 (2006). https://doi.org/10.1016/j.neucom.2005.12.126
11. Peng, Y., Wang, S., Long, X., Lu, B.: Neurocomputing discriminative graph regularized extreme learning machine and its application to face recognition. Neurocomputing **149**, 340–353 (2015). https://doi.org/10.1016/j.neucom.2013.12.065
12. Soleymani, M., Asghari-esfeden, S., Member, S.: Analysis of EEG signals and facial expressions for continuous emotion detection. IEEE Trans. Affect. Comput. **7**(1), 17–28 (2016)

Treble Band RF Energy Harvesting System for Powering Smart Textiles

B. Naresh, V. K. Singh, V. K. Sharma, Akash Kumar Bhoi and Ashutosh Kumar Singh

Abstract A flexible rectenna has been studied to provide power to wearable electronic systems. Prototype antenna has a compact size, which has been studied from −10 to 10 dBm power levels, that converts RF into DC supply. The measured impedance bandwidth of anticipated antenna is 6500 MHz, i.e. from 6.50 to 12.80 GHz. The anticipated rectenna is having measured efficiency is 50.24% at 8.53 GHz, 42% at 10.32 GHz and 29.78% at 12.34 GHz.

Keywords Rectification · RF energy harvesting · Conversion efficiency · Wearable electronics

1 Introduction

In the present decade, most of the electronic devices are consuming low power and this power is supplied by means of batteries. However, these batteries are rechargeable ones but powering them is a difficult task. Even though power consumption is low only solar cells could power them, no other sources provide a great deal of energy. On the other side energy harvesting will provide a better solution to powering the wearable electronics and wireless sensors.

B. Naresh · V. K. Singh (✉) · V. K. Sharma
Department of Electrical Engineering, Bhagwant University, Ajmer, Rajasthan, India
e-mail: singhvinod34@gmail.com

B. Naresh
e-mail: nareshbangari@gmail.com

A. K. Bhoi
Department of Electrical and Electronics Engineering, Sikkim Manipal Institute of Technology, Sikkim Manipal University, Gangtok, Sikkim, India
e-mail: akash730@gmail.com

A. K. Singh
Indian Institute of Information Technology, Allahabad, India
e-mail: ashutosh_singh@iiita.ac.in

© Springer Nature Singapore Pte Ltd. 2020
P. K. Mallick et al. (eds.), *Cognitive Informatics and Soft Computing*,
Advances in Intelligent Systems and Computing 1040,
https://doi.org/10.1007/978-981-15-1451-7_8

The flexible items where electronic gadgets and sensors are coordinated into apparel to move toward becoming body-worn are eluded as wearable innovation [1]. As of now, use of electronic devices is quickly growing and furthermore electronic gadgets are scaled-down in size, control utilization is impressively diminished by new electronic assembling advances [2, 3]. Wearable innovation is a mix of both material and electronic innovation, which makes simple access to the ordinary exercises. Wearable gadgets could be incorporated into frill and attire, for example, shirts, glasses, watches, caps and so forth [4–6]. These wearable gadgets are controlled by a battery framework which required successive charging. To charge these electronic gadget power supply isn't accessible in a simple way; it a misfortune for the regular clients. This issue can be overwhelmed by utilizing vitality reaping procedures which catch vitality from in normally accessible sources like sun-powered, wind, warm, and radio recurrence. The principle capacity of the gathering circuit is catching vitality with least return loss and putting away the vitality to future use [7–9].

A novel Microstrip-line-fed parachute-shaped textile antenna with integrated cross slit in first and second quadrants has been proposed in this paper. The antici-pated antenna has good bandwidth (6.5–13 GHz) covering the C-band and X-band applications. The X-band is very valuable because it is widely used by military forces for dismounted soldier communication because of high-speed data rates, resilience to rain fade, remote coverage and interference.

2   Antenna Design

2.1   Antenna Description and Manufacturing

The typical design of the parachute is shown in Fig. 1. The upper section of the parachute is known as the canopy which is the main radiating element for the proposed antenna. A set of lines connects the canopy to the backpack are called suspension lines; radiating element and feed line are tying together with these lines. Patch radius is computed by Eq. (1)

$$a = \frac{87.94}{f_r \sqrt{\varepsilon_r}} \tag{1}$$

2.2   Antenna Topology and Simulation

The geometry of CST model of cross-slot loaded parachute patch antenna is shown in Fig. 2. The textile substrate dimension is 42 * 42 mm^2 whereas the dimension of ground is 42 * 20 mm^2. The overall thickness of this design mentioned was 1.076 mm.

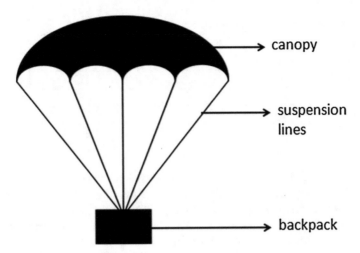

Fig. 1 Typical model of parachute

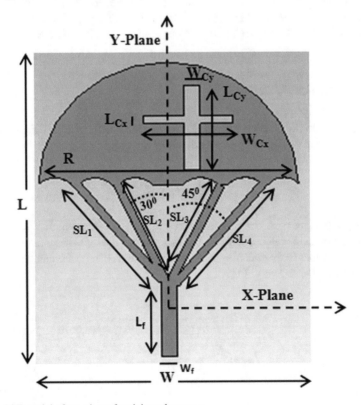

Fig. 2 CST model of top view of anticipated antenna

Table 1 Dimensions of proposed antenna

Parameter	L	W	R	L_f	W_f
Units (mm)	42	42	32	10	2
Parameter	SL_1	SL_2	SL_3	SL_4	W_{cy}
Units (mm)	17	15	15	17	2
Parameter	L_{cy}	W_{cx}	L_{cx}	L_g	W_g
Units (mm)	11	11	1	20	42

Table 1 shows the characteristics of textile material. The radius of patch is 16 mm which has four suspensions lines in which outer two lines SL_1 and SL_4 are at an angle $\alpha = 45°$ to (Y-Plane) the non-radiating edge of patch antenna. The two inner lines SL_2 and SL_3 are having angle $\alpha = 30°$, width of all suspension lines was 2 mm. The snapshot of anticipated antenna is depicted in Fig. 2.

The antenna resonates at frequencies $f_1 = 7.65$ GHz, $f_2 = 8.53$ GHz, $f_3 = 10.32$ GHz and $f_4 = 12.34$ GHz with impedance bandwidth of 57.92% at dual-band (6.5–11 GHz and 12–12.8 GHz). These four resonance frequencies have the $-22, -37, -33$ and -18 dB good return loss values at respective frequencies.

3 Rectification Circuit

HSMS2850 Schottky diode is utilized for RF to DC transformation with a low edge voltage of 150 mV. The voltage amplitude at the receiving antenna is calculated by Eq. (2)

$$V_s = 2\sqrt{2R_s P_{av}} \tag{2}$$

where R_s is the source series resistance in ohms, P_{av} is the available power at the receiving antenna in watts. The receiving antenna has the three resonance operating frequencies so that three matching L circuits designed at these frequencies. Basic L matching circuit is shown in Fig. 3.

$$C_m = \frac{1}{\omega_r R_s}\sqrt{\frac{R_s}{R_{in} - R_s}} \tag{3}$$

$$L_m = \frac{R_{in}}{\omega_r}\frac{1}{\omega_r R_{in} C_{in} + \frac{1}{\sqrt{\frac{R_s}{R_{in} - R_s}}}} \tag{4}$$

$$G = \frac{V_{in}}{V_s} = \frac{1}{2}\cdot\sqrt{\frac{R_{in}}{R_s}} \tag{5}$$

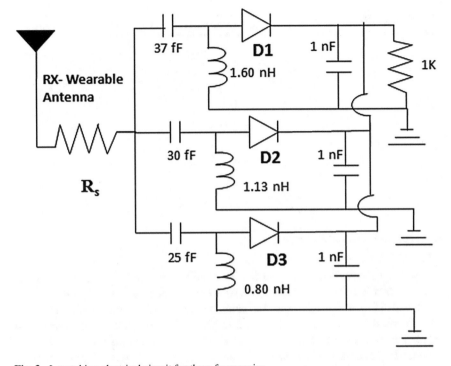

Fig. 3 *L*-matching electrical circuit for three frequencies

$$Q = \sqrt{\frac{R_{in}}{R_s} - 1} \qquad (6)$$

The calculation of *L*-matching circuit components as followed [10, 11]; First, Gain (*G*) has to select in such a way that the voltage available at the diode should overcome its dropping voltage. Then, for series resistance of the receiving antenna $R_s = 50 \ \Omega$ calculate the R_{in} by using Eq. (6). Finally, at resonance frequencies calculate the ω_r and C_{in} is obtained from diode manufacturing data sheet; L_m and C_m are obtained from Eqs. (3), (4) and (5).

The designed half-wave rectifier circuit is optimized for 100 μw which is equivalent to −10 dBm. Textile antenna has source resistance of 50 Ω, for these values available power at the antenna is $V_s = 0.2$ V. HSMS2850 Schottky diode is utilized for RF to DC transformation with a low edge voltage of 150 mV, so that gain (*G* = 5) is selected to overcome the voltage drop. Equations (4), (5) and (6) are used to calculate the values of C_m and L_m at three resonant frequencies which are given in Table 2. The output value of capacitor is taken as 1 nf.

Table 2 C_m and L_m values at three resonant frequencies

Resonant frequency GHz	L_m (nH)	C_m (fF)
8.53	1.60	37
10.32	1.13	30
12.34	0.80	25

$$\eta = \frac{P_{\text{load}}}{P_{\text{av}}} \tag{7}$$

4 Results and Discussions

The series vector network analyzer has been utilized to determine the return loss of proposed textile antenna. The fabricated prototype of antenna with defected ground is displayed in Fig. 4. The measured and simulated return–loss curves of the parachute-shaped cross-slot antenna are displayed in Fig. 5. The measured 10 dB return loss bandwidth is around 6500 MHz (6.5–12.8 GHz). The measured return loss values are reduced in magnitude but the resonance frequencies are almost unaffected; the first resonance point ($f_1 = 7.65$ GHz) is the effected resonance frequency point. So that, harvesting circuit is designed for upper frequencies only. The power transformation efficiency of the rectifier is expanded up to 1 kΩ thereafter drop essentially from 1 to 5 kΩ as for change in load and it appears in Fig. 6. The power tranformation efficiency of the rectifier is calculated by Eq. (7)

The measured maximum efficiency for the rectenna is 50.24% at 8.53 GHz, 42% at 10.32 GHz and 29.78% at 12.34 GHz which is shown in Fig. 7. Maximum efficiency

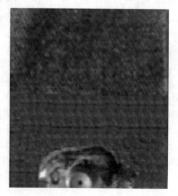

Fig. 4 CST model of top and back view of anticipated antenna

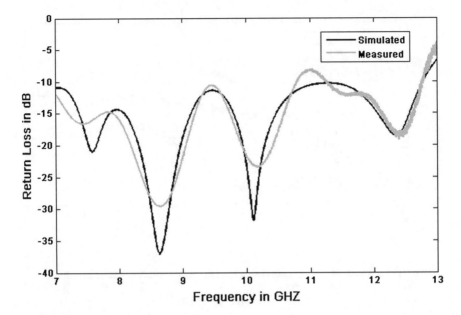

Fig. 5 Measured and simulated return loss plot

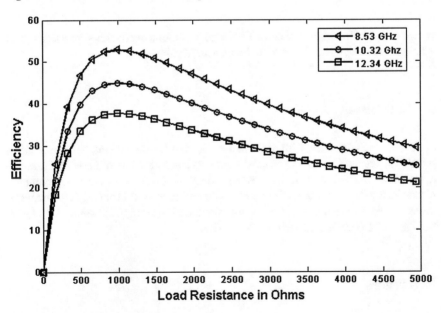

Fig. 6 Efficiency versus load resistance at 10 dBm power level

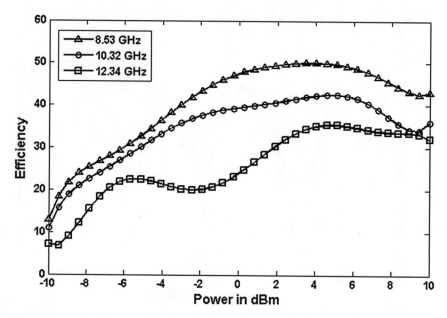

Fig. 7 Measured efficiency versus power in dBm

is received at 5 dBm input power for 1 kΩ load. In simulated efficiency, plot maximum efficiency is somewhat higher than the measured values.

5 Conclusion

A novel parachute shape with cross-slot embedded textile antenna has been antici-pated. The results proved that proposed antenna has bandwidth of 6500 MHz ranges from 6.5 to 12.80 GHz with four resonant frequencies. Antennas are very compact in size hence the integration of antenna into clothing is useful for body area network systems. Maximum measured efficiency for the rectenna is 50.24% at 8.53 GHz, 42% at 10.32 GHz and 29.78% at 12.34 GHz.

References

1. Vallozzi, L., Van Torre, P., Hertleer, C., Rogier, H.: A textile antenna for off-body communica-tion integrated into protective clothing for fire fighters. IEEE Trans. Antennas Propag. **57**(4), 919–925 (2009)
2. Kaivanto, E.K., Berg, M., Salonen, E., de Maagt, P.: Wearable circularly polarized antenna for personal satellite communication and navigation. IEEE Trans. Antennas Propag. **59**(12), 4490–4496 (2011)

3. Singh, N., Singh, A.K., Singh, V.K.: Design & performance of wearable ultra wide band textile antenna for medical applications. Microw. Opt. Technol. Lett. **57**(7), 1553–1557 (2015)
4. Masotti, D., Costanzo, A., Adami, S.: Design and realization of a wearable multi-frequency RF energy harvesting system. In: Proceedings of the 5th European Conference on Antennas Propagation, pp. 517–520 (2011)
5. Roundy, S.J.: Energy scavenging for wireless sensor nodes with a focus on vibration to electricity conversion. PhD Thesis, University of California, Berkeley, USA (2003)
6. Shukla, S.S., Verma, R.K., Gohir, G.S.: Investigation of the effect of Substrate material on the performance of Microstrip antenna. IEEE (2015). 978-1-4673-7231-2/15
7. Mane, P., Patil, S.A., Dhanawade, P.C.: Comparative study of microstrip antenna for different subsrtate material at different frequencies. Int. J. Emerg. Eng. Res. Technol. **2**(9) (2014)
8. Daya Murali, S., Narada Maha Muni, B., Dilip Varma, Y., Chaitanya, S.V.S.K.: Development of wearable antennas with different cotton textiles. Int. J. Eng. Res. Appl. **4**(7)(Version 5), 08–14 (2014). ISSN: 2248-9622
9. Agilent, H.S.M.S.: 285x Series Surface Mount Zero Bias Schottky Detector Diodes-Data Sheet 5898-4022EN, Agilent Technologies, Inc., Sept 2005
10. Bowick, C., Blyler, J., Ajluni, C.: RF Circuit Design, 2nd edn. Elsevier Inc., Burlington (2008)
11. Soltani, H.N., Yuan, F.: A high-gain power-matching technique for efficient radio-frequency power harvest of passive wireless microsystems. IEEE Trans. Circ. Syst. **57**(10), 2685–2695 (2010)

Performance Analysis of Data Communication Using Hybrid NoC for Low Latency and High Throughput on FPGA

C. Amaresh and Anand Jatti

Abstract The perception of the theory of communication network has led to enormous supplanting committed to simplex, duplex system in terms of various scales of interconnecting systems and switches. Each data interactive system needs to have a smart way for the design of efficient and tolerant system in terms of adaptability, versatility, execution, and effective data delivery. In this peculiarity, the system architects demand for a novel network on-chip router which is error free and minimized circuit path with greater packet delivery ratio, minimum delay, and better bandwidth utilization. In this paper, we propose the modeling of network architecture in consideration with 8 × 8 switch router which indulges the suitable algorithm for shortest path finder, i.e., minimum spanning tree, for efficient routing in run-time. We have demonstrated warmhole technique and virtual cut-through mechanisms for automatic correction with validating errors. So, we have selected verilog HDL for development under the environment of VIVADO Xilinx 2018-1 and demonstrated on Nexys DDR-4 Artix-7 Field Programmable Gate Array family bearing part number XCA7CGS100t comprised of 324 pins with the results it is noticed that better reliability and minimized latency of 36.5% with enhanced throughput reaching 40% than the existing router. The proposed design is acceptable in terms of better performance in terms of area, delay, and resource allocation.

Keywords Communication switch · Error · FPGA · HDL · NoC · Router · Shortest path algorithm · Topology · Switching · Power reduction schemes and encoder and decoder

C. Amaresh (✉) · A. Jatti
R.V. College of Engineering, Bangalore, India
e-mail: amareshapj@gmail.com

A. Jatti
e-mail: anandjatti@rvce.edu.in

© Springer Nature Singapore Pte Ltd. 2020
P. K. Mallick et al. (eds.), *Cognitive Informatics and Soft Computing*,
Advances in Intelligent Systems and Computing 1040,
https://doi.org/10.1007/978-981-15-1451-7_9

1 Introduction

The day to day growing number of IP-Modules in System-on-chip's (Soc's) Conventional interconnection architecture through bus, maybe one of the Design parameters to satisfy the requirements in the application. In any Digital system of successive parallel communication with bus communication will fail in Regard of Area, Speed, Delay and Bandwidth. A solution for above-mentioned shortfall of communication is to incorporate the switching network using embedded system identified as network on-chip (NoC) for interconnection of IP modules with respect to SOC [1, 2]. The interesting feature of NoC is it provides a wide space for design when compared with bus-based technique, as various arbitration strategies and various routings can be tested and demonstrated with multiple organizations of communication architecture. The notable feature of NoC's is its resistive redundancy that narrates to withstand faults and errors of communication network. This increases the interest of the SOC designer to enhance suitable design for required constraint and system characteristics. Figure 1 shows the functional and architectural view of a typical router, thereby indicating the four different ways of routing and a line input and output for the communication.

A typical NoC is comprised of trilateral functional blocks, viz., network adapter router followed by links. The efficiency of any NoC can be evaluated by bandwidth, speed, area consumption, and packet delivery ratio. The bandwidth of NoC refers to maximum amount of data propagated over the network in the topology designed by considering the complete packet with its header bit tail and payload [3–5]. This parameter of measurement of bandwidth is bits per second. Pocket delivery ratio is the ratio of maximum amount of information content delivered per time unit to the amount of information sent over the channel. This ratio is the normalized function of dividing it by the initial division by size of message and by its size of network in our proposed system; it is 8 × 8; that is 64 possible nodes switch the considerable unit for analysis bits per node per cycle. The speed of operation of the system is the

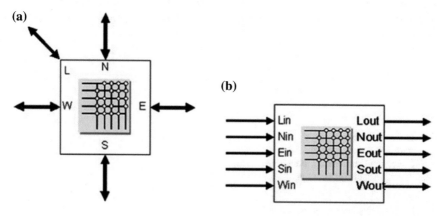

Fig. 1 Interface views of a typical router: **a** functional view and **b** architectural view

maximum amount of data described per clock cycle from source to destination node by its shortest part usually measured interest of second. The other parameter for the evaluation is the area consumption that depends on the number of LUT and gates and gate utilization for error trace information delivery [6].

Network Adapter
This functional block can be broken down into two subparts, namely back-end and front-end adapters. The handling of network protocols on occasion of packetization, and shuffling and reshuffling of buffers dealing with synchronization protocols will be done by back-end adapter; the front-end adapter deals with core request as it is defined in software as a socket. It will be unaware of existing NoC [7].

Collectively front-end and back-end adapters are responsible for the network interface, logically between network and IP cores. Each IP will have significant interface protocol in view of its network. This functional block is critical in NoC as it separates the operation such as computation and communication for its reusability [8].

Router The demonstration of a router demands for a bunch of policies to handle: packet collision, rerouting, and other routing-related issues. A NoC router can be viewed as the combination of set of input and output ports unified with a switching matrix and a local port to provide avenue for IP core with its connected router. The concept of router does not stop with the connection part but also extend to logic block that furnishes the various strategies for movement of data with its flow control policies [9]. Figure 2 shows the architecture of typical router with necessary signals.

The flow control policy is characterized by various packet movements globally and locally. This can ensure dilemma of predicament for resources by considering suitable flow control policy with its optimization of available resources guaranteeing between qualities of service [10]. The governance of data packets can be distributed or centralized; in centralized governance, the decision of routing was made globally and is intended to all the nodes with the assuming strategy of zero congestion. Their by avoiding the need for arbitration but it requires the synchronized operation of all nodes [11]. In case of distributed governance, each NoC can make its decision locally with a systematic design of virtual channels, in which virtual channel is the designing of single physical channel with multiplexing over an available channel locally with individual buffer queue. The implementation of virtual channel is to minimize the number of predicament resources and optimization of wire usage [12].

Links The execution of links is generally the combination of set of wires and connection approach between two routers in a network. Link may consist of logical connection which is supportive to full duplex mode of operation any routers. Link corresponds to operation with synchronization protocol between destination node and transmitter node; ultimately, links are responsible for overall performance of NoC in terms of speed, reliability, and delay [13, 14].

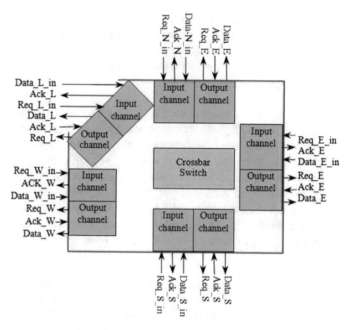

Fig. 2 Architecture of typical NoC router

2 Problem Definition

NoC design is to have the better establishment among the locus of interest in terms of speed, area, packet delivery ratio. Over the 8 × 8 network defined providing the better efficiency and accuracy according to routing algorithm and various switching schemes which will suffer with the starvation of resources and ending with an dilemma of predicament which may result in failure of entire embedded system, but this criteria can be best suited and redefined if the embedded system develop and shortest path algorithm to find its path, reconfiguration of router upon the requirement of wormhole switching and virtual cut-through switching.

3 Proposed System and Methodology

The proposed architecture of NoC router switch is purely based on the virtual channel routing intern divided into wormhole and virtual cut-through switching on an 2D mesh topology design comprising of mainly three various blocks, viz., input channel, output channel, and an crossbar. The NoC router comprises five ports as shown in Fig. 2, to be specific east, west, north, south, and neighborhood port and a focal cross-point lattice. Every one of the ports has an information channel furthermore, a particular channel to get and transmit information transmission separately.

Information packets travel through the information channel of one port of the switch and after that are sent to the corresponding channel of other port. Each information and a yield channel have unraveling rationale which plays out the elements of the switch. Cushions are consolidated at all ports to store the information briefly. The stop and wait sort buffering technique is utilized. Control rationale settles on discretion choices, and communication is built up among information and corresponding ports. The association of communication ports is made among ports and focal crosspoint grid. As per the goal, way of information bundle and control bit lines of the cross-point framework is set. Coordinating the development of information from source to destination is called switching mechanism.

The Input Channel The architecture of input channel at each port has its own control logic, and flowchart of operation is shown in Fig. 4. Each input channel has a first-input first-output (FIFO) of depth 16, data width of 8 bits, and a control logic which is implemented based on the model of finite state machine (FSM) (Fig. 3).

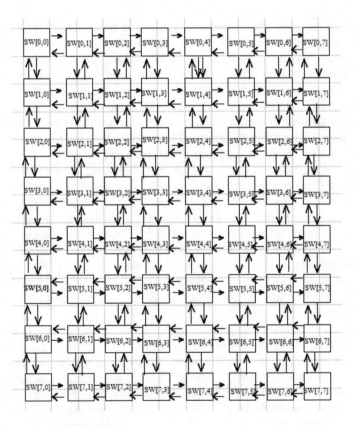

Fig. 3 Proposed 8 × 8 NoC router

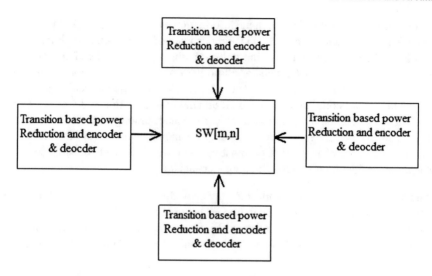

At the point when the information channel gets demand from neighboring switches, on the off chance that it is free quickly it sends the recognition of the solicitation. In the parcel position, first is header and the following bounces are information bits; as long as the solicitation signal is held high, it acknowledges the information for transmission. Info channel stream outline is shown in Fig. 4b. The yield channel of mentioning switch guarantees that the solicitation line is held high until it discharges the information bundle, which is being acknowledged by the information channel. The information channel keeps the affirmation line at high, as long as the exchange of information is proceeded with which it is shown by the solicitation line. Once the exchange information is cultivated, the solicitation and the affirmation lines go low in an arrangement. The information parcels got from the mentioning switch are put away locally in the FIFO for store and forward dataflow style. Next, the control rationale peruses the header of the parcel and chooses which yield channel is to be mentioned for conveying information bundles from the switch. On choice, it sends the solicitation to distinguished yield channel. It is to be noticed that every one of the info channels is running a free FSM and thus can start five conceivable parallel associations all the while. When the information channel gets an allow from the mentioned yield channel, the control bits of cross point framework are et as needs be to set up information stream process.

The functioning of the configuration buffer initiates as soon as the input signal is triggered from the input channel and various VC activates, and its RTL representation is shown in Fig. 5. The input block is the active participation of initial data reception and various VCs with the multiplexer and demultiplexer for data handling; in regard to packet switching operation, it follows the condition of routing computation and dataflow control for determining packet delivery ratio and to establish the shortest path for the packet delivery in terms of shortest path algorithm. Routing control is designed, and the output of this block is channelized for the crossbar.

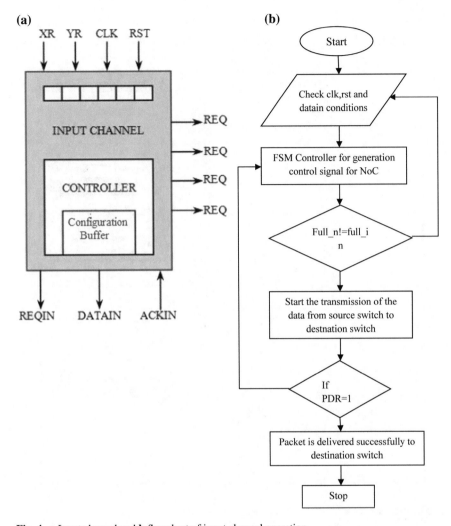

Fig. 4 **a** Input channel and **b** flowchart of input channel operation

Crossbar Crossbar switch comprises a lot of multiplexers and demultiplexers. These are interconnected so that every single imaginable association between the information and the yield channels is set up whenever required. The crossbar switch stream outline is shown in Fig. 6. The yield channel while giving the solicitation to an information channel arranges the multiplexers and demultiplexers of accessible info and yield channels. Hence, the associations are built up between channels for the exchange of the information parcels. A cross-point switch likewise called as lattice switch or crossbar switch interfaces various info and yields in a network plot. The crossbar switch configuration has five data sources and five yields, and the multi-plexer crossbar switch is shown in Fig. 6; also, its RTL portrayal of crossbar is shown

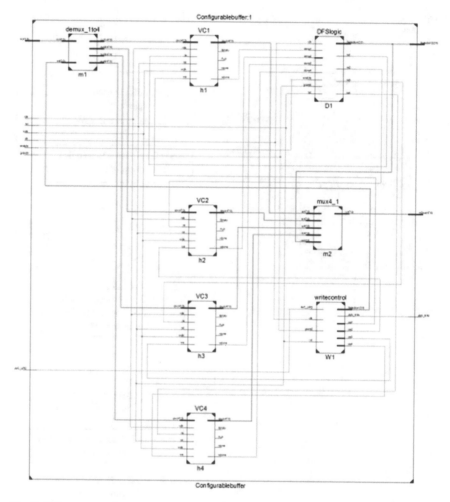

Fig. 5 RTL representation of configuration buffer

in Fig. 7. Every one of the multiplexers is associated with all contributions with select lines of 3 bits long, out of accessible select lines which will be chosen dependent on the authority rationale at once. Yields of multiplexers are the yield ports of the switch.

Output Channel In light of the determination bits, the multiplexer works as indicated by the solicitation from the information channel. Further dependent on the info, the comparing yield will get information from the crossbar. Each yield channel of the considerable number of ports has input bit FIFO and allotment rationale to settle on mediation choices. The yield channel gets demand from various info channels and allows consent to anybody. Further it sets the control bit lines of grid cross focuses as appeared in Fig. 8. With the assistance of a solitary deciphering rationale into its

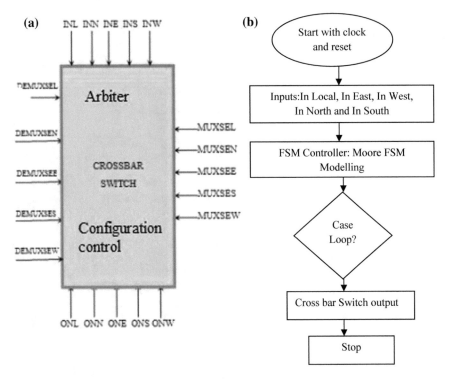

Fig. 6 **a** Crossbar switch and **b** flowchart of crossbar

FIFO, the yield channel acknowledges the parcel as long as the sending inputs FIFO isn't vacant. In the event that the information exchange is finished, at that point the framework cross-point controls are reset. FSM then starts the procedure forward the information into the close-by switch utilizing handshake calculation, void status of its FIFO starts the following between channel exchange. The VC allotment rationale assumes an indispensable job in the information transmission and portion of channel for the transmission of required information. The RTL portrayal of VC square is shown in Fig. 9.

Channel routing algorithm: Channel routing is used to find the shortest path between the interconnected sub-modules of architectures. Both area and channel routing algorithm establish proved and improved efficiency network data traffic of the NoC. The sample result after channel routing for the data movement over the channel is shown in Figs. 10 and 11.

Routing topology: Routing topology depends on the stream control strategies and various accessible steering calculations for advancement. The stream control module centers around cradle size and channel transfer speed prerequisites. The versatile reconfigurable NoC can be utilized to wipe out the support unit of engineering. This diminishes on-chip territory and accordingly less power utilization. Wire delays are typically affected by channel transmission capacity assignment plans.

Fig. 7 RTL diagram of
crossbar switch

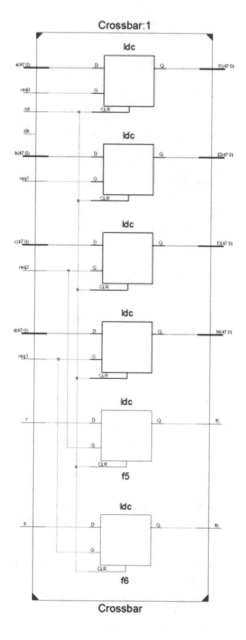

Routing is cultivated in two stages. The initial step is called worldwide or free
steering, and it centers around wiring channels. The second step is called definite
directing, and it guarantees exact wire courses for various channel layers. Reconfiguration and configuration stream and warm gap steering are shown in Fig. 12.
Virtual cut-through switching dataflow is shown in Fig. 13 in which it represents the

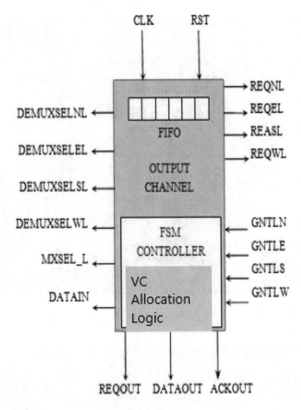

Fig. 8 Output channel

four different stages of cut-through data segment movement from one node to other node initially. Figure 13a represents the path allocation from the router, and after allocation from the router data packet has been subdivided into various segments as observed in Fig. 13b and immediately jumps over to the next node which was found out by shortest path algorithm as represented in Fig. 13c; a virtual path is established from source to destination for the data movement as depicted in Fig. 13d.

Committed virtual channel (VC) for bypassing the bustling channel way is to structure a versatile VC switch with diminished low-stack inactivity. Bypassing should be very much intended to give VC similarity, in the mean time continuing the proficiency of intra-measurement bypassing in switch. An approaching dance may have a place with a subjective VC. Picking whether a move can avoid the present switch, at first the VC must be decoded, and thereafter, the openness of contrasting credits for the downstream switch must be checked. Here, we acknowledge that the bob holds its VC ID while bypassing. Accept the VC ID of a gotten bob is VC, and the yield port of Dimension course of action mentioned is method of reasoning o. In case the going with two conditions are met, the gotten move can evade the present switch in one cycle. Directly off the bat, bypassing must not cause overshooting

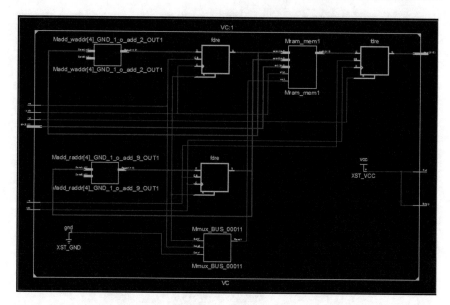

Fig. 9 RTL representation of VC block at output channel

Name	Value		2,640 ns	2,650 ns	2,660 ns	2,670 ns
v74[47:0]	XXXXXXXXXXXX		XXXXXXXXXXX		efaccbadadca	
v75[47:0]	XXXXXXXXXXXX		XXXXXXXXXXX		efaccbadadca	
clk	1					
rst	0					
in[47:0]	efaccbadadca					efaccbadadca
enable0	1					
enable1	0					
enable2	0					
enable3	0					
enable4	0					
enable5	0					
enable6	0					
enable7	0					
enable8	0					
enable9	1					
enable10	0					
enable11	0					

Fig. 10 Simulation results for the feeding of data to the channel

to the objective (irrelevant coordinating). Besides, the VC at yield o must be inert (ensuring a successful VA). Executing this bypassing justification requires using the VC ID as the commitment to list relating information. This control method of reasoning will unavoidably construct the fundamental path length of bypassing justification appeared differently in relation to the one [11] in view of VC unraveling. For example, the execution on the X estimation is according to the accompanying:

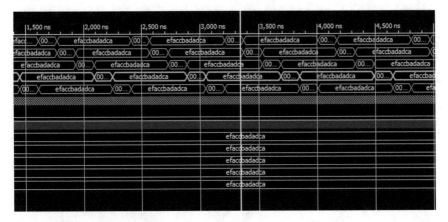

Fig. 11 Simulation of result after data routing on to the channel

(a)   **(b)**

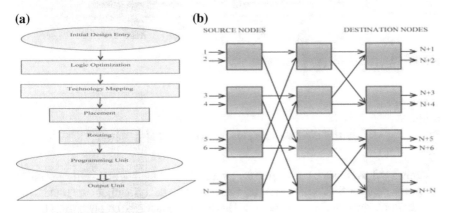

Fig. 12 **a** Data flow representation in wormhole switching and **b** $N \times N$ representation of nodes

$$bypassing \leq (destination.x! = present.x)\&vc\_idle[0][vc]$$

To quicken this system, we present a submitted VC for bypassing. Expect the uncommon VC introduced is known as the slide virtual channel (SVC). We now simply perform bypassing for moves having a spot with SVC. To check if a SVC bob can avoid the present switch, a change simply needs to check if the SVC of yield o is inactive. Avoiding essential administration is snappier in light of the way that we do not need to use VC ID as the record to ingest credit information or other information. The planning speed is invariant to the amount of VCs.

If we use SVC for bypassing, the decision making on the X dimension is as follows:

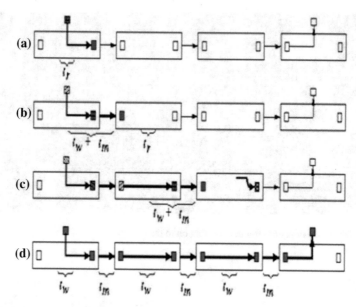

Fig. 13 Virtual cut-through representation

If (svc = 1)
begin
Bypassing $<=$ (destination.x! = present.x)&svc_idle[0]
end

Only SVC packets are considered for bypassing, and there is also dedicated buffer space reserved for SVC in each router. To avoid extra delays without the routers, the SVC buffer is restricted to only one slot. This design reduces the complexity of bypass decision making. Bypassing with SVC is faster and, more importantly, invariant to the number of VCs. Adding an extra VC does not necessarily increase buffer space in the router because most NoC routers use a shared buffer between VCs.

Design of Microarchitecture of the NoC Router

As the need for the SoC has provided an ideal pathway for the designers to test and evaluate an innovative NoC which satisfy the operation of packet switching from end to end application. In our proposed scheme of network on-chip, we simulated the desired system for 64 switches. Through out the discussion, we have considered data in terms of various packets. These packets are rerouted upon the employing of spanning tree algorithm with is switching technique, thereby reducing the unavoidable redundancy of movement of data over the network. The total system is viewed as three major sub-blocks, namely input channel, crossbar switch, and output channel.

The input channel block is internally equipped with write control, a pair of multiplexer and demultiplexer, dataflow control (DFC), virtual channels connecting between multiplexer and demultiplexer. The conflict sensing (CFS) requests the arbiter for the data transfer from input channel, when a grant signal is generated from the arbiter. The data which has to be transmitted over the crossbar switch will be uploaded from the receiver buffer by finding the shortest path depending on the minimum routing technique as soon as the data flits arrive at the input port at 8 × 8 switch with suitable control signals from the configuration controller. The data with various switching techniques will be loaded to output buffer, and the transmitter block sends a signal to the virtual channel allocation logic block; meanwhile, VC allocation logic and arbiter will be in half duplex communication mode for the exchange of data in handshaking mode. VC allocation logic enables the transmission by allocating the virtual channel dynamically for data uploading operation.

Power Reduction in NoC

Transition-Based Power Reduction Algorithm (TBPRA): Encryption Process

1: Start.
2: Provide "w" bit information alongside the already encoded dance to the encoder.
3: At that point, contrast current bounce and past encoded dance.
4: At that point, check the condition $2(T_2 - T_4**) < 2T_y - w + 1, T_y > (w - 1)/2$.
5: In this condition, we check for two conditions, i.e., odd transform and no transform conditions.
6: Contingent upon that condition, we are performing reversal on odd bits.
7: At that point for various information, the procedures will be rehashed.

Example

The input packet size is 32 bits and its value say $x = 10101011101010111010101110101011$, and it has 24 transitions. The power reduction process is followed step by step as given below

x=01 00 11 10 01 00 11 11 01 00 11 10 01 00 11 11 01 00 11 10 01 00 11 11 01 00 11 10 01 00 11 11

y=00 01 00 11 10 01 00 11 11 01 00 11 10 01 00 11 11 01 00 11 10 01 00 11 11 01 00 11 10 01 00 11

Ty=1 1 2 1 2 1 2 0 1 1 2 1 2 1 2 0 1 1 2 1 2 1 2 0 1 1 2 1 2 1 2 0

$T_y = 39 = 00101000, \quad T_y = \frac{w-1}{2}, \quad T_2 > T_4$
$T_2 = 00 = 00000000$
$T_4 = 00 = 00000000$

$$T_y = \frac{32 - 1}{2} = 15.5$$

For that reason, half invert (HI) = 1 and full invert (FI) = 0

$Z_0 = X_0 \oplus \mathrm{FI} = 1 \oplus 0 = 1$

$Z_1 = X_1 \oplus \mathrm{FI} \oplus \mathrm{HI} = 1 \oplus 0 \oplus 1 = 0$

$Z_2 = X_2 \oplus \mathrm{FI} = 0 \oplus 0 = 0$

$Z_3 = X_3 \oplus \mathrm{FI} \oplus \mathrm{HI} = 1 \oplus 0 \oplus 1 = 0$

$Z_4 = X_4 \oplus \mathrm{FI} = 0 \oplus 0 = 0$

$Z_5 = X_5 \oplus \mathrm{FI} \oplus \mathrm{HI} = 1 \oplus 0 \oplus 1 = 0$

$Z_6 = X_6 \oplus \mathrm{FI} = 0 \oplus 0 = 0$

$Z_7 = X_7 \oplus \mathrm{FI} \oplus \mathrm{HI} = 1 \oplus 0 \oplus 1 = 0$

$Z_8 = X_8 \oplus \mathrm{FI} = 1 \oplus 0 = 1$

$Z_9 = X_9 \oplus \mathrm{FI} \oplus \mathrm{HI} = 1 \oplus 0 \oplus 1 = 0$

$Z_{10} = X_{10} \oplus \mathrm{FI} = 0 \oplus 0 = 0$

$Z_{11} = X_{11} \oplus \mathrm{FI} \oplus \mathrm{HI} = 1 \oplus 0 \oplus 1 = 0$

$Z_{12} = X_{12} \oplus \mathrm{FI} = 0 \oplus 0 = 0$

$Z_{\_(w-1)} = \mathrm{FI} \oplus \mathrm{HI} = 0 \oplus 1 = 1$

$Z_{13} = X_{13} \oplus \mathrm{FI} \oplus \mathrm{HI} = 1 \oplus 0 \oplus 1 = 0$

$Z_{14} = X_{14} \oplus \mathrm{FI} = 0 \oplus 0 = 0$

$Z_{15} = X_{15} \oplus \mathrm{FI} \oplus \mathrm{HI} = 1 \oplus 0 \oplus 1 = 0$

$Z_{16} = X_{16} \oplus \mathrm{FI} = 1 \oplus 0 = 1$

$Z_{17} = X_{17} \oplus \mathrm{FI} \oplus = 1 \oplus 0 \oplus 1 = 0$

$Z_{18} = X_{18} \oplus \mathrm{FI} = 0 \oplus 0 = 0$

$Z_{19} = X_{19} \oplus \mathrm{FI} \oplus \mathrm{HI} = 1 \oplus 0 \oplus 1 = 0$

$Z_{20} = X_{20} \oplus \mathrm{FI} = 0 \oplus 0 = 0$

$Z_{21} = X_{21} \oplus \mathrm{FI} \oplus \mathrm{HI} = 1 \oplus 0 \oplus 1 = 0$

$Z_{22} = X_{22} \oplus \mathrm{FI} = 0 \oplus 0 = 0$

$Z_{23} = X_{23} \oplus \mathrm{FI} \oplus \mathrm{HI} = 1 \oplus 0 \oplus 1 = 0$

$Z_{24} = X_{24} \oplus \mathrm{FI} = 1 \oplus 0 = 1$

$Z_{25} = X_{25} \oplus \mathrm{FI} \oplus \mathrm{HI} = 1 \oplus 0 \oplus 1 = 0$

$Z_{26} = X_{26} \oplus \mathrm{FI} = 0 \oplus 0 = 0$

$Z_{27} = X_{27} \oplus \mathrm{FI} \oplus \mathrm{HI} = 1 \oplus 0 \oplus 1 = 0$

$Z_{28} = X_{28} \oplus \mathrm{FI} = 0 \oplus 0 = 0$

$Z_{29} = X_{29} \oplus \mathrm{FI} \oplus \mathrm{HI} = 1 \oplus 0 \oplus 1 = 0$

$Z_{30} = X_{30} \oplus \mathrm{FI} = 0 \oplus 0 = 0$

$Z_{31} = X_{31} \oplus \mathrm{FI} \oplus \mathrm{HI} = 1 \oplus 0 \oplus 1 = 0$

The encryption yield is the recorded bits from Z_{31} to Z_0, i.e., $Z = 10000000100000001000000100000001 = 8$ propels; the MSB bit exhibits half steamed or full alter. In this manner, the number advances have been diminished from 24 changes to just 8 propels; consequently, the power usage is generally reduced to 75% when appeared differently in relation to any other cryptography systems.

Transition-Based Power Reduction Algorithm (TBPRA): Decryption Process

The output of the encryption process is the input to the decryption process for recovering the original data, and its steps are explained as follows.

$z = 10000000100000001000000010000000$

z=<u>11</u> <u>01</u> <u>00</u> <u>00</u> <u>01</u> <u>01</u> <u>00</u> <u>00</u> <u>11</u> <u>01</u> <u>00</u> <u>00</u> <u>01</u> <u>01</u> <u>00</u> <u>00</u> <u>11</u> <u>01</u> <u>00</u> <u>00</u> <u>01</u> <u>01</u> <u>00</u> <u>00</u> <u>11</u> <u>01</u> <u>00</u> <u>00</u> <u>01</u> <u>01</u> <u>00</u> <u>00</u>

y=<u>00</u> <u>11</u> <u>01</u> <u>00</u> <u>00</u> <u>01</u> <u>01</u> <u>00</u> <u>00</u> <u>11</u> <u>01</u> <u>00</u> <u>00</u> <u>01</u> <u>01</u> <u>00</u> <u>00</u> <u>11</u> <u>01</u> <u>00</u> <u>00</u> <u>01</u> <u>01</u> <u>00</u> <u>00</u> <u>11</u> <u>01</u> <u>00</u> <u>00</u> <u>01</u> <u>01</u> <u>00</u>

Ty=2 1 1 0 1 0 1 0 2 1 1 0 1 0 1 0 2 1 1 0 1 0 1 0 2 1 1 0 1 0 1 0

$$T_y = 24 = 00011000 \quad T_y = \frac{w-1}{2}, \quad T_2 > T_4$$

$$T_y = \frac{24 - 1}{2} = 11.5$$

In this manner, the half modify (HI) = 1 and full alter (FI) = 0 and the decoder activity can execute as it takes after utilizing just XOR task and the first data be unscrambled.

$X_0 = Z_0 \oplus FI = 1 \oplus 0 = 1$
$X_1 = Z_1 \oplus FI \oplus HI = 0 \oplus 0 \oplus 1 = 1$
$X_2 = Z_2 \oplus FI = 0 \oplus 0 = 0$
$X_3 = Z_3 \oplus FI \oplus HI = 0 \oplus 0 \oplus 1 = 1$
$X_4 = Z_4 \oplus FI = 0 \oplus 0 = 0$
$X_5 = Z_5 \oplus FI \oplus HI = 0 \oplus 0 \oplus 1 = 1$
$X_6 = Z_6 \oplus FI = 0 \oplus 0 = 0$
$X_7 = Z_7 \oplus FI \oplus HI = 0 \oplus 0 \oplus 1 = 1$
$X_8 = Z_8 \oplus FI = 1 \oplus 0 = 1$
$X_9 = Z_9 \oplus FI \oplus HI = 0 \oplus 0 \oplus 1 = 1$
$X_{10} = Z_{10} \oplus FI = 0 \oplus 0 = 0$
$X_{11} = Z_{11} \oplus FI \oplus HI = 0 \oplus 0 \oplus 1 = 1$
$X_{12} = Z_{12} \oplus FI = 0 \oplus 0 = 0$
$X_{13} = Z_{13} \oplus FI \oplus HI = 0 \oplus 0 \oplus 1 = 1$
$X_{14} = Z_{14} \oplus FI = 0 \oplus 0 = 0$
$X_{15} = Z_{15} \oplus FI \oplus HI = 0 \oplus 0 \oplus 1 = 1$
$X_{16} = Z_{16} \oplus FI = 1 \oplus 0 = 1$
$X_{17} = Z_{17} \oplus FI \oplus HI = 0 \oplus 0 \oplus 1 = 1$
$X_{18} = Z_{18} \oplus FI = 0 \oplus 0 = 0$
$X_{19} = Z_{19} \oplus FI \oplus HI = 0 \oplus 0 \oplus 1 = 1$
$X_{20} = Z_{20} \oplus FI = 0 \oplus 0 = 0$
$X_{21} = Z_{21} \oplus FI \oplus HI = 0 \oplus 0 \oplus 1 = 1$
$X_{22} = Z_{22} \oplus FI = 0 \oplus 0 = 0$
$X_{23} = Z_{23} \oplus FI \oplus HI = 0 \oplus 0 \oplus 1 = 1$
$X_{24} = Z_{24} \oplus FI = 1 \oplus 0 = 1$
$X_{25} = Z_{25} \oplus FI \oplus HI = 0 \oplus 0 \oplus 01 = 1$
$X_{26} = Z_{26} \oplus FI = 0 \oplus 0 = 0$
$X_{27} = Z_{27} \oplus FI \oplus HI = 0 \oplus 0 \oplus 1 = 1$
$X_{28} = Z_{28} \oplus FI = 0 \oplus 0 = 0$
$X_{29} = Z_{29} \oplus FI \oplus HI = 0 \oplus 0 \oplus 1 = 1$

Table 1 Summary of various resources available and its utilization percentage

Resource	Utilization	Available	Utilization %
LUT	9405	303,600	3.10
FF	39,488	607,200	6.50
BRAM	256	1030	24.85
IO	100	600	16.67
MMCM	1	14	7.14

$X_{30} = Z_{30} \oplus \mathrm{FI} = 0 \oplus 0 = 0$

$X_{31} = Z_{31} \oplus \mathrm{FI} \oplus \mathrm{HI} = 0 \oplus 0 \oplus 1 = 1$

Record the resultant bits from decoder yield X_{31} to X_0 to get back the primary data, for example $X = 10101011101010111010101110101011$; consequently, the least intricate cryptography structure for control decline for both encryption and unscrambling task is TBPRA best system (Table 1).

4 Conclusion

In this paper, we have aimed at developing and demonstration of efficient routing scheme and enumerating the application of wormhole and virtual cut-through switching architecture with the avoidance of congestion control by insight modification of the packet switching for 2D mesh topology in the arrangement of 8×8 passion. In case of efficient traffic handling with the existing buffer on the principle of FIFO technique, but primary goal and concern of the paper is to optimize the architecture design in terms of area, latency and packet delivery ratio in lieu of available resources which was achieved by 24.85% utilization of BRAM and its utilization as represented in Figs. 14 and 15, the interesting achievement in the paper is the

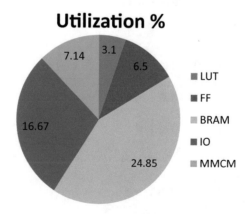

Fig. 14 Graphical representation of the utilization in % in Artix-7 FPGA

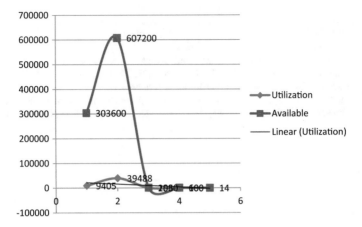

Fig. 15 Comparative graph of available resources versus utilization (%)

almost of minimization of usage of LUT by just 3.1% over the available resources their by supporting the designers for the desired employment of LUT for the other applications.

References

1. Modarressi, M., et.al.: A hybrid packet-circuit switched on-chip network based on SDM. 978-3-9810801-5-5/DATE09, 2009 EDAA
2. Kuma, S.: A network on chip architecture and design methodology. In: Proceedings of the IEEE Computer Society Annual Symposium on VLSI (ISVLSI'02). IEEE (2002). 0-7695-1486-3/02
3. Lotfi-Kamran, P.: An efficient hybrid-switched network-on-chip for chip multiprocessors. IEEE Trans. Comput. **65**(5), 1656–1662 (2016)
4. Shin, K.G., et al.: Investigation and implementation of hybrid switching. IEEE Trans. Comput. **45**(6) (1996)
5. Agyeman, M.O., et al.: Broadening the performance of hybrid NoCs past the limitations of network heterogeneity. J. Low Power Electron. Appl. **7**, 8 (2017). https://doi.org/10.3390/jlpea7020008
6. FallahRad, M., et al.: In: 2016 Euromicro Conference on Digital System Design. IEEE (2016). 978-1-5090-2817-7/16, https://doi.org/10.1109/dsd.2016.87
7. Aghdai, A., et al.: Structure of a Hybrid Modular Switch. arXiv:1705.09999v1 [cs.NI], 28 May 2017
8. Biswas, A.K.: Proficient timing channel protection for hybrid (packet/circuit-switched) network-on-chip. IEEE Trans. Parallel Distrib. Syst. (2017)
9. Ezhumalai, P., et al.: Superior hybrid two layer router architecture for FPGAs using network-on-chip. (IJCSIS) Int. J. Comput. Sci. Inf. Secur. **7**(1) (2010)
10. Paramasivam, K.: System on-chip and its research challenges. ICTACT J. Microelectron. **01**(02) (2015)
11. Tsai, W.-C., et al.: Systems on chips: structure and design methodologies. J. Electr. Comput. Eng. **2012**, 15 (2012). Article ID 509465, https://doi.org/10.1155/2012/509465. (Hindawi Publishing Corporation)

12. Mikkel, B., et al.: ReNoC: a network-on-chip architecture with reconfigurable topology. In: Second ACM/IEEE International Symposium on Networks-on-Chip. IEEE (2008). 978-0-7695-3098-7/08, https://doi.org/10.1109/nocs.2008.13
13. Lusala, A.K., et al.: Joining SDM-based circuit switching with packet switching in a router for on-chip networks. Int. J. Reconfigurable Comput. **2012**, 16. Article ID 474765, https://doi.org/10.1155/2012/474765. (Hindawi Publishing Corporation)
14. William, J., et al.: Course packets, not wires: on-chip interconnection networks. DAC 2001, 18–22 June 2001, Las Vegas, Nevada, USA. Copyright 2001 ACM 1-58113-297-2/01/0006

Detection and Prediction of Schizophrenia Using Magnetic Resonance Images and Deep Learning

S. Srivathsan, B. Sreenithi and J. Naren

Abstract Researchers are continuously making breakthroughs on the impact of deep learning in the medical industry. This approach of Deep Learning (DL) in neuroimaging creates new insights in modification of brain structures during various disorders, helping capture complex relationships that may not have been visible otherwise. The main aim of proposed work is to effectively detect the presence of schizophrenia, a mental disorder that has drastic implications and is hard to spot, using Magnetic Resonance Image Features from fMRI Database. Dataset is then fed to a Neural Network classifier, which learns to predict and give indications for preventing the onset of Schizophrenia using regression models.

Keywords Schizophrenia · Neural network · Deep learning · MRI · FNC · SBM

1 Introduction

1.1 Schizophrenia

Schizophrenia is a chronic and severe mental disorder that affects 1% of the entire population. Known recently to be caused both by genetic inheritance and biological factors, it causes abnormal brain activity which has been documented by brain imaging techniques such as Computer Tomography (CT) and Magnetic Resonance Imaging (MRI). Overall volume of the brain reduced by 3%, Temporal Lobe (6% Left, 9.5% Right); largest being in the hippocampus (14%, 9%) [1]. It is observed that ventricular enlargement is one of the earliest and consistent findings in brains

S. Srivathsan · B. Sreenithi · J. Naren (✉)
SASTRA Deemed University, Tirumalaisamudram, Thanjavur, Tamil Nadu, India
e-mail: naren@cse.sastra.ac.in

S. Srivathsan
e-mail: 120014052@sastra.ac.in

B. Sreenithi
e-mail: 120014050@sastra.ac.in

© Springer Nature Singapore Pte Ltd. 2020

P. K. Mallick et al. (eds.), *Cognitive Informatics and Soft Computing*,
Advances in Intelligent Systems and Computing 1040,
https://doi.org/10.1007/978-981-15-1451-7_10

of people with Schizophrenia. The ventricles of a schizophrenic are approximately 130% of the normal controls [2]. It is also known to have 25% less cortical grey matter especially in the frontal and temporal lobes, regions known for controlling thinking and judgement capacity of a person [3].

1.2   FNC and SBM

Functional Network Connectivity (FNC) are parameters that quantize the overall co-relation and communication between independent brain regions over fixed periods of time. FNC thus gives an overview of the connectivity pattern over time between individual brain maps. Connectivity analysis is continuously disregarding localizing activations, deactivations and increasingly using characterization of co-activation patterns [4]. Thus, resting-state functional connectivity features can be used to differentiate healthy controls and schizoaffective patients with high accuracy [5].

Source-based Morphometry (SBM) loadings correspond to the weights assigned on the brain maps collected from ICA on the grey matter concentration maps of all subjects. Grey matter concentration refers to the brain's outermost layer where most of the brain signal processing occurs. Thus, higher concentration of grey matter in an area refers to a more working part of the brain [6]. SBM previously identified up to five grey matter clusters that are present in most of the schizophrenia affected patients [7].

1.3   Deep Learning

Processing data in its raw form is one of the major shortcomings of machine learning. Deep Learning is a family of Machine Learning paradigms which is vaguely based on information processing and communication patterns in biological systems. It learns features in multiple levels of abstractions from the data, forming a hierarchy of concepts. Each layer of perceptrons tries to learn from the input data and change it into a more composite representation [8]. Deep neural networks, specifically convolutional neural networks are derived from one of the earliest forms of neural networks—the multi-layer perceptron (MLP). CNN's were designed specifically, to handle visual data. All neural networks includes a set of layers, namely—input layer, convolutional net, activation node, pooling layer and a fully connected layer [9].

2 Architecture

1. **(Training/Test) Data Input Layer**: It takes input from the data which is equal to the number of features available for each subject (Fig. 1).
2. **Hidden Layer**: Layer's aim is to extract features from the vector of reduced dimension. There can be more than one hidden layer, which leads to improved performance of the network. The above-mentioned layers contain neurons (less than or more than the size of the input layer).
3. **Activation Layer**: typically **Sigmoid, tanh or ReLU**, is an element-wise activation function, which fits values into a certain range. The most commonly used function is ReLU which turns negative values to zeros and keeps the positive values as it is.

$$\text{ReLU: } f(x) = \max(0, x) \tag{1}$$

4. **Pooling Layer**: is a layer added for dimensionality reduction and also to reduce the computational complexity of the network. MaxPooling is a popular pooling function, which takes only the value of the brightest pixel as its input volume.
5. **Fully Connected Layer**: connects all neurons in one layer, to all neurons in the next layer. The final fully connected layer uses a SoftMax activation for classifying the generated features of the input image, into various classes. The SoftMax function turns values into the range 0–1. Here, there are two classes—schizophrenic and not schizophrenic.
6. **Other Layers**: If the dimensions of a vector need to be changed, a dense layer can be used. Application of rotation, scaling and transformation to the vector is done. Another method to avoid overfitting is by adding a dropout layer, which takes a value between 0 and 1 and signifies the fraction of neurons to turn off/dropout.

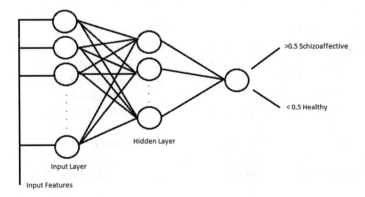

Fig. 1 Architecture of the proposed model

3   Literature Survey

Sharma and Ramkiran [10], the research approach seems to have the most accuracy when compared to CNN and DNN using Meta cognitive neural network architecture. But failed to offer an accurate representation of the data which lead to low efficiency.

Zhang and Zong [11] and Schmidhuber [12] articles provided an overview of the use of deep neural networks for translation of machine data. It compiles the various neural networks currently in usage and various architectures.

Kingma and Ba [13], introduced the Adam optimizer in 2015, which was selected for use in the model among other optimizers. It implements adaptive gradient algorithm and root mean square propagation rather than stochastic gradient descent. This allows the optimizer to handle sparse gradients on noisy problems.

Allen et al. [14], measured the resting state characteristics of brain which could then be compared to active states of the brain between healthy controls and schizoaffective patients using a new multivariate analytic approach from which the dataset was recorded.

Ulloa et al. [15], focused on generating synthetic realistic training data, which improved deep learning predictions of schizophrenia in the brain. The neural network's output data used in the model continuously produced artificial data that gave the best area under the curve as compared to real data collected from patients.

Nimkar and Kubal [16] used a machine learning technique (SVM) and the aim was to auto diagnose schizophrenia using multimodal features extracted from an MRI. It had an accuracy score of 94%, one of the best accuracies in detecting schizoaffective patients. It also elucidates on data cleaning, pre-processing the data to improve accuracy.

Talo et al. [17] was a more generalized approach as it focused on all types of brain abnormalities. The technique used deep transfer learning to classify images into normal and abnormal brain scans. Using a CNN on 613 MRI Scans, it also used techniques such as optimal learning rate, learning rate finder, data augmentation and fine-tuning to train the model.

de Moura et al. [18] studied the similarities of brain pattern of first-episode psychosis patients to chronic schizophrenia and healthy controls. First episode psychosis patients already had a brain volumetric pattern similar to schizoaffective patients.

Bordier et al. [19] studied the abnormal brain resting-state functional connectivity in patients affected by Schizophrenia as compared to normal resting state of brain in Allen E.'s article [14]. It helped in setting clear difference between the two states of human brains.

De Pierrefeu et al. [20] identified the key sites in human brains which undergoes most changes when a person is affected by schizophrenia. It helped identify various networks in the brain from which features could be extracted by an fMRI scan.

4 Method and Analysis

The dataset used in the research article was collected from a contest of IEEE International Workshop on Machine Learning for Signal Processing, compiled at the Mind Research Network under an NIH NIGMS Centres of Biomedical Research Excellence (COBRE) grant 5P20RR021938/P20GM103472 to Vince Calhoun (PI). It contained data from 46 healthy controls and 40 schizoaffective patients. Correlation values between pairs of brain maps and the activation level of ICA brain maps recorded from their grey matter concentration where present in all the cases (Fig. 2).

The Given Data was first classified using traditional Machine Learning techniques such as *K* Neighbours Classifier, Support Vector Machine (SVM), Decision tree Classifier, Random Forest Classifier, AdaBoost Classifier, Gradient Boosting Classifier, Gaussian NB, Linear Discriminant Analysis and Quadratic Discriminant Analysis and their Accuracy and Log Loss are plotted in Figs. 3 and 4.

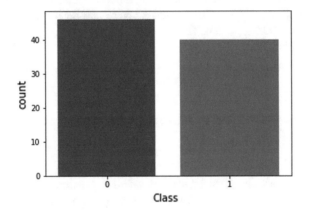

Fig. 2 Distribution of the dataset used

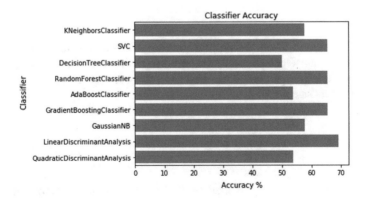

Fig. 3 Accuracy of the methods

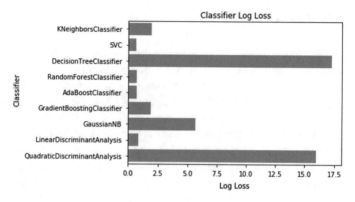

Fig. 4 Average Log Loss of the models

The Neural Network trained in the research paper contained 411 Input Neurons which was reduced to 200 Neurons in hidden layer, enabling it to learn from the input in a more constrained approach. The Output Node was activated through a sigmoid activation function, i.e.

$$S(x) = \frac{1}{1 + e^{-x}} \tag{2}$$

While the other layers used a rectifier activation function, i.e.

$$f(x) = \max(0, x) \tag{3}$$

The Data was first standardized using standard scaler function. It scaled the data points between -1 and 1., i.e. with mean as 0 and the standard deviation as 1. Generally, standard scaler function uses

$$z = \frac{(x - u)}{s} \tag{4}$$

where u is the mean of the training samples and s is the standard deviation of the training set. Adam Optimizer [10] was used to optimize the model training over 150 epochs. The Trained model was evaluated using K-Fold Cross Validation with number of splits as 5.

5 Result

The Neural Network model exhibited decreasing loss on the progress of consecutive epochs which was plotted, with final loss being 2.31×10^{-7} which is quite negligible (Fig. 5).

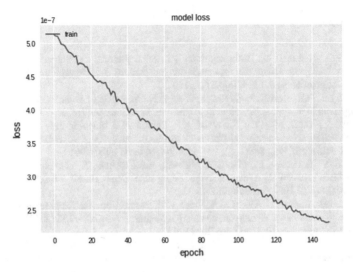

Fig. 5 Epoch-loss graph

The Resultant Neural Network finished training with over 82% accuracy and a small standard deviation of ±3.74%.

6 Conclusion

The paper presented a system to classify patient's healthy controls versus schizoaffective using a deep learning approach. The Neural Network model outperformed the highest accuracy of the traditional machine learning model by about 13% which shows that the neural network approach for classification of patients into healthy controls and schizoaffective is quite accurate. The scope of future work is to increase the model's accuracy above 90% which could imply employing the model to mainstream applications in hospitals, clinics to assist doctors in diagnosing the disease. The above-mentioned issue could be done by tweaking layers and optimizers used in the algorithm which could increase accuracy or reduce loss further. Other areas of the brain could also be mapped and features are extracted, implications currently being unknown.

References

1. Lawrie, S., Abukmeil, S.: Brain abnormality in schizophrenia: a systematic and quantitative review of volumetric magnetic resonance imaging studies. Br. J. Psychiatry **172**(2), 110–120 (1998). https://doi.org/10.1192/bjp.172.2.110
2. Weinberger, D.R., Torrey, E.F., Neophytides, A.N., Wyatt, R.J.: Lateral cerebral ventricular enlargement in chronic schizophrenia. Arch. Gen. Psychiatry **36**(7), 735–739 (1979)
3. Fornito, A., Yücel, M., Patti, J., Wood, S.J., Pantelis, C.: Mapping grey matter reductions in schizophrenia: an anatomical likelihood estimation analysis of voxel-based morphometry studies. Schizophr. Res. **108**(1–3), 104–113 (2009). ISSN 0920-9964
4. Allen, E.A., Damaraju, E., Plis, S.M., Erhardt, E.B., Eichele, T., Calhoun, V.D.: Tracking whole-brain connectivity dynamics in the resting state. Cereb. Cortex **24**(3), 663–676 (2014). https://doi.org/10.1093/cercor/bhs352
5. Mohammad, A., Kent, K., Godfrey, P., Vince, C.: Classification of schizophrenia patients based on resting-state functional network connectivity. Front. Neurosci. **3**, 133 (2013). https://www.frontiersin.org/article/10.3389/fnins.2013.00133
6. Segall, J.M., Allen, E.A., Jung, R.E., Erhardt, E.B., Arja, S.K., Kiehl, K.A., Calhoun, V.D.: Correspondence between structure and function in the human brain at rest. Front. Neuroinf. **6**, 10 (2012). https://www.frontiersin.org/article/10.3389/fninf.2012.00010
7. Xu, L., Groth, K.M., Pearlson, G., Schretlen, D.J., Calhoun, V.D.: Source-based morphometry: the use of independent component analysis to identify gray matter differences with application to schizophrenia. Hum. Brain Mapp. **30**, 711–724 (2009). https://doi.org/10.1002/hbm.20540
8. Lecun, Y., Bengio, J., Hinton, J.: Deep learning. Nature **521**, 436 (2015). https://doi.org/10.1038/nature14539
9. Lee, J.G., Jun, S., Cho, Y.W., Lee, H., Kim, G.B., Seo, J.B., Kim, N.: Deep learning in medical imaging: general overview. Korean J. Radiol. **18**(4), 570–584 (2017). https://doi.org/10.3348/kjr.2017.18.4.570
10. Sharma, A., Ramkiran, S.: MRI based schizophrenia patient classification: a meta-cognitive approach. In: 2015 International Conference on Cognitive Computing and Information Processing (CCIP), Noida, pp. 1–6, 2015. https://doi.org/10.1109/ccip.2015.7100719
11. Zhang, J., Zong, C.: Deep neural networks in machine translation: an overview. IEEE Intell. Syst. **30**(05), 16–25 (2015). https://doi.org/10.1109/MIS.2015.69
12. Schmidhuber, J.: Deep learning in neural networks: an overview. Neural Netw. **61**, 85–117 (2015). ISSN 0893-6080, https://doi.org/10.1016/j.neunet.2014.09.003
13. Kingma, D., Ba, J.: Adam: a method for stochastic optimization. In: International Conference on Learning Representations (2014)
14. Allen, E.A., Erhardt, E.B., Damaraju, E., Gruner, W., Segall, J.M., Silva, R.F., Havlicek, M., Rachakonda, S., Fries, J., Kalyanam, R., Michael, A.M.: A baseline for the multivariate comparison of resting-state networks. Front. Syst. Neurosci. **5**, 2 (2011)
15. Ulloa, A., Plis, S., Erhardt, E., Calhoun, V.: Synthetic structural magnetic resonance image generator improves deep learning prediction of schizophrenia. In: 2015 IEEE 25th International Workshop on Machine Learning for Signal Processing (MLSP), Boston, MA, pp. 1–6, 2015. https://doi.org/10.1109/mlsp.2015.7324379
16. Nimkar, A.V., Kubal, D.R.: Optimization of schizophrenia diagnosis prediction using machine learning techniques. In: 2018 4th International Conference on Computer and Information Sciences (ICCOINS), Kuala Lumpur, pp. 1–6, 2018. https://doi.org/10.1109/iccoins.2018
17. Talo, M., Baloglu, U.B., Yıldırım, Ö., Acharya, U.R.: Application of deep transfer learning for automated brain abnormality classification using MR images. Cogn. Syst. Res. **54**, 176–188 (2019). ISSN 1389-0417, https://doi.org/10.1016/j.cogsys.2018.12.007
18. de Moura, A.M., Pinaya, W.H.L., Gadelha, A., Zugman, A., Noto, C., Cordeiro, Q., Belangero, S.I., Jackowski, A.P., Bressan, R.A., Sato, J.R.: Investigating brain structural patterns in first episode psychosis and schizophrenia using MRI and a machine learning approach. Psychiatry Res. Neuroimaging **275**, 14–20 (2018). ISSN 0925-4927

19. Bordier, C., Nicolini, C., Forcellini, G., Bifone, A.: Disrupted modular organization of primary sensory brain areas in schizophrenia. NeuroImage Clin. **18**, 682–693 (2018). ISSN 22131582, https://doi.org/10.1016/j.nicl.2018.02.035
20. De Pierrefeu, A., Löfstedt, T., Laidi, C., Hadj-Selem, F., Bourgin, J., Hajek, T., Spaniel, F., Kolenic, M., Ciuciu, P., Hamdani, N., Leboyer, M.: Identifying a neuroanatomical signature of schizophrenia, reproducible across sites and stages, using machine learning with structured sparsity. Acta Psychiatr. Scand. 1–10 (2018) (Wiley)

Intrusion Detection Systems (IDS)—An Overview with a Generalized Framework

Ranjit Panigrahi, Samarjeet Borah, Akash Kumar Bhoi and Pradeep Kumar Mallick

Abstract The greatest challenge in the present era of Internet and communication technologies is the identification of spiteful activities in a system or network. An Intrusion Detection Systems (IDS) is the traditional approach that they use to follow to minimize such kind of activities. This paper provides an overview of an IDS highlighting its basic architecture and functioning behavior along with a proposed framework. It also provides a classification of threats. The IDSs are classified into various categories based on many criteria. A generalized framework of intrusion detection has been proposed to be implemented by the future network engineer.

Keywords Intrusion detection · Network · Host-based IDS · Network security · Anomaly detection · Misuse detection

1 Introduction

A generalized meaning of intrusion detection is the unlawful usage of computing resources [1, 2]. The computing resources can be network resources, system resources or resources of a computing environment. From the system point of view, an intrusion

R. Panigrahi (✉) · S. Borah
Department of Computer Applications, Sikkim Manipal Institute of Technology (SMIT), Sikkim Manipal University, Majitar, Sikkim 737136, India
e-mail: ranjit.p@smit.smu.edu.in

S. Borah
e-mail: samarjeet.b@smit.smu.edu.in

A. K. Bhoi
Department of Electrical and Electronics Engineering, Sikkim Manipal Institute of Technology, Sikkim Manipal University, Majitar, Sikkim 737136, India
e-mail: akash730@gmail.com; akash.b@smit.smu.edu.in

P. K. Mallick
School of Computer Engineering, Kalinga Institute of Industrial Technology (KIIT) University, Bhubaneswar, India
e-mail: pradeepmallick84@gmail.com

© Springer Nature Singapore Pte Ltd. 2020
P. K. Mallick et al. (eds.), *Cognitive Informatics and Soft Computing*,
Advances in Intelligent Systems and Computing 1040,
https://doi.org/10.1007/978-981-15-1451-7_11

107

or threat undertake deliberate activities that lead to security breaches, viz., integrity, confidentiality, availability, and authenticity of system resources [3–6]. This unauthorized access of system resources is intended to disrupt normal working flow of the system. Therefore, building a broad and integrated catalogue for security threats is essential. This will not only help the researchers to build tools capable of identifying various security breaches, but also motivates the cybersecurity professionals to undertake appropriate security measures.

One of the earliest classes of threats [7] was proposed by Kendall [8]. According to the research carried out by Kendall threats were classified into Denial of Service, Remote to Local (R2L), Probing and User to Root (U2R). Kendall successfully grouped threats based on the similarity of behavior and severity. For instance, all the threats belong to the group Denial of Service (DoS) tend to stop the service receivers from accessing a given resource or service. Further, in case of U2R and R2L type of threats, the attacker gains access to the target system, either by impersonating the user to the root user or by gaining local access. Similarly, in a probing attack, the attackers aggressively reconnaissance a system for vulnerabilities. Babar et al. [9] categorized threats into identification, communication, storage supervision, implanted security and physical threats in an Internet of Thing (IoT) network. As per Welch and Lathrop [10], there are seven attack categories of common threats, viz., Man-in-the-middle, Replay, Session High-Jacking Traffic Analysis, Unauthorized Access and finally, Active and Passive Eavesdropping.

2 Threats Classification

A compressive classification of threats associated with various IDS have been undertaken by Hindy et al. [11]. According to the authors, threats were classified according the cause of the threats and behavior. It can further be classified based on OSI model. A detailed schema of threat has been presented in Table 1. Similar to the guideline provided by Hindy et al. [11], threats have been classified as the source of origination and the location of threats. A threat can be originated from hosts, software, and human which are propagated through one or more conceptual layers of OSI model.

2.1 Threats Originated by Hosts

In this class of attacks, a host itself plays a crucial role in propagating attacks in a computer network. Malicious packets are triggered by the compromised nodes, which ultimately negotiate the normal functionalities of the system.

The Denial of Service (DoS) and Distributed Denial of Service (DDoS) are two basic examples of threats that originates from a compromised host to overflows the network and consume the available bandwidth unnecessarily.

Table 1 Classification threats

Layers	Source		
	Hosts	Software	Human
Application	• Impersonate (cloning) • Spoofing • Trojans • Remote access (sending, proxy, FTP, SSD, DoS) • Malwares (worms, virus, adware, spyware, ransomware)	• Information gathering • SQL injection • Cross-site scripting • Fingerprinting	• Masquerade • Phishing • U2R • R2L • Repudiation • Fraud
Presentation	• Man in the browser • Man in the middle	• Fake certificate	
Session			• Session hijacking
Transport	• Packet forging • Non-tor traffic		

Network	• Dos/DDoS
	  – Flood (Smurf)
	  – Flood (buffer overflow)
	  – Flood (SYN)
	  – Amplification
	  – Protocol exploit (teardrop)
	  – Protocol Exploit
	  – Malformed packets (ping to death)
	• Impersonate
	  – Rough access point
	  – IP spoofing
	• Scanning (content scanning, SYN scanning, FIN scanning, ACK scanning)
	• Flooding
	• Probing		

Datalink	• ARP spoofing • VLAN hooping (switch spoofing, double tagging)		
Physical			• Backdoor • Misconfiguration • Physical damage

Scanning is a form of attack originated by hosts [12]. Many attacks are found initiated from a host and propagated to other hosts through the communication medium or storage devices. Most of these threats are categorized as malwares, which includes adware, viruses, worms, Trojans, spywares, and ransomware. Adware is used for displaying advertisements to users during web surfing as well as software installation. Adware is not generally used to run malicious code but, slows down the system performance. Spyware is used to obtain information from a system, such as browsing history, cookies, emails, etc. It can also be used to track and monitor user activities.

2.2   *Threats Originated by Software*

Software originated threats are such as cross-site scripting (XSS), SQL injection, fingerprinting, misconfiguration, etc. SQL injection acquires confidential data, data manipulation by dropping database objects, viz., tables, rows, columns or indexes, rows or tables. XSS executes malicious code to take control of cookies or credentials.

2.3   *Threats Originated by Human*

This includes human action based attacks such as user masquerade, phishing, session hijacking or sniffing, etc. In phishing attack, the attacker deliberately uses emails or other chat services in order to acquire identifications or personal data of the user. U2R and R2L are also human-based attacks with higher privileges. In repudiation attack, a user is denied an action. Again, in session hijacking or sniffing, the attacker takes control of an active session, which ultimately provides access to cookies and query strings.

3   Classes of IDS

Intrusion Detection Systems can be classified into various groups depending upon many aspects such as operational area, detection approaches, and deployment location. Various logical classes of network IDS are presented in Fig. 1.

3.1   *Based on Detection Methods*

An IDS can be categorized into unsupervised and supervised detection based on the detection strategies it employs.

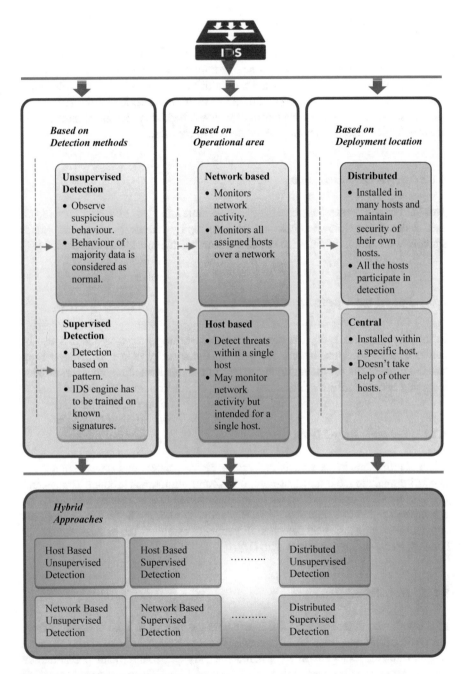

Fig. 1 Classes of intrusion detection

3.1.1 Unsupervised Detection

The intrusion detection systems based on unsupervised learning detects attacks by recognizing abnormal behavior. In case of anomaly-based methods, behavior of the attacker is found different from the normal traffic or activity, which helps in detecting the same. But, generation of false alarms is a major concern for all the IDS due to varying separation line between normal and attack behaviors. Similarly, new attacks are usually detected by anomaly-based detection engines.

3.1.2 Supervised Detection

This kind of detection takes place based on available knowledge base. Therefore, it is used to detect only known attacks and used mostly by commercial products. Attack signatures are stored in database which are searched against traffic data available.

3.2 Based on Operational Area

In this divisional category an, considering the working periphery an IDS can be network-based—responsible for maintaining the security of a network and host-based—responsible for maintaining security of a single host along with scrutinizing incoming network information.

3.2.1 Network Based Detection

NIDS identify attacks by seizing and analyzing the packets that traverse in a given network connection. It is the most commonly found commercial product. The systems involved in NIDS use to sniff the network for any kind of abnormality and malicious activity and report to a central console. Basically, it monitors the features of network data and performs intrusion detection. It generally examines the packet contents, IP headers, transport layer headers, etc. [13].

3.2.2 Host-Based Detection

In a host-based detection model the detectors are placed in host and dedicatedly responsible for the security of that host only. It analyses every activity of a host against possible attacks. If any abnormal or malicious activity is found (being performed by users or processes) an alert is generated immediately. However, a host-based IDS can be designed as a network IDS to detect network events incoming to the specific host where it is deployed [13–17].

A typical host-based detection engine holds one or more of the following characteristics [15]:

- *System Calls*: Observation of system call are important to any IDS to keep track of behavior of any program or application.
- *Network Events*: An IDS can trace the network communications to view data which is to be received by end-users. It can inspect network stack before passing on to the user-level processes.
- *File System*: This data is quite important for an IDS. Changes reflected in the file system of a host may indicate the activities performed, where irregular patterns may be suspicious. This can be viewed carefully for possible discovery of attacks.

A host-based IDS can be implemented considering one or more of these characteristics.

3.3 Based on Deployment Location

An IDS can be centrally deployed in a system or can be spanned over many systems in a distributed fashion; thus, categorizing the IDS into:

3.3.1 Distributed IDS

In a distributed IDS, the IDS is deployed over many systems to monitors those systems or network of systems. The only characteristic that make a distributed IDS unique is its ability of team work, i.e. all the IDSs in a distributed IDS take help of each other in the detection process of threats. It should be noted that both network-based and host-based IDS can be replicated to other systems in order to design a distributed IDS.

3.3.2 Centrally Deployed IDS

A centrally deployed IDS is similar to host-based IDS with a minor difference in detection process. The difference is that the centrally deployed IDSs are deployed centrally in a server, which is responsible for monitoring the security aspect of the same server as well as connected clients.

4   A Generalized IDS Framework

By going through literature various research gaps are encountered, which are not
addressed properly in most of the systems. As a result, most of these existing IDS
are not achieving high detection rate with low false-positive rate.

The research gaps found in the literature are summarized below:

- Datasets are not fully normalized.
- Lack of feature selection techniques.
- Absence of an alert categorization system.
- Most of the systems are not using feedback approach about the types and degree
  of threats.

In typical IDS, the system stimulates a vast amount false positives, which is
bottleneck for the security manager to decode potential attacks and to take preventive
measures accordingly. This is because most of the datasets used by the IDS are
not properly normalized in the preprocessing stage. As a result, attacks have not
been classified in accordance with their severity. Therefore, this classification and
normalization of intruder messages at the preprocessing stage of any IDS is necessary
for reaching a final conclusion. So it is necessary to develop an efficient process with
normalization of intruder messages and developing an effective system for detecting
intrusion with high detection rate and low false-positive rate.

Therefore, based on the above discussion, the problem statement for any research
work in the field of intrusion detection should focus more on designing and developing host and network-based intrusion detection models to estimate attack severity.
Further classification of detected alerts will also reduce the false alarms significantly.

The proposed framework explained in this section will contain three modules:
Data Normalization Module (DNM), which will be responsible for collection and
normalization of traffic messages at the preprocessing stage, Alert Generation Module (AGM), which will be responsible for detecting alerts from the incoming normalized messages from DNM stage, Alert Generation Module (AGM), which will be
responsible for detecting alerts from the incoming normalized messages from DNM
and Threat Score Processing Module (TPM), which will be responsible for evaluating threat out of the alert messages of AGM. The TPM process includes assigning
a threat score to individual threats for automated classification of IDS alert. The
proposed system will have three different databases namely: Data Normalization
Module Database (DNMDB), Alert Generation Module Database (AGMDB) and
Threat Score Processing Module Database (TPMDB).

A generalized algorithm describing the entire intrusion detection process is
described in Table 2.

All three modules mentioned above are briefly discussed below:

Table 2 Intrusion detection process

Intrusion detection process	
DMM Module	**Step 1**: Read data from a standard dataset **Step 2**: Pass the data to the normalization process **Step 2.1**: Use a standard normalization algorithm to normalize unprocessed data **Step 2.2**: Save normalized data to the DNMDB **Step 3**: Generate random sample from the normalized data for AGM
AGM Module	**Step 4**: Read normalized sample data from DNM **Step 5**: Extract the best feature of sample data using a new feature selection algorithm **Step 5.1**: Assign rank to each feature and arrange features accordingly **Step 5.2**: Remove redundant data using a new aggregation algorithm **Step 6**: Pass the aggregated features to the signature detection engine for known attacks **Step 7**: Pass the aggregated features to the anomaly detection engine for unknown attacks **Step 8**: Update AGMDB with the features, known and unknown attacks
TPM Module	**Step 9**: Read attacks and its feature from AGMDB **Step 10**: Compare and group alerts having feature similarity using an Apriori Algorithm **Step 11**: Provide score to each threat type **Step 12**: Extract the predefined rules from the TPMDB **Step 13**: Update TPMDB with threats, threat type, and threat scores **Step 14**: Pass threats, threat type and rules to the Visualization Engine for follow up action

4.1 Data Normalization Module (DNM)

The DNM module receives data from IDS datasets or from live traffics [18]. The received data are sent for normalization and sample generation. The DNM will be responsible for carrying out two major activities. The first activity involves collection and storing of traffic messages to a DB file. Instead of live traffic messages, an identical dataset would solve the purpose. But in a nutshell this activity ends with normalization or standardize data. The second and most important activity will be Sample Generation (SG). It will be designed to extract a random sample from the normalized dataset.

4.2 Alert Generation Module (AGM)

The main function of AGM is to generate alerts from the normalized dataset received from DNM. This phase is passed through four stages. The first stage involves feature selection. Many feature selection or feature ranking mechanisms are suitable in this case. Feature ranking, in fact, doesn't generate subset of features, rather it ranks the

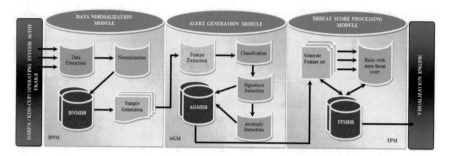

Fig. 2 An architecture of intrusion detection

features by allocating weights to each feature [19]. The second stage classifies the dataset. The supervised classification is the core of any signature detection engine. The signature detection identifies known threats. Further, the undetected instances of signature detection are passed to the anomaly detection for detecting unknown attacks. In this way, the detection process will be faster and effective for detecting threats.

4.3   Threat Score Processing Module (TPM)

The TPM Module allots weight to each threat type in order to identify seriousness of threats. It has two main components: the generating feature sets and the generate rules feature sets with auto threat score. The generating feature sets can be used to correlate similar threats [20]. The correlation mechanism ultimately allocates scores to each threat type. The threat score can be sent to visualization engine or to the administrators for follow up action (Fig. 2).

5   Conclusion

Since attacks on information and communication technology are growing day-by-day, it is seen that intrusion detection system is an essential part of such technology. Nowadays, several algorithms are available that can successfully detect various types of computer intrusions. But, it is seen that majority of the intrusion detection techniques are not being able to differentiate between attacks and the modified user behavior. Intrusion detection systems generally employ some classification techniques for differentiating attacks from normal user behavior. In such cases, selection of proper classification techniques plays a crucial role in the detection process. Again, most such systems are developed targeting some specific attacks. It will not provide a standard technique for various types of attack detection.

References

1. Abraham, T.: IDDM: Intrusion Detection Using Data Mining Techniques. Defense Science and Technology Organization. http://www.dtic.mil/cgi-bin/GetTRDoc?AD=ADA392237. Accessed on Feb 2016
2. Kemmerer, R.A., Vigna, G.: Intrusion detection: a brief history and overview. Comput. Soc. **35**(4) (2002). https://doi.org/10.1109/mc.2005
3. Kemmerer, R.A., Vigna, G.: Intrusion detection: a brief history and overview. Comput. Soc. **35**(4) (2002). https://doi.org/10.1109/mc.2002.1012428
4. Anjum, F., Mouchtaris, P.: Intrusion detection systems. In: Security for Wireless Ad Hoc Networks. Wiley (2007). https://doi.org/10.1002/9780470118474.ch5
5. Douligeris, C., Serpanos, D.N.: Intrusion detection versus intrusion protection. In: Network Security: Current Status and Future Directions. IEEE (2007). https://doi.org/10.1002/9780470099742.ch7
6. Ovaska, S.J.: Intrusion detection for computer security. In: Computationally Intelligent Hybrid Systems: The Fusion of Soft Computing and Hard Computing. IEEE (2005). https://doi.org/10.1002/9780471683407.ch8
7. Viegas, E.K., Santin, A.O., Oliveira, L.S.: Toward a reliable anomaly-based intrusion detection in real-world environments. Comput. Netw. **127**, 200–216 (2017). https://doi.org/10.1016/j.comnet.2017.08.013, ID: 271990
8. Kendall, K.R.: A database of computer attacks for the evaluation of intrusion detection systems. Ph.D. Dissertation (1999)
9. Babar, S., Mahalle, P., Stango, A., Prasad, N., Prasad, R.: Proposed security model and threat taxonomy for the Internet of Things (IoT). In: International Conference on Network Security and Applications, pp. 420–429. Springer, Berlin (2010)
10. Welch, D., Lathrop, S.: Wireless security threat taxonomy. In: Information Assurance Workshop, 2003, pp. 76–83. IEEE Systems, Man and Cybernetics Society. IEEE (2003)
11. Hindy, H., Brosset, D., Bayne, E., Seeam, A., Tachtatzis, C., Atkinson, R., Bellekens, X.: A taxonomy and survey of intrusion detection system design techniques. Netw. Threats Datasets (2018)
12. McClure, S., Scambray, J., Kurtz, G.: Hacking Exposed: Network Security Secrets and Solutions, 6th edn. McGraw-Hill Osborne Media (2009). https://www.amazon.com/Hacking-Exposed-Network-Security-solutions/dp/0071613749?SubscriptionId=0JYN1NVW651KCA56C102&tag=techkie-20&linkCode=xm2&camp=2025&creative=165953&creative ASIN=0071613749
13. Dubrawsky, I.: Topologies and IDS, Book: How to Cheat at Securing Your Network, pp. 281–315. Syngress (2007). ISBN 9781597492317, https://doi.org/10.1016/B978-159749231-7.50010-1, http://www.sciencedirect.com/science/article/pii/B9781597492317500101
14. Beal, V.: Intrusion Detection (IDS) and Prevention (IPS) Systems. URL: https://www.webopedia.com/DidYouKnow/Computer_Science/intrusion_detection_prevention.asp, 15 July 2005
15. Kumar, N., Angral, S., Sharma, R.: Integrating intrusion detection system with network monitoring. Int. J. Sci. Res. Publ. **4**(5) (2014). ISSN 2250-3153
16. Parande, V., Kori, S.: Host based intrusion detection system. Int. J. Sci. Res. (IJSR) (2015). ISSN (Online): 2319-7064
17. Kumar, B.S., Chandra, T., Raju, R.S.P., Ratnakar, M., Baba, S.D., Sudhakar, N.: Intrusion detection system-types and prevention. Int. J. Comput. Sci. Inf. Technol. **4**(1), 77–82 (2013)
18. Debar, H., Curry, D., Feinstein, B.: The Intrusion Detection Message Exchange Format (IDMEF). http://www.ietf.org/fc/rfc4765.txt, No. 4765, Mar 2007
19. Hee, C., Jo, B., Choi, S., Park, T.: Feature selection for intrusion detection using NSL-KDD. Recent Adv. Comput. Sci. 184–187 (2013). ISBN: 978-960-474-354-4
20. Singh, J., Ram, H., Sodhi, J.S.: Improving efficiency of Apriori algorithm using transaction reduction. Int. J. Sci. Res. Publ. **3**(1) (2013)

Hybrid Machine Learning Model for Context-Aware Social IoT Using Location-Based Service Under Both Static and Mobility Conditions

D. P. Abhishek, Nidhi Dinesh and S. P. Shiva Prakash

Abstract In recent times, the data traffic generated by the Internet of Things (IoT) devices is increasing. As these data are informative and useful, the categorization of these data helps in its effective and efficient use. Social Internet of Things (SIoT) generates huge data traffic compared to IoT devices and these data need to be manipulated so that the data are used productively. There exist many approaches in the reported work using machine learning models to use data effectively. In this paper, a Hybrid Machine Learning (HML) model is proposed to detect the device movement (location) based on X-, Y-coordinates on private mobile devices and private static devices, by applying Naive Bayes classifier on static data sets and K-Means clustering on mobile data sets. The result shows that the proposed model exhibit more accurate solution compared to the existing approaches.

Keywords Hybrid machine learning (HML) · Machine learning (ML) · Supervised machine learning (SML) · Unsupervised machine learning (UML) · Naive Bayes algorithm (NB) · K-Means algorithm (KM)

1 Introduction

Science and Technology brings in new advances and surprises each day. With the advent of IoT, a large scale of automation is creeping. An advance in IoT is Social IoT (SIoT), SIoT is a paradigm that involves interaction among the IoT devices. SIoT makes interaction as argumentation as discussed in [1] among IoT devices to

D. P. Abhishek (✉)
Cobalt Labs, Mysuru, India
e-mail: abhishek.dp263@gmail.com

N. Dinesh
DoS in Computer Science, UoM, Mysuru, India
e-mail: nidhidinesh9@gmail.com

S. P. Shiva Prakash
JSS Science and Technology University (Formerly SJCE), Mysuru, India
e-mail: shivasp@sjce.ac.in

© Springer Nature Singapore Pte Ltd. 2020
P. K. Mallick et al. (eds.), *Cognitive Informatics and Soft Computing*,
Advances in Intelligent Systems and Computing 1040,
https://doi.org/10.1007/978-981-15-1451-7_12

Fig. 1 Devices in SIoT system

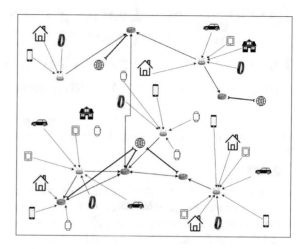

achieve common goal. SIoT devices control and coordinate among IoT devices to make decisions based on the analysis of the collected data and interactions.

One of the features of SIoT devices is context-aware; devices know the environment in which it is placed such as the users, owner, time, location, role, tasks, neighboring devices, as shown in Fig. 1 and these data are collected by each SIoT device in real time and analyzed to guide responses to attain common goal. With a lot of features in single device and interaction among the devices causes huge surge in the data traffic that in turn causes congestion leading to delays, errors in transmit/receive of data by devices. Ultimately degrading the performance and efficiency of the device, and making the device unfeasible, incapable to work in the complex data-driven world.

To address these issues many methods are proposed, one such method is machine learning approach. The use of machine learning approaches in IoT helps devices gather, understand, and actuate the data with minimal time complexity, minimal storage, and minimal network resources, thereby increasing system performance. Many surveys are made [1, 2] to choose a better model to scrutinize generated data of IoT devices. Many approaches have led to improve and create a new algorithm to increase the overall performance of the system.

The rest of the paper is organized as follows. The overview of the related works is discussed in Sect. 2. In Sect. 3, problem statement is defined. The proposed model design is presented in Sect. 4. The proposed algorithm is described in Sect. 5. The results obtained are discussed in Sect. 6, and Sect. 7 presents conclusion.

2 Related Works

In this section, work related to IoT, machine learning approaches are shown.

Authors Chandra and Yao [3] propose a framework that treats diversity and accuracy as evolutionary pressures that are exerted at multiple levels of abstraction.

Authors Tu and Sun [4] propose a framework to combine class-separate and domain-merge objectives to achieve cross-domain representation learning. Authors Bengio et al. [5] reviewed recent work in the area of deep learning and unsupervised feature learning, covering advances in deep networks, manifold learning, auto-encoders, and probabilistic models. Authors Qiu et al. [6] reviewed machine learning techniques and emphasize learning methods, such as deep learning, representation learning, parallel and distributed learning, active learning, transfer learning, and kernel-based learning. Authors Gubbi et al. [7] proposed the technologies and application domains that can be applied to drive radical change in the capabilities of IoT research in the near future. Authors Tsai et al. [8] discussed on the IoT and gave analysis on the features of "data from IoT" and "data mining for IoT". Authors Gil et al. [9] reviewed many surveys that are connected to IoT to provide well-integrated and context-aware intelligent services for IoT. Authors Atzori et al. [10] have proposed as follows: (i) recognize suitable policies for establishing and managing social relationships between devices so the developing social networks are traversable. (ii) Details a viable IoT architecture including the functionalities that are required to integrate things into a social network. (iii) Study the behavior of the SIoT network structure from simulations. Authors Lippi et al. [1] demonstrated a natural form of argumentation for conversational coordination between the devices through practical examples and case study scenarios. Authors Perera et al. [2] presented the survey that addresses a wide range of methods, techniques, functionalities, models, applications, middleware, and systems solutions connected to IoT and context-awareness. Authors Athmaja et al. [11] presented a literature review of different machine learning techniques. Authors Alturki et al. [12] explored a hybrid approach in cloud-level and network-level processing that work together to build effective IoT data analytics. Authors Bengio et al. [13] gave a better representation of the model that can produce Markov chains mixing faster between modes. Mixing between modes would be more efficient at higher levels of representation.

3   Problem Statement

The exponential increase in data traffic from the SIoT devices leads to cumbrous analyzation. Several models are proposed to interpret IoT data by applying machine learning techniques. Supervised machine learning algorithms perform better on categorized training data set, but on uncategorized training data set this approach breaks down as the accuracy will be infeasible. Similarly, unsupervised machine learning algorithms perform better on uncategorized training data. But, this approach fails on categorized training data; as the result, comprehension of data cannot be ascertained. Hence, there is a need to develop a model that increases the performance of SIoT devices.

4 Proposed Model Design

The proposed model is a hybrid approach; the combination of both supervised and unsupervised machine learning algorithms in analyzing the data traffic generated from the SIoT devices that are context-aware. The model simulates the prediction of the SIoT device movement (location) based on X- and Y-coordinates by applying machine learning algorithms, namely Naive Bayes and K-Means on SIoT device data. Figure 2 depicts the model that will review both the performance of Naive

Fig. 2 Model design of hybrid algorithm

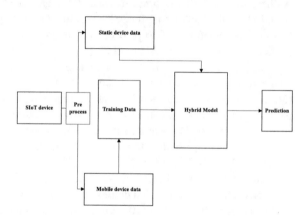

Bayes and K-Means on SIoT device data. Data generated from the SIoT devices are preprocessed, and mobile device data are trained. These data are then processed with hybrid algorithms that produce better performance on data. The model will give the accuracy chart between the applied algorithms that help in understanding the approaches in handling static and highly autonomous data (mobile data) generated by SIoT devices.

5 Algorithm

The model proposes a hybrid approach with multiple machine learning algorithms applied on the SIoT devices data.

Algorithm 1: Hybrid Algorithm

Input: SIoT device data
Result: Performance prediction

1) begin
2)   preprocessing
3)   if *data = static data set* then
4)   call Modified   Naive Bayes Static Algorithm
5)   else
6)   call Modified   K Means Mobility Algorithm
7)   end
8) end

Algorithm 2: Modified Naive Bayes Static Algorithm

Input: T // static data set
Result: Performance prediction

1) begin
2)   for $i := 1n$ do
3)   if $T[i]$ *exist* then
4)   *sum* = (sum of data points in T)
5)   end
6)   end
7)   *mean* = sum/*length* (T);
8)   for $i := 1n$ do
9) $V = V + (T[i] - mean)^2$; // V is variance
10)   end
11) $sd = sqrt((V)/length(T))$; // sd is standard deviation
12)   for $i := 1n$ do
13) $P(T[i], mean, sd) = (1/(sqrt(2*pi*sd)))*exp(-((T[i]-(mean)^2)/(2-(sd)^2)))$;
14)   end
15) $P(Dt | T) = P(T | Dt) * P(Dt)/P(T)$; // Dt is device   type   in static   data set
16) end

Algorithm 3: Modified *K*-Means Mobile Algorithm

Input: T // mobile data set
Result: Performance prediction

1) begin

2)   *normalize* (T) = (T [i] − min(T [n]) / max(T [n] − min(T [n]);

3) TM = T ; // TM is matrix of T

4) m = n * 0.7 ; // 70% out of n total data points are taken for sampling

5)   for i := 1 n do

6)   if *sizeof* (KS > m) then

7) $KS[i]$ = sample(TM, 1, FALSE); // KS is K Means sample

8)   end

9)   end

10)   for i := 1 n do

11)   for i := 1 n do

12) S = S + TM [i] − ((sum of data points in TM) / *length* (T)) ^{2} ;

13)   end

14) V = V + S / *length* (T) ; // Variance V is sum of each column of data

15)   end

16) WSS = *sizeof* (KS) * V ; // W SS is within sum of squares in cluster

17)   for i := 1 n do

18) WSS [i] = *sum* (*Kmeans* (KS , i)$ withinss ; // sum of W SS within centers i

19)   end

20)   for i := 1 n do

21)   if *length* (K [i] < $\dfrac{T}{4}$) then

22) K [i] = sample(KS , 24 , *FALSE*);

23)   end

24)   end

25) end

In Algorithm 1, initially SIoT device data are preprocessed for features extraction. Depending upon the features, data are categorized into mobile data and static data. For static data, the modified Naive Bayes algorithm is called, and for mobile data, the modified *K*-Means algorithm is called. In Algorithm 2, mean, variance, standard deviation are calculated on the static data and Gaussian distribution is used in prediction, this algorithm produces an output with better performance on static data. Algorithm 3 defines the normalization of mobile data and converting it to data matrix. Then, 70% of mobile data is sampled to train data set. Further, variance and WSS (within sum of squares) are calculated by which K value is determined by Elbow method and K clusters are created. Prediction is made using test data on the trained data clusters; this algorithm produces an output with a better performance on mobile data.

6 Experiment, Results, and Discussion

This section contains the experimental setup, and results obtained from the experiment.

Table 1 SIoT device data set features

Parameters	Values
Total users	4000
Total private devices	14,600
Total device type	8
Total mobile data points	290,706
Total static data points	6080
Total data points	296,786

Table 2 SIoT device types

Device	Mobility	Device type id
Smart phone	Mobile	1
Car	Mobile	2
Tablet	Mobile	3
Smart fitness	Mobile	4
Smart watch	Mobile	5
Pc	Static	6
Printer	Static	7
Home sensors	Static	8

6.1 Experimental Setup

The data taken for the simulation of the model are SIoT private devices from [14].

Each user has many devices associated with them. Private devices shown in Fig. 1 have features mentioned in Table 2. Table 1 lists the features that are part of the prediction analysis.

6.2 Results and Discussion

Hybrid model; the combination of both Algorithm 2 and Algorithm 3, produces the overall result as shown in Fig. 3. In this hybrid model, Naive Bayes is applied on static device data sets as it shows an accuracy of 0.9781 and K-Means is applied on mobile device data sets as it shows 0.7448 of accuracy, and the average cumulative accuracy of 0.8615 is obtained by hybrid model.

It is seen that the computation of Algorithm 2 on huge data volume takes more time compared to Algorithm 3. Algorithm 2 produces 0.9781 accuracy on static devices data, and on mobile device data, it produces 0.0417 accuracy. The performance of Algorithm 2 is shown in Fig. 4, and it shows low accuracy for mobile data sets. Overall performance of Algorithm 2 on static and mobile data produces 0.5099. Due to its low accuracy, it is not preferred to apply in complex and huge data analysis.

Fig. 3 Accuracy chart of Naive Bayes, *K*-Means, and hybrid model

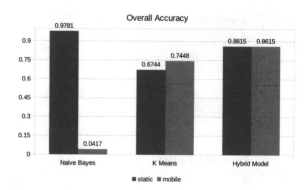

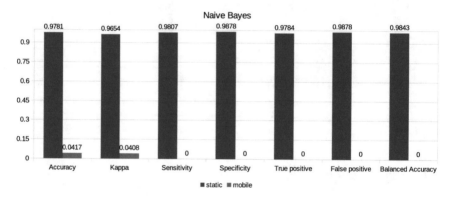

Fig. 4 Performance metrics of Naive Bayes

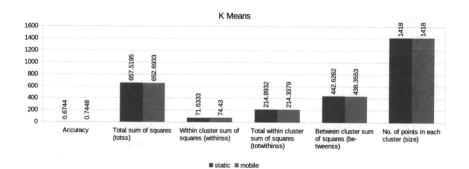

Fig. 5 Performance metrics of *K*-Means

Algorithm 3 produces initial data distributions and scaling for number of clusters needed with cluster distributions as shown in Figs. 6 and 7 on static and mobile data. Performance of Algorithm 3 shows difference in performance on static and mobile data sets as shown in Fig. 5. The combined result of accuracy on static and

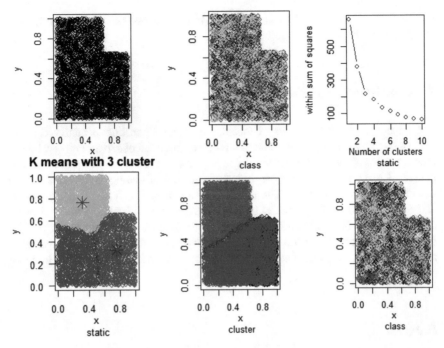

Fig. 6 *K*-Means algorithm initial results. Top left: *x* and *y* data points distributed in static clusters. Top middle: *x* and *y* data points distributed originally as per "class" attribute in static data set. Top right: Elbow graph on static data set. Down left: 3 cluster formation on static data sets. Down middle: *x* and *y* data points distributed in mobile clusters. Down right: *x* and *y* data points distributed originally as per "class" attribute in mobile data set

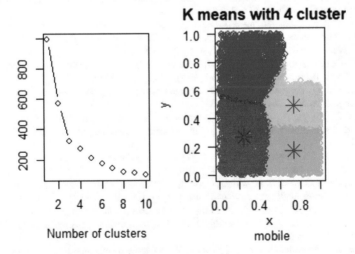

Fig. 7 *K*-Means algorithm initial results. Left: Elbow graph of mobile data clusters. Right: cluster formation of mobile data points based on Elbow's result

mobile data sets by Algorithm 3 produces an average of 0.7096, and this outcome is comparatively better than Algorithm 2.

7   Conclusion and Future Work

The impact of the combination of machine learning algorithms is analyzed under both static and mobile data sets. Based on the experimental observations, new hybrid model proposed helps in achieving high degree of accuracy and improves performance of machine learning algorithms on data sets of distinguished features and occurrences in SIoT devices. The hybrid model can be tested by applying various algorithms for the data generated by SIoT devices with different patterns. Hence, analyzing SIoT data on algorithms such as hierarchical clustering and OPTICS, fuzzy C Means, and SVM combinations in the hybrid model is considered necessary along with artificial neural network (ANN) algorithms as the hybrid models in future works.

References

1. Lippi, M., Mamei, M., Mariani, S., Zambonelli, F.: An argumentation-based perspective over the social IoT. IEEE Internet Things J. **5**(4), 2537–2547 (2018)
2. Perera, C., Zaslavsky, A., Christen, P., Georgakopoulos, D.: Context aware computing for the internet of things: a survey. IEEE Commun. Surv. Tutorials **16**(1), 414–454 (2014)
3. Chandra, A., Yao, X.: Evolving hybrid ensembles of learning machines for better generalization. Neurocomputing **69**(7–9), 686–700 (2006)
4. Tu, W., Sun, S.: Cross-domain representation-learning framework with combination of class-separate and domain-merge objectives. In: Cross Domain Knowledge Discovery in Web and Social Network Mining (CDKD'12), pp. 18–25. ACM (2012)
5. Bengio, Y., Courville, A., Vincent, P.: Representation learning: a review and new perspectives. IEEE Trans. Pattern Anal. Mach. Intell. **35**(8), 1798–1828 (2013)
6. Qiu, J., Wu, Q., Ding, G.: A survey of machine learning for big data processing. EURASIP J. Adv. Signal Process. **2016**(1), 67 (2016)
7. Gubbi, J., Buyya, R., Marusic, S., Palaniswami, M.: Internet of Things (IoT): a vision, architectural elements, and future directions. Future Gener. Comput. Syst. **29**(7), 1645–1660 (2013)
8. Tsai, C., Lai, C., Chiang, M., Yang, L.T.: Data mining for internet of things: a survey. IEEE Commun. Surv. Tutorials **16**(1), 77–97 (2014)
9. Gil, D., Ferrández, A., Mora-Mora, H., Peral, J.: Internet of things: a review of surveys based on context aware intelligent services. Sensors **16**(7), 1069, 1424–8220 (2016)
10. Atzori, L., Iera, A., Morabito, G., Nitti, M.: The Social Internet of Things (SIoT)—when social networks meet the internet of things: concept, architecture and network characterization. Comput. Netw. **56** (2012)
11. Athmaja, S., Hanumanthappa, M., Kavitha, V.: A survey of machine learning algorithms for big data analytics. In: International Conference on Innovations in Information, Embedded and Communication Systems (ICIIECS), pp. 1–4 (2017)

12. Alturki, B., Reiff-Marganiec, S., Perera, C.: A hybrid approach for data analytics for internet of things. In: Proceedings of the Seventh International Conference on the Internet of Things (IoT'17), pp. 7–1. ACM (2017)
13. Bengio, Y., Mesnil, G., Dauphin, Y., Rifai, S.: Better mixing via deep representations. In: International Conference on Machine Learning (ICML'13), vol. 28, pp. I-552–I-560 (2013)
14. Data set. http://social-iot.org/Downloads website. Last accessed 3 Feb 2019

An Efficient Fuzzy Logic Control-Based Soft Computing Technique for Grid-Tied Photovoltaic System

Neeraj Priyadarshi, Akash Kumar Bhoi, Amarjeet Kumar Sharma, Pradeep Kumar Mallick and Prasun Chakrabarti

Abstract Here, manuscript presents an efficient fuzzy logic controller (FLC)-based soft computing for grid-tied photovoltaic (PV) schematic as a maximum power point tracking (MPPT). An inverter controller for unity power factor operation of grid-tied PV system is achieved using space vector pulse width modulation (SVPWM) technology. Zeta chopper has been kept as an interface between inverter and utility grid. Under steady, dynamic and different loading situations have been presented by captured simulation estimations.

Keywords FLC · MPPT · PV · SVPWM · Zeta converter

N. Priyadarshi (✉) · A. K. Sharma
Department of Electrical Engineering, Birsa Institute of Technology (Trust), Ranchi 835217, India
e-mail: neerajrjd@gmail.com

A. K. Sharma
e-mail: ermaxamar@gmail.com

A. K. Bhoi
Department of Electrical & Electronics Engineering, Sikkim Manipal Institute of Technology, Sikkim Manipal University, Gangtok, Sikkim, India
e-mail: akash730@gmail.com

P. K. Mallick
School of Computer Engineering, Kalinga Institute of Industrial Technology (KIIT) University, Bhubaneswar, India
e-mail: pradeepmallick84@gmail.com

P. Chakrabarti
Department of Computer Science and Engineering, ITM University, Vadodara, Gujarat 391510, India
e-mail: drprasun.cse@gmail.com

© Springer Nature Singapore Pte Ltd. 2020
P. K. Mallick et al. (eds.), *Cognitive Informatics and Soft Computing*,
Advances in Intelligent Systems and Computing 1040,
https://doi.org/10.1007/978-981-15-1451-7_13

131

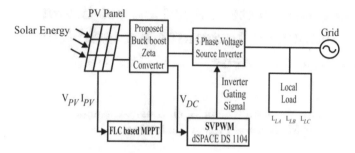

Fig. 1 Schematic diagram of FLC-based MPPT system for PV grid-tied system

1 Introduction

Because of industrial revolution, the requirement of electrical power is increasing day by day. Moreover, recently energy crisis issues have been appeared as fossil fuels are in depletion state [1–3]. Renewable energy sources provide green, clean, and environment-friendliness origination of power. Amidst all renewable sources, photovoltaic (PV) system has considered most promising technology because of less maintenance in lower fuel cost [4–6]. Maximum power point tracking (MPPT) behaves as an imperative component which provides optimal utilization efficiency of PV power system. MPPT trackers are designed by considering solar insolation and temperature in which module voltage should be close to maximum power point (MPP) region. Several MPPT methods such as perturbation and observation, incremental conductance, feedback voltage, and open-circuit voltage implemented for maximum output power generation from PV module [7–9]. The above MPPT methods have design complexities with low convergence speed under abrupt solar insolation variations. In this manuscript, fuzzy logic controller (FLC) is implemented as a MPPT trackers because of simpler design and not required mathematical analysis under abrupt weather conditions which extracts optimal PV power from PV system. Figure 1 depicts the schematic diagram of FLC-based MPPT system for PV grid-tied structure. SVPWM mechanism as an inverter controller has been employed which provides sine wave output and injects low harmonic current to the utility grid. Figure 2a, b shows P–V and I–V properties of PV scheme during varying environmental temperature and solar irradiance values.

2 FLC-Based MPPT Algorithm

There are numerous MPPT algorithms which have been implemented in past. In this manuscript, a fuzzy logic-based MPPT controller has been discussed. This method provides optimal PV power tracking with zero oscillations in voltage magnitude. Compared to genetic and neural network methods, the fuzzy logic controller provides solution to the problems as per human perception. The fuzzification, inference rules, and defuzzifications are the major steps of FLC design. Membership functions

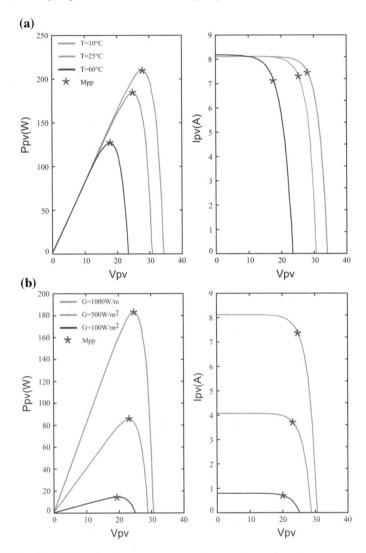

Fig. 2 PV cell's P–V and I–V nature during varying **a** environmental temperature and **b** solar irradiance values

are used to transform crisp values to fuzzy values as far as fuzzification method is concerned. Fuzzy outcomes can be decided by combining inference rule base as well as membership functions. Centroid methods can be employed to estimate the output in case of defuzzifications. In this paper, change in PV power and voltage are considered as an input of FLC and duty cycle is treated as output of FLC. Seven linguistic variables with triangular membership functions are employed in this method. Again crisp output is produced using defuzzification method. Figure 3 presents fuzzy logic design block diagram. Figure 4 describes membership functions of (a) duty ratio, (b) error, (c) change in error for fuzzy logic controller design.

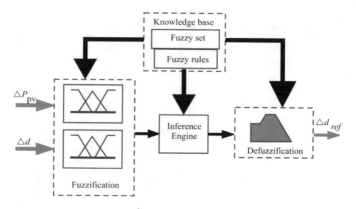

Fig. 3 Fuzzy logic design block diagram

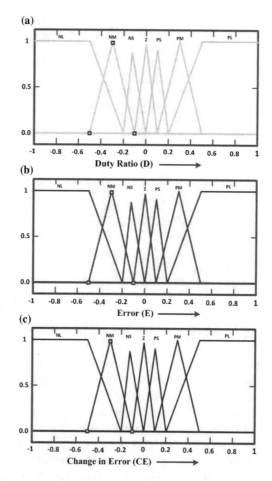

Fig. 4 Membership values **a** duty ratio, **b** error, and **c** change in error

Fig. 5 Switching sectors of
SVPWM control

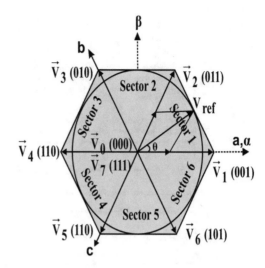

3 SVPWM-Based Inverter Controller

The SVPWM method for inverter control is best suitable pulse width modulation technique to generate sine wave for power inverter. It provides reduction in harmonics distortion, DC-link voltage utilization, and smooth control with minimized switching losses which is based on instantaneous 3-phase reference values. The 3-phase voltage output can be obtained by rotating reference vector by angular velocity $2\pi f$. Figure 5 depicts the switching sector in which 6 nonzero and 2 zero vectors are presented in which angle between two active/nonzero vectors is kept $\frac{\pi}{3}$. The active/nonzero vectors inject electric power to the loads, while zero vectors inject zero voltage to the loads. The employed SVPWM method approximates reference vectors to 8 switching patterns.

4 Simulation Results and Discussions

MATLAB/Simulink-based simulation estimation has been done under steady as well as dynamic situations. The steady-state analysis has been performed with sun irradiance value of 1000 W/m². Simulated PV parameters with improved performance of grid-connected system have been realized. Figure 6 depicts the steady behavior of grid-connected PV structure under 1000 W/m² solar irradiant.

The dynamic responses are analyzed for grid-tied PV system by applying solar insolation levels 1000–400 W/m² at $t = 0.37$ s. Corresponding variations in grid voltage/current, DC-link voltage, active/reactive power, PV voltage/current, and power are noticed. Figure 7 presents the transient nature of grid-tied PV in unstable solar insolation.

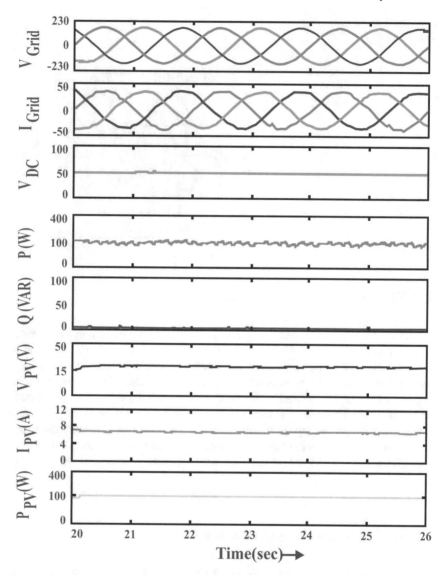

Fig. 6 Steady behavior of grid-connected PV structure under standard solar irradiant

Behavior of grid-fixed PV structure has been evaluated by connecting 10 KW load with variations in sun insolation level at $t = 0.35$ s from 1000 to 400 W/m^2. Figure 8 presents the simulated responses under loading and varying sun irradiance level. Due to decrement in sun irradiant, current of PV and corresponding PV power declined. Despite this, load requirement remains same and thus additional power has been supplied by grid which results in increment in AC current. Because of employment of SVPWM inverter controller, the behavior of grid-tied PV system is efficient and controlled.

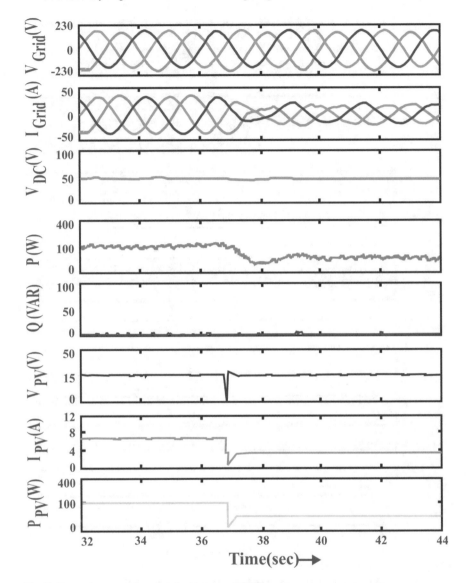

Fig. 7 Dynamic state responses of grid-connected PV system

5 Conclusions

This paper presented a soft computing-based fuzzy logic controller as a MPPT tracker for grid-tied PV system. SVPWM method as an inverter control has been used for unity power factor operation. MATLAB simulation-based simulated responses have

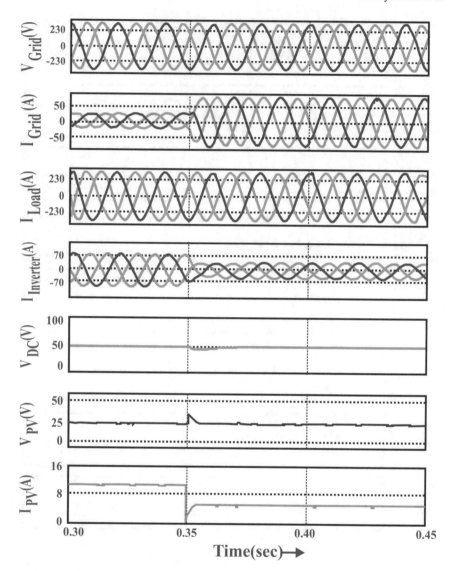

Fig. 8 Simulated responses under loading and varying sun irradiance level

been realized for steady, dynamic, and changing loading conditions. The behavior of grid-tied PV system has been evaluated by providing changing loading as well as varying sun insolation level.

References

1. Priyadarshi, N., Padmanaban, S., Maroti, P.K., Sharma, A.: An extensive practical investigation of FPSO-based MPPT for grid integrated PV system under variable operating conditions with anti-islanding protection. IEEE Syst. J. 1–11 (2018)
2. Priyadarshi, N., Padmanaban, S., Bhaskar, M.S., Blaabjerg, F., Sharma, A.: A fuzzy SVPWM based inverter control realization of grid integrated PV-wind system with FPSO MPPT algorithm for a grid-connected PV/wind power generation system: hardware implementation. IET Electr. Power Appl. 1–12 (2018)
3. Priyadarshi, N., Anand, A., Sharma, A.K., Azam, F., Singh, V.K., Sinha, R.K.: An experimental implementation and testing of GA based maximum power point tracking for PV system under varying ambient conditions using dSPACE DS 1104 controller. Int. J. Renew. Energy Res. 7(1), 255–265 (2017)
4. Nishant, K., Ikhlaq, H., Bhim, S., Bijaya, K.P.: Self-adaptive incremental conductance algorithm for swift and ripple free maximum power harvesting from PV array. IEEE Trans. Ind. Info. 14, 2031–2204 (2018)
5. Mahdi, J., Seyed, H.F.: A new approach for photovoltaic arrays modeling and maximum power point estimation in real operating conditions. IEEE Trans. Ind. Elec. 64, 9334–9343 (2017)
6. Priyadarshi, N., Sharma, A.K., Azam, F.: A hybrid firefly-asymmetrical fuzzy logic controller based MPPT for PV-wind-fuel grid integration. Int. J. Renew. Energy Res. 7(4) (2017)
7. Priyadarshi, N., Sharma, A.K., Priyam, S.: Practical realization of an improved photovoltaic grid integration with MPPT. Int. J. Renew. Energy Res. 7(4) (2017)
8. Prasanth, R.J., Rajasekar, N.: A novel flower pollination based global maximum power point method for solar maximum power point tracking. IEEE Trans. Power Elec. 32, 8486–8499 (2017)
9. Bhim, S., Chinmay, J., Anmol, B.: An improved adjustable step adaptive neuron based control approach for grid supportive SPV system. IEEE Trans. Ind. Appl. 54, 563–570 (2018)

Prediction of Academic Performance of Alcoholic Students Using Data Mining Techniques

T. Sasikala, M. Rajesh and B. Sreevidya

Abstract Alcohol consumption by students has become a serious issue nowadays. Addiction to alcohol leads to the poor academic performance of students. This paper describes few algorithms that help to improve the efficiency of academic performance of students addicted to alcohol. In the paper, we are using one of the popular Data Mining technique—"Prediction" and finding out the best algorithm among other algorithms. Our project is to analyze the academic excellence of the college professionals by making use of WEKA toolkit and *R* Studio. We implement this project by making use of alcohol consumption by student datasets provided by kaggle website. It is composed of 395 tuples and 33 attributes. A classification model is built by making use of Naïve Bayes and ID3. Comparison of accuracy is done between *R* and WEKA. The prediction is performed in order to find out whether a student can be promoted or demoted in the next academic year when previous year marks are considered.

Keywords Data mining · Prediction · Naïve Bayes · ID3 · WEKA · *R* studio · Confusion matrix

1 Introduction

Data mining, Prediction, Naïve Bayes, ID3, WEKA, *R* Studio, Confusion MatrixA large amount of data is being produced from different fields every day in order to pull out a valid and useful data which is used in a decision-making process. The

T. Sasikala · M. Rajesh (✉) · B. Sreevidya
Department of Computer Science and Engineering, Amrita School of Engineering, Bengaluru, India
e-mail: m_rajesh@blr.amrita.edu

T. Sasikala
e-mail: t_sasikala@blr.amrita.edu

B. Sreevidya
e-mail: b_sreevidya@blr.amrita.edu

Amrita Vishwa Vidyapeetham, Coimbatore, India

© Springer Nature Singapore Pte Ltd. 2020
P. K. Mallick et al. (eds.), *Cognitive Informatics and Soft Computing*,
Advances in Intelligent Systems and Computing 1040,
https://doi.org/10.1007/978-981-15-1451-7_14

141

decision-making process is performed by using different Data Mining techniques. Various mining Techniques are Clustering, Classification, Prediction, and Association. Classification is the process of arranging the data based on similarities. It is a supervised learning technique as we build the model by making use of training data which consists of class labels. Clustering is the grouping of objects based on the principle of increasing the intra-cluster distance and decreasing the inter-cluster distance. Prediction is the process of digging out the information from a huge amount of data and helps to predict the outcome irrespective of the past, present, and future events. The prediction data mining technique is used in order to predict the performance of the students. Using Naïve Bayesian and ID3 algorithms we built a model and found out the accuracy both in WEKA and R and the comparison between them is performed.

2   Related Works

In [1] they have used different data mining technologies to analyze students' performance in the courses. They have used classification techniques to assess the student's performance.

Among all the classification techniques, they have used decision tree method. The data they have used is of students belonging to Yarmouk University of the year 2005 who took C++ course. CRISP-DM a methodology is used to build a classification model. Among the 20 attributes, only 12 conditional attributes are considered which affects the performance of the students.

In [2], they have created a web-based application making use of Naïve Bayesian algorithm. The data consists of 19 attributes like student details, course details, admission details, attendance details, etc., from 700 students studying at Amrita Vishwa Vidyapeetham, Mysore. Among all the algorithms they have used, Naïve Bayesian has got the highest accuracy. Here, students, academic history is taken as input and the output is student's performance on the basis of a semester.

In [3], they have mainly focused on the data mining techniques that help in studying the educational data mainly in higher learning institutions. This shows how data mining helps in decision-making in order to maintain university reputation. It predicts the student's performance at the end of their bachelor's degree and found out the students who are at risk in the early years of their study and provides measured to improve the quality. They have collected different data from two different batches of the years 2005–06 and 2006–07 from 214 undergraduate students belonging to the civil engineering department at NEDUET, Pakistan. They have used decision tree with Gini index, with information gain, with accuracy, Naïve Bayesian, Neural Networks, Random forest.

In [4], the data has been collected from 300 students of computer science department for all three years. The attributes are related to different subjects like English, Maths, and programming language. They have used algorithms like Neural Networks,

Naïve Bayesian, J48 and SMO. According to the output, we can say that Multi-Layer Perception algorithm is more efficient for predicting student's performance.

In [5], they have collected two different sets of data of the students belonging to second year and third year of Amrita School of Engineering, Bangalore. The dataset constitutes 20 attributes like gender, 10, 12%, father education, mother education, etc. Naïve Bayesian classifier is used for predicting the student's excellence and according to the performance they have suggested a learning style for underperformed students.

In [6], they have used alcohol consumption by student's data set of age 10–14. They have used SVM method, decision trees, and Naïve Bayes algorithms and they found out that SVM is more efficient than other algorithms.

In [7], classification of data mining techniques illustrates few techniques to classify data along with their applications to health care. IF-THEN prediction rules are one of them which is a popular technique in data mining. This scheme present discovered knowledge at highest level of abstraction.

In [8], they have used datasets related to education and they have used various data mining techniques to predict and evaluate their performance.

In [9], they have analyzed the college student performance for Villupuram district. They have used clustering technique and k-means clustering algorithm. They have used Gaussian mixture model in order to improve the accuracy.

In [10], the authors have collected student information from B. J College and analyzed them using K-means clustering. The data set used for this analysis was obtained by surveying the B.C.A students from B. J College.

In [11], critical relationship between variables from a large data set is analyzed. The authors propose a mechanism which can be used by the teachers to examine the academic growth of their students.

In [12, 13], they have large amount of data from medical industry which consists of attributes like about the patient including the details, diagnosis, and medications. Using this data we train the model and find a pattern that helps in prediction.

3 Proposed System

See Fig. 1.

Dataset: The data is collected from the internet which consists of 34 attributes. The dataset is obtained from kaggle website. Name of the school, gender, age, qualification of parents, occupation of parents, weekly study time, internet access, alcohol consumption levels on working day as well as weekends, health condition and grades are some of the major attributes listed.

Preprocessing: After collecting the data we need to preprocess it. The preprocessing includes four steps. They are

Data Cleaning: In this step, we remove the noisy data and fill the missing data.

Data Integration: Combing different forms of data into single form.

Data Transformation: In this step data of different forms is made into a single form.

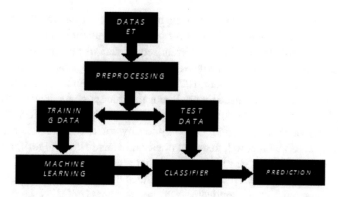

Fig. 1 Working model

Data Reduction: This technique is applied to obtain reduced representation of data.
Training Data: We split and consider the first 300 tuples as training data. The training data consists of class labels. Using this train data we build a classifier.
Machine Learning: In order to build a classifier we use machine learning techniques like prediction. In prediction, we have many data mining algorithms. Using these algorithms we build a classifier.
Classifiers: We have implemented Naïve Bayesian and ID3 algorithm in *R* and WEKA to build a classifier.

Naïve Bayesian

Input: dataset
Output: confusion matrix and predicted class labels

   Do
   For each value of the class label (Ci,Cj) find probability
   For each attribute belonging to the class label (either Ci or Cj) find probability
   Compare probabilities of each attribute of different class labels
   If p(Ci) > p(Cj)
   Class label will be Ci else Cj
   *P(Ci|X) = P(X|Ci)*P(Ci)/P(X)*
   Confusion matrix: It is a tool for finding the accuracy.

ID3

Input: dataset
Output: confusion matrix and predicted class labels

   Do

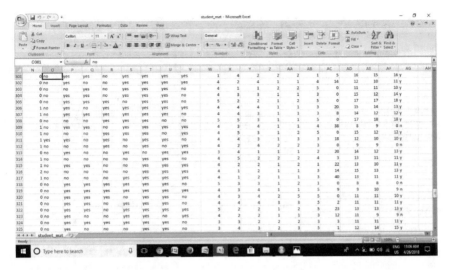

Fig. 2 Test data

Calculate information gain of all the attribute
The attribute with highest information gain value will be taken as the root node
and according to the outcomes the tree will be further extended till all the leaf
node becomes the class labels.

Test Data

The left out dataset is taken as test data and this test data is given as input to the
classifier and the prediction is performed (Fig. 2).

Prediction: It is a mining technique used to predict the outcome of the data tuple. We
perform prediction by training the classifier with trained data that consists of class
labels and then the test data without class label is given to the model and class label
is predicted. This class label indicates the performance of the student in the next
academic year i.e., yes = pass, no = fail.
yes = pass, no = fail.

4 Experimental Results

Table 1 the comparison between accuracy in *R* and WEKA for all the attributes.

Table 2 describes the comparison between accuracy in *R* and WEKA for the
attributes which effects student's performance.

Table 3 represents confusion table for Naïve Bayes algorithm in *R*. The accuracy
obtained is 94.73%.

Table 1 Accuracy comparison for R and WEKA for all attributes

Algorithm	R (%)	WEKA (%)
Naïve Bayes	96.8	95.95
ID3	94.9	92

Table 2 Comparison of accuracy in R and WEKA for few attributes

Algorithm	R (%)	WEKA (%)
Naive Bayes	94.7	87
ID3	100	100

Table 3 Confusion table for naïve bayes

Actual class label	Predicted class label	
	Y	N
Y	61	2
N	3	29

Table 4 Confusion matrix of ID3

Actual class label	Predicted class label	
	Y	N
Y	0	31
N	64	0

Table 4 represents confusion matrix of IDE algorithm in R. The accuracy is 100%. The Pass percentage will be 67.36 and fail percentage will be 32.63.

Figure 3 represents the predicted class label that explains whether a student is pass or fail. y represents pass and n represents fail.

Figure 4 represents performance of the attributes and usefulness of the attributes. Each histogram in the above figure indicates distribution of each attributes and each color indicates different classes. The attributes used travel time, study time, gout, G1, G2, G3, Dalc, Walc, performance.

5 Conclusion

Using Naive Bayesian and ID3 classifier we got the highest accuracy than many other algorithms. The attributes that have got highest priority got less accuracy when we have used ID3 algorithm than the other attributes when considered. When all the attributes are considered naïve Bayes when implemented in R has got more accuracy than in WEKA. The accuracy obtained in R and WEKA for ID3 when all the attributes are considered is same as the accuracy obtained in R and WEKA for ID3 when only some of the attributes are considered. From this, we can infer that the time required

Fig. 3 Predicted class labels

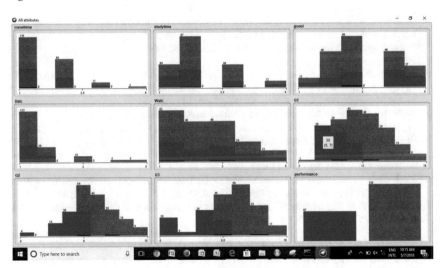

Fig. 4 Visualization of attributes

to implement the classifier for only a few attributes is less than the time when all attributes are considered. The class label is predicted and this indicates that whether a student is pass or fail by considering previous year's marks.

6 Future Scope

Using these classifiers one can build an application in order to predict the performance of the students and provide the necessary guidance.

References

1. Al-Radaideh, Q., Al-Shawakfa, E., Al-Najjar, M.I.: Mining student data using decision trees. The Int. Arab J. Inf. Technol.—IAJIT (2006)
2. Devasia, T., Vinushree T.P., Hegde, V.: Prediction of students performance using educational data mining. In: *International Conference on Data Mining and Advanced Computing (SAPIENCE)*, pp. 91–95 (2016)
3. Asif, R., Hina, S., Haque, S.I.: Predicting student academic performance using data mining methods. Int J Comput. Sci. Netw. Secur. (IJCSNS) **17**(5), 187–191 (2017)
4. Ramesh, V., Parkavi, P., Yasodha, P.: Performance analysis of data mining techniques for placement chance prediction. Int. J. Sci. Eng. Res. **2**, 2229–5518 (2011)
5. Krishna, K.S., Sasikala T.: Prognostication of students performance and suggesting suitable learning style for under performing students. In: International Conference on Computational Systems and Information Technology for Sustainable Solutions (CSITSS—2018), December 2018
6. Fabio, M.P., Roberto, M.O., Ubaldo, M.P., Jorge, D.M., Alexis, D.L.H.M., Harold, C.N.: Designing A Method for Alcohol Consumption Prediction Based on Clustering and Support Vector Machines. Res. J. Appl. Sci., Eng. Technol. **14**, 146–154
7. Sreevidya B., Rajesh M., Sasikala T.: Performance analysis of various anonymization techniques for privacy preservation of sensitive data. In: Hemanth J., Fernando X., Lafata P., Baig Z. (Eds.) International Conference on Intelligent Data Communication Technologies and Internet of Things (ICICI) 2018. ICICI 2018. Lecture Notes on Data Engineering and Communications Technologies, vol 26. Springer (2019)
8. Krishnaiah, V., Narsimha, G., Subhash Chandra, N.: Diagnosis of lung cancer prediction system using data mining classification techniques. Int. J. Comput. Sci. Inf. Technol. **4**, 39–45 (2013)
9. Shelke, N.: A survey of data mining approaches in performance analysis and evaluation. Int. J. Adv. Res. Comput. Sci. Softw. Eng. (2015)
10. Jyothi, J.K. Venkatalakshmi, K.: Intellectual performance analysis of students by using data mining techniques. Int. J. Innov. Res. Sci. Eng. Technol. **3**, (2014)
11. Sreevidya, B.: An enhanced and productive technique for privacy preserving mining of association rules from horizontal distributed database. Int. J. Appl. Eng. Res. (2015)
12. Bhise, R.: Importance of data mining in higher education system. IOSR J. Hum. Soc. Sci. **6**, 18–21 (2013)
13. Sumitha Thankachan, Suchithra, Data mining warehousing algorithms and its application in medical science. IJCSMC, **6** (2010)

Decision Support System for Determining Academic Advisor Using Simple Additive Weighting

M. Sivaram, S. Shanmugapriya, D. Yuvaraj, V. Porkodi, Ahmad Akbari, Wahidah Hashim, Andino Maseleno and Miftachul Huda

Abstract In every college there is always an Academic Advisor. Every student studying in a tertiary institution must have an Academic Advisor. Higher education is also a continuation of secondary education which is organized to prepare students to become members of the community who have the ability, not only in the academic field but also in all fields where students or student forms a form of agent of change, agent of control, and iron stock. The problem that occurs in determining the Academic Advisor is the need for Academic Advisors for students or students at STMIK Pringsewu Lampung. Where the parties concerned are still confused in determining the Academic Advisor, who is eligible to be an Academic Advisor? This problem can be solved by the method of saw in determining criteria and decision-making. This method will give an alternative weighting where the biggest weight is the alternative choice who will be determined to be Academic Supervisor in STMIK Pringsewu Lampung College.

Keywords Decision support system · Simple additive weighting · Academic advisor

M. Sivaram · V. Porkodi
Department of Computer Networking, Lebanese French University, Erbil, Iraq

S. Shanmugapriya
Department of Computer Science Engineering, M.I.E.T Engineering College, Trichy, India

D. Yuvaraj
Department of Computer Science, Cihan University – Duhok, Kurdistan Region, Iraq

A. Akbari
Department of Information Systems, STMIK Pringsewu, Lampung, Indonesia

W. Hashim · A. Maseleno (✉)
Institute of Informatics and Computing Energy, Universiti Tenaga Nasional, Kajang, Malaysia
e-mail: andimaseleno@gmail.com

M. Huda
Universiti Teknologi Malaysia, Skudai, Malaysia

© Springer Nature Singapore Pte Ltd. 2020
P. K. Mallick et al. (eds.), *Cognitive Informatics and Soft Computing*,
Advances in Intelligent Systems and Computing 1040,
https://doi.org/10.1007/978-981-15-1451-7_16

1 Introduction

1.1 Background

In order to help students complete their studies, universities are expected to provide Academic Advisors [1]. Academic advisors are lecturers appointed and assigned to the task of guiding a group or individual students who aim to help students complete their studies as quickly and efficiently as possible in accordance with the conditions and potential of individual students [2].

Academic Guidance is guidance efforts carried out by Academic Supervisors, Bahasa: Pembimbing Akademik (PA) for students of STMIK PRINGSEWU LAMPUNG who are their mentors, during semester I to done. These activities are to assist students in planning learning programs, solving specific problems, education problems and developing their potential towards the achievement of optimal development potential and learning outcomes [3]. The word "help" contains an element of direction/influence of the Academic Advisor to his students.

The level of direction of PA lecturers to students depends on the type of mentoring activities carried out, ranging from the briefing of student problems to the heavy ones [4]. The types of activities can be identified as follows: providing relevant scientific information, research dissertation orientation, developing skills to access scientific references, giving consideration or suggestions in the process of preparing a dissertation supporting scientific work, giving correction, approval or rejection of concepts the scientific writing submitted by the student's guidance based on the provisions.

Applicable provisions on the campus of STMIK Pringsewu Lampung must be put into a process framework that includes the stages of preparation, monitoring, and follow-up of student programs. Carrying out all of these steps thoroughly requires a long and continuous time, and cannot be completed within an hour or two.

1.2 Problem Formulation

In accordance with the background above, the formulation of the problem that will be resolved is to help the related parties or the STMIK Pringsewu campus to determine the Academic Advisor on the STMIK Pringsewu campus.

1.3 Research Objectives and Benefits

To make it easier for related parties or the STMIK Pringsewu campus to determine the Academic Advisor while studying at the STMIK Pringsewu College. In order not to be manual in determining the Academic Advisor in the STMIK Pringsewu college.

2 Literature Review

2.1 Decision Support System

Decision Support System is a portion regarding a computer-based information rule which includes knowledge-based systems and competencies management so that it is aged in accordance with aid decision erection of an enterprise yet employer [5]. It performs additionally keep mentioned as like a pc rule as procedures records between data to redact decisions beyond particular semi-structured problems [6].

Decision Support System (DSS) does stand described as much a dictation successful regarding assisting information adhoc analysis, choice modeling, decision-oriented, time-planning orientation the front is chronic at uncommon instances [7]. Decision Support System is additionally a mixture of sources over individual genius including the potential of elements to improve the virtue concerning selections then grow to be a computer-based statistics system because decision-making management that offers with semi-structural troubles [8]. Decision assist law, recognized as the Decision Support System (DSS), is a rule so is in a position according to grant problem-solving services then verbal exchange services because problems along semi-structured yet unstructured stipulations [9]. This provision is used to assist selection erection of semi-structured conditions yet unstructured situations, the place no certain knows because of certain or a selection should stay made [10].

Decision support regulation namely a regulation ancient in imitation of support and help management make decisions into semi-structured and unstructured conditions. Basically, the thought over DSS is only restricted in conformity with activities to that amount assist managers redact an evaluation yet replace the function or function of managers.

2.2 System Definition

Information explains that the system is a collection of elements that interact to achieve certain goals.

Two approaches are used to understand the system: procedure approach and component/element approach.

a. Understanding the system with a procedure approach is a sequence of activities that are interconnected, gathered together to achieve certain goals.
b. Understanding the system with an element approach is a collection of interrelated components and work together to achieve a specific goal.

2.3 Decision Definition

Decision-making is a process of choosing actions (among various alternatives) to achieve a goal or several goals, where decisions must include the main components, which are given as follows:

a. Data management subsystem
b. Model administration subsystem
c. User interface subsystem
d. Knowledge-based administration subsystem.

3 Research Methodology

3.1 Data Collection Method

Data series is an activity to locate records between the field so choice stands chronic in accordance with reply lookup problems.

a. Literature study

Literature lesson is an endeavor concerning gathering information yet information beyond a variety of sources, such as many books so contain a variety over theoretical research so are wished by means of researchers, magazines, texts, historical stories, and documents.

b. Observation

Observations are often interpreted as a narrow activity, particularly paying attention in imitation of something solely including the bare sight.

3.2 Data Analysis

3.2.1 Simple Additive Weighting

The SAW method or Simple Additive Weighting is a method hourly recognized as a weighted addition method. The reason for the weighted content is to locate the content weighted beyond the rating of every alternative on every attribute/criteria. The result/total rating present because of an choice is after assemble every the multiplication consequences among ranking/in contrast in conformity with the move attributes or weights regarding each attribute. Ratings regarding everything concerning the preceding attributes ought to have been through the normalization process. The SAW technique requires the method of normalizing the choice mold x after a range that can stand compared in accordance with the existing alternative ratings. SAW Method is repeatedly additionally known as much the weighted volume method. The primary thought regarding the SAW method is in accordance with find a weighted extent regarding performance scores of every choice of whole attributes.

3.2.2 Criteria Determination

Table 1 shows criteria, Table 2 shows level of education, Table 3 shows expertise, Table 4 shows length of position, Table 5 shows status of lecturers, and Table 6 shows many students.

Table 1 Criteria

Criteria	Description
C1	Level of education
C2	Expertise
C3	Duration of office
C4	Lecturer status
C5	Many students

Table 2 Level of education

Education	Criteria	Value
S1	Very less	0.10
S2	Moderate	0.30
S3	Very high	0.60
Amount of		100

Table 3 Expertise

Expertise field	Criteria	Value
Less computer technician	Meet	0.10
Multimedia	Meet	0.25
Programming	Very meet	0.60

Table 4 Length of position

Length of position (years)	Description	Value
2	Very less	0.10
3	Medium	0.25
4	Height	0.60

Table 5 Status of lecturers

Status	Description	Value
LB lecturers	Very less	0.10
Permanent lecturer	Medium	0.20
Civil servant lecturer	Very high	0.65

Table 6 Many students

Many students	Description	Value
45	D1 (very less)	0.10
500	D3 (moderate)	0.30
50	S1 (very high)	0.60

3.2.3 Normalization of Each Criteria

$$\text{Criteria Benefit}(B1, B2, B3)$$
$$R_{ij} = (X_{IJ}/\text{Max}\{X_{ij}\})$$

$$X = \left\{ \begin{array}{ccc} 0.30 & 0.30 & 0.75 \\ 0.30 & 0.60 & 0.75 \\ 0.30 & 0.60 & 0.75 \end{array} \right\}$$

3.2.4 Calculation

$$\begin{aligned} PA1 &= (0.10 \times 0.30) + (0.30 \times 0.30) + (0.60 \times 0.75) \\ &= 0.3 + 0.09 + 0.45 \\ &= 0.84 \end{aligned}$$

$$\begin{aligned} PA2 &= (0.30 \times 0.30) + (0.30 \times 0.60) + (0.30 \times 0.75) \\ &= 0.09 + 0.18 + 0.22 \\ &= 0.55 \end{aligned}$$

$$\begin{aligned} PA3 &= (0.60 \times 0.30) + (0.60 \times 0.60) + (0.60 \times 0.75) \\ &= 0.18 + 0.36 + 0.45 \\ &= 100 \end{aligned}$$

From the above calculation, the values are as follows:

$$PA1 = 0.84$$
$$PA2 = 0.55$$
$$PA3 = 1.0$$

Then the alternative has the highest value of PA3 criteria and an alternative can be chosen with a value of 1.0.

4 Conclusion

From the design of the SAW method above, the decision support system determines the Academic Advisor through an alternative weighting as a solution to determine who is eligible to become an Academic Advisor in the STMIK Pringsewu college.

References

1. Kamenez, N.V., Vaganova, O.I., Smirnova, Z.V., Bulayeva, M.N., Kuznetsova, E.A., Maseleno, A., Experience of the use of electronic training in the educational process of the Russian higher educational institution, International Journal of Engineering and Technology(UAE), Vol. 7, No. 4, pp. 4085–4089, 2018
2. Vaganova, O.I., Zanfir, L.N., Smirnova, Z.V., Chelnokova, E.A., Kaznacheeva, S.N., Maseleno, A., On the linguistic training of future teachers of unlike specialties under the conditions of Russian professional education, International Journal of Engineering and Technology(UAE), Vol. 7, No. 4, pp. 4090–4095, 2018
3. Vaganova, O.I., Kamenez, N.V., Sergeevna, V.I., Vovk, E.V., Smirnova, Z.V., Maseleno, A., Possibilities of information technologies to increase quality of educational services in Russia, International Journal of Engineering and Technology(UAE), Vol. 7, No. 4, pp. 4096–4102, 2018
4. Smirnova, Z.V., Zanfir, L.N., Vaganova, O.I., Bystrova, N.V., Frolova, N.V., Maseleno, A., WorldSkills as means of improving quality of pedagogical staff training, International Journal of Engineering and Technology(UAE), Vol. 7, No. 4, pp. 4103–4108, 2018
5. Aminin, S., Dacholfany, M.I., Mujib, A., Huda, M., Nasir, B.M., Maseleno, A., Sundari, E., Masrur, M., Design of library application system, International Journal of Engineering and Technology (UAE), Vol. 7, No. 2.27, 2018, pp. 199–204
6. Aminudin, N., Huda, M., Kilani, A., Embong, W.H.W., Mohamed, A.M., Basiron, B., Ihwani, S.S., Noor, S.S.M., Jasmi, K.A., Higher education selection using simple additive weighting, International Journal of Engineering and Technology (UAE), Vol. 7, No. 2.27, 2018, pp. 211–217
7. Maseleno, A., Huda, M., Jasmi, K.A., Basiron, B., Mustari, I., Don, A.G., and Ahmad, R. Hau-Kashyap approach for student's level of expertise. *Egyptian Informatics Journal*, 2019
8. Maseleno, A., Tang, A.Y.C., Mahmoud, M.A., Othman, M., Shankar, K., Big Data and E-Learning in Education, International Journal of Computer Science and Network Security, Vol. 18, No. 5, pp. 171–174

9. Amin, M.M., Sutrisman, A., Stiawan, D., Maseleno, A., Design Restful WebService of National Population Database for supporting E-health interoperability service, Journal of Theoretical and Applied Information Technology, vol. 96, issue 15, 2018
10. Maseleno, A., Tang, A. Y., Mahmoud, M. A., Othman, M., Negoro, S. Y., Boukri, S., ... & Muslihudin, M. The Application of Decision Support System by Using Fuzzy Saw Method in Determining the Feasibility of Electrical Installations in Customer's House. International Journal of Pure and Applied Mathematics, 119(16)

Analysis of Human Serum and Whole Blood for Transient Biometrics Using Minerals in the Human Body

N. Ambiga and A. Nagarajan

Abstract Transient biometrics is the new concept that can be applied to biometric characteristics that do change over a period of time and that are constant as possible. We came with the elements in the body which is a transient biometric with a lifetime of approximately two months. For healthy people, this can be a quiet good biometric recognition system because for healthy people, the mineral concentration of whole blood and serum is constant over a period of time. The better understanding of the range and variability of the content of these minerals in the biological samples can provide knowledge about the relationship between the mineral content and the health of individuals. This paper determines the mineral content of an individual which varies from one person to another, and concentration remains unchanged for healthy people. So for other people with disease, it will be a transient biometrics. This paper describes the analysis of mineral content for the determination of biometric authentication. In this work, we used the open data set of human serum and whole blood available from National Institute of Health Monitoring (NIHM).

Keywords Biometrics · Whole blood · Serum · Minerals

1 Introduction

By using a conventional biometric system such as smart card and passwords, the biometric authentication systems have a unique advantage that the subject need not carry anything and need not know the password or any id. There is a risk of loss of data or disclosure of the recognition taken. Biometric authentication attracts the government or any other large organizations that gain support and acceptance in critical recognition situations. The major drawback of biometric recognition system

N. Ambiga (✉) · A. Nagarajan
Department of Computer Applications, Alagappa University, Karaikudi, Tamilnadu, India
e-mail: ambiga2876@gmail.com

A. Nagarajan
e-mail: nagarajana@alagappauniversity.ac.in

© Springer Nature Singapore Pte Ltd. 2020
P. K. Mallick et al. (eds.), *Cognitive Informatics and Soft Computing*,
Advances in Intelligent Systems and Computing 1040,
https://doi.org/10.1007/978-981-15-1451-7_17

is a misuse of biometric data and it is a major concern of individuals. An individual's privacy may be compromised or discrimination may be enabled.

In this paper, we introduced transient biometric which is defined as biometric recognition technologies which depend on the biometric characteristics that are proven to change over time. Thus, they automatically cancel after a known period of time. To support the body function and biological functions, minerals or inorganic are needed.

The level of these minerals is constantly maintained by a process of the human body called homoeostasis. The excess of minerals or shortage of minerals can be associated with a variety of illnesses. The diseases such as hemochromatosis and Wilson's disease are a result of elevated iron and copper levels in the human body. The study about all the metals (inorganic compounds) is called metallomics and developed to study about all the toxic and micro, macro, trace inorganic compounds in the human body and their interactions in biological systems. A subfield called as mineralomics is the subfield of metallomics, which focuses the study of essential minerals in the human body and their relationship to a wide variety of health of individuals.

Mineralomics studies can be combined with metabolomics studies to provide a view of the minerals or chemicals contained in the body and their relationship to a variety of health states including obesity and heart disease.

2 Transient Biometrics Using Whole Blood and Serum

Despite the advantages of biometric recognition systems, a major concern of individuals is the possibility of misuse of their biometric data. A card or password can be cancelled, but what happens if your biometric data falls into the wrong hands, then the individual's privacy may be compromised. However, cancellable biometrics requires that the subject trusts the biometrics capture point and also that the misuse is detected in order to activate a transform change. For instance, the people will not happily provide their fingerprints just to have a check-in or check-out to their hotel room. In this paper, we introduce transient biometrics. Transient biometrics is defined as biometric recognition technologies which rely on biometric characteristics that are proven to change over time. Thus, they automatically cancel themselves out after a known period of time. A transient biometric approach for the verification task is shown. In contrast to cancellable biometrics, it is the actual biometric data that are naturally changing over time. As a consequence, it will presumptively help in the creation of more sociable acceptable recognitions systems. We show that the concentration of blood and serum constitutes a transient biometric with a lifetime of two months.

3 Data Set Creation

We used the open data set of the **National Institute of Health (NIH)** and is available in (www.ncbi.nim.nih.gov). NIHMS603989-SUPPLEMENT-12011-33-MOESM1-ESM.PDF.

4 Previous Work

The use of elements in the human body for biometric authentication is a new topic in the field of research. The elements in the human body differ from one person to another person and the concentration of each individual varies. So this is possible to take this for biometric authentication.

5 The Proposed Approach

Identifying an individual can be done by using a data set of previously collected samples. For the analysis of mineral content, a small volume of blood (250 µL) and serum (250 µL) samples for eight essential minerals sodium (Na), Calcium (Ca), magnesium (Mg), potassium (K), iron (Fe), zinc (Zn) and selenium (Se) was measured by plasma spectrometric methods.

A comparison of the concentration of the elements or minerals is for ten serum samples, and six whole blood samples are carried out. The results describe that the digestion and analysis method can be used to measure the concentration of the minerals.

6 Identification Performance Analysis

The minerals selected for this study are sodium, molybdenum, cobalt, chromium, manganese, selenium, iron and magnesium. Inductively coupled plasma mass spectrometry (ICP-MS) was used for quantification of analytes that are present in the ng/ML or low microgram/mL concentration in samples.

ICP-OES was used for quantification for finding the ubiquitous elements that were present in the high microgram/mL concentration range such as potassium and sodium.

Table 1 Subject information for samples used in this study

Study sample ID	Sex	Age
Serum samples		
1	Male	54
2	Male	45
3	Male	39
4	Male	24
5	Male	52
6	Female	48
7	Female	35
8	Female	20
9	Female	18
10	Female	44
Whole blood samples		
1	Male	31
2	Male	47
3	Male	46
4	Female	59
5	Female	31
6	Female	30

7 Methods and Discussion

7.1 Samples and Materials

Ten serum samples and six whole blood samples are taken from different subjects. Samples were collected from a healthy adult. Basic information about the donors is given in Table 1.

8 Results

8.1 Serum Elemental Analysis Results

The elemental composition of the digested serum sample and the inter-day analysis of the samples are measured and tabulated in Table A2. The values tabulated in Tables 2, 3 and 4 are the average of three preparation within a day and analyses ($n = 3$).

Table 2 Serum intraday analysis results (ng element/mL sample), Day 1

Sample		Ca[a]	K[a]	Mg[a]	Na[a]	Cr	Mn	Fe	Co	Cu	Zn	Se	Mo
1	Re 1	92.5	213	14.8	2420	ND	0.943	ND	0.291	1394	698	107	3.17
	Re 2	90.9	210	14.3	2315	ND	0.943	125	0.251	1429	729	108	3.33
	Re 3	97.8	228	16.6	2542	ND	1.54	22.9	0.171	1438	735	107	2.65
	Ave.	93.7	217	15.2	2426	ND	1.14	74.1	0.238	1420	721	107	3.05
	%RSD	3.13	3.63	6.38	3.82	NA	24.7	97.8	21.0	1.33	2.28	0.583	9.51

Table 3 Serum intraday analysis results (ng element/mL sample), Day 2

Sample		Ca[a]	K[a]	Mg[a]	Na[a]	Cr	Mn	Fe	Co	Cu	Zn	Se	Mo
1	Re 1	92.5	213	14.8	2420	ND	0.943	ND	0.291	1394	698	107	3.17
	Re 2	90.9	210	14.3	2315	ND	0.943	125	0.251	1429	729	108	3.33
	Re 3	97.8	228	16.6	2542	ND	1.54	22.9	0.171	1438	735	107	2.65
	Ave.	93.7	217	15.2	2426	ND	1.14	74.1	0.238	1420	721	107	3.05
	%RSD	3.13	3.63	6.38	3.82	NA	24.7	97.8	21.0	1.33	2.28	0.583	9.51

Table 4 Serum intraday analysis results (ng element/mL sample), Day 3

Sample		Ca^a	K^a	Mg^a	Na^a	Cr	Mn	Fe	Co	Cu	Zn	Se	Mo
1	Re 1	92.9	210	14.6	2350	4.27	2.38	687	ND	1370	750	115	3.75
	Re 2	96.0	217	14.8	2330	ND	ND	ND	0.160	1416	808	120	3.35
	Re 3	97.0	217	14.9	2351	ND	ND	544	0.160	1403	785	113	3.03
	Ave.	95.3	214	14.8	2344	4.27	2.38	615	0.160	1396	781	116	3.38
	%RSD	1.82	1.54	0.736	0.413	NA	NA	16.5	0	1.39	3.05	2.59	8.71

Table 5 Sample 1 serum analysis result in 3 days

	Ca[a]	K[a]	Mg[a]	Na[a]	Cr	Mn	Fe	Co	Cu	Zn	Se	Mo
Day 1	93.4	213	14.8	2073	3.06	2.07	424	0.329	1430	768	105	4.41
Day 2	93.7	217	15.2	2426	ND	1.14			1420	721	107	3.05
Day 3	**95.3**	**214**	**14.8**	**2344**	**4.27**	**2.38**	**615**	**0.160**	**1396**	**781**	**116**	**3.38**

The final average concentration is an average of three days. The intraday concentration analyses of samples determined from precision were taken as the relative standard deviation of the replicate preparations ($n = 3$) on each of three days.

8.2 Whole Blood Elemental Analysis Results

The inter- and intraday analyses of the elemental composition of the digested whole blood samples are shown in Tables 3, 4 and 5.

8.3 Mineral Concentration Reference Ranges and Consideration of Biometric Authentication

The minerals show a slight variation in Table 5.

The serum on the first day, second day and third day on a first sample shows slight variation which is 93.7, 93.4 and 95.3. The average of this to be taken is 94.76. This is the concentration of calcium in serum of sample which is 94.76. This is recorded for Sample 1. For all the elements or minerals, the level of three days is to be calculated and recorded. This value can be taken for biometric authentication. Table 5 Show this.

9 Conclusion

Biometrics research has produced significant results in terms of universality. This work records the transient data that do change over time. A transient biometric recognition exploits the mineral concentration extracted from the whole blood and serum which has been taken from different intervals. This indicates that an element concentration in human body is a valid transient biometric authentication.

Acknowledgements This research work has been supported by RUSA PHASE 2.0, Alagappa University, Karaikudi, India.

Ethical Statement This manuscript does not contain samples that were obtained from clinical studies, and no personally identifiable patient data is included.

Conflict of Interest The authors declare that they have no conflict of interest.

Comparison of Data

See Tables 2, 3, 4 and 5.

References

1. Barbosa, I.B., Theoharis, T., Schellewald, C., Athwa, C.: Transient biometrics using finger nails. In: 2013 IEEE Sixth International Conference on Biometrics: Theory, Applications and Systems (BTAS), 29 Sept–2 Oct 2013, Arlington, VA, USA. Date Added to IEEE Xplore: 16 January 2014, IEEE. Electronic ISBN: 978-1-4799-0527-0, INSPEC Accession Number: 14042228. https://doi.org/10.1109/btas.2013.6712730
2. Harrington, J.M., Young, D.J., Essader, A.S., Sumner, S.J., Levine, K.E.: Analysis of human serum and whole blood for mineral content by ICP-MS and ICP-OES: development of a mineralomics method. Biol. Trace Elem. Res. **160**(1), 132–142 (2014). https://doi.org/10.1007/s12011-014-0033-5. Epub 2014 Jun 11
3. Warda Hussain, A.M., Yasmeen, F., Khan, S.Q., Butt, T.: Reference range of zinc in adult population (20–29) years of Lahore, Pakistan. Pak. J. Med. Sci. **30**(3), 5445–5548 (2014). https://doi.org/10.12669/pjms.303.4027
4. Jantzen, C., Jorgensen, H.L., Duss, B.R., Sporring, S.L., Lauritzen, J.B.: Chromium and cobalt ion concentrations in blood and serum following types of metal-on-metal hip arthoplasties. Acta Orthop. **84**(3), 229–236 (2013)
5. Slotnick, M.J., Nriagu, J.O., Johnson, M.M., Linder, A.M., Savoie, K.L., Jamil, H.J., Hammad, A.S.: Profiles of trace elements in toenails of Arab-Americans in the Detroit area. Michigan. Biol. Trace Elem. Res. **107**, 113 (2005)
6. Boogaard, P.J., Money, C.D.: A proposed framework for the interpretation of biomonitoring data. Environ. Health **7**(Suppl 1), S12 (2008). https://doi.org/10.1186/1476-069X-7-S1-S12
7. *Metrology of Nail Clippings as Trace Element Biomarkers*, Proefschrift, ISBN 978-1-61499-287-5
8. Wee, B.S., Ebihara, M.: Neutron activation analysis and assessment of trace elements in finger-nail from residents of Tokyo, Japan (Analisis Pengaktifan Neutron dan Penilaian Unsur Surih dalam Kuku Penduduk di Tokyo, Jepun). Sains Malaysiana **46**(4), 605–613 (2017). http://dx.doi.org/10.17576/jsm-2017-4604-13
9. Derived from the Canadian Health Measures Survey 2007–2013. Int. J. Hygiene Environ. Health **220**, 189–200 (2017). www.elsevier.com/locate/ijheh
10. Versick, J., Mccall, J.T.: Trace elements in human body fluids and tissues. CRC Crit. Rev. Clin. Lab. Sci. **22**(2), 97–184 (1985)
11. Dermiencea, M., Lognaya, G., Mathieub, F., Goyens, P.: Effects of thirty elements on bone metabolism. J. Trace Elem. Med. Biol. **32**, 86–106 (2015). www.elsevier.com/locate/jtemb. https://doi.org/10.1016/b978-0-12-811353-0.00006-3

12. Zhu, Y., Wang, Y., Meng, F., Li, L., Wu, S., Mei, X., Li, H.: *Distribution of Metal and Metalloid Elements in Human Scalp Hair in Taiyuan*, China
13. Zhangd, G., Wua, D.: Ecotoxicol. Environ. Saf. **148**, 538–545 (2018)
14. Camina Martín., M.A., de Mateo Silleras, B., Redondo del Río, M.P.: *Body Composition in Older Adults*. Cons handbook of Models for Human Aging (2016)
15. Mazzoccoli, G.: Body composition: where and when. Eur. J. Radiol. http://dx.doi.org/10.1016/j.ejrad.2015.10.020. PII: S0720-048X(15)30140-6
16. Karatelad, S., Ward, N.I., Zeng, I.S., Paterson, J.: Status and interrelationship of toenail elements in Pacific children. J. Trace Elem. Med Biol. **46**, 10–16 (2018)
17. Usuda, K., Kono, K., Dote, T., Atanabe, M., Shimizu, H., Tanimoto, Y., Yamadori, E.: An overview of boron, lithium, and strontium in human health and profiles of these elements in urine of Japanese. Environ. Health Prev. Med. **12**(November), 231–237 (2007)
18. Liu T.: The scientific hypothesis of an "energy system" in the human body. J. Tradit. Chin. Med. Sci. **5**, 29e34 (2018)
19. Bel'skaya, L.V., Kosenok, V.K., Sarf E.A.: Chronophysiological features of the normal mineral composition of human saliva. Arch. Oral Biol. PII:S0003 9969(17)3020 http://dx.doi.org/10.1016/j.archoralbio.2017.06.024 Reference: AOB 3927
20. Jha, S.K., Hayashi, K.: Body odor classification by selecting optimal peaks of chemical compounds in GC–MS spectra using filtering approaches. Int. J. Mass Spectrom. 8-3-2017, Oct 2015. PII: S1387-380 6(16)30290 http://dx.doi.org/10.1016/j.ijms.2017.03.003 Reference: MASPEC 15769
21. Chang, R.: *Chemistry*, 9th Edn, p. 52. McGraw-Hill (2007)
22. Poddalgoda, D., Macey, K., Jayawardene, I., Krishnan, K.: Derivation of biomonitoring equivalent for inorganic tin for interpreting population-level urinary biomonitoring data. Regul. Toxicol. Pharmacol. **81**, 430–436 (2016)
23. Hu, Z., et al. (eds.): Advances in Artificial Systems for Medicine and Education. Detection of Hidden Mineral Imbalance in the Human Body by Testing Chemical Composition of Hair or Nails. Advances in Intelligent Systems and Computing 658, https://doi.org/10.1007/978-3-319-67349-3_20
24. . Kaur, K., Gupta, R., Saraf, S.A., Saraf, S.K.: The metal of life. Compr. Rev. Food Sci. Food Saf. **13**, 358–376 (2014). Toxicological importance of human biomonitoring of metallic and metalloid elements in different biological samples
25. Saiki, M., Vasconcellos, M.B.A., de Arauz, L.J., Fulfaro, R.: Determination of trace elements in human head hair by neutron activation analysis. J. Radioanal. Nucl. Chem. **236**(12), 25–28 (1998)
26. Christensen, J.M., Ihnat, M., Stoeppler, M., Thomassen, Y., Veillon, C., Wolynetz, M., Fresenius, Z.: Human body fluids—IUPAC proposed reference materials for trace elements analysis. Anal. Chem. **326**, 639–642 (1987)
27. Michalke, B., Rossbach, B., Göen, T., Schäferhenrich, A., Scherer, G.: Saliva as a Matrix for Human Biomonitoring in Occupational and Environmental Medicine. Springer, Berlin, Heidelberg (2014)

Design and Implementation of Animal Activity Monitoring System Using TI Sensor Tag

Shaik Javeed Hussain, Samiullah Khan, Raza Hasan and Shaik Asif Hussain

Abstract Wireless sensor networks are used to monitor the movement of animals by exact location in either conscious or unconscious state. This work initiates the details for the fundamental environmental research. The designed animal activity monitoring system employs the use of TI sensor tag attached to the animal body to provide the status and exact location of the animal. Security systems employed for animals these days are not so smart enough to find the exact location and state of animals; hence, the integration of traditional methods with the modem can help sensors to monitor the status of animals. The proposed method uses Raspberry Pi board, TI sensor tag, Pi camera and Wi-Fi dongle for real-time monitor. Here, TI sensor tag is tagged on the animal body which also provides temperature, humidity, pressure, light and motion sensor readings too. The gyro sensor monitors the movement, and their readings are measured based on the threshold value. After determining the location of animal, the camera built captures the image and sends the image to the Gmail account registered for real-time activity monitoring of animal. The prototype designed can also be built through an embedded device for stand-alone mode operation as well.

Keywords TI sensor tag · Pi camera · Wireless sensor networks · Gmail · Raspberry Pi 2

S. J. Hussain
Global College of Engineering and Technology, Muscat, Oman
e-mail: s.javeedhussain@gcet.edu.om

S. Khan · S. A. Hussain (✉)
Middle East College, Muscat, Oman
e-mail: sah.ssk@gmail.com

S. Khan
e-mail: akhan@mec.edu.om

R. Hasan
Department of Information Technology, School of Science and Engineering, Malaysia University of Science and Technology, Petaling Jaya, Selangor, Malaysia
e-mail: raza.hasan@pg.must.edu.my

© Springer Nature Singapore Pte Ltd. 2020   167
P. K. Mallick et al. (eds.), *Cognitive Informatics and Soft Computing*,
Advances in Intelligent Systems and Computing 1040,
https://doi.org/10.1007/978-981-15-1451-7_18

1   Introduction

The information technology and communication sector have been abundantly improved in recent past for monitoring extinct animals. Animal activity monitoring is to monitor the lifestyle of some particular animals, to prevent some of the species from being attacked by their superior races or to prevent some endangered species from becoming extinct. Several technologies have to be integrated to implement these monitoring activities [1]. The significant challenge faced in monitoring of animals is through latest technology to enhance their functionality in various fields.

Wireless sensor network (WSN) easily monitors the animal movements and their behavior and also provides exact status or information about them. Hence, this technology acts as a basic platform to create more impact on environmental species [2]. This work is dedicated for welfare of animals and also to help the researchers to analyze the data on this species. Internet is also a medium to efficiently monitor the status through wireless or wired communication. WSN is a collection of local and remote communication to process, monitor and transmit the information generated by sensors [3].

2   Literature Survey

In the present work the monitoring of climatic conditions [3] such as temperature, humidity in chicken farm sheds using smartphone and also to improve the climatic conditions prevailing inside the sheds of an air quality detection sensor used to control the fan switch using raspberry pi and Arduino. The implementation of automatic drip irrigation system [4] to the 50 water pots with the email. This implementation can be carried by using the Raspberry Pi, Arduino microcontroller with Zigbee, ultrasound sensors and the solenoid values. The work presents [2] environmental monitoring such as temperature, humidity and carbon dioxide; these data can be sensed by the sensors, and it can uploaded to the Web server for the real-time application purpose by using the Raspberry Pi for exhibiting environmental sensors. In the present work [8] the Robotic arm controlling using the smart phone. In the remote areas, robotic arm can be used same as the human hand; these can be controlled from the remote place, and delay can be reduced by using the Raspberry Pi, smartphone and Wi-Fi. These can be written in the Python scripting language. The author states that [9] presented the IOT-based temperature tracking system, in the environment using the Raspberry Pi and SOC, and this can be carried out by sensing the data for uploading to the Internet through the local area network; these readings can be displayed to the mobile clients with android applications for getting the alert messages. The implementation of e-health [10] smart networked system, In this system reduces the delay coming patients in emergency condition formalities such entering the data increasing capacity of beds in the hospital all these can be carried out by using the medical sensor for data transferring and patient data can be transferred to cloud environment over a wireless networks the data can be monitored in real time purpose these can be support in the medical staff. The author [10] presented Raspberry Pi and Zigbee environmental monitoring applications. In this system, presented Raspberry Pi collects the data from different sensors; the data can be collected from sensor nodes via Zigbee; these data can be supplied to multiple clients through the Web site or Ethernet. The work implemented by the author [10] is Internet of things-based monitoring and control system for home automation. In this raspberry pi can be act as central part of the system with these the home appliance like fan or light can be control on or off with the help of web server and also buzzer when smoke exploits. The implemented work of the real time vehicle monitoring system. In this system, they used the sim900A module leakage sensor, temperature sensor and Raspberry Pi; with these, owner of vehicle can observer the vehicle without any restrictions. Sim900A module can be placed in the vehicle tracking in at a certain location and also travel the people securely or safely with a LPG. The work [8] presented human–robot cloud architecture-based security and surveillance systems where it uses Raspberry Pi's placed at different places. Raspberry Pi's were connected with cameras, and different sensors, security and surveillance module include person detection, motion detection and object tracking.

3   System Design and Implementation

Raspberry Pi is the central part of the system. Raspberry Pi and sensor tag can interface with the help of upgraded version of Bluetooth Dongle (4.0). Sensor tag readings can be collected by the Raspberry Pi with the help of Bluetooth Dongle: temperature, humidity, pressure, light and motion sensor. Motion sensor consists of accelerometer, magnetometer and gyroscope. Thus the readings of sensor tag we make use of the gyroscope sensor to monitor the animal in the cages or zoos by getting the changes in the gyro sensor readings camera is captured the position of the animal, activity of the animal like is it sleeping, eating and also sensor tag the surrounding environmental conditions like temperature, humidity, pressure etc. Wi-Fi adapter is used to get the Internet connection to the Raspberry Pi with these readings, and the captured image can be send to the Gmail for real-time monitoring purpose with the support of Wi-Fi adapter; hence, we can observe the readings of sensor tag in the Internet from anywhere in the world (Fig. 1).

Hardware consists of sensor tag, Raspberry Pi, Bluetooth Dongle, Wi-Fi adapter and Pi camera; these are the hardware used in this system. Sensor tag is a latest updated TI component (Texas component). Sensor tag is a device consisting of 32-bit ARM Cortex-M3 processor which runs at 48 MHz as the original processor with rich device of features. It is a Bluetooth low-energy peripheral device based on cc26xx hardware family including different sensors.

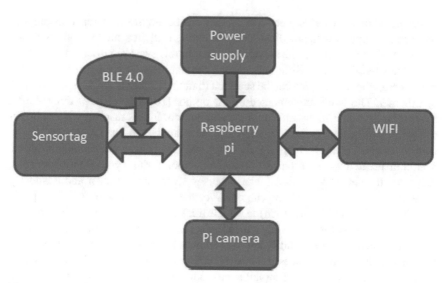

Fig. 1 Proposed block diagram

3.1 Hardware Components

They are temperature, humidity, pressure, motion (accelerometer, magnetometer and gyroscope) and light. In this paper, we use gyroscope sensor to monitor the animal monitoring system in the cages, security, etc. Sensor tag consists of five sensors: IR temperature sensor consists in both object and ambient temperature of a size 2×16 bits; moment sensor consists in totally nine axes (accelerometer, gyroscope and magnetometer) of a data size 9×16 bits; the humidity sensor consists in both relative and temperatures of a size 2×16 bits; barometer consists in both pressure and temperature of a size 2×16 bits; optical sensor consists of a size 1×16 bits. We use the gyroscope sensor to monitor the animal activity in the cages. They are several generations of Raspberry Pi by upcoming the technology in A, B, A+B+ model can be released. Broadcom series chips can be inserted in Raspberry Pi for the late of version of Raspberry Pi 2 consisting of BCM 2836 Broadcom socket on chip with a speed of 32-bit quad core ARM Cortex-A7 processor with a cache memory of L2 shared with 236 Kb.

Raspberry Pi exhibits 6 times than the processing capacity of previous models. Raspberry Pi 2 consists of 2 USB ports, Ethernet port and also GPIO pins for hardware extending purpose, SD card slot. It can support 32 GB SD card into the Raspberry Pi, display connector slot, camera connector slot, HDMI display connector, audio and REA video jack in this Raspberry Pi (Fig. 2).

Latest Raspbian Pi camera has a board of Sony 1MX219 with 8-megapixel sensor. This camera can be used in both either in image capture or video capturing. It is very much efficient for the home security and also for wildlife traps in the camera with

Fig. 2 Major hardware components used

v2 series of Pi camera. This can be connected to the Raspberry Pi with the help of 15-pin camera strip.

3.2 Software Requirements

Python programming language is the suitable language for the Raspberry Pi. It is the default programming language to Raspberry Pi. By using Python language, camera is capturing the image, sensor tag readings can be collected into the Raspberry Pi and those values can be sent to the Gmail with the help of Python language.

3.3 Flowchart

Initially, read the sensor tag readings from Raspberry Pi, observe the gyro sensor readings and apply the threshold value for gyro sensor readings. If gyro sensor values are greater than threshold, camera gets turn on and captures the image. The image captured by the camera. It will store and send to the Gmail for real-time purpose; if the gyro sensor value is less than the threshold, it will again read the sensor tag readings and gyro sensor value (Fig. 3).

4 Experimental Results

The experimental results shown here define how the location of the animal is identified by Raspberry Pi and sensor tag. The values are shown in the Raspberry Pi Python language with the correct location, and that information is sent to your email for appropriate verification.

Fig. 3 Flowchart of the
proposed system

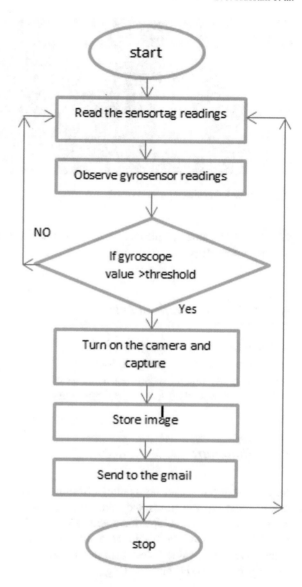

Figure 4 shows that by applying above command it scans the Bluetooth devices with the MAC address which are located near to the Raspberry Pi to pair the device: sudo python sensortag.py B0:B4:48:BC:78:03–t 0.01—all.

Figure 5 shows that by applying above command it addresses the code to get the readings of sensors which are presented in the sensor tag in the lx terminal. This code contains only about sensor tag readings: sudo python sensortagemail.py B0:B4:48: BC:78:03–t 0.01—all.

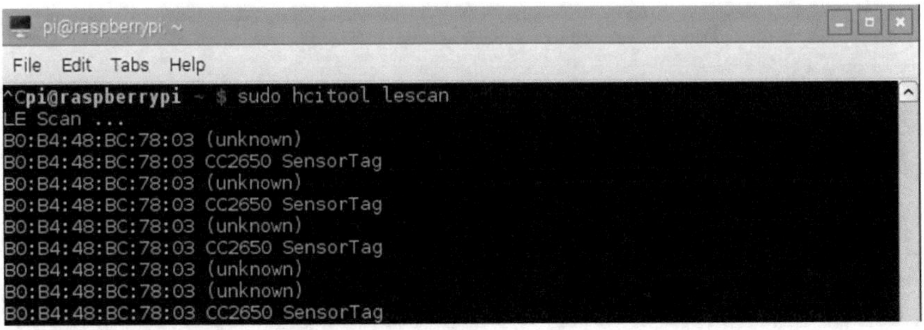

Fig. 4 Obtaining the MAC address for sensor tag

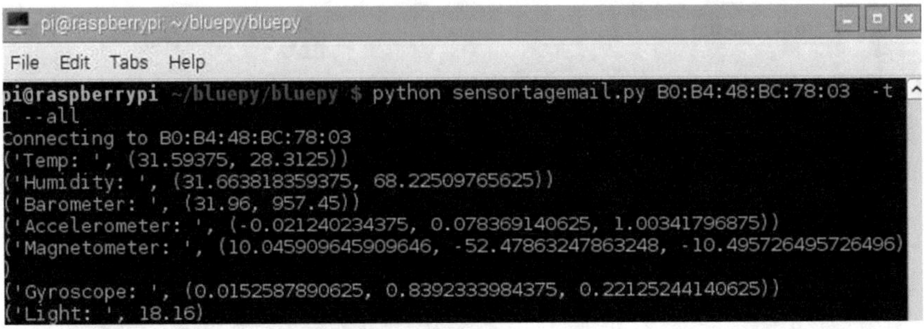

Fig. 5 Obtaining all sensor tag readings

Figure 6 shows the notification of image sent to the Gmail successfully when gyro sensor values exceed the threshold value and also it shows the sensor readings which are presented in the sensor tag: sudo python sensortagemail.py B0:B4:48:BC:78:03–t 0.01—all (Fig. 7).

Receiving the email of image captured by the camera with the code written in the python language greater than applied threshold to the sensor tag values mainly for gyroscope sensors.

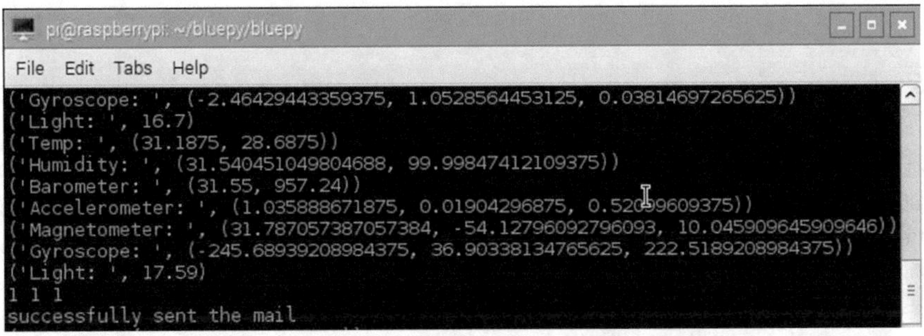

Fig. 6 Sending the Gmail after crossing the threshold

```
pi@raspberrypi: ~/bluepy/bluepy                                    _ □ ×
File  Edit  Tabs  Help
('Gyroscope: ', (-2.46429443359375, 1.0528564453125, 0.03814697265625))
('Light: ', 16.7)
('Temp: ', (31.1875, 28.6875))
('Humidity: ', (31.540451049804688, 99.99847412109375))
('Barometer: ', (31.55, 957.24))
('Accelerometer: ', (1.035888671875, 0.01904296875, 0.5205909609375))
('Magnetometer: ', (31.787057387057384, -54.12796092796093, 10.045909645909646))
('Gyroscope: ', (-245.68939208984375, 36.90338134765625, 222.5189208984375))
('Light: ', 17.59)
1 1 1
successfully sent the mail
```

Fig. 7 If readings are greater than the threshold, then send the email

Fig. 8 Image sent to the email

@gmail.com

to me ▾

Figure 8 shows the image which is captured by the camera when the gyro sensor exceeds the threshold and it is sent to the Gmail for real-time monitoring purpose. Like these, we can monitor the activity of the animals which are presented in the cages or enclosures of zoo.

5   Conclusion

In this paper, the proposed system is the non-invasive method if animal activity monitoring by deploying sensor tag on the body of the animal which is placed in a cage or an enclosure. The proposed system is based on intelligent watching devices with the camera interfaced to the Raspberry Pi. This system can consume low energy, is easily portable and works with stand-alone mode. Pi camera will capture the image when animal moves. The motion of the animal determined by the values of gyroscope sensor is present in the sensor tag. In this paper, we used threshold for only one motion is captured like slightly moving. The system can be further improved by providing threshold for different movements. The captured image can be sent for real-time monitoring purpose.

Acknowledgements Firstly, I would like to thank Middle East College, Muscat, for providing me the facilities and equipment to perform this work. I would like to extend my sincere thanks to pixabay.com for providing such high-quality HD images for free to perform my work. Also, my deepest thanks to all the co-authors for constant support and dedication.

References

1. Abaya, W.F., Abad, A.C., Basa, J., Dadios, E.P.: Low cost smart security camera with night vision capability using Raspberry Pi and open CV. In: International Conference on Humanoid, Nanotechnology, Information Technology, Communication and Control, Environment and Management (HNICEM), pp. 1–6. IEEE, Philippines (2014)
2. Atabekov, A., Starosielsky, M.: Internet of Things-based temperature tracking system. In: International Conference in Computer Software and Applications Conference (COMPSAC), vol. 3, no. 1, pp. 493–498. IEEE (2015)
3. Balakrishnan, R., Pavithra, D.: IoT based monitoring and control system for home automation. In: Global Conference in Communication Technologies (GCCT), pp. 169–173. IEEE, Thuckalay (2015)
4. Ben Abdelkader, C., Krstikj, A., Karaiskos, C., Mavridis, N., Pierris, G.: Smart buildings and the human-machine cloud. In: GCC Conference Exhibition, pp. 1–6. IEEE, Muscat (2015)
5. Jadhav, G., Jadhav, K., Nadlamani, K.: Environment monitoring system using Raspberry-Pi. Int. Res. J. Eng. Technol. (IRJET) **3**(04), 1168–1172 (2016)
6. Jassas, M.S., Mahmoud, Q.H., Qasem, A.A.: A smart system connecting e-health sensors and the cloud. In: International Conference in Electrical and Computer Engineering (CCECE), pp. 712–716. IEEE, Canada (2015)
7. Jindarat, S., Wuttidittachotti, P.: Smart farm monitoring using Raspberry Pi and Arduino. In: International Conference in Computer, Communications and Control Technology (I4CT), pp. 284–288. IEEE, Malaysia (2015)
8. Mane, Y.B., Shinde, P.A., Tarange, P.H.: Real time vehicle monitoring and tracking system based on embedded Linux board and android application. In: International Conference on Circuit, Power and Computing Technologies (ICCPCT), pp. 1–7. IEEE, Nagercoil (2015)
9. Nigel, K.G., Prem Kumar, K.: Smart phone based robotic arm control using raspberry pi, android and Wi-Fi. In: International Conference in Innovations in Information, Embedded and Communication Systems (ICIIECS), pp. 1–7. IEEE, Coimbatore (2015)
10. Nikhade, S.G.: Wireless sensor network system using Raspberry Pi and zigbee for environmental monitoring applications. In: International Conference in Smart Technologies and Management for Computing, Communication, Controls, Energy and Materials (ICSTM), pp. 376–381. IEEE, Chennai (2015)
11. Singhal, S.: Smart drip irrigation system using raspberry pi and arduino. In: International Conference in Computing, Communication and Automation (ICCCA), pp. 928–932. IEEE, Noida (2015)

Improved Filtering of Noisy Images by Combining Average Filter with Bacterial Foraging Optimization Technique

K. A. Manjula

Abstract Biologically inspired algorithms have attracted a large number of researchers and found to have application in many areas of computer science including image processing. This paper presents the application of a biologically inspired algorithm, bacterial foraging optimization algorithm (BFOA), to optimize the filtering of images for denoising and its performance comparison with existing noise reduction techniques like average filter. Usually, images suffer from noises which will corrupt image quality and appearance. Hence, denoising of images plays a great role in the image processing and the frequently used filters are average filter, median filter, etc., to remove noises. This research paper explores the suitability of applying BFOA on the filtered image produced by average filter to result a further denoised image. In this proposed method of bacterial foraging-based optimization, peak signal-to-noise ratio (PSNR) is used as fitness function to denoise the noisy images. The implemented code is tested for noisy images (Gaussian noise and salt–pepper noise) filtered with average filter, and results show the optimization capability of BFOA-based method and that it improves the denoised images produced by average filter.

Keywords Image processing · Denoising · Bacterial foraging optimization · Average filter

1 Introduction

Digital image processing is a significant technique applied in many areas like biomedical science, defence, etc. Noise is some random signal, and its presence may destroy some of the image contents. Image distortion is treated as one of the most crucial issues in image processing, and filters are found to be useful in removing noises. The pixels containing noise are detected and replaced with their estimated values, leaving the other pixels unaffected. Hence, the filtering process plays a crucial role in getting the accurate data from images.

K. A. Manjula (✉)
Department of Computer Science, University of Calicut, Malappuram, Kerala, India
e-mail: manjulaka@gmail.com

© Springer Nature Singapore Pte Ltd. 2020
P. K. Mallick et al. (eds.), *Cognitive Informatics and Soft Computing*,
Advances in Intelligent Systems and Computing 1040,
https://doi.org/10.1007/978-981-15-1451-7_19

177

Several researchers have been trying to improve the denoising process of images by introducing new techniques. K. M. Bakwada et al. in their work proposed a new approach to improve peak signal-to-noise ratio of images corrupted by salt–pepper noise [1]. The proposed work used images with adaptive median filter and the bacterial foraging optimization (BFO) technique based on the fitness function mean square error. And, it has been observed from the results that the proposed method is better than conventional methods in terms of image quality, clarity, and smoothness especially in edge regions of the final image. This technique is found to be suitable for eliminating the salt-and-pepper noise of corrupted images with a high noise level such as ninety percentage. S. Yaduwanshi and J. S. Sidhu describe application of BFO as an optimizing technique to improve the denoising of medical images like CT scan and MRI images [2]. This research work is dealt with the optimization of images filtered by adaptive median filter on the basis of fitness function mean square error. V. Sharma et al. conducted a study providing a comprehensive review of the algorithms applying bacterial foraging optimization along with their potential in engineering applications like scheduling of networks, forecasting of electric load, designing of antenna, etc. [3]. S. Gholami et al. put forward a system with an evolutionary algorithm applying the foraging behaviour of *E. coli* bacteria to improve peak signal-to-noise ratio of images [4]. In 2013, Beenu et al. worked on image segmentation by applying algorithm based on BFO. Entropy, standard deviation and PSNR were used to evaluate the methods of thresholding. The results of this work had yielded satisfactory results [5]. Om Prakash Verma et al. discussed a fuzzy system for edge detection suitable for noisy images applying bacterial foraging optimization in 2011. The results were compared with the results of applying some of the standard edge detection operators like Sobel, Canny, ACO and GA. The results showed the superior performance and reduced computational complexity of the new method [6]. According to S. Binitha and S. S. Sathya, bio-inspired algorithms are opening up new opportunities in computer science and they will be extremely useful in the next-generation computing, modelling and algorithm engineering [7].

Bio-inspired computing optimization techniques based on ant colony, bacterial foraging, firefly population and bee colony are considered to be effective in many stages of image processing and literature survey points to the suitability of BFO in improving the filtering of images. In most of the works reviewed, the fitness function of BFO was taken with respect to mean square error and only one type of noise was considered for experimenting the performance of BFO-based smoothening of images. The improvement that can be done for a popular filter like average filter is not well studied yet. The review of literature suggests the possibility of exploring the methods for improving the performance of all commonly used filters including average filter.

Fitness function plays a crucial role in applying bio-inspired optimization algorithms, and hence, this research work proposes peak signal-to-noise ratio as a fitness function in deciding the optimization of BFO in image filtering. Also, it is planned to test two different noises—salt–pepper and Gaussian—for the average filter, to confirm the suitability of applying BFO to filtering and for confirming its capability for improved smoothening of images.

In this research paper, the whole work is arranged in the coming sections as follows: Section 2 discusses filtering of images, salt–pepper noise, Gaussian noise and average filter in general. Section 3 discusses BFO. Section 4 discusses the proposed method for image smoothening by applying BFO-based method along with average filter. Section 5 discusses simulation and result of experiments and the final section presents the conclusion and future directions.

2 Noises in Images and Filtering

One of the very relevant tasks in image processing is to suppress the noise from corrupted images, and it is usually done by filtering of images. One simple and popular filter, known as average filter, is considered in this research work.

2.1 Gaussian Noise

This type of noise is usually found associated with devices like detectors or amplifiers and is commonly known as electronic noise. Natural sources such as discrete nature of radiation from warm objects or thermal vibration of atoms are found to cause this type of noise. This noise usually corrupts the grey values in digital images.

2.2 Salt-and-Pepper Noise

The original pixel values of image data are found to drop, if affected with salt–pepper noise, and such an affected image will probably have bright pixels in dark regions and dark pixels in bright regions. This type of noise corruption can be caused by dead pixels, errors arising from analog to digital conversion or bit transmission. This can be done away mostly by applying dark frame subtraction and by interpolating around dark or bright pixels.

Filters are noise reduction techniques applied on corrupted images to reduce the effect of noises, and there are many commonly used filters like average filter, median filter, adaptive median filter, etc. This research paper deals with smoothening of the result generated by average filter.

2.3 Average Filter

Average filters, also known as mean filters, have a general less complicated structure compared to other filters. To eliminate corruption, replacement of the value of every pixel in an image with the average value of its neighbours is performed. This filter performs well in the case of signal-dependent noises too. This filter is applied to suppress the minute details in an image and also to bridge the small gaps present within the curves or lines. This filter can be applied successfully in removing grain noise from a photograph.

3 Bacterial Foraging

Bacterial foraging optimization algorithm is developed based on the foraging strategy of *Escherichia coli* (*E. coli*) bacteria which is largely found living in human intestine [8]. This optimization method finds a significant role in many domains due to its advantages like lesser mathematical complexity, faster convergence and more accuracy. The foraging aim of each *E. coli* bacteria is to find places with high nutrient level, at the same time avoiding noxious locations using certain patterns in its motion. A chemical substance called attractant is released while heading to a nutritious place and repellent is released while happening to be at a noxious place. The type of motion pattern of *E. coli* in finding nutrient is called chemotaxis. When the bacteria reach higher nutrient level locality than their previous position, they start moving forward continuously and this is known as swimming or running. But, if bacteria are reaching a locality with a lower nutrient level than their previous position, the movement they make is known as tumbling.

3.1 Process of Bacterial Foraging Technique

There are four key steps in bacterial foraging technique—chemotaxis, swarming, reproduction and elimination–dispersal. These steps are to be simulated in implementing the optimization technique for problem-solving based on bacterial foraging.

3.1.1 Chemotaxis

This step denotes the movement of an *E. coli* applying swimming and tumbling.

3.1.2 Swarming

E. coli develop complex and stable spatio-temporal patterns called swarms in semisolid nutrient medium. A group of *E. coli* cells are found to arrange themselves in a travelling ring by moving up the nutrient gradient.

3.1.3 Reproduction

The least healthy bacteria finally die when each of the healthier bacteria (with lower value of the objective function) asexually splits into two bacteria, which are then placed in the same location. This keeps the swarm size constant.

3.1.4 Elimination and Dispersal

This step denotes the elimination and dispersal process found in bacteria life. If there comes any change like sudden high temperature in the local environment of *E. coli* population, the bacteria in that region may be killed or a group is dispersed into a new part of the environment. If this dispersal places the bacteria near better food sources, it may assist in the chemotaxis.

3.2 General Algorithm of Bacterial Foraging Optimization Technique

The general algorithm of BFO can be represented as follows [8].

1. Start with initial population of bacteria
2. Perform chemotaxis for N1 times (steps 3–5)
3. Move to new position
4. Calculate nutrient gradient
5. If better nutrient gradient than previous

   a. Swim in same direction else
   b. tumble to new direction

6. If chemotaxis ended, perform reproduction N2 times (steps 7–9)
7. Sort ascending order of fitness of bacteria, divide into two halves
8. Let better nutrient half reproduce
9. Let other half die
10. If reproduction ends, perform elimination and dispersal N3 times (steps 11–12)
11. Sustain bacteria in nutritious environment
12. Disperse/eliminate bacteria in noxious environment.
13. End.

4 Proposed Method of BFO-Based Optimized Filtering

In our proposed method of smoothening images by applying bacterial foraging optimization algorithm, PSNR is used as fitness function, to denoise the noisy images which are filtered by average filter.

4.1 Steps for the Proposed Denoising Technique

(1) Step 1: Images (with resolution 128×128, 256×256 and formats jpg, png, bmp and tiff) are added with noises—salt-and-pepper noise/Gaussian noise with different measures (noise density/standard deviation such as 1, 2, 10, ..., 90%).

(2) Step 2: These two sets of noisy images are separately passed through a filter—average filter. And the parameters MAE and PSNR of the filtered images are computed.

(3) Step 3: The proposed BFO Algorithm is applied to both image sets resulted in the previous step. BFOA works on the basis of fitness function PSNR. This step efficiently reduces noises which are not fully filtered by the average filter. The parameters PSNR and MAE of the optimised image are computed.

(4) Step 4: Compare the parameter values obtained from the filtered images (step 2) and further optimised images (step 3).

4.2 Proposed BFO Algorithm for Improving the Filtered Images

Different parameters considered are as follows—number of bacteria in population used for searching, dimension of search space, number of chemo-tactic steps N1, number of swimming steps, number of reproduction steps N2, number of elimination and dispersal steps N3, probability of elimination and dispersal.

1. Start with initialization of relevant parameters; filtered image and noisy image set as input matrix. Initialize a matrix to store the result.
2. Evaluate fitness of initial population (PSNR).
3. Perform chemotaxis (swim/tumble) (steps 4–5) N1 times.
4. Evaluate PSNR (fitness function).
5. If better fitness value is resulted than previous, then swim (run in the same direction), else tumble (change the direction).
6. If chemotaxis is ended, perform reproduction (steps 7) N2 times.
7. Sustain half of population (higher fitness value), and remove the other half.
8. If reproduction is ended, perform elimination and dispersal N3 times (Step 9).
9. Preserve higher fitness values and disperse lower fitness values.
10. Update result matrix's corresponding values (pixel intensity).
11. Output: result matrix (new filtered image).
12. Stop.

5 Experimental Results

The proposed noise reduction optimization technique along with average filter is simulated on MATLAB 2017a, experiments are performed on 50 greyscale images, and a few test results have been presented here. To analyse the performance of various denoising techniques in the noisy environment, a set of 50 images is added with salt-and-pepper noise of noise density varying from 0.01 to 0.9. Another set of 50 images is added with Gaussian noise of standard deviation varying from 0.01 to 0.9. Both these sets are filtered by average filter and then applied by the newly proposed algorithm. The PSNR and MAE are taken as performance measures in both the cases. Results of performance evaluation of both image sets have been shown qualitatively as well as quantitatively through table (Table 1) and figures (Fig. 1 and Fig. 2).

Table 1 Quantitative analysis of the effect of applying average filter as well as BFO-based method on two images w.r.t. PSNR and MAE

Image	Resolution	Type of noise, noise level, SD			Average filter		BFO	
					PSNR	MAE	PSNR	MAE
Shoe.jpg	128 × 128	Salt–pepper	0.07	–	16.180	0.051	16.933	0.040
Shoe.jpg	256 × 256	Salt–pepper	0.08	–	23.519	0.019	24.776	0.011
Lotus.bmp	128 × 128	Gaussian	–	0.07	13.805	0.164	15.295	0.128
Lotus.bmp	256 × 256	Gaussian	–	0.08	13.421	0.171	14.865	0.135
Leaf.bmp	128 × 128	Salt–pepper	0.5	–	8.791	0.294	9.808	0.249
Leaf.bmp	128 × 128	Salt–pepper	0.9	–	6.797	0.425	7.844	0.370
Leaf.bmp	128 × 128	Gaussian	–	0.5	8.876	0.310	9.970	0.263
Leaf.bmp	128 × 128	Gaussian	–	0.9	8.170	0.343	9.231	0.294

Fig. 1 Salt–pepper noisy image—shoe.jpg—average filter versus BFO-based method

Original image

Salt & pepper noise added image

Average filtered image

BFO optimised image

Fig. 2 Gaussian noisy
image—lotus.bmp—average
filter versus BFO-based
method

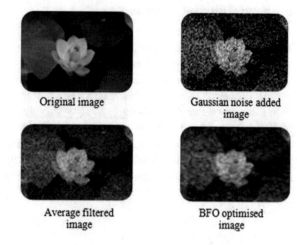

Original image

Gaussian noise added
image

Average filtered
image

BFO optimised
image

From the results of 50 test cases, it is observed that the proposed denoising technique yields better results at all noise level with respect to parameters—MAE and PSNR. The high value of PSNR of images resulted after applying the proposed denoising technique for all images indicates that generally peak signal-to-noise ratio is raised as compared to the case of applying the denoising technique average filter alone. The low value of MAE resulted after applying proposed denoising technique for all images indicates that generally the mean absolute error of the proposed denoising technique is less than the denoising technique average filter considered in this study.

In short, the proposed denoising technique is found to be more accurate, stable and better on noisy images in terms of all quantitative parameters as well as qualitative judgement (visual judgement) when compared to the standard method of applying average filter. The noise suppression capability of the new method is observed to be better.

6 Conclusion

In this work, the basic issues regarding noises, problems in noise reduction and criteria for good noise reduction were analysed. A new method for optimized noise reduction using average filter combined with bacterial foraging optimization algorithm is introduced in this study, and this new method is based on applying bacterial foraging optimization algorithm, BFOA, based on the fitness function, PSNR of images, in combination with average filter and the same is tested with 50 test cases. The test images are noised using salt–pepper/Gaussian noise, and then they are passed through average filter. The filtered images are then optimized applying BFOA. The image quality of resulted images after applying BFOA is compared with those resulted after applying, average filter alone, with respect to parameters PSNR and MAE. From the results obtained, it is evident that BFOA combined with average filter is more efficient for reducing noises in images than compared to applying average filter alone. In the filtered images after applying BFOA smoothening, PSNR gets maximized and MAE gets minimized as compared to those images which were only filtered using average filters. Image gets smoothened and clear after the application of BFOA.

From the quantitative as well as qualitative (visual judgement) results of denoising techniques on noisy images which are corrupted with salt-and-pepper noise/Gaussian noise, it is concluded that the proposed technique works well in images with higher noise densities too. It is concluded that the proposed denoising technique as a combination of average filter and BFOA is efficient on noisy images and performs better than the average filter, with respect to salt–pepper and Gaussian noises.

There is enough scope for improvement in the proposed denoising technique based on BFOA. The proposed technique can be extended to cover the denoising of colour images also. And, the applicability of BFOA method can be tested for images corrupted with other noises like speckle or Poisson noises too. More parameters can be tested for establishing the accuracy and potential of this proposed method.

References

1. Bakwad, K.M., Pattnaik, S.S., Sohi, B.S., Devi, S., Gollapudi, S.V., Sagar, C.V., Patra, P.K.: Fast motion estimation using small population-based modified parallel particle swarm optimisation. Int. J. Parallel Emergent Distrib. Syst. 26(6), 457–476 (2011)
2. Yaduwanshi, S., Sidhu, J.S.: Application of bacterial foraging optimization as a de-noising filter. Int. J. Eng. Trends Technol. 4(7), 3049–3055 (2013)
3. Sharma, V., Pattnaik, S.S., Garg, T.: A review of bacterial foraging optimization and its applications. Int. J. Comput. Appl. (2012)
4. Gholami-Boroujeny, S., Eshghi, M.: Active noise control using bacterial foraging optimization algorithm. In: IEEE 10th International Conference Signal Processing (ICSP), pp. 2592–2595 (2010)
5. Beenu, S.K.: Image segmentation using improved bacterial foraging algorithm. Int. J. Sci. Res. (IJSR) (2013)
6. Verma, O.P., Hanmandlu, M., Sultania, A.K., Parihar, A.S.: A novel fuzzy system for edge detection in noisy image using bacterial foraging. Multidimens. Syst. Signal Process. 24(1), 181–198 (2013)
7. Binitha, S., Sathya, S.: A survey of bio inspired optimization algorithms. Int. J. Soft Comput. Eng. 2(2), 137–151 (2012)
8. Passino, K.M.: Biomimicry of bacterial foraging for distributed optimization and control. IEEE Control Syst. 22(3), 52–67 (2002)

Accessing Sensor Data via Hybrid Virtual Private Network Using Multiple Layer Encryption Method in Big Data

Harleen Kaur, Ritu Chauhan, M. Afshar Alam, Naweed Ahmad Razaqi and Victor Chang

Abstract Enormous data generation takes place with the increase in the number of systems and enhanced technology. This data generation occurs exponentially with time. Sensor networks are one of the main generators of big data. The data/information received from the sensor network is transferred to the other end user/s via hybrid virtual private network (HVPN) using multiple layer encryption through Internet of things (IoT). The communication between sensors and end user/s takes place through the server with the help of inter-networking procedure called Internet of things (IoT). With the help of IoT, the objects can be traced and controlled from large distances over the network for a nonstop assimilation of the world into computer-based systems resulting in superior efficiency, accuracy, and reduced human intervention. The HVPN using multiple layer encryption strategy will permit the association between the initiator and responder to remain unidentifiable. We have used multiple layer encryption technique to scramble the information to secure IoT. Furthermore, we experimented the contrast of the speed for information transmission and multiple layer encryption and without multiple layer encryption with IoT.

Keywords Hybrid virtual private network (HVPN) · Internet of things (IoT) · Virtual private network (VPN) · Multilayer encryption

H. Kaur (✉) · M. Afshar Alam · N. A. Razaqi
School of Engineering Sciences and Technology, Jamia Hamdard, New Delhi, India
e-mail: harleen.unu@gmail.com

M. Afshar Alam
e-mail: aalam@jamiahamdard.ac.in

N. A. Razaqi
e-mail: naweed.razaqi@gmail.com

R. Chauhan
Centre for Computational Biology and Bioinformatics, Amity University, Noida, India
e-mail: rituchauha@gmail.com

V. Chang
School of Computing, Engineering and Digital Technologies, Teesside University, Middlesbrough, UK
e-mail: victor.chang@xjtlu.edu.cn

© Springer Nature Singapore Pte Ltd. 2020
P. K. Mallick et al. (eds.), *Cognitive Informatics and Soft Computing*,
Advances in Intelligent Systems and Computing 1040,
https://doi.org/10.1007/978-981-15-1451-7_20

1 Introduction

With the passing time, there is an increase in the number of systems and enhanced technology, which results in the huge data accumulation exponentially. This data accumulation is useful only if it is used further to generate information. Since the data sets are so huge, it is wayward to maintain and process it using traditional method. Big data deals with the behavior analytics, predictive analytics, and other analytic methods. "Big data volumes are relative and vary by factors such as time and type of data" [1]. Due to the increase in the storage power which results in capturing of mammoth data sets, what appeared big data may not be even enough to reach the limen in the future [1–3]. Scientific instruments, social media and networks, mobile devices, sensor technology and networks are the main generators of big data. In this paper, we are focusing on sensor networks which build considerable percentage of big data.

Sensors can be used to observe physical or environmental conditions. The data produced by the sensors can be used for many purposes and can be sent to the remotely situated users. Along with this, one sensor node can communicate with other node/s, thus forming sensor networks. Internet of things (IoT) allows different physical devices, vehicles, and buildings that are using electronic devices, software, sensors, and actuators in order to connect with the network so that they can be able to gather and swap data. To make this communication safe and attack free, we can send the sensor data via hybrid virtual private network (HVPN) and encrypting the data using multilayer encryption technique with IoT devices [4].

To secure IoT, HVPN using multilayer encryption provides a method for two parties' communication connection sender and connection acceptor to communicate with each other unidentifiably. Virtual private network (VPN) guards its communications against traffic analysis attacks with IoT security. For VPN observers (such as crackers, companies), it seems so hard in order to easily recognize the entities that are going to communicate with each other (i.e., sender and receiver) and what is the purpose of communicating each other, by checking data packets that are flowing over the VPN [5]. The main focus of VPN observers is on the process of encrypting the address where from the data packets had come and the address where the packet has to reach, rather than the complacent of the packet. With the help of any kind of cryptography, the encryption of the complacent of the packet is possible before sending. It is the main function of HVPN to offer very powerful private communications which is in real time over a public VPN at a lesser cost and advanced efficiency. Correspondences are proposed to be consecrated as in a busybody on the general population VPN cannot decide either the substance of messages spilling out of Akram and Karim or significantly whether Akram and Karim are speaking with each other. There are various types of information exchange where Internet plays a very important role, e.g., VoIP and email, multimedia services like Online Music and Online Movie, VPN activities such as Online Ads and Social VPN (the anonymizer—http://www.Anonymizer.com).

2 Proposed Hybrid Virtual Private Network (HVPN) Architecture in IoT

The overall architecture consists of sensor network, sensor server at one end, and end user/s at another end as shown in Fig. 1. The data or information is generated by the sensor/s and passed on to the other side. This information is passed via HVPN. HVPN will help in secure access to the sensor data/information. Sensor/s are directly connected to the server. The information is actually passed by this server to the end user/s. The end users do not have direct contact with the sensor network. An introduction of server between the sensor/s and end user/s helps to conserve bandwidth, reduces system cost, and provides better protection of data/information as no direct contact is allowed. Whatever data/information passed has to pass this server just like in spoke–hub distribution paradigm. Server can also help in easy scalability. The connection made for data/information transfer over the network is not socket connection; instead, it is unassigned connection made using multiple layer encryption model. The security is provided in such a way that the data/information originator and receiver are hidden, thus who is connected to whom remains secret.

3 IoT Security by Hybrid Virtual Private Network (HVPN) Using Multiple Layer Encryption Model

Multiple layer encryption method is the machines which initiate applications to make connections through them. This multiple layer encryption method is used in multiple layer encryption method for making connection instead of making socket connection directly in the IoT. The connection between originator and responder will remain undisclosed if multiple encryption method is used. This is specifically called as unassigned plug connection or unassigned connection. Which is will not clarify who is connected to whom and for what purpose is hidden through the unassigned connection for both outside eavesdroppers and inside attacker in HVPN. HVPN is called hybrid VPN multiple layer encryption routing topology that we will be using in this study. Basic information about the configuration is illustrated in Fig. 3.

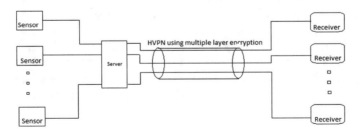

Fig. 1 Secure architecture of hybrid virtual private network (HVPN) to IoT

In the basic composition, this hybrid VPN multiple layer encryption method works in the process of being an interface between engines at the back of the firewall and the external VPN in IoT. Using multiple layer encryption method, security by other means (e.g., physical security) is given for connection from machines behind the firewall to the multiple layer encryption method [4, 6, 7]. To increase the complication of traffic in a site, this multiple layer encryption should also route data between other HVPN using multiple layer encryptions routers. Otherwise, even basic examination Different layer encryption technique in the VPN are associated by changeless attachment association. Undisclosed associations through the VPN utilizing different layer encryption strategy are multiplexed over the longstanding associations. For any undisclosed connection, the sequence from begin to the finish of various layer encryption strategies in a course is entirely characterized at the connection setup. In any case, each different layer encryption technique is permitted to just distinguish the past and next switch along a course. The information is passed along the undisclosed connection which seems distinctive at each layer of numerous layer encryption techniques, so nobody will have the capacity to track the date in route. The HVPN utilizing multiple layer encryption technique is constantly gotten to by means [5, 8–10]. That intermediary defines a route through the VPN with the assistance of multiple layer encryption by developing a layered information structure called a multiple layer encryption strategy and sending that multiple layer encryption technique through the VPN using multiple layer encryption technique. Next bounce in a route is defined by each layer of the multiple layer encryption strategy. Multiple layer encryption technique that gets multiple layer encryption peels off its layer, distinguishes the following bounce, and sends the installed multiple layer encryption to that various layer encryption strategies [11–13].

The last multiple layer encryption strategy router advances the information to another kind of intermediary on a similar machine, known as responder's intermediary, whose undertaking is to pass information between the HVPN using multiple layer encryption technique and the responder. An outline VPN using multiple layer encryption technique and undisclosed attachment association is likewise depicted in Fig. 2. Each multiple layer encryption contains scratch seed material from which keys are produced for encoding the information is to be sent forward or in reverse along the unassigned association, notwithstanding conveying next jump data. (We characterize forward to be the bearing in which the different layer encryption ventures and in reverse as the other way.) Information can be conveyed once the undisclosed connection is built up. Before sending information over an undisclosed connection, the initiator's multiple layer encryption technique includes a layer of encryption for different layer encryption strategies in the course. Each multiple layer encryption strategy evacuates one layer of encryption as information travels through the undisclosed connection, so it comes to the collector side as plaintext. So all information that has gone in reverse through the unassigned connection must be over and again postscript to pick up the plaintext.

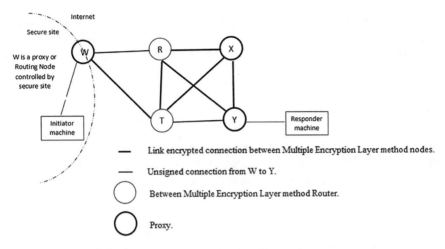

Fig. 2 Routing topology of HVPN using multiple layer encryption model in IoT

4 Hybrid Virtual Private Network (HVPN) Using Multiple Layer Encryption

Intermediaries

An intermediary is a perfect administration between two application clients that would ordinarily make an immediate socket connection to each other yet cannot. For instance, a firewall may safeguard coordinate socket connection among inside and outer machines. An intermediary dealing with the firewall may empower that sort of associations.

Our objective has been to outline a design for private correspondence that would interface with unmodified applications, so we utilized intermediaries as the interface among applications and multiple layer encryption steering's unknown connection in sensor network. Since the application goes between interfaces among applications and the VPN, it must comprehend both application conventions and different layer encryptions coordinating conventions. To energize the blueprint of utilization specific mediators, we fragment the middle person into two sections.

The intermediary interfaces between an attachment association from an application and an attachment association with the middle delegate. In particular, the client mediator must prepend to the information stream a standard structure that recognizes a conclusive objective by either hostname/port or IP VPN building the mysterious connection with the responder proxy. After getting a new demand, the center proxy uses the prepended standard structure as an indication in building a numerous layer parcel characterizing the route of an unknown connection with that goal. It then passes the multiple layer packets to the mysterious VPN. The principal datum got will be the standard structure determining a definitive goal. The responder mediator makes an attachment (socket) association with that IP or port, reports a one-byte status message back to VPN (and along these lines back to the center go-between which thusly drives it back to the client middle person), and hence moves information between the VPN and the new connection to the customer proxy [13].

4.1 Procedure Used

The HVPN setup, the durable connection between multiple layer encryption switches; which builds up mysterious connection through the multiple layer encryption router; information development over an unknown connection.

Client Proxy

The interface between an application and the customer mediator is application express. The interface between the customer go-between and the inside mediator is unmistakable as takes after. In the event that rejected, it reports an application particular blunder message and after that closures the attachment and sits tight for the following solicitation. In the event that acknowledged, it will make a socket connection to the client proxy's important port.

Retry check stipulates how often the responder intermediary ought to wound to retry association with the conceivable objective. At long last, the Addr arrangement field demonstrates the kind of a legitimate target address: 1 for an invalid completed ASCII string with the hostname straightaway taken after by another invalid wrapped up as shown in Figs. 3 and 4.

Core Proxy

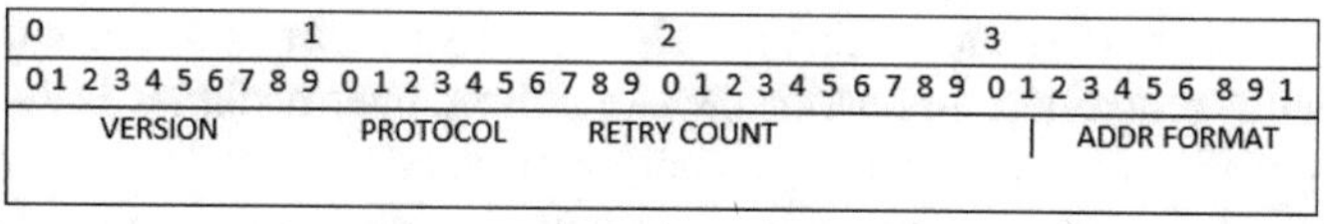

0	1	2	3
0 1 2 3 4 5 6 7 8 9	0 1 2 3 4 5 6 7 8 9	0 1 2 3 4 5 6 7 8 9	0 1 2 3 4 5 6 8 9 1
VERSION	PROTOCOL	RETRY COUNT	ADDR FORMAT

Fig. 3 Client proxy structure

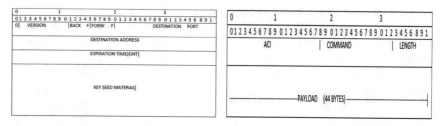

Fig. 4 Layer and cell structure

The core proxy can choose whether to acknowledge or dismiss the demand. In the event that rejected, it sends a sensible blunder code back to the customer intermediary, closes the attachment, and sits tight for the accompanying requesting. It keeps on building the baffling association with the responder intermediary utilizing the standard structure, sends the standard structure to the responder intermediary over the obscure association and a while later passes each and every normal information and from the client delegate and obscure association [9].

Multiple Layer Encryption Packet Structure

In order to develop the obscure association with the responder go-between, the inside middle person makes a package. The package is a multi-layered information structure that epitomizes the unassigned affiliation course will started with the responder middle person and working verse to the inside delegate as shown in Fig. 4.

The Back F field indicates as far as possible will be related to information moving in the retrogressive bearing (depicted as information moving the other way that the package voyaged, by and large around the initiator's end of the unsigned affiliation association) using key: depicted underneath. The FORW F field proposes the cryptographic power to move the data in the forward course (depicted as information moving in a practically identical heading that the multi-layer packet voyaged in rendition 1, packet has five layers. Before talking about how packet and information are sent between multiple layer encryption routers, we will define multiple layer encryption interconnection.

VPN using Multiple Layer Encryption Interconnection

In the process of VPN setup (not to be mistaken for mysterious connection setup), revered connection between neighboring multiple layer encryption is set up and keyed. The VPN topology is predefined and each multiple layer; to stay associated with each of its neighbors, multiple layer encryption must both tune in for connection from neighbors and endeavor to start associations with neighbors. In order to deny the impasses or crush issues between many neighbors, multiple layer encryption method makes connection with IP or ports and start with lower IP or ports.

5 Results

Security of interchanges against activity investigation does not oblige bolster for unknown correspondence. By scrambling information to be sent over a movement examination safe association, for instance, endpoints may group themselves to each other without meager the nearness of their correspondence to whatever remains of the system. However, activity investigation is an energetic instrument for scanty gatherings in discussion, in this way trading off a correspondence that was proposed to be indistinct.

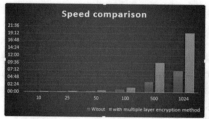

Size of data	10Mb	25Mb	50Mb	100Mb	500Mb	1024Mb
Without multiple layer encryption method.	00:06 seconds	00:10 seconds	00:23 Seconds	00:50 seconds	03:31 seconds	11:10 seconds
With multiple layer encryption method	00:10 Seconds	00:14 Seconds	00:38 Seconds	01:30 Seconds	05:48 seconds	19:35 seconds
Percentage of speed test for individual data	66.67%	70.00%	65.21%	66.00%	64.92%	66.41%

Fig. 5 Speed comparison of two types of VPN in IoT

In this way, we consider objectives for mysterious and additionally private, correspondence. These are the essential strongholds with which we will be uneasy. We will likewise be worried about more theoretical obscurity security.

Our work for the speed trial of information downloading with multiple layer encryption and without numerous layer encryption is distinctive, and we test this procedure for different information on the Web in IoT. We have taken different sizes of data which are 10, 25, 50, 100, 500, and 1024 Mb; we have tested all different data with the speed of 20 Mbp/s and got a very fair result; each data is tested separately; the percentage of speed for each data is also shown separately; and then we have shown the average speed of using VPN with multiple layer encryption approach.

The average speed percentage between VPN with multiple layer encryption and VPN without multiple layer encryption is 66.53%. In this approach since there are multiple times of encryption of every packet at every node, this encryption is done separately so the speed of data transfer decreases but still a very fair speed, which is 66.53%. The above speed test has been experimented, with 20 Mbp/s Internet speed, Windows 10, and all different nodes which had the same speed of 20 Mbp/s. Figure 5 the analysis of VPN with multiple layer encryption and without multiple layer encryption have been shown graphically.

6 Conclusion

Providing secure access to the data/information produced by the sensor/s is a critical issue. Sensor/s produce significant volume of data/information which results in formation of big data. HVPN not only secures system from unauthorized user but also helps from eavesdropping for the IoT. HVPN using multiple layer encryption gives stronghold against both listening stealthily and movement examination. In spite of the fact that our attention is on unassigned connection and not unassigned correspondence, unassigned correspondence is likewise conceivable by destroying recognizing data from the information stream. HVPN using multiple layer encryption mysterious connection are application self-governing and can interface with essential Web applications by methods for intermediaries. Our execution of VPN using multiple layer encryption involves intermediaries for Web perusing, email, and remote login with IoT devices.

References

1. Sherwood, R., Bhatacharjee, B., Srinivasan, A.: P^5: a protocol for scalable anonymous communication. In: Proceedings of the IEEE Symposium on Security and Privacy, pp. 58–70. 12–15 May 2002
2. Chauhan, R., Kaur, H., Lechman, E., Marszk, A.: Big data analytics for ICT monitoring and development. In: Kaur, H., et al. (eds.) Catalyzing Development Through ICT Adoption: The Developing World Experience, pp. 25–36. Springer, New York (2017)
3. Kaur, H., Tao, X.: ICT and Millennium Development Goals: A United Nations Perspective, p. 271. Springer, New York (2014)
4. Diffie, W., Oorschot, P.C.V., Wiener, M.J.: Authentication and authenticated key exchanges. Des. Codes Cryptogr. **2**, 107–125 (1992)
5. Goldschlag, D., Reed, M., Syverson, P.: Hiding routing information. In: Proceedings of the 1st International Workshop on Information Hiding, 30 May–01 June, pp. 137–150. Springer (1996)
6. Gandomi, A., Haider, M.: Beyond the hype: big data concepts, methods, and analytics. Int. J. Inf. Manag. **35**(2), 137–144 (2015)
7. Claessens, J., Diaz, C., Goemans, C., Preneel, B., Vandewalle, J.: Revocable anonymous access to the internet? Internet Res. **13**, 242–258 (2003)
8. Fielding, R.T., Taylor, R.N.: Principled design of the modern web architecture. ACM Trans. Internet Technol. (TOIT) **2**(2), 115–150 (2002)
9. Hoang, N.P.: Anonymous communication and its importance in social networking. In: Proceedings of the 16th International Conference on Advanced Communication Technology, 16–19 Feb, pp. 34–39. IEEE Xplore Press (2014)
10. Kristol, D.M.: HTTP cookies: Standards, privacy and politics. ACM Trans. Internet Technol. **1**, 151–198 (2001)
11. Reed, M.G., Syverson, P.F., Goldschlag, D.M.: Proxies for anonymous routing. In: Proceedings of the 12th Annual Computer Security Applications Conference, 9–13 Dec, pp. 95–104. IEEE Xplore Press (1996)
12. Rubin, M.K., Reiter, A.D.: Crowds: anonymity for web transactions. ACM Trans. Inf. Syst. Secur. **1**, 66–92 (1998)
13. Schneier, B.: Applied Cryptography: Protocols, Algorithms and Source Code in C, 2nd edn. Wiley, New York (1996)

Statistical Features-Based Violence Detection in Surveillance Videos

K. Deepak, L. K. P. Vignesh, G. Srivathsan, S. Roshan and S. Chandrakala

Abstract Research over detecting anomalous human behavior in crowded scenes has created much attention due to its direct applicability over a large number of real-world security applications. In this work, we propose a novel statistical feature descriptor to detect violent human activities in real-world surveillance videos. Standard spatiotemporal feature descriptors are used to extract motion cues from videos. Finally, a discriminative SVM classifier is used to classify violent/non-violent scenes present in the videos with the help of feature representation formed out of the proposed statistical descriptor. Efficiency of the proposed approach is tested on crowd violence and hockey fight benchmark datasets.

Keywords Violence detection · Statistical features · Histogram of gradients · SVM (support vector machines)

1 Introduction

Automated recognition of human activities has been widely applied to many real-time security applications, including violence detection in public and private spaces. The Violence detection system mainly aims at recognizing and localizing of violent

K. Deepak · L. K. P. Vignesh · G. Srivathsan · S. Roshan · S. Chandrakala (✉)
Intelligent Systems Group, School of Computing, SASTRA Deemed to be University, Thanjavur, Tamil Nadu, India
e-mail: chandrakala@cse.sastra.edu

K. Deepak
e-mail: deepak@sastra.ac.in

L. K. P. Vignesh
e-mail: lkpvignesh@gmail.com

G. Srivathsan
e-mail: njsg69@gmail.com

S. Roshan
e-mail: mailroshan.8@gmail.com

© Springer Nature Singapore Pte Ltd. 2020
P. K. Mallick et al. (eds.), *Cognitive Informatics and Soft Computing*,
Advances in Intelligent Systems and Computing 1040,
https://doi.org/10.1007/978-981-15-1451-7_21

197

behavior in a target scenario, associated with a surveillance camera. Sample violent scenes from hockey fight dataset are shown in (Fig. 1) Since manual monitoring is highly tedious along with the rapid installation of thousands of surveillance cameras in public places, there exists a wide scope in automating the surveillance event detection to seek intelligent solutions through machine learning algorithms. But the main challenge lies in extracting and representing useful information from thousands of cameras recording huge volumes of videos.

The applications such as video event detection generally include three phases: detection of interest points in a spatiotemporal space, robust representation of the detected points, and performing classification with a sequential or discriminative classifier. Another key challenge comes from the feature selection part, trajectory analysis seems to be most natural approach to capture high-level features from a video, and it extremely performs well on un-crowded scenes, but the approach fails to detect anomalies on crowded scenes due to complex foreground motion and occlusions. A recent work [1] by Yang Zian et al. stated that low-level spatiotemporal features are more robust to illuminations, occlusions, and inconsistency than the global features. Thus, in this work we have tried to explore a robust low-level feature [1] to discriminate the violent behavior from non-violent scenes in a video. We have used the well-known HOF features to represent motion in our spatiotemporal video volume. In the past, a lot of research over activity recognition experimented on very simple and clean datasets, which are simulated by actors performing explicit actions [2–5].

Our goal is to precisely represent the input data which is being fed to the classifier and to detect the violent behavior with a less false alarm rate in unconstrained scenarios. The proposed global statistical feature descriptor gradually reduces the dimension of training data. Instead of concentrating on the signal-based perspectives like [5], we have chosen to explore the representation learning approach in order to take advantage of discriminative classifier like SVM, which accepts input data in the form of set of vectors. In the proposed framework, we first extract dense HOF features from the input videos using a descriptor of $3 \times 3 \times 2$ block size. The histograms are varying length feature vectors obtained from a 2D vector field computed from the famous Horn–Schunck optical flow algorithm [6]. The same configuration is followed for extracting MBH features proposed by Uijilings et al. [6]. Then, the proposed global statistical feature descriptor is used to represent a varying length feature vector to a fixed length pattern, and then, we use a discriminative model for classification. From now, the paper follows with the following order in explaining the details of our work, (I) system model, (II) proposed methodology, (III) experimental results over 2 benchmark datasets and conclusion.

2 Statistical Features-Based Violence Detection

The proposed system consists of three elementary components as shown in Fig. 2. (1) extraction of motion features; (2) representing the extracted features using the proposed global statistical descriptor; and (3) learning the violence detection model and prediction with a SVM classifier.

Optical flow of a video volume is computed by estimating the direction and velocity of the moving pixels from two consecutive video frames. The computed optical flow is a 2D vector field, which describes the motion in a video. The HOF is calculated by voting the flow magnitude of each pixel in to X bins by their flow direction. In the proposed work, the magnitude and flow gradient responses of the optical flow fields are quantized in to 8 bins. Before learning and the classification process, each spatiotemporal cube is converted to a sequential feature vector and further it is converted to a fixed length pattern by using the proposed global statistical descriptor for precise definition of the training data.

Classifiers such as SVM focus on modeling the discrimination between the activities, and it is proven to be effective for classifying static data. It is also suitable for carrying out classification tasks with less training data. These models accept inputs represented as feature vectors of fixed size. First, the HOF feature descriptors are extracted for alternative frames in a video and represented as a motion sequence per activity. Then, a set of statistical features are extracted along the frames of feature sequences and combined as a single feature vector. The statistical features such as mean, median, mode, quantiles, and skewness are used to form the statistical feature descriptor. This statistical feature descriptor achieves better classification results among the violent and non-violent classes with an added advantage of representing the training data in a compact form, which in turn reduces the modeling time.

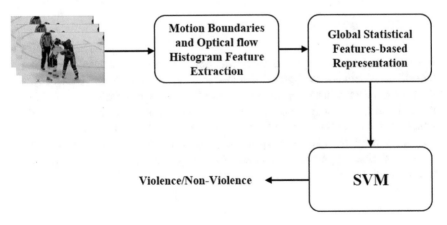

Fig. 2 Overview of the proposed violence detection system

$$S = \frac{1}{N} \sum_{i=1} \frac{(y_i - \mu)^3}{\sigma^3} \tag{1}$$

The asymmetry in the data distribution is measured from the third central moment using the skewness measure. Skewness of a varying length video data S is given by Eq. (1), where μ is the mean, σ is the standard deviation, and N is the number of examples. One way to characterize the skewness is that if it is said to be negative, it indicates that the distribution of data points is more to the left side of mean.

Similarly, features such as mean, mode, median, min, and max are computed over the 144-dimensional feature vectors extracted from the training data and combined to obtain the statistical vector embedding of activities. These activity descriptors are of fixed length, and it is used to exploit the discriminative power of the support vector machine (SVM) classifier. Finally, the representations obtained out of each feature vector are combined to model the activities.

3 Experimental Study

To evaluate the efficiency of the statistical feature descriptor, we compare it against the BoW methods cited from the literature. We employ fivefold cross-validation on each dataset to avoid biased results based on the training set. The reported results are presented based on the mean prediction accuracy. Next, we briefly explain about the two benchmark datasets used in our work and proceed with the experimental setup, results, and discussion.

3.1 Datasets

Experiments over the proposed approach were conducted on datasets for violence detection, namely hockey fights dataset [4] and the crowd violence dataset [4]. The hockey fight dataset has a total of 1000 video clips of category fight and no fight from the national hockey leagues (NHL). It is a collection of real hockey game videos, which consists of 500 violent and 500 non-violent clips. Each clip exactly consists of 50 frames (resolution: 360×288 pixels). The crowd violence dataset totally consists of 246 real video clips, which includes 123 violent and 123 non-violent clips with a resolution of 320×240 pixels.

3.2 Experimental Setup

The proposed work generates about 144 descriptors for a video clip, which is formed by the product of total number of video blocks and number of orientation bins as, for both the datasets, the average size of a sequence matrix formed is about 7200×144, the lesser detailing is due to the 12×12 block size. The obtained sequence vectors are then converted in to a set of fixed dimensional feature representation using the proposed global statistical feature descriptor. Finally, with the help of LibSVM the prediction and classification of the violent clips are performed.

3.3 Results and Discussion

Existing methods based on existing feature encoding techniques are usually combined with the traditional bag-of-words approach since it has shown some promising performance over violence detection tasks. Table 1 shows the results over the hockey and crowd violence dataset compared with the standard BoW approaches; however, the baseline BoW methods purely depend on the distinguishing ability of the local space–time descriptors; the results are directly based on how often these descriptors occur in a video.

Another important disadvantage of a BoW model is the performance can be significantly degraded by the quantization error. The performance is evaluated using two different sets of statistical features; the first set is tested with only four moments. The second set is tested with min, max, median, and mode measures. Our studies show that the proposed statistical feature descriptor with 8 features has shown slightly improved performance. We have empirically demonstrated the performance

Table 1 Results over hockey fight and crowd violence dataset	Algorithm	Hockey fight dataset accuracy (%)	Crowd violence accuracy (%)
	HOG + Bag of words [4]	88.77	57.98
	HOF + Bag of words [4]	86.07	58.71
	HNF + Bag of words [4]	89.27	57.05
	MoSIFT + Bow [5]	88.81	57.09
	HOF + Statistical	**91.50**	**84.00**
	MBH + Statistical	**86.40**	**79.87**

Fig. 3 False alarms in crowd violence dataset

of our proposed idea over the existing popular HOF and MBH feature descriptors, as presented in Table 1 appearance-based HOG + BoW comparatively performs better than HOF + BoW for hockey fight dataset, but our proposed method outperforms all other methods with the proposed global statistical feature representation. As shown in Fig. 3 most of the false alarms are caused because of the scenes like people clapping the hands vigorously and waving the flags, which are similar to violent actions. All the other BoW approaches failed to perform because of the optical flow noise, since our proposed method is highly discriminative; it has managed to outperform all the other BoW methods with almost 15% increase in the average accuracy by exploiting the feature representation.

4 Conclusion

Detecting violent human activities in crowded scenes is considered of core importance among all the computer vision applications. In this work, a statistical feature descriptor to represent the violent activities in surveillance videos has been proposed. The proposed approach reduces the dimension of the training data which explicitly reduces the training time. Experiments over the two benchmark datasets have also proven the efficiency and performance of proposed feature-based representation over the existing bag-of-words approaches cited in the literature. From the experimental results, we conclude that the proposed statistical feature descriptor is efficient and robust for detecting violent actions in complex scenes.

References

1. Xian, Y., et al.: Evaluation of low-level features for real-world surveillance event detection. IEEE Trans. Circuits Syst. Video Technol. **27**(3), 624–634 (2017)
2. Barrett, D.P., Siskind, J.M.: Action recognition by time series of retinotopic appearance and motion features. IEEE Trans. Circuits Syst. Video Technol. **26**(12), 2250–2263 (2015)
3. Rodriguez, M., et al.: One-shot learning of human activity with an MAP adapted GMM and simplex-HMM. IEEE Trans. Cybern. **47**(7), 1769–1780 (2017)
4. Zhang, T., et al.: Discriminative dictionary learning with motion weber local descriptor for violence detection. IEEE Trans. Circuits Syst. Video Technol. **27**(3), 696–709 (2017)
5. Wang, S., et al.: Anomaly detection in crowded scenes by SL-HOF descriptor and foreground classification. In: 2016 23rd International Conference on Pattern Recognition (ICPR). IEEE (2016)
6. Uijlings, J., et al.: Video classification with densely extracted hog/hof/mbh features: an evaluation of the accuracy/computational efficiency trade-off. Int. J. Multimed. Inf. Retr. **4**(1), 33–44 (2015)

Encapsulated Features with Multi-objective Deep Belief Networks for Action Classification

Paul T. Sheeba and S. Murugan

Abstract Human action classification plays a challenging role in the field of robotics and other human–computer interaction systems. It also helps people in crime analysis, security tasks, and human support systems. The main purpose of this work is to design and implement a system to classify human actions in videos using encapsulated features and multi-objective deep belief network. Encapsulated features include space–time interest points, shape, and coverage factor. Initially, frames having actions had been separated from the input videos by means of structural similarity measure. Later, spatiotemporal interest points, shape and coverage factor are extracted and combined to form encapsulated features. To improve the accuracy in classification, MODBN classifier was designed by combining multi-objective dragonfly algorithm and deep belief network. Datasets such as Weizmann and KTH are used in MODBN classifier to carry the experimentation. Accuracy, sensitivity, and specificity are measured to evaluate the classification network. This proposed classifier with encapsulated features can produce better performance with 99% of accuracy, 97% of sensitivity, and 95% of specificity.

Keywords Action recognition · SSIM · STI · DBN · DA · MODBN

1 Introduction

The world is moving toward fully automated systems, in which there is a high influence of interaction between human and machine. In human–machine interacting system, computer vision is the major part in processing the visual inputs and producing automated results. In human–computer interactive automation systems, human

P. T. Sheeba (✉)
Faculty of Computer Science & Engineering, Sathyabama Institute of Science and Technology, Chennai, India
e-mail: sheebajames8082@gmail.com

S. Murugan
Department of Computer Science and Engineering, Sathyabama Institute of Science and Technology, Chennai, India
e-mail: snmurugan@gmail.com

© Springer Nature Singapore Pte Ltd. 2020
P. K. Mallick et al. (eds.), *Cognitive Informatics and Soft Computing*,
Advances in Intelligent Systems and Computing 1040,
https://doi.org/10.1007/978-981-15-1451-7_23

205

action recognition is important to make system more interactive with human. Hence, human action classification becomes vigorous research area in the field of human–computer interactive systems. Even though many algorithms and techniques are being available, action recognition is still challenging and a difficult research area due to its intra- and interclass variations.

Many previous works have been carried out for the representation of actions. Two different categories of representations are possible: global feature representation and local feature representation. Global feature representation follows top–down method and encodes visual observation as a whole, and it is powerful, since it encodes much of information. Local representations follow bottom–up method which detects STI points and forms local feature representation. Local representations in videos are less sensitive to noise and partial occlusion. Global representations were adopted in [1–3], to describe an action. But global feature representation methods cannot detect silhouettes from realistic videos. Local representation methods extract spatial–temporal points like [4–6] and show more improvement in accuracy and recognition. Human posture estimation methods distinguish the actions based on the spatiotemporal action features. STI features give detailed features such as shape, swiftness, and scale for the human action classification [7]. STI feature extraction does not undergo any segmentation or action units tracking step. These local representation methods are robust to speed and resolution. Even though it shows improvement, it causes loss of information, since it detects few points from videos.

Recently, in the field of computer vision, human activity recognition uses dynamic texture patterns like local binary patterns (LBP) [8] for the actions to retrieve moving texture patterns.

The work presents a technique with encapsulated features for human action classification in video frames using multi-objective deep networks (MODDBN). In order to select the key frames, structural similarity (SSIM) measure is calculated in between frames and frame with high threshold value will be selected. Then, these key frames are used to extract the features. To retain the spatial information, grid-based shape feature and coverage factor are extracted and space–time interest (STI) features were extracted to retain temporal information. These features are encapsulated and used to define feature vector. Feature vector is used as input over MODBN, which has been projected by incorporating MODA into DBN, for the classification of human actions. The projected MODBN model recognizes human actions in video frames by picking the appropriate weights in the deep belief network using MODA.

The framework includes the following two tasks to recognize the actions

- Design of extraction process for encapsulated features, where the features like shape, coverage value, and temporal features are extracted for an efficient human actions classification.
- Design of MODBN by incorporating DBN with MODA optimization method, to perform efficient classification of human actions in the videos.

2 Literature Review

In human action recognition, there are many works have been evolved in local and global representations. Local representation based on spatiotemporal features has conquered in human action classification. Spatial–temporal feature extraction was proposed in [9], with a size adopted descriptor for a dense representation of videos. Grounded on this knowledge, many categorization and recognition tasks were proposed in [4, 5]. Beyond normal spatial interest points, Bregonzio et al. [9] proposed holistic features calculated on these spatial temporals spreading of interest points which intervent the ratio of the interest points at different scales. To improve more, instead of local features, mid-level features are learned and used spatial feature pyramid to feat temporal details in [10].

High-level feature representation has also been performed well in various recognition tasks. In [11], it is exploited semantic attributes like "hairy" and "four legged" to classify conversant objects and to describe unfamiliar objects when images and boundary box annotation are provided. Wu et al. [12] proposed novel scheme in representing the actions. They developed compact action units to signify human actions in video.

Rodriguez et al. [13] proposed two different techniques such as video encoding and time flexible kernel for human activity classification. Video encoding has been used to retain the temporal structure features of the frames, and time flexible kernel has been used to compare the sequences to identify the difference in lengths to randomly arrange them. Through these methods, accuracy has been improved but failed with short duration complex activities like blinking eyes, finger movements, etc.

Li et al. [14] proposed video feature representation using convolution neural networks. In this work, gray images and optical flow maps of the videos were taken as input to one layer of CNN to extract the details of human motion. For the second layer, color frames were given to retrieve the context information. As a next stage, linear support vector machine has been used to train the classifier. Time complexity of this method was improved because of feature maps.

Wang et al. [15] proposed a video representation method and implemented it in human action detection and classification. SURF descriptor was used to extract features, and camera motion estimation was done. A human detector has been used to improve the robustness in homography estimation. Moreover, bag of words (BOW) has been replaced by Fisher vector to encode the features. Still this performs poor, when irrelevant information is available.

Li and Vasconcelos [16] has designed a model to represent the videos, based on the action attribute dynamics. In this approach, binary dynamic system (BDS) has been utilized for an efficient video representation. To improve the accuracy in complex actions, BDS has been embedded with bag of visual words (BoVW). But, still this method of representation failed to perform well on lengthy videos.

Nigam and Khare [17] used holistic feature vector to describe the human actions. Holistic feature vector incorporates local video descriptors to retain the details about the actions. Moment invariants and uniform local binary patterns are combined to present the local motion descriptor. Finally, support vector machine has been used for classification. Time complexity of this method has pulled down the performance of the entire framework.

A continuous activity learning approach has been proposed by Hasan and Roy-Chowdhury [18] by merging deep hybrid networks and active learning. Deep neural network model is a unsupervised learning model for segmented video frames. Active learning is a training technique to manually train the labeled videos. This can be applied only when there is countable numbers of instances and it cannot deal with large instances.

Even though there are different algorithms exist to classify the human actions, still it is a challenging task to classify the human actions

- Intra-class variation is the one of the major challenges, due to difference in the style of actions.
- Camera motion and occlusions make a challenging situation to classify the actions.
- Action detection and classification are made difficult because of poor quality of video frames captured with camera jitter.
- Time complexity in algorithms due to high dimensional features.

3 Encapsulated Features with Multi-objective Deep Belief Networks for Action Classification

In this section, the overall structure of the proposed system and its functions are presented in detail. The system has four different phases, such as key frame extraction, feature extraction, training phase, and action classification. A video containing action may have multiple numbers of frames, from which required frames must be selected based on the similarities between the frames. Thus, key frames are extracted from the given input video using structure similarity measure.

Selected key frames undergo feature extraction process to select the features and to form the feature vector. These features are combined as encapsulated features and feature vectors are generated.

To classify the actions, the system is trained with large number of videos by the MODBN network and finally classification is done to recognize the human actions (Fig. 1).

3.1 SSIM-Based Key Frame Extraction

Redundant features affect the complexity of the algorithms; hence, key frames are selected form the input videos and given for feature extraction. To select the unique key frames, structure similarity measure is used. Such key frames are the subset

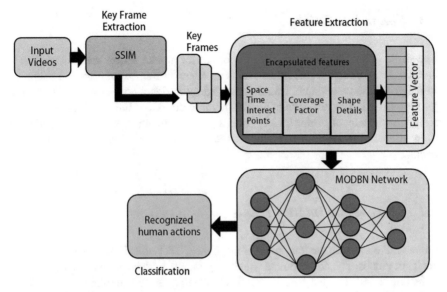

Fig. 1 MODBN for action recognition

of frames of the input video representing the contents. Thus, it improves the space complexity the system.

Consider a video "V" containing "F_n" frames and represented as

$$V = \{F_1, F_2, F_3, \ldots, F_n\} \tag{1}$$

where n represents the number of frames in the video.

A predefined threshold has been used to select the key frames. All the pairs of frames are compared to calculate the SSIM index, and frames having SSIM less than the predefined threshold are selected as key frames. Key frames are represented as

$$K = \{K_1, K_2, K_3, \ldots, K_m\} \tag{2}$$

3.2 Encapsulated Features

Features are the important elements used for classification. Once the key frames are ready, different features are extracted to form the encapsulated features. Encapsulated features include space–time interest points, shape, and coverage factor.

Feature extraction process uses selected key frames for extracting the features required for classification. The proposed technique of action classification extracts both spatial and temporal features, which includes shape extraction, space–time interest point extraction, and coverage factor calculation.

3.3 Shape Extraction

The shape details of the human are important for human action recognition. Grid-based technique can retrieve shape details based on the number of rows and columns. In the grid-based technique, frame is divided into grids with size 4 × 4. Then, the gird is perused from left to right and from top to bottom of the frame. Every grid is assigned with a binary value 0 or 1. Thus, a matrix with binary values zeros and ones is constructed for the frames. Finally, a vector of size 1 × 16 is obtained for shape details.

3.4 Space–Time Interest Points

It is essential to retain the temporal details; hence, STI points [19] are extracted from the input videos. The active points which vary locally with its intensities are selected as feature points, and they are very informative to retain the temporal details. Such active points are named as "interest points" [20].

With selected features, feature vector is formed and its size can be adjusted with respect to the structure of the frames [21]. The variance of objects in the input video frames can be calculated based on the scale and velocity. The interest points of neighborhoods have information based on the appearance actions, and it can be calculated as below.

$$x = (K_a, K_b, K_n, K_{aa}, \ldots, K_{nnnn}) \tag{3}$$

With respect to the calculated velocity, the features of neighborhoods can be adjusted to estimate the invariance based on the motions of the camera [20]. Finally, the feature vector of size 1 × 40 will be obtained with the spatiotemporal features.

3.5 Coverage Factor

By using the interest points extracted, the coverage factor is calculated. At first, the mean value of the interest points will be calculated. Then, the center point is identified, which is used to specify the location of the object. With the center point, vertical length and horizontal length are determined. Using this information, a feature vector is formed with the size of 1 × 2.

4 Multi-objective Deep Belief Networks for Action Classification

The proposed network in Fig. 1 is designed by combining deep belief network (DBN) and multi-objective dragonfly optimization algorithm. DBN [22] has multiple layers of restricted Boltzmann machine (RBM) and multilayer perceptron (MLP). RBM accepts and learns probability distribution of input features. Since MLP has multiple layers, it maps input videos to a set of output videos. Each neuron in the MLP is handled by means of activation function.

By default, RBM in the MODBN uses unsupervised learning and MLP learns using supervised learning with back propagation technique. MODBN network selects weights based on the weight update process in the MODA algorithm. Using MODA, most appropriate weight can be selected for each neuron in the MLP layer.

5 Results and Discussion

This segment describes the outcomes of the projected MODBN classifier used for action classification.

5.1 Experimental Setup

To train the network, two datasets, namely KTH [23] and Weizmann [24], are used. KTH contains actions such as clapping, running, hand waving, walking, boxing, and jogging of 25 persons. Ten classes of actions are available in Weizmann dataset, such as run, walk, jump, bend, gallop sideways, jumping jack, one-hand wave, two-hands wave, skip, and jump in place.

To measure the performance, there are three different metrics such as accuracy, specificity, and sensitivity.

Accuracy: Accuracy is calculated [25] as given below,

$$Acc = \frac{TN + TP}{TN + TP + FN + FP} \tag{4}$$

where, TP is true positive, TN is a true negative, FP is false positive, and FN is a false negative.

Sensitivity: Sensitivity [25] is calculated as below,

$$Sen = \frac{TP}{TP + FN} \tag{5}$$

Specificity: Specificity [25] is calculated as below,

$$\text{Spc} = \frac{\text{TN}}{\text{TN} + \text{FP}}. \tag{6}$$

5.2 *Experimental Results*

Here, the experimental results of the MODBN classification system for action classification are explained with two datasets, namely KTH and Weizmann datasets. Figures 2 and 3 prove the results of the projected multi-objective dragon deep belief network for action classification using KTH and Weizmann datasets, respectively.

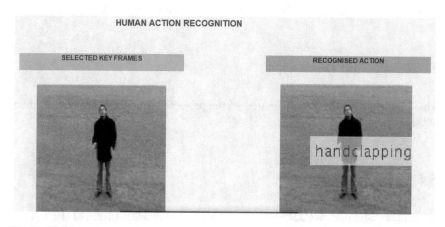

Fig. 2 KTH dataset

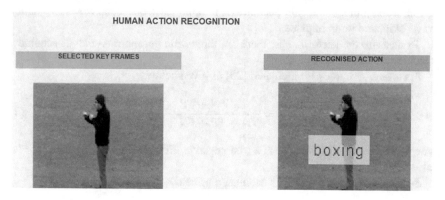

Fig. 3 Weizmann dataset

6 Conclusion

An action classification method is proposed to classify the human actions in the video frames by designing MODBN. The proposed work selects key frames using structure similarity measure, and selected frames are given as input for the feature extraction process. Feature extraction process extracts, features like shape, coverage factor, and interest points. Later, these features are combined to form encapsulated features. In this work, it also proposed a novel MODBN classifier by combining DA and DBN with feasible weights. The MODBN could categorize different actions in the video frames based on the encapsulated features. From the performance evaluation conducted, the proposed MODBN classifier can achieve improved performance than the other methods compared with 0.99 accuracy, 0.97 sensitivity, and 0.95 specificity. Thus, an enhanced result is accomplished for identifying the human actions using the MODBN. In future, the algorithm can be enhanced to the other complex actions, and the efficiency can be analyzed.

References

1. Blank, M., Gorelick, L., Shechtman, E., Irani, M., Basri, R.: Actions as space-time shapes (2007)
2. Tong, M., Li, M., Bai, H., Ma, L., Zhao, M.: DKD–DAD: a novel framework with discriminative kinematic descriptor and deep attention-pooled descriptor for action recognition. Neural Comput. Appl. 1 (2019)
3. Jia, C.C., et al.: Incremental multi-linear discriminant analysis using canonical correlations for action recognition. Neurocomputing 83, 56–63 (2012)
4. Dollar, P., Rabaud, V., Cottrell, G., Belongie, S.: Behavior Recognition via Sparse Spatio-Temporal Features, pp. 65–72. IEEE (2005)
5. Schuldt, C., Barbara, L., Stockholm, S.: Recognizing human actions: a local SVM approach. In: Proceedings of 17th International Conference, vol. 3, pp. 32–36 (2004)
6. Moussa, M.M., Hemayed, E.E., El Nemr, H.A., Fayek, M.B.: Human action recognition utilizing variations in skeleton dimensions. Arab. J. Sci. Eng. 43, 597–610 (2018)
7. Huynh-The, T., Le, B.V., Lee, S., Yoon, Y.: Interactive activity recognition using pose-based spatio–temporal relation features and four-level Pachinko Allocation model. Inf. Sci. (NY) 369, 317–333 (2016)
8. Kong, Y., Jia, Y.: A hierarchical model for human interaction recognition. In: Proceedings of IEEE International Conference Multimedia Expo, pp. 1–6 (2012)
9. Bregonzio, M., Gong, S., Xiang, T.: Recognising action as clouds of space-time interest points. In: 2009 IEEE Computer Society Conference on Computer Vision and Pattern Recognition, pp. 1948–1955 (2009)
10. Liu, J., Shah, M.: Learning human actions via information maximization. In: 26th IEEE Conference Computer Vision and Pattern Recognition, CVPR (2008)
11. Farhadi, A., Endres, I., Hoiem, D., Forsyth, D.: Describing objects by their attributes. In: CVPR 2009. IEEE Conference, pp. 1778–1785 (2009)
12. Wu, D., Shao, L.: Silhouette analysis-based action recognition via exploiting human poses. IEEE Trans. Circuits Syst. Video Technol. 23, 236–243 (2013)
13. Rodriguez, M., Orrite, C., Medrano, C., Makris, D.: A time flexible kernel framework for video-based activity recognition. Image Vis. Comput. 48–49, 26–36 (2016)

14. Li, H., Chen, J., Hu, R.: Multiple feature fusion in convolutional neural networks for action recognition. Wuhan Univ. J. Nat. Sci. **22**, 73–78 (2017)
15. Wang, H., Yuan, C., Hu, W., Ling, H., Yang, W., Sun, C.: Action recognition using nonnegative action component representation and sparse basis selection. IEEE Trans. Image Process. **23**(2), 570–581 (2014)
16. Li, W.X., Vasconcelos, N.: Complex activity recognition via attribute dynamics. Int. J. Comput. Vis. **122**, 334–370 (2017)
17. Nigam, S., Khare, A.: Integration of moment invariants and uniform local binary patterns for human activity recognition in video sequences. Multimed. Tools Appl. **75**, 17303–17332 (2016)
18. Hasan, M., Roy-Chowdhury, A.K.: A continuous learning framework for activity recognition using deep hybrid feature models. IEEE Trans. Multimed. **17**, 1909–1922 (2015)
19. Meng, H., Pears, N., Bailey, C.: Human action classification using SVM_2K classifier on motion features, pp. 458–465 (2006)
20. Everts, I., Van Gemert, J.C., Gevers, T.: Evaluation of color spatio-temporal interest points for human action recognition. IEEE Trans. Image Process. **23**, 1569–1580 (2014)
21. Laptev, I., Lindeberg, T.: Velocity adaptation of space-time interest points. In: Proceedings of International Conference on Pattern Recognition, vol. 1, pp. 52–56 (2004)
22. Vojt, J.: Deep neural networks and their implementation (2016)
23. KTH dataset from, http://www.nada.kth.se/cvap/actions/
24. Weizmann dataset from, http://www.wisdom.weizmann.ac.il/~vision/SpaceTimeActions.html
25. Sopharak, A., Uyyanonvara, B., Barman, S., Williamson, T.H.: Automatic detection of diabetic retinopathy exudates from non-dilated retinal images using mathematical morphology methods. Comput. Med. Imaging Graph. **32**, 720–727 (2008)

LU/LC Change Detection Using NDVI & MLC Through Remote Sensing and GIS for Kadapa Region

A. Rajani and S. Varadarajan

Abstract This research letter considered land use/land cover (LU/LC) changes in Kadapa region from 2001 to 2016 by using GIS (Geographical Information Systems) LANDSAT 7 and 8 images level-1 data of ETM+ sensor were collected for the years 2001 and 2016 to estimate the NDVI values. Based on index values study area is classified into five classes like water bodies, built-up area, barren land, sparse vegetation and dense vegetation. The proposed method performs two-stage classification. From the results it is observed that, increase in barren land of 6.23%, built-up area of 24.74% and decrease of water resources of 2.87%, sparse vegetation of 2.62% and dense vegetation/forest land of 25.47% in the Kadapa region. Finally, the classification algorithm was assessed by the confusion matrix method. As NDVI is used basis for classification process vegetation and water resources classification is 100% and 86% built up and barren lands are classification, i.e., some of built-up areas classified as barren land and barren land as built-up area. Compared to existing classification algorithms based on Principal Component Analysis (PCA) and False Color Composite (FCC) the proposed NDVI histogram and Maximum Likelihood Classifier (MLC) multiple classification gives the best results to estimate the change detection. The work was carried out by using ArcGIS software.

Keywords Normalized Difference Vegetation Index (NDVI) · Maximum Likelihood Classifier (MLC) · Change detection · Remote sensing · GIS · Multispectral image

A. Rajani (✉) · S. Varadarajan
Department of ECE, S V University College of Engineering, Tirupati, Andhra Pradesh, India
e-mail: rajanisvu2015@gmail.com

S. Varadarajan
e-mail: svaradajan@svuniversity.edu.in

© Springer Nature Singapore Pte Ltd. 2020
P. K. Mallick et al. (eds.), *Cognitive Informatics and Soft Computing*,
Advances in Intelligent Systems and Computing 1040,
https://doi.org/10.1007/978-981-15-1451-7_24

215

1 Introduction

Remote sensed information and their dynamics from terrestrial vegetation are extremely useful for applications like environmental monitoring, agriculture, forestry, urban green infrastructures and other related fields can be studied and analyzed. Specially remote sensing information obtained from earth observation satellite sensors can be applied to agriculture for micro- and macro-analysis of yield estimation of crops [1]. However, the applicability of remote sensing and its different vegetation indices estimated using various techniques usually relies heavily on the instruments and its platforms to determine which solution is best to analyze particular issue. These innovations give information to study and screen the dynamic parameters of natural assets for practical maintenance [2].

GIS is a coordinated environment comprising of computer software programming and hardware equipment utilized for analysis, recovering, controlling, investigating and showing geologically referenced (spatial) data with the end goal of advancement arranged administration and management. The major vital uses of satellite images are for land use/land cover (LU/LC) [2–4] mapping. Land covers are the customary benefactors of the physical state of the ground surface. Satellite-based remote detecting has helped in the investigation of land spread elements. The quantity of remote detecting stages expanded amid the previous decades and delivered progressively gritty topographical information [5].

Specialists utilizing strategies including reflected sun-based radiation have utilized either a single band or a proportion of two bands information to depict water highlights, vegetation and soil parameters, etc. [6]. It has been observed that NIR radiation was strongly absorbed by the water and it strongly reflected by vegetation and dry soil. In NDVI images, vegetation [7] appears as white and water surfaces appear as dark black color.

Multispectral remote sensing images contain the vital spatial and spectral highlights of the different objects [8, 9]. Grouping of the objects is done through spectral and spatial investigation of vitality reflected radiant energy discharged by the objects. Multispectral, spatial and transient images are utilized for finding the Normalized Difference Vegetation Index (NDVI) values and arranging [10] distinctive land spread sorts over the chose investigation zone. The NDVI depends on the distinction between spectral radiance of the sensor in Visible Red band (Band4) and the Near-Infrared band (Band5) of LANDSAT 7 and 8 pictures. Hypothetically the estimations of the NDVI changes between −1.0 and +1.0, however, are usually positive for soil and vegetation and negative for water.

The remainder of the paper is organized as: Sect. 2 gives portrayal about the district which is chosen for the investigation to perform. The focal point of Sect. 3 is on the methodology utilized for evaluating NDVI histogram and MLC-based supervised classification and furthermore about confusion matrix method utilized for precision assessment of methodology used for classification. Section 4 examines about outcomes acquired during the time spent change location. At long last in last Sect. 5, conclusions drawn from the images and information examination are introduced.

2 Study Area

Kadapa (on the other hand spelled Cuddapah) is a city in the Rayalaseema of the south-focal piece of Andhra Pradesh, India. It is the area home office of Kadapa region. Starting at 2011 Census of India, the city had a populace of 344,078. It is found 8 km (5.0 mi) south of the Penna River. The city is encompassed on three sides by the Nallamala and Palkonda Hills lying on the structural scene between the Eastern and Western Ghats. Dark and Red ferrous soils involve the region. The city is nicknamed "Gadapa" ("limit") since it is the passage from the west to the holy slopes of Tirumala.

Spatial and ephemeral changes of LU/LC Kadapa in the district of YSR Kadapa, Andhra Pradesh, India are picked for the change analysis purpose. Kadapa Latitude and Longitude are 14.4702 (14° 28′ N) North and 78.8206 (78° 52′ E) East, respectively. Normalized Difference Vegetation Index (NDVI) [7, 8] is evaluated considering multispectral band images of LANDSAT 8 Operational Land Imager/Thermal Infrared Sensor (OLI/TIRS) C1 Level and LANDSAT 7 Enhanced Thematic Mapper Plus (ETM+). The reference datum is World Geodetic System 1984 (WGS84) the pixel spatial resolution is 30 meters for both LANDSAT 7 and LANDSAT 8. One image captured on 24-02-2001 using LANDSAT 7 and the other image captured on 10-02-2016 using LANDSAT 8, here 15 years duration is considered for observing the changes. Table 1 indicates satellite information. The present work considered the Row-50 and path 143 of the LANDSAT picture for the spatial and fleeting elements of Land use and Land Cover.

The multispectral remote sensing images of Kadapa district for 2 unique years were gathered from USGS (United States Geological Survey) earth explorer. LANDSAT 8 satellite pictures the whole earth once every 16 days. The sensors called Operational Land Imager (OLI) and Thermal Infrared Sensor (TIRS) give information in multispectral groups. Band data of LANDSAT 7 and 8 are as given in Table 2. Figure 1 shows the location of chosen area (Kadapa which is a place in the YSR Kadapa District in Andhra Pradesh State in the Country India) for study purpose in the map.

Table 1 Spatial image sources

Data source	Sensor	Date	Spatial resolution (m)	Bands considered
LANDSAT 7	ETM+	24-02-2001	28.5	3 & 4
LANDSAT 8	OLI/TIRS	10-02-2016	30	4 &5

Table 2 LANDSAT 7 and 8 band information

LANDSAT 7 enhanced thematic mapper plus (ETM+)	Bands	Wavelength (μm)	Resolution (m)
	Band 3—Red	0.63–0.69	30
	Band 4—Near-infrared (NIR)	0.77–0.90	30
LANDSAT 8 Operational Land Imager (OLI) and Thermal Infrared Sensor (TIRS)	Band 4—Red	0.636–0.673	30
	Band 5—Near-infrared (NIR)	0.851–0.879	30

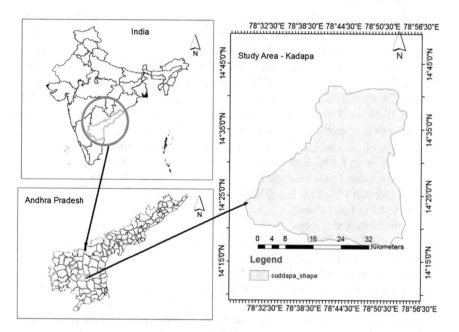

Fig. 1 Geographic location of study area—Kadapa, Kadapa District, Andhra Pradesh in India

3 Methodology

The method used to estimate changes occurred during 2001–2016 is two-stage classification. Proposed method, after performing preprocessing, by applying NDVI formula NDVI image is obtained. NDVI image is classified using histogram and index values as threshold parameter to perform classification. Features are selected based on training samples chosen and then later Maximum likelihood supervised classifier is used to perform two-stage classification. As the classification done in two stages, accuracy is improved compared to existing methods using PCA and FCC. Accuracy assessment is done using confusion matrix. Normalized Difference Vegetation Index (NDVI) is estimated for 2001 and 2016 images using the following Eq. (1):

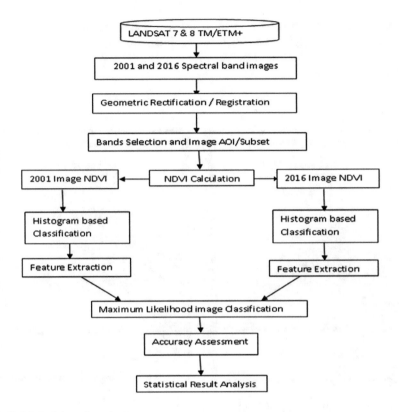

Fig. 2 Methodology flow chart

$$\text{NDVI} = \frac{(\text{NIR} - \text{RED})}{(\text{NIR} + \text{RED})} \tag{1}$$

where

NIR is the Near-Infrared band pixel value
RED is the Red band pixel value

The process and steps followed for finding the change detection of Kadapa region over 2001 to 2016 is represented as flowchart form in Fig. 2.

4 Results and Discussions

The after effects of proposed procedure, two-stage classification method using NDVI histogram and Maximum Likelihood Classifier (MLC) images of 2001 and 2016 year, are shown in Figs. 3 and 4, respectively. By visually observing the NDVI histogram-based Maximum Likelihood Classifier images of 2001 and 2016 the progressions in various parameters is plainly obvious. The image classification success is based on number factors like, accessibility of quality images, secondary data and based on the algorithm chosen for classification process and also based on expertise of person who does the analysis.

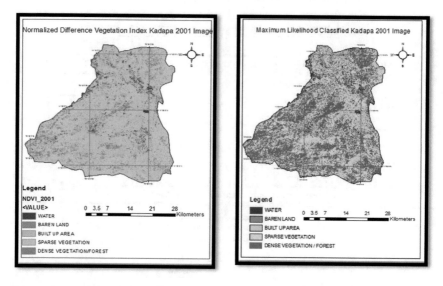

Fig. 3 NDVI and MLC images of Kadapa region 2001

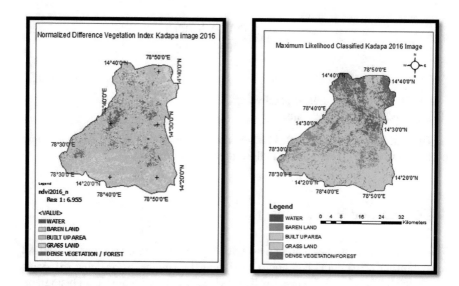

Fig. 4 NDVI and MLC images of Kadapa region 2016

The resultant images after performing NDVI histogram classification and Maximum Likelihood classification for the years 2001 and 2016 of Kadapa region are shown in Figs. 3 and 4, respectively. By performing two-stage classification the areas are classified well. The statistical analysis is done considering the number of pixels in each class. Table 3 presents data for the change detection in terms of number of pixel count. Using this data, it is estimated that out of total area 3.21%—Water, 7.03%—barren land, 8.88%—built-up area, 38.79%—sparse vegetation and 42.09%—dense vegetation occupied for the year 2001. Percentage of each parameter occupied compared to overall area is given in Table 4. Change detected in the Kadapa region for the various classified parameters is tabulated in Table 5.

From the results, it is observed that there is decrease (negative change) in water—2.87%, sparse vegetation—2.62% and Dense forest reduced to—25.47%, whereas increase in built-up area 24.74% and barren land 6.23% is observed.

Accuracy assessment: The categorize results were additionally assessed utilizing confusion matrix method. The exactness of the pixel-based land use/land cover change evaluation can be surveyed quantitatively by translating an error matrix-based statistical analysis called confusion matrix. Table 6 shows the accuracy estimation of classification algorithm using two-stage classifier for Kadapa region images over the years 2001 and 2016.

Table 3 Classified parameters in terms of no. of pixels count

Each parameter occupancy in terms of no. of pixels					
	Water	Barren land	Built-up area	Sparse vegetation	Dense vegetation
2001	83,143	181,919	229,768	1,003,484	1,088,946
2016	7,989	309,714	785,450	844,905	388,248

Table 4 Classified parameters in terms of percentage

Percentage of each parameter pixels with respect to total pixels in the total area					
	% Water	% Barren land	% Built-up area	% Sparse vegetation	% Dense vegetation
2001	3.21	7.03	8.88	38.79	42.09
2016	0.34	13.26	33.62	36.16	16.62

Table 5 Change detected from 2001 to 2016

Change from 2001 to 2016 in terms of percentage value				
Water	Barren land	Built-up area	Sparse vegetation	Dense vegetation
−2.87	6.23	24.74	−2.62	−25.47

Table 6 Accuracy assessment of classified images of Kadapa region for 2001 and 2016 year

Land cover classes	Land cover map 2001		Land cover map 2016	
	Users accuracy (UA %)	Producers accuracy (PA %)	Users accuracy (UA %)	Producers accuracy (PA %)
Water	100	100	100	100
Barren land	94	96	93	95
Built-up area	92	94	89	91
Sparse vegetation	100	100	100	100
Dense vegetation	100	100	100	100

By taking 50 random samples for each class, the accuracy is estimated using confusion matrix. From the results, it is observed that, water, sparse and dense vegetation are classified 100% and barren land and built-up area are misclassified ~10% remaining 90% correctly classified.

5   Conclusions

Through this result, analysis shows that multispectral and transient LANDSAT time arrangement has extraordinary potential for breaking down vegetation changes in Kadapa, Andhra Pradesh, India. Not withstanding that, NDVI histogram and Maximum Likelihood-based supervised classifier can give best outcomes contrasted with existing calculations utilizing PCA Principal Component Analysis (PCA) algorithm presented and tested.

Two-stage classification using NDVI along with histogram-based classification and Maximum Likelihood classification shows the efficiency to produce high accuracy vegetation maps over Kadapa region over the years 2001 and 2016. The area of Kadapa region was classified into five classes like water, built-up area, barren land, sparse vegetation and finally into dense vegetation/forest land. Compared to False Color Composite (FCC)- and PCA-based classification techniques specific index and histogram-based technique gives the best accurate results observed in this work.

Acknowledgements   Authors would like to thank USGS Earth Explorer for providing LANDSAT images to carry out the present research work.

References

1. Das, K.: NDVI and NDWI based change detection analysis of Bordoibam Beelmukh Wetland-scape, Assam using IRS LISS III data. ADBU-J. Eng. Technol. **6**(2), 17–21 (2017). ISSN: 2348-7305
2. Xue, J., Su, B.: Significant remote sensing vegetation indices: a review of developments and applications. J. Sensors **2017**, Article Id 1353691 (Hindawi) (2017)
3. Taufik, A., Ahmad, S.S.S., Ahmad, A.: Classification of LANDSAT 8 satellite data using NDVI thresholds. J. Telecommun. Electron. Comput. Eng. **8**(4), 37–40 (2016). ISSN: 2180-1843 e-ISSN: 2289-8131
4. Jeevalakshmi, D., Narayana Reddy, S., Manikiam, B.: Land cover classification based on NDVI using LANDSAT 8 time series: a case study Tirupati region. In: International Conference on Communication and Signal Processsing (ICCSP) (2016)
5. Aburas, M.M., Abdullah, S.H., Ramli, M.F., Ash'aaria, Z.H.: Measuring land cover change in Seremban, Malaysia using NDVI index. In: International Conference on Environmental Forensics 2015 (iENFORCE 2015), Proc. Environ. Sci. **30**, 238–243 (2015)
6. Mallupattu, P.K., Reddy, J., Reddy, S.: Analysis of land use/land cover changes using remote sensing data and GIS at an urban area, Tirupati, India. Sci. World J. (Hindawi Publishing corporation) (2013)
7. Bhandari, A.K., Kumar, A., Singh, G.K.: Feature extraction using normalized difference vegetation index (NDVI): a case study of Jabalpur city. In: 2nd International Conference on Communication. Computing and Security [ICCCS-2012], Procedia Technol. **6**, 612–621 (Elsevier) (2012)
8. Hayes, D.J., Sader, S.A.: Comparison of change detection techniques for monitoring tropical forest clearing and vegetation regrowth in a time series. Photogramm. Eng. Remote Sens. **67**, 1067–1075 (2001)
9. McFeeters, S.K.: The use of the normalized difference water index (NDVI) in the delineation of open water features. Int. J. Remote Sens. **17**(7), 1425–1432 (1996)
10. Singh, A.: Digital change detection techniques using remotely sensed data. Int. J. Remote Sens. **10**, 989–1003 (1989)

Text Region Extraction for Noisy Spam Image

Estqlal Hammad Dhahi, Suhad A. Ali and Mohammed Abdullah Naser

Abstract In this paper, the problem of spam filtering for images, a type of fast-spreading spam where the text is included in images to overcome the text-based spam filter. One common method for detecting spam is the optical character recognition system (OCR) that detecting and recognizing the text embedded, following by a classifier which distinguishes spam from ham. Nevertheless, the spammers begin hiding image text for preventing OCR from detecting spam. To recompense for the shortages of the OCR system, a method based on the detection algorithm is proposed for the text region. To estimate the performance of the projected system, the methodology was applied to a group of unwanted images Dredze (available to the public) to check the efficiency of our method which outperforms the initial OCR system in sensible use with a complex background in spam. The test results indicated that the new method gives good text regions detection even for noisy images.

Keywords OCR · Text localization · Spam image · Text-based spam filtering · Text features

1 Introduction

The growing use of the Internet has led to the promotion of an easy and fast way to communicate electronically, the example is known for this email. Now, send and receive email as a commonly used means of communication. Today, spammers will evade images in three methods, implanting in email body appending an image document to messages, put a link within the email's body and therefore the objective

E. H. Dhahi · S. A. Ali (✉) · M. A. Naser
Department of Computer Science, College of Science for Women, University of Babylon, Babylon, Iraq
e-mail: Suhad_ali2003@yahoo.com

E. H. Dhahi
e-mail: estqlal87@gmail.com

M. A. Naser
e-mail: wsci.mohammed.abud@uobabylon.edu.iq

© Springer Nature Singapore Pte Ltd. 2020
P. K. Mallick et al. (eds.), *Cognitive Informatics and Soft Computing*,
Advances in Intelligent Systems and Computing 1040,
https://doi.org/10.1007/978-981-15-1451-7_25

message is sent by the image. Text-based spam filtering techniques are not success and careless in recent to recognize the new spammers' methods. Image spam is rich in content and can be a variety of unwanted images or a singular image can include various images inside a unique image. So image spam filters use distinct technology and different filters to get the ability to recognize image spam and to fight the planning of image spammers. The most straight forward strategy is to utilize an optical character recognition (OCR) tool to find. In response to the OCR-based discovery, spammers introduced new obfuscation spamming techniques, which prevent OCR reading text embedded within images. So, good text detection and extraction method can enhance the result of text spam filtering. There are numerous processes have been presented through many researchers and professionals to detect and extract of text regions in advance. Depending on the processes used, they are classified in many ways, like the method based on the connected component, the edge-based strategies, the region-based methodology, the texture-based methodology, and therefore the mathematical morphology methodology [1]. All techniques which want to extract the text image have their reign benefits and determination supported completely various criteria like accuracy average, the speed of restoration, accuracy, etc. [2].

2 Related Works

This area reviews several ways associated with the suggested approach:

In 2017, Mathur and Rikhari [3] suggest an algorithm for extracting the text regions from images based on FAST algorithm. It used for detecting many writing and languages, in additional to blurry and non-blurry images. But this technique fails in the large lines and for some specific images that respond to the corners a lot. Despite these problems, they are fast (and can be parallel) and are less complex than other OCR tools and can be improved in the future.

In 2017, Kulkarni and Barbadekar [4] proposed completely approaches used for text extraction from color images. Two usually used ways for this downside are stepwise strategies and integrated strategies within the method of text detection and recognition and analyses.

In 2018, Natei et al. [2] proposed an algorithm for extracting the text through collecting the edge-based and connected components algorithms. OCR is used to recognize the extracted text recognition and then will produce audio as output. This method does not apply on handwritten and complex font text.

In 2018, Dai et al. [5] suggested increased multi-channels MSER method for detecting a scene text. For detecting text in complex background images, a honing and separating different channels of the image is applied which may introduce non-text regions. The local differentiation and limit key focus high lights are used to filter these non-text regions. The support vector machine is employed to classify the candidate text regions.

3 The Suggested Method

The suggested method is implemented in two phases. Firstly, we find candidate text regions of an input spam image by using the lifting wavelet transform. Secondly, a set of trusted spam-indicative features from detected text regions are defined.

3.1 Text Regions Extraction

Firstly, the text regions are extracted from the image. Our method for text regions detection is consists of several steps as shown in Fig. 1.

3.1.1 Detection of Text Pixels

Text pixels have a high contrast around adjacent pixels, so the edge-based method will be applied. The proposed method used lifting Haar wavelet (LHW) [6] to detect edge pixels. The foreground text is also of any color and so the edges are also visual in one or a lot of these three color channels. Therefore, the edge is detected in each color bar separately.

As shown in Fig. 2, LHW is applied on each band of color image to generate four sub-bands (LL, LH, HL, and HH). The edges found in high frequencies, therefore, a wavelet edge array (WE) is found from combinations of three-wavelet edge arrays (WER, WEB, and WEG) of three color bands. For the red band, the WER is computed as follows:

$$\text{WER}(x1, y1) = \begin{cases} 1 \ \text{DAE}(x1, y1) \geq T \\ 0 \ \text{otherwise} \end{cases} \tag{1}$$

where DAE means details an average array which is computed according to the following equation:

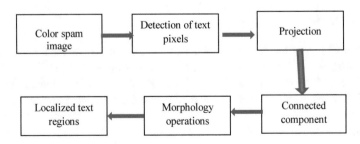

Fig. 1 Steps of the proposed method

(a)   (b)

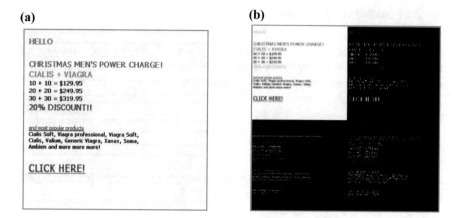

Fig. 2 Applying LHW **a** initial image, **b** after applying LHW an original image

$$DEA(x1, y1) = \frac{LH(x1, y1j)^2 + HL(x1, y1)^2 + HH(x1, y1)^2}{3} \tag{2}$$

And T is a threshold value, the value of which is determined by WER $(x1, y1)$ as follows:

$$T = k \times (m + \sigma) \tag{3}$$

The parameters (m and σ) represented the mean and standard deviation of the image, respectively, and k represents the parameter whose value is set to 1.2.

In the same way, the other edge arrays (WEB and WEG) are computed. To construct wavelet edge array (WE), the logical OR (V) operator is applied between three edge arrays (WER, WEG, and WEB) as given by Eq. (4)

$$WE = (WER) \vee (WEG) \vee (WEB) \tag{4}$$

3.1.2 Projection

The obtained binary image WE( ) contains text and non-text regions. So, to remove the non-text region, a horizontal and vertical projection is used [7]. For an $M \times N$ image, the horizontal projection $\mathbf{P_{hor}}$ of the row (i_0) is computed accordingly the total of element values there in row and each one the columns among the image as follow:

$$P_{hor}(i_0) = \sum_{j=0}^{N} WE(i_0, j) \quad \text{for } 0 \leq i_0 \leq M \tag{5}$$

While the vertical projection $\mathbf{P_{ver}}$ of the column (j_0) is calculated by the Eq. (6) as the total of pixel values there in column and each one the rows among the image.

$$P_{ver}(j_0) = \sum_{i=0}^{M} WE(i, j_0) \quad \text{for } 0 \leq j_0 \leq N \tag{6}$$

Equations (7 and 8) are used to find two thresholds $(\mathbf{T_h}, \mathbf{T_v})$ as horizontal and vertical thresholds, respectively. These thresholds are used to separate non-text regions from text regions.

$$T_h = \frac{\text{mean}(P_{hor}) + \text{min}(P_{hor})}{2} \tag{7}$$

$$T_v = \frac{\text{mean}(P_{ver}) + \text{min}(P_{ver})}{2} \tag{8}$$

Each value in horizontal projection $\mathbf{P_{hor}}$ $(\mathbf{i_0})$ is seeing as a section of an elect text region if its value is greater than $\mathbf{T_h}$; otherwise, this row is suppressed. Also, if each value in $\mathbf{P_{ver}}$ (j_0) is larger than $\mathbf{T_v}$ then column (j_0) can be seen according to a section of an elect text region.

3.1.3 Connected Components

At this stage, a single image will be produced from merge the horizontal and vertical images, and non-text areas detected using geometric features are ignored. Initially, it applied four connected components [8] to the embedded projection image. Secondly, the area (A) is calculated from each region and the largest area (LA) is found. Finally, the area of each area is compared to the threshold $(T2)$. If the space of every area is a smaller amount than the threshold, it will be considered a non-text area and its pixels must become ignored. The threshold value $(T2)$ is calculated as follows:

$$T2 = \frac{LA}{25} \tag{9}$$

3.1.4 Morphology Operations

The resulting image might contain non-text pixels. For that the dilation morphology process is applied using a 9×9 structure elements followed by the opening process using a 9×9 structure elements to remove the non-text pixels.

3.2   Features Extraction

In this step, a set of features are extracted from detected text regions, which foreseeable to be discriminatory between spam and non-spam images. These features are [9]:

1. **Occupy Rate (OR)**: It measures the text area to the total text region size according to the following equation

$$OR = \frac{\text{text region area}}{T_m \times T_n} \tag{10}$$

    where T_m and T_n are the no. of rows and columns of the text region, respectively.
2. **Aspect ratio (AR)**: It measures the rate between width and height of text regions. It can be defined as the following equation:

$$AR = \frac{\max(T_m, T_n)}{\min(T_m, T_n)}, \tag{11}$$

3. **Perimetric Complexity (PC)**: It is known as the square length of the border between the black and white pixels ("ocean P") within the whole image, divided by the black area (A), according to the following equation:

$$PC = \frac{P^2}{A}, \tag{12}$$

4. **Compactness (C)**: It measures the ratio between the area of the text region to the perimeter of the text region. It can be defined as follows:

$$C = \frac{A}{P * P}, \tag{13}$$

5. **The Extent of the Text Region (ETR)**: This feature measures the amount of text regions to the total size of images. It can be defined as follows:

$$ETR = \frac{\text{area of the extracted regions}}{M \times N} \tag{14}$$

    where M, N representing the rows and columns of the image, respectively.

4   Experimental Results

Our system method is checked on the public Dredze dataset [10]. This dataset contains email images of various sizes. Figure 3 shows samples of original spam images and images of detected text regions.

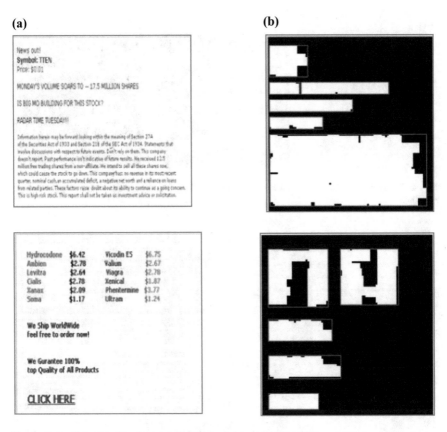

Fig. 3 Extraction of text region **a** spam images, **b** labeled text regions

The proposed method can deal with the noise that causes confusion of OCR. Figure 4a shows the original spam image which contains noise pixels that are not belong to text pixels. Figure 4b shows detected text regions and Fig. 4c shows the perfect localization of text regions without noise pixels.

The text region may be found in natural images such as the name of building, the signs of the road, street names, and others. So, the text features extraction as explained previously is helping in distinguish between non-spam and spam images. Figure 5 illustrated the allocation of features for each spam and ham images in different data domains. For all features, it gives good discernment between both types of images. For example, it can be seen from Fig. 5b that more than 61% of the spam images are distributed within 0.25–0.99. And more than 89% of the important images are within 0.25–0.99.

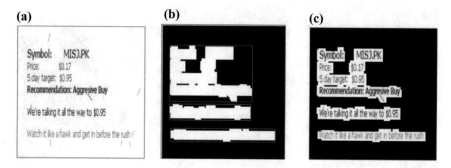

Fig. 4 Noisy text regions detection **a** span image, **b** labeled text regions, **c** the extraction of text regions

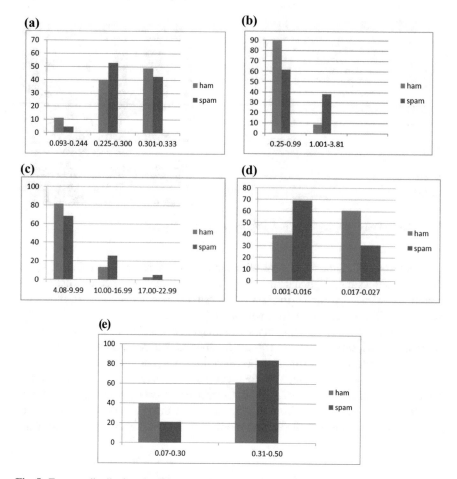

Fig. 5 Features distributions in all images **a** feature 1, **b** feature 2, **c** feature 3, **d** feature 4, **e** feature 5

5 Conclusions

In this suggested system, a new method depends on connected components (CC) and edge detection method is proposed. Firstly, it can be noted from experimental results that using lifting Haar transform gives good method for text detection. Secondly, the suggested method is more powerful in handling low-contrast images, images of different sizes, fonts, and text images that containment noise. Lastly, the features that are extracted from the detected text regions give good indication to separate spam images from non-spam images as shown in Fig. 5. For future works, these features can be used for classifying images in spam-based text filtering either it is used singly or combined with other image features to enhance the classification results.

References

1. Gupta, Y., Sharma, S.H., Bedwal, T.: Text extraction techniques. Int. J. Comput. Appl. NSFTICE, 10–12 (2015)
2. Natei, K.N., Viradiya, J., Sasikumar, S.: Extracting text from image document and displaying its related information. J. Eng. Res. Appl. **8**(5), 27–33 (Part-V) (2018). ISSN: 2248-9622
3. Mathur, G., Rikhari, S.: Text detection in document images: highlight on using FAST algorithm. Int. J. Adv. Eng. Res. Sci. (IJAERS) **4**(3) (2017). ISSN: 2349-6495(P)|2456-1908(O)
4. Kulkarni, C.R., Barbadekar, A.B.: Text detection and recognition: a review. Int. Res. J. Eng. Technol. (IRJET) (2017). e-ISSN: 2395-0056, p-ISSN: 2395-0072
5. Dai, J., Wang, Z., Zhao, X., Shao, S.: Scene text detection based on enhanced multi-channels MSER and a fast text grouping process. Int. J. Comput. Linguist. Res. **9**(2) (2018)
6. Lee, H.: Wavelet analysis for image processing. Institute of Communication Engineering, National Taiwan University, Taipei, Taiwan. On http://disp.ee.ntu.edu.tw/henry/wavelet_analysis.pdf
7. Javed, M., Nagabhushan, P., Chaudhuri, B.B.: Extraction of projection profile, run-histogram and entropy features straight from run-length compressed text documents. In: IAPR Asian conference on Pattern Recognition, IEEE proceedings, pp. 813–817 (2013)
8. Burger, W., Burge, M.J.: Principles of digital image processing. Cor Algorithms. Springer Publishing Company (2009)
9. Gonz´alez, A., Bergasa, L.M., Yebes, J.J., Bron, S.: Text location in complex images. In: Proceedings of the 21st International Conference on Pattern Recognition (ICPR 2012), pp. 617–620, Tsukuba, 11–15 Nov 2012
10. From https://www.cs.jhu.edu/~mdredze/datasets/image_spam/

Noise Reduction in Lidar Signal Based on Sparse Difference Method

P. Dileep Kumar and T. Ramashri

Abstract Lidar is the only remote sensing device used to measure the dynamic properties of the atmosphere from stratosphere through the mesosphere. The range of Rayleigh lidar is affected due to different noises present in the atmosphere. In this paper, different types of noises in lidar signal are interpreted, and distinct denoising methods such as wavelets, Empirical Mode Decomposition (EMD) and Sparsity are tested on signal received from Rayleigh lidar receiver at National Atmospheric Research Laboratory (NARL), Gadanki. The proposed denoising using sparsity achieves better signal-to-noise ratio at higher altitudes, and the temperature profile also matches good with the SABER instrument in TIMED satellite and NRLMSISE-00 model data.

Keywords Rayleigh lidar · Denoising techniques · Sparse difference · Signal-to-noise ratio · Temperature

1 Introduction

In recent decades, light detection and ranging (Lider) is an effective remote sensing tool used to determine temperature of the atmosphere and vertical profiles of aerosol layers. The vertical atmosphere density profile is calculated from the Rayleigh scattering molecules. Rayleigh scattering lidar is an efficient ground-based instrument used to measure the temperature of atmosphere at an altitude of 30–80 km. The experimental denoising techniques are applied on Rayleigh lidar situated in National Atmospheric Research Laboratory (NARL), Gadanki (13.8°N, 79.2°E). The backscattered light intensity from the atmosphere with respect to different altitudes is measured by lidar, and the measured profile is called as photon count profile. The backscattering light intensity from the atmosphere is due to both particles and molecules,

P. D. Kumar (✉) · T. Ramashri
Department of ECE, Sri Venkateswara University College of Engineering, Tirupati, India
e-mail: dileepk38@gmail.com

T. Ramashri
e-mail: rama.jaypee@gmail.com

© Springer Nature Singapore Pte Ltd. 2020
P. K. Mallick et al. (eds.), *Cognitive Informatics and Soft Computing*,
Advances in Intelligent Systems and Computing 1040,
https://doi.org/10.1007/978-981-15-1451-7_26

but the presence of particles is negligible above 30 km. Therefore, the backscattered photon counts from the Rayleigh lidar are proportional to molecular number density of the atmosphere. Rayleigh lidars are so-called as they use Rayleigh scatter from air molecules. There are primarily three different noises present in the backscattered Rayleigh lidar signal. The noises are Quantum noise, Background noise and dark current noise occurs [1]. The effect of these noises increases with increase in the height of the lidar signal. Various signal denoising techniques are proposed like wavelet and EMD are proposed to reduce the noise by improving signal-to-noise ratio at higher altitudes. Wavelets are used for joint time-frequency analysis. Scaling and shifting are the major concepts in wavelets [2]. Symlet 4 (Sym 4) wavelet is used for denoising lidar signals. There are some limitations in Wavelet transforms like oscillations, shift variance, aliasing and lack of directionality. Due to mode mixing problem in EMD [3], the significant part of the signal also denoised. In order to overcome these drawbacks, baseline correction and denoising of lidar signal based on sparse technique is proposed in this paper.

This paper is organized as follows. The block diagram and description of Rayleigh lidar explained in Sect. 2. In Sect. 3, proposed denoising technique using baseline correction sparsity algorithm system model is explained. Simulations for baseline correction of lidar signals with various algorithms are provided in Sect. 4. Temperature profile retrieval using Chanin-Hauchecorne (CH) method is interpreted, and the comparison of denoised temperature profile with TIMED satellite data and NRLMSISE-00 model is illustrated in Sect. 5.

2 Indo-Japanese Rayleigh Lidar

The block diagram of Indo-Japanese Rayleigh lidar is shown in Fig. 1. The transmitter of Rayleigh scatter lidar consists of a laser, beam expander, steering mirror and pulse detector. The system laser is an Nd:YAG laser with an oscillator section succeeded by an amplifier performing at 1064 nm and a extra harmonic generator that bring about the output light at 532 nm. The receiver of main mirror diameter is 760 mm, and the F-ratio is 3:1. The telescope's FoV is limited to 1.0 mm rad since wide FoV simply allows most of the sky's background noise and allows high axis regulation tolerance. The laser light from the aperture is correlated as well as passes through a 532 nm wavelength narrow band-pass filter and a 1.07 nm bandwidth full width Half Maximum (FWHM) [4]. The high sensitive R-channel (90% channel) collects backscattered lidar signals only from high altitudes (35–90 km). The backscattered lidar signal from altitudes well below 5 km is denied mostly by aperture as the telescope receiver is placed about 2 m distant from steering mirror. The current signal from PMT is intensified by the Pre-amplifier (PA) and then transmitted to Multichannel scalar (MCS) unit [5]. The raw lidar signal from Rayleigh lidar is shown in Fig. 2 with the number of photon count backscattered with respect to time. The round time is transformed into altitude (h) using speed of light (c) as $h = ct/2$.

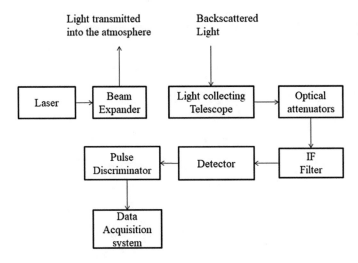

Fig. 1 Block diagram of lidar system

Fig. 2 Raw lidar signal

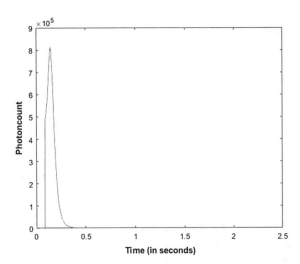

The time bins are 2 µs for a 300 m resolution profile. The range-corrected lidar signal with respect to height is shown in Fig. 3.

The basic lidar signal is formed by accumulating the data for 250 s duration for each measurement corresponding to 5000 laser shots. The background noise is predicted for each measurement and finally subtracted from the signal for correction. A few of such noise corrected signals are combined further for lidar signal analysis. The main technique used for temperature retrieval from density profiles was based on the theory described by Chanin and Hauchecorne method. In Rayleigh lidar, laser light pulse directed vertically into the atmosphere is backscattered by aerosols and

Fig. 3 Range-corrected
lidar signal

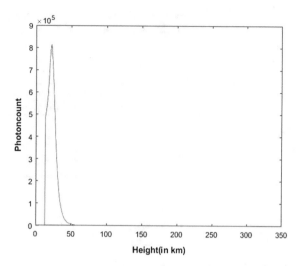

Height(in km)

air molecules. The detected altitude profile of backscattered photon accommodates n
channels of range bins of each with a thickness of Δz. Then, the measured backscat-
tered signal in the zth altitude layer $\left(z_i - \frac{\Delta z}{2}\right) - \left(z_i + \frac{\Delta z}{2}\right)$ is given by lidar equation
in Eq. (1).

$$N(z_i) = \left[\frac{P_L \Delta t}{hc/\lambda}\right][o(z)]\left[\frac{c\Delta t}{2}\right][\beta_\pi(z)]\left[\frac{A_R}{(z_i - z_a)^2}\right][T_a^2(z)][\eta_T] + N_B \quad (1)$$

3 Proposed Baseline Correction Denoising Using Sparsity

The proposed approach represents a noisy lidar signal as sparse with a very sparse
derivative. The slopes of interest in the lidar signal are of second derivative and
named as sparse. This sparsity concept has been an effective and beneficial ride in
signal processing since several decades. The upper and lower case bold represent the
matrices and vectors, respectively. The lidar signal X with N-point is defined through
a vector X in Eq. 2.

$$X = [x(0), ---- x(N-1)]^T \quad (2)$$

All derivatives indicated as finite differences. The first-order difference is represented by $(N - 1)X \ N$. The matrix length $(N - k)X \ N$ by an order like k is
expressed as D_k. D_0 is represented as identity matrix $D_0 = I$. The Rayleigh lidar
signal is represented as the raw lidar signal, noise signal and baseline interference
[6]. The N-point noise-free lidar data is represented as

$$z = g + n, \quad z \in R^N \tag{3}$$

The vector g consists of more peaks and represented as sparse derivative signal. The vector n represents the baseline and represented by a low-pass lidar signal. The recognized noisy data in lidar signal is further modeled in Eq. 4.

$$h = z + s \tag{4}$$

$$h = g + n + s, \quad h \in R^N \tag{5}$$

where s is a stationary Gaussian process, and the aim must assessment baseline with the peaks "g" together against examination h. When the peaks in lidar signal are missing, then the baseline can be relatively recovered by low-pass filtering and represented in Eq. 6.

$$n \approx \text{LPF}(n + s) \tag{6}$$

Therefore, given an estimate of $\hat{g}$ of peaks of lidar signal, we can obtain a estimate $\hat{n}$ based on baseline through filtering h in (4) with a low-pass filter.

$$\hat{n} = LPF(h - \hat{g}) \tag{7}$$

We can obtain an estimate $\hat{z}$ by adding $\hat{g}$

$$\hat{z} = \hat{n} + \hat{g} \tag{8}$$

$$\hat{z} = \text{LPF}(h - \hat{g}) + \hat{g} \tag{9}$$

$$\hat{z} = \text{LPF}(h) + \text{HPF}(\hat{g}) \tag{10}$$

where HPF is the high-pass filter

$$\text{HPF} = I - \text{LPF} \tag{11}$$

To calculate estimate $\hat{g}$ of the lidar signal peaks from observed signal data h, we will formulate inverse problem with the quadratic data fidelity term $\left\| h - \hat{z} \right\|_2^2$.

$$\left\| h - \hat{z} \right\|_2^2 = \left\| h - \text{LPF}(h) - \text{HPF}(\hat{g}) \right\|_2^2 = \left\| \text{HPF}(h - \hat{g}) \right\|_2^2 \tag{12}$$

Consequently, information loyalty term does not rely upon the baseline assessment $\hat{n}$ and only depends on the high-pass filter. The low-pass and high-pass filters are to be noncasual, zero phase and recursive filters. Therefore, high-pass and low-pass filters filter peaks in lidar signal without recommending any shifts in the peak locations.

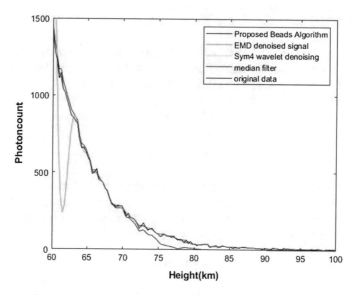

Fig. 4 Comparison of denoising techniques on 05/04/07

In this paper, HPF can be defined as HPF $= BA^{-1}$, where A along with B are matrices represented by linear time-invariant system. Sparse performance in a signal is obtained over the usage of non-quadratic regularization terms. Accordingly, to access the estimation of $\hat{g}$, successive optimization effect is suggested and expressed in Eq. 13.

$$\hat{g} = \arg\min_{g} \left\{ F(g) = \frac{1}{2} \|\mathrm{HPF}(h - g)\|_2^2 + \sum_{i=0}^{M} \lambda_i R_i(D_i g) \right\} \tag{13}$$

The output g can be obtained with less memory and more efficiency. The proposed denoising technique using sparsity is compared with existing denoising techniques like EMD and Wavelet on 05/04/07 Rayleigh lidar data. The concluding results are presented in Fig. 4.

4 Simulation Results

The signal-to-noise ratio of proposed sparsity algorithm compared with existing techniques which are shown in Table 1. The range of the lidar signal also increased by reducing noise at higher altitudes.

Table 1 Signal-to-noise ratio on 05/04/2007 at different altitudes in NARL

Method	SNR (dB) at different altitudes			
	40 km	50 km	70 km	80 km
Proposed sparse	50.59	42.15	28.11	19.84
EMD	47.18	38.74	24.70	16.44
Sym 4 wavelet	42.75	34.30	20.27	12.00
Median filter	42.88	34.43	20.40	12.13
Original signal	39.33	30.89	16.86	8.59

5 Temperature Retrieval Using Chanin-Hauchecorne Method

Temperature is expressed as $T(Z_i)$ using Chanin method [7], and temperature at an altitude is expressed in Eq. 14.

$$T1(z_i) = \frac{M.g1(z_i).\Delta z}{R \log\left[\frac{P1\left(z_i - \frac{\Delta z}{2}\right)}{P1\left(z_i + \frac{\Delta z}{2}\right)}\right]} \tag{14}$$

The atmospheric temperature is expressed as $T1(z_i) = \frac{M.g1(z)dz}{R \log(1+X)}$.

5.1 SABER Instrument

Sounding of atmosphere using broadband emission radiation (SABER) is an instrument in NASA's TIMED satellite. SABER begins to operate from January 2002. SABER attains the profiles from 13.8°N, 79.2°E 0N 05/04/2007. As SABER uses the approach of limb infrared radiometry, it brings near global radiance data with good vertical resolution daily [8].

5.2 Nrlmsise-00

Naval Research Laboratory Mass Spectrometer Incoherent Scatter (MSIS) Radar Extended Model (NRLMSISE-OO) is an atmospheric model database maintains dependent parameters such as mass density, oxygen, nitrogen, total mass density and neutral temperature with respect to independent variables such as different locations, time and data of particular year. The MSIS model represents better results compared with Jacchia-70 model and used for different operational and scientific communities. The temperature profiles of original and denoised signals using proposed method and compared with SABER instrument in TIMED satellite data and NRLMSISE-OO model are shown in Fig. 5.

Fig. 5 Comparing
temperature profiles on
05/04/07

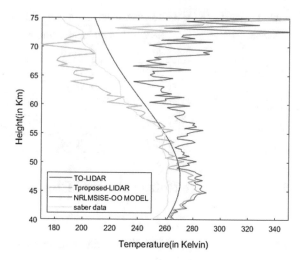

6 Conclusion

In this paper, sparse difference method is proposed for denoising nonlinear, nonstationary Rayleigh lidar signal. The experimental results show that sparse difference method performs better than other methods like EMD, Sym 4 Wavelet and median filter with an improvement in signal-to-noise ratio from 10 to 12 dB. The range of lidar signal is increased by improving signal-to-noise ratio. The temperature retrieved using proposed sparse difference method better matches with SABER instrument used in TIMED satellite and NRLMSISE-00 data.

References

1. Xu, F., Wang, J., Zhu, D., Tu, Q.: Speckle noise reduction technique for lidar echo signal based on self adaptive pulse matching independent component analysis. Opt. Lasers Eng. **103**, 92–99 (2018). (Elsevier Publications)
2. Xu, F., Zhang, X., Zhu, D.: Generalized wavelet thresholding technique for optimal noise reduction of lidar echo signals. In: Proceedings of the 9th International Conference on Signal Processing Systems, pp. 140–145 (2017)
3. Li, H., Chang, J., Xu, F., Liu, Z., Yang, Z., Zhang, L., Zhang, S., Mao, R., Duou, X., Liu, B.: Efficient lidar signal denoising algorithm using variational mode decomposition combined with whale optimization algorithm. Remote Sens. **11**(126), 01–15 (2019). (MDPI Publisher)
4. Chang, J.H., Zhu, L., Li, H., Xu, F., Liu, B., Yang, Z.: Noise reduction in lidar signal using correlation based EMD combined with soft thresholding and roughness penalty. Opt. Commun. **407**, 290–295 (2018). (Elsevier Publications)
5. Sharma, S., Vaishnav, R., Shukla K.K., Lal, S., Chandra, H., Acharya, Y.B., Jayaraman, A.: Rayleigh lidar observed atmospheric temperature characteristics over a western indian location: intercomparison with satellite observations and models. Eur. Phys. J. D. **71**(187), 01–08 (2017)
6. Rai, S., Tripathy, P., Nayak, S.K.: Using sparsity to estimate oscillatory mode from ambient data. Indian Acad. Sci. **44**(90), 01–09 (2019). (Saadhana Publishers)

7. Wing, R., Hauchecorne, A., Keckhut, P., Godin, BS., Khaykin, S., Mccullogh, E., Mariscal, J.F., Almedia, E.D.: Lidar temperature series in the middle atmosphere as a reference data set—Part I: improved retrievals and a 20 year cross validation of two co-located French lidars. Atmos. Meas. Tech. **11**, 5531–5547 (2018)
8. Kulikov, M.Y., Nechaev, A.A., Belikovich, M.V., Vorobeva, M.V., Grygalashvyly, M., Sonnemann, G.R., Feigin, A.M.: Boundary of night time ozone chemical equilibrium in the mesopause region from saber data. Implications for deviation of atomic oxygen and atomic hydrogen. Geophys. Res. Lett. **10**, 01–08 (2018)

Iris Recognition System Based on Lifting Wavelet

Nada Fadhil Mohammed, Suhad A. Ali and Majid Jabbar Jawad

Abstract At present, the need for a precise biometric identification system that provides reliable identification and individual verification has rapidly increased. Biometric recognition system based on iris is a reliable human authentication in biometric technology. This paper proposed a new iris system using on lifting wavelet transform to recognize persons using low-quality iris images. At first, the iris area is localized. Then, it converted to the rectangular area. For discrimination purpose, a set of features are determined from the lifting wavelet subbands, where the iris area is analyzed to three levels. Also, the new method depends on using quantizing the two subbands (LH3 and HL3) and the average values for the two high-pass filters areas (HH1, HH2) to build the iris code. CASIA V1 dataset of iris images is used to measure the performance of proposed method. The test results indicated that the new method gives good identification rates (i.e., 98.46%) and verification rates (i.e., 100%) for CASIA V1 dataset.

Keywords Biometric · Iris recognition · Lifting wavelet · Identification · Verification

1 Introduction

Nowadays, biosecurity is the basic condition in the world. Different types of biometric security systems are deployed based on the requirements of applications designed primarily to identify and authenticate people. Biometrics is the field of automatic

N. F. Mohammed (✉) · S. A. Ali · M. J. Jawad
Department of Computer Science, College of Science for Women, University of Babylon, Hillah, Iraq
e-mail: Nfm.computers@gmail.com

S. A. Ali
e-mail: suhad_ali2003@yahoo.com

M. J. Jawad
e-mail: wsci.majid.jabbar@uobabylon.edu.iq

© Springer Nature Singapore Pte Ltd. 2020
P. K. Mallick et al. (eds.), *Cognitive Informatics and Soft Computing*,
Advances in Intelligent Systems and Computing 1040,
https://doi.org/10.1007/978-981-15-1451-7_27

identification of a person who depends on the physiological and behavioral characteristics of individuals. As is known, the behavioral property is a reflection of the physiological composition of the individual such as gait, signature, and speech patterns while the physiological property is a relatively stable physical property such as the face and fingerprints, iris, signatures, and sounds. Nowadays iris, face, signature, and fingerprint biometric are from the best authentication system. These systems are selected depended on the security level required and type of biometric is used in the system to fulfill the desired requirements.

Iris recognition system is from the most significant systems for verification and identifying individual. The iris is a thin circular structure of color in the eye responsible for controlling the diameter of pupil size and thus the amount of light that reaches the retina. The iris pattern is highly unique. In addition, two irises are not similar even if they are twins [1]. The structure of the iris is fixed approximately one year and remains constant over time and is strong against damage caused by external factors. In addition, it has several properties such as high level of singularity, availability, and easy to use. So that using iris patterns in the authentication technique is suitable to provide high level of security.

The suggested work deals with using the iris patterns for identification and verification the person based on the iris. Both identification and verification procedures are passing through two phases: training phase and testing phase.

2   Related Works

In this section, some previously proposed schemes related to iris recognition will be listed.

Ukpai et al. were proposed a new method of iris features extraction. The proposed method depended on the conversion of dual-tree complex wavelet transform (DT-CWT) and the Principal Texture Pattern (PTP). The evaluating of the principal direction (PD) of the iris tissue is done by via the principal component analysis and obtains the PD angle. Later, complex wavelet filters are created and rotated CWFs in the PD direction as well as in the opposite direction—to analyze the image into 12 sub-domains using DT-CWT [2].

In [3], Umer and Dhara display a rapid segmentation of the iris part from the image of the eye into the iris recognition system. The circular Hough transform is implemented on the inner boundary. To determine the location of the outer border image, the first time is passed by inversion transform after that by using the circular Hough conversion method. The using of circular Hough method in addition to the invention transformation will minimize the search area of the circular Hough transform.

Dhavale in [4] presented an approach for detecting iris boundaries using both Canny Edge Detection and Hough transform. Statistical properties of (DCT) are used for extracting features from iris picture.

In [5], Mohamad et al. suggested a multimodal biometric identification system. In this system, the characteristics and characteristics of the cytoskeleton sequentially in the identification process are combined. In the step of iris encoding, the most discriminating features of the iris region are extracted only and are encoded in a compact form to improve the recognition rate of accuracy. Extraction features of the iris are extracted using a Log-Gabor. The encoding process creates a mono-focal template for the iris region with phase-analysis information.

Radu et al. [6] suggest a scheme to improve multiple sets of 2D Gabor filter parameters for enhancing the iris recognition system. In order to obtain a binary traits from the iris texture, a methodology for finding the 2D Gabor filter bank parameters is employed.

3 The Presented System

The presented consists of two stages: identification and verification. Figure 1 depicts the presented system. The presented system is done by five stages: pre-processing, iris localization, iris enhancement, features extraction, and pattern matching.

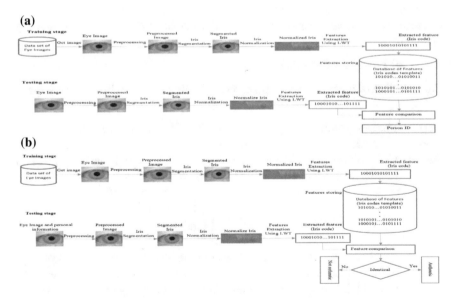

Fig. 1 Proposed system for iris recognition. **a** identification, **b** verification

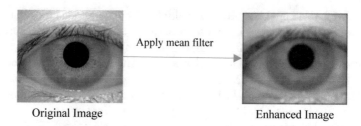

| Original Image | Enhanced Image |

Fig. 2 Pictures enhancement

3.1 Pre-processing Stage

In this stage, the picture is enhanced by using mean filter. In addition, its size is resized to required size. Figure 2 shows the original and enhanced iris images.

3.2 Iris Segmentation Stage

In this stage, an automatic segmentation algorithm was used for localize the iris part in an eye picture by applying two steps detection namely: **inner** and **outer boundary detection** of the iris region.

A. Inner boundary detection

The detection of inner boundary (pupil region) is done by applying a set of steps as follows:

Step 1:   Image smoothing using mean filter with window size (9×9).

Step 2:   Compute histogram of a smoothed image.

Step 3:   Convert the smoothed image to binary. Since the pupil is the largest darkest region. So a threshold is selected based on histogram.

$$\text{Bimg}(i, j) = \begin{cases} 1 \text{ if } O(i, j) \geq t \\ 0 \text{ otherwise} \end{cases} \tag{1}$$

where $O(i, j)$ and $\text{Bimg}(i, j)$ refer to the intensity value at location (i, j) of the original image and the changed image, respectively, and t refer to the value of threshold (this value usually is determined by test).

Step 4:   The pupil region may contain reflection points, so the connected component method is used to extract these points.

Step 5:   Extract the largest region which represents the pupil region.

Step 6:   Compute the center coordinates according to the following equations:

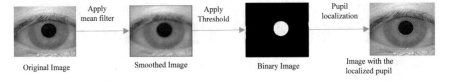

Fig. 3 Steps of inner boundary detection

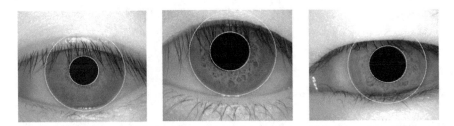

Fig. 4 Examples of iris localization

$$x_p = \frac{1}{N} \sum_{i=1}^{N} x_i, \quad y_p = \frac{1}{N} \sum_{i=1}^{N} y_i \tag{2}$$

Figure 3 shows the above steps for determining the pupil region.

B.   **Outer boundary detection**

The determined parameters of pupil, i.e., radius of pupil R_p, center of pupil (x_p, y_p) are used for localization of outer boundary of iris image through using a circular Hough transform. Figure 4 shows samples of localized iris images.

3.3   Iris Normalization Stage

Due to several reasons such as the size of pupil, lighting and distance from the camera, the iris size different from person to other, even for the same person. The performance of iris matching can be effected by these factors. So, to obtain accurate results, it is important to remove or eliminate the effect of these factors. To accomplish that, the region of the iris is plotted to be flat, rectangular form rather than circular. For this mapping task, conversion from polar coordinates (r, θ) to coordinates (x, y) is applied to each point within the iris region. Equations (3, 4) are used for accomplishing mapping process.

$$I(x, y) \rightarrow I(r, \theta) \rightarrow I(x_f, y_f) \tag{3}$$

Such that,

Fig. 5 Normalized iris image

$$x_f = \frac{NW \times \theta}{2\pi}, y_f = \frac{NW \times (r - R_p)}{R_i - R_p} \tag{4}$$

Figure 5 shows a sample of normalized iris image.

3.4 Feature Extraction Stage

This stage aims at extracting the characteristics of iris picture texture. The distinguishing features of the iris are the basis of comparison (matching) of any two images. The resulting template is an iris code, which is presented as an input to the matching unit. Several methods have been presented for extraction the features. In the proposed method, a lifting wavelet method is exploited on the iris template and generates iris code.

3.4.1 Lifting Scheme Haar Transform

LWT is the second generation of wavelets that rely on traditional wavelets. It is proposed first by Sweldens. The resulting odd element $(j + 1, i)$ is obtained by evaluating the subtracts the existing odd element (j, i) from the predicted (even) element (j, i) according to Eq. (5).

$$\text{odd}_{j+1,i} = \text{odd}_{j,i} - \text{even}_{j,i} \tag{5}$$

The new even element $(j + 1, i)$ is obtained by evaluating the average of the (odd and even values):

$$\text{even}_{j+1,i} = \frac{\text{even}_{j,i} + \text{odd}_{j,i}}{2} \tag{6}$$

For the next recursive step, the averages (even elements) will be the input for the forward transform as illustrated in Fig. 6.

3.4.2 Iris Code Generations

To extract the feature (iris code) from iris, the following steps are done:

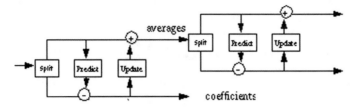

averages

coefficients

Two steps in the wavelet Lifting Scheme forward transform

Fig. 6 Predict and update lifting steps block diagram

001111001001000000000000000000000101101000001000110100100100001011100000100011111111111
010111111110110000010010000000110100001010100111010111111111111111111111010111101011010
111111101000011000111111110111111111111111111111110111101111111101111110111111111011111
111101111111111111111111111111111111101111110111111111101101111001100101111111111111111111
111011001101001011111111111111101011000111101001111111111111111111111010000001101111111001
011000001101110111000001111111111111111111111111110000000000000000011111110000000000000010

Fig. 7 Example of iris code

Step 1: Lifting wavelet transform is applied on the normalized iris image by decomposing the picture into four subbands. The first one (LL) represents the approximation picture which processed in the next step (level). The second one (LH) represents the horizontal. The third one (HL) represents the vertical. The fourth one represents the detail components.

Step 2: Lifting wavelet transform is applied 3 times on the 64×512 iris image in polar coordinates to obtain the 8×64 sub-images (this means 512 features).

Step 3: Compute the average of (LH3 and HL3) subbands. Also, the average values for the two high-pass filters areas namely HH1 and HH2 are computed. The length of the resulting feature vector is 514. Each value of 514 has a real value (between -1.0 and 1.0). By applying quantization to these real values of the feature vector into binary form by using average of (LH3 and HL3) subbands as a threshold value. Figure 7 shows an example of iris code generation.

Finally, all featured from enrolled iris images from different persons are saved as templates in the database of features.

3.5 Feature Comparison Stage

The comparison of iris code metric means computing of a Hamming Distance (HD) the distance between the iris features code extracted from the input iris image and template of iris code stored in the databases. If the input iris code is X and iris code template in the database is Y, the difference between these two codes will be computed by using the Eq. (7):

$$\text{Humming Distance} = \frac{\sum_1^{514} X \text{ xor } Y}{514} \tag{7}$$

4 Result of the Experiments

Iris images obtained from CASIA V1.0 iris image database which are used to measure our system performance. The proposed system was applied on iris images for 98 person and each person has seven images (total 686 images) divided into five images for training and two images for testing.

4.1 Identification Stage

In this stage, the comparison is done between presented iris image and all templates of iris code saved in the database. Then, return the person information of the person forms the database which has minimum distance. The percentage of correct classification is 98.46%.

4.2 Verification Stage

Verification stage will have same procedure as identification stage except feature comparison stage. In the identification stage, test feature vector is compared with all feature vectors stored in a database. Whereas in the verification phase, the test feature vector is compared with only the template feature vector of the person who claimed stored in a database. In verification stage, the name of user is known, so that the matching is done one by one, the presented iris code is compared only with iris code of that person stored in the database, if the computed distance is less than the determined threshold, the system verifies that the user is authentic.

4.2.1 Determination the Optimal Threshold for Verifying Stage

The pictures for the same person do not match 100% since these pictures were captured under different status. So, that there is need to determine the threshold in order to achieve the best values for performance parameters such as False Positive Rate (FPR) and True Negative Rate (TNR). In order to find the values of TPR and TNR, the distances between the code of the presented iris and iris codes that are

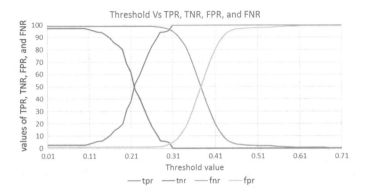

Fig. 8 TPR, TNR, FPR, and FNR values for the suggested system

stored in the database for the same person are computed then determine the range of threshold between (average of these distances -0.05, average of these distances $+ 0.05$) and if the distance less this threshold, it means that two codes are match and TPR incremented otherwise TNR incremented. To find the values of False Positive Rate (FPR) and False Negative Rate (FNR), the distances here are computed between the presented iris code belongs to a user with all iris codes in the database to the other users and if the distance less this threshold, it means that unauthorized user is classified as authorized one and FPR incremented otherwise mean unauthorized user is classified as unauthorized one and FNR incremented. When the value of threshold is very low, the TNR is increasingly high. With increasing the threshold, FRR is decreased, and overall accuracy is improved. Figure 8 shows the dependency between the four verification rates on the similarity distance threshold value. It can be noted that the optimal threshold is 0.31 that gives best TPR and TNR.

5 Summary/Conclusion

A method for iris recognition using lifting wavelet transform features has been proposed in our paper. Wavelet transform has proved to be good iris representation media to provide discriminating features lead to good recognition accuracy. In iris localization step, the experimental results show that the suggested method achieved a well results. The mechanism of generated iris code by quantizing lifting wavelet coefficients achieved good results in terms of recognition. Weighted Euclidean distance is used for matching operation. It was found that the suggested method achieved a good results and this proves that the accuracy of each stage affects the accuracy of recognition rates.

References

1. Agarwal, V., et al.: Human identification and verification based on signature, fingerprint and iris integration. In: 2017 6th International Conference on Reliability, Infocom Technologies and Optimization (Trends and Future Directions) (ICRITO). IEEE (2017)
2. Ukpai, C.O., Dlay, S.S., Woo, W.L.: Iris feature extraction using principally rotated complex wavelet filters (PR-CWF). In: 2015 International Conference on Computer Vision and Image Analysis Applications (ICCVIA). IEEE (2015)
3. Umer, S., Dhara, B.C.: A fast iris localization using inversion transform and restricted circular Hough transform. In: 2015 Eighth International Conference on Advances in Pattern Recognition (ICAPR). IEEE (2015)
4. Dhavale, S.V.: DWT and DCT based robust iris feature extraction and recognition algorithm for biometric personal identification. Int. J. Comput. Appl. **7** (2012)
5. Elhoseny, M., et al.: Cascade multimodal biometric system using fingerprint and Iris patterns. In: International Conference on Advanced Intelligent Systems and Informatics. Springer, Cham (2017)
6. Radu, P., et al.: Optimizing 2D gabor filters for iris recognition. In: 2013 Fourth International Conference on Emerging Security Technologies (EST). IEEE (2013).

Classification of Abusive Comments Using Various Machine Learning Algorithms

C. P. Chandrika and Jagadish S. Kallimani

Abstract In the past decade, we have seen an increasing surge in the popularity of social networking sites, with Twitter and Facebook being some of the most popular ones. These sites allow you to share your expressions and views. However, as of now, there is no particular restriction applied by them to control the kind of content that is being uploaded. These uploaded contents may have obnoxious words, explicit images which may be unsuitable for social platforms. There is no predefined method for restricting unpleasant texts from publishing on social sites. To solve this problem, we propose a method that can aid human moderators as well as work independently. In this approach, logistic regression, multinomial Naïve Bayes, and random forest techniques are used to extract features like term frequency–inverse document frequency features, text features, and frequency features of the comments, respectively, to obtain a weak prediction model. Gradient boosting is applied to this model to obtain the final prediction model. We also applied a neural network using bidirectional long short-term memory and compared the accuracy rate of the two models. We believe that these models can help human moderators on various online platforms to filter out abusive comments.

Keywords Logistic regression · Multinomial Naïve Bayes · Random forest · Gradient descent rule · Long short-term memory

1 Introduction

We have tried to develop prediction models, which will help a comment moderator to find comments that are potentially abusive to a person or an organization using machine learning. Today, on many news feed and social media, users post thousands

C. P. Chandrika (✉) · J. S. Kallimani
Department of Computer Science and Engineering, M. S. Ramaiah Institute of Technology, Bangalore, India
e-mail: Chandrika@msrit.edu

J. S. Kallimani
e-mail: jagadish.k@msrit.edu

© Springer Nature Singapore Pte Ltd. 2020
P. K. Mallick et al. (eds.), *Cognitive Informatics and Soft Computing*,
Advances in Intelligent Systems and Computing 1040,
https://doi.org/10.1007/978-981-15-1451-7_28

of comments in matter of seconds. Web sites may have their own content policy, and to adhere with them, it may get necessary to have a moderator who may restrict the display of comments.

Flagging of abusive comments on online platforms has been around for a long time. The comments were filtered out by site owners or moderators who had to constantly look out for incoming comments put by the user. As the online community grew, it became an extremely tedious job to do with thousands of comments coming in every second. The only way that this could be handled was to add more moderators or automate the process using computers. In the recent years, machine learning has gained traction and it is now being used to deal with this problem. It can be trained for a wide range of words and phrases and even add multi-lingual filtering. Several classification algorithms and neutral network can be used to do the required task. This can be deployed to either aid the moderators or even work as an independent system with a high accuracy rate.

2 Background Work

Some of the research works in this area are summarized as follows.

Developing tools for validating whether a verbal aggression offense report refers to a real offense or not was discussed in [1]. They have analyzed player behavior of online social game, called Okey. Bayes Point Machine (BPM) was adopted as primary classifier. Cyberbullying using supervised learning techniques were discussed in [2] and authors focus on two feature extraction algorithms based on capturing pronouns and skip-grams. A distributed low-dimensional demonstration of remarks using neural language models was discussed in [3]. Authors have discussed paragraph2vec and words with lowest text embeddings approaches in their work.

Detection and prevention of offensive words to get published on social media using a string pattern matching algorithm called AHO-Corasick is discussed in [4]. Cyberbullying on the Twitter was recognized by an emergent machine algorithm based on set of unique features like network, activity, user and tweet content was discussed in [5]. A private attack at scale is considered in [6], the authors proposed a method that includes both crowdsourcing and machine learning techniques. A method of identification of hate speech that was published online is discussed in [7]. This approach includes a development of corpus of user comments annotated for abusive language. In [8], the authors wanted to determine whether a comment should be moderated. Two baselines—best-fitting straight line with a least squares loss function and feedforward neural network classifier baseline are used in their work. In [9], research focused on developing a methodology for accumulating text, user, and network-based attributes.

Abusive comments uploaded on the Twitter can be detected by single-step and double-step classification methods that are based on CNN-based models are explored in [10]. A Mallet Learning for language routines to assess the performance of various classification algorithms are discussed in [11]. Tests conducted on these three different data sets—English tweets, German tweets, and Wikipedia talks.

3   Design

The design part is one of the most crucial parts of this proposed work. The key is to keep the design simple yet functional and incorporate all the intended functionality under a simple scalable design. The following modules constitute Architecture module:

- Data collection and reading: For both the models, the comment data is taken from an open-source dataset and kept under an extension called comma separated values (.csv) which a popular extension for keeping data for use in a script. Care should be taken while considering units from various standards, as taking the unit of a particular value in SI and another in CGS might give us bogus results.
- Feature extraction: From the dataset, features are extracted which will be later used in implementing various algorithms. For the first model, features directly extracted while for the second model, the features are extracted as word embeddings.
- Classification: We obtained a weak model after applying algorithms like LR, MNB, and RF for the extracted features. To this model, gradient boosting is applied to get the final refined model. In the second model, bidirectional LSTM is applied with dropouts with two different activation methods for different density equations.
- Results: For both the models, the conclusive results are obtained and tabulated which is shown in the result section.

4   Implementation

Due to its simplicity, its object-oriented nature, combined with its ability to support multiple programming paradigms, Python is the programming language generally preferred for the development of prediction models. Classification was done in two phases/models.

4.1   Phase I Classification Using LR, RF, and GDR

Phase I begins with reading the train data stored in .csv file. The features like text, frequency, and TF-IDF. After extracting the features, the dataset is trained using— LR, MNB, and RF algorithms, and this leads to weak prediction model because accuracy is less so to improve it the same dataset is classified using GDR algorithm. Using GDR accuracy is improved up to 70%. The same steps are repeated for test data also. The details of this algorithm implementation are as follows.

(a)   Logistic regression: LR basically uses two sets of variables—an independent and a dependent one, where the independent variable is used to estimate the value of the dependent variable. The plot obtained using the logit function provides us with the probability estimate. However, to obtain the coefficients, the logit function needs to be optimized using the standard likelyhood and log likelyhood functions. The second derivation is called the Hessian matrix. The second derivative of the log likelihood function which along with the derivative is considered and the following equation is obtained on which we can apply Newton–Raphson method to get the optimal solution (Eq. 4.1).

$$\Delta_k = \left(x w_k x^t\right)^{-1} x (y - p_k) \qquad (4.1)$$

In order to tokenize an assemblage of text documents and to construct a vocabulary of familiar words, an effortless approach is delivered by the CountVectorizers. For the frequency of each word that has transpired in the document, an integer count was returned along with an encoded vector comprising the extent of the complete vocabulary.

(b) Multinomial Naïve Bayes: Multinomial Naïve Bayes is a stratification approach that assumes conditional independence, i.e., the affect of one attribute is independent of affect on other attributes. Using the Eq. 4.2, posterior probability which is the prior requirement can be found:

$$P(c|x) = \frac{\left(p\left(\frac{x}{c}\right) p(c)\right)}{p(x)} \qquad (4.2)$$

To do the classification, three simple steps are followed—(1) Convert the dataset as a frequency table. (2) Form a likelihood table by discovering the probabilities. (3) Implement the above equation in order to compute the posterior probability for every class. The given tuple from the dataset will be assigned with the class label which has the maximum posterior probability proposes by the given class label in the dataset. Here, term frequency–inverse document frequency of sample comments is determined. The Tf–idf Vectorizer is said to tokenize documents, comprehend the vocabulary along with inverse document frequency weightings.

Random forests: Random forest is a technique in machine learning that ensembles the ventures of both regression and classification. It is a bagging method that constitutes several learning models to frame an enhanced model. Numerous supplemental concerns that the random forest addresses include dimensional reduction techniques, handle missing values, outlier values, additional vital procedures of data exploration. In random forest, the feature extraction function extracts all possible features which are encoded as an integer. The random forest algorithm takes these 7 binary digit array and give the corresponding output after performing supervised learning similar to the decision tree.

(c) Gradient boosting: Machine learning employs gradient boosting to aid in classification and regression problems. The decision trees, by way of explanation; the weaker prediction models, structures a ultimate prediction model by gradient boosting technique. Similar to other boosting techniques, the model is constructed in a stagewise architecture and later conceptionalized by optimizing an arbitrary differentiable loss function. It is found to have two strategies to boost the precision of a predictive model. It can be achieved by engulfing feature engineering or promptly seeking boosting algorithms. Boosting algorithms can be categorized to be gradient boosting, XGBoost, AdaBoost, Gentle Boost, etc. Every algorithm has its own underlying mathematics and a slight variation is observed while applying them. To increase the accuracy, we assign weights to the observations. The steps to do this can be summarized like this—start with a uniform distribution assumption. Call it as $D1$ which is $1/n$ for all n observations (1) assume an alpha(t). alpha is simply the weights value. It can be calculated with the following Eq. 4.3:

$$\alpha t = \frac{1}{2} \ln\left(\frac{1 - \varepsilon t}{\varepsilon t}\right) \tag{4.3}$$

(2) Get a weak classifier $h(t)$. (3) Change the population distribution with new value for the next step using Eq. 4.4:

$$Dt + 1(i) \equiv \frac{Dt(i) \exp(-\alpha_t y_i h_t(x_i))}{Zt} \tag{4.4}$$

(4) Employ fresh population distribution to again perceive a new learner. (5) Iterate from Step 1–4 up until a surpassable hypothesis is no longer procurable. (6) At the hand of all the learners used till now, a weighted median of the frontier is attained. This technique is deployed to effectuate top efficiency and boost the learners rapidly. Equation 4.5 to calculate Z is given below

$$Zt = \sum_{i=1}^{m} Dt(i) \exp(-\alpha_t y_i h_t(x_i)) \tag{4.5}$$

4.2 Phase II: Neural Networks (Long Short-Term Memory Cells or LSTM)

"Memory" in LSTM, the three gates sparingly can be considered to be a "conventional" artificial neuron, implying a multi-layer also going by the name feedforward neural network. By means of which, with the usage of an activation function, an activation of a weighted sum is enumerated. Connections are planted between these gates and the cell.

Here, the output is calculated at each node of the hidden and the output layers by calculating the sum of weighted input ($w_i * x_i$). A best fit nonlinear hypothesis can be achieved by nonlinear transformation by using the output calculated at each node of hidden and output layers. The below illustration portrays the steps in the implementation of LSTM. For both training and testing, embedding size of 128 has been taken. This makes sure that in input layer of arbitrary length, for each node, we give a vector of size 128. These vectors consist of real numbers which are mapped to higher dimension in a continuous vector space.

Bidirectional recurrent neural networks (BRNN) focus on maximizing the quantifiable input data existing in the network. BRNNs work on variable data.

The current state of the network is used to access the future input data.

5   Results

For both the classification models, the comments data were took from an open-source Web site kaggle and is known as a popular dataset provider for research and competitions. For training and testing, 1.6 Lakhs and 6000 comments were taken, respectively.

From the dataset, features are extracted which will be used in implementing various algorithms. For the first classification model, features are extracted by getting those words, which fall under seven categories like personal reference, common insults, common swear words, exaggeration, words with all letters in upper case, and number of words in comments.

Sample example of comments which has the above features:

Another dumb comment. Put America first before your idiotic views (common insult and personal reference).

In this work, the identified class labels are toxic, severe toxic, obscene, insult, and identity hate. Attributes are extracted from the comments and classified using the machine learning algorithms into one or more of the above-mentioned group.

The algorithms mentioned in the phase I read text features, frequency features, and TF-IDF features which are extracted from the feature extraction module. The algorithms put the value 1 if the features in the comments match to one or more groups. Even though a comment has some abusive words, whole comment cannot be classified as abusive, following example makes it clear.

My sister, I have never trolled this site. I am here only because I enjoy giving my unbiased opinions, asking questions, sparking open debates, and trying to get a feel of other people's perspective. Please stop calling me a troll because I feel like you are insulting me and I have not called you out of your name at all.

For comments like above, the algorithms put the value 0 for all the groups.

For the second classification model, the features are extracted as word embeddings using Keras library. It requires that the input data in the training set be normalized by removing noise and computes the missing data.

After the classification is done, results are stored in a .csv file which contains the probability values between 0 and 1 in each group against each comment. For

Table 1 Classification using phase I

Id	Toxic	Severe toxic	Obscene	Threat	Insult	Identity hate
00025465	0	0	0	0	0	0
0002bcb3	1	1	1	0	1	0
00031b1e	0	0	0	0	0	0

Table 2 Classification using phase II

Id	Toxic	Severe toxic	Obscene	Threat	Insult	Identity hate
00001	0.99717	0.503591	0.963368	0.105894	0.874642	0.331935
00002	0.000201	1.29E−07	3.36E−05	7.83E−07	8.60E−06	8.81E−07

Table 3 Accuracy rate of the models

Algorithm	Accuracy rate (%)
LR	67.9
MNB	60.4
RF	66.0
GDR	69.7
LSTM	97.8

example, a comment has been classified as insult if value in the insult group has the highest probability than other. The sample Tables 1 and 2 show the result obtained from different classifiers. The accuracy rate in Table 3 indicates that neural network is best suited for natural language processing applications. It has the highest rate due to the effective vectorization technique.

Note: Since the comments taken for testing was very abusive/vulgar and sensitive in nature, ID is assigned to comments.

Figure 1 shows the accuracy rate obtained using various classification algorithms.

6   Conclusion and Future Work

From the comparison we found that for the given dataset, neural network's performance is far better than the other classical machine learning algorithms like MNB, LR, and RF (even after applying GDR). Using the first model, we achieve a score of ~69% after applying gradient boosting. For the second model, the neural network achieves a high score of ~98% accuracy using bidirectional LSTM. Future work can include better webpage UI, better security on the web page, much more robust model, handling of censored and jumbled comments, multi-lingual support, and more.

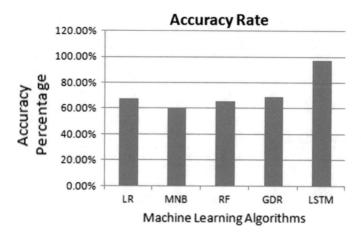

Fig. 1 Chart showing accuracy rate for algorithm

References

1. Balci, K., Salah, A.A.: Automatic analysis and identification of verbal aggression and abusive behaviors for online social games. Comput. Human Behav. **53**, 517–526 (2015)
2. Chavan, V.S., Shylaja, S.S.: Machine learning approach for detection of cyber-aggressive comments by peers on social media network. In: International Conference on Advances in Computing, Communications and Informatics, pp. 2354–2358 (2015)
3. Djuric, N., Zhou, J., Morris, R., Grbovic, M., Radosavljevic, V., Bhamidipati, N.: Hate speech detection with comment embeddings. In: 24th International Conference on WWW '15, vol. 9, pp. 29–30 (2015)
4. Yadav, S.H., Manwatkar, P.M.: An approach for offensive text detection and prevention in social networks. In: International Conference on Innovations in Information, Embedded and Communication Systems, pp. 1–4 (2015)
5. Al-garadi, M.A., Varathan, K.D., Ravana, S.D.: Cybercrime detection in online communications: the experimental case of cyberbullying detection in the twitter network. Comput. Human Behav. **63**, 433–443 (2016)
6. Wulczyn, E., Thain, N., Dixon, L.: Ex machina: personal attacks seen at scale. In: Proceedings of 26th International Conference on World Wide Web—WWW '17, pp. 1391–1399 (2016)
7. Nobata, C., Tetreault, J., Thomas, A., Mehdad, Y., Chang, Y.: Abusive language detection in online user content. In: 25th International Conference on World Wide Web—WWW '16, pp. 145–153 (2016)
8. Chatzakou, D., Kourtellis, N., Blackburn, J., De Cristofaro, E., Stringhini, G., Vakali, A.: Mean birds: detecting aggression and bullying on twitter. In: Proceedings of ACM Web Science Conference—WebSci '17, pp. 13–22 (2017)
9. Park, J.H., Fung, P.: One-step and two-step classification for abusive language detection on twitter. In: Proceedings of First Workshop on Abusive Language, pp. 41–45 (2017)
10. Bourgonje, P., Moreno-Schneider, J., Srivastava, A., Rehm, G.: Automatic classification of abusive language and personal attacks in various forms of online communication. In: LNAI, vol. 10713, pp. 180–191 (2018)
11. Yadav, S.H., Manwatkar, P.M.: An approach for offensive text detection and prevention in social networks. ICIIECS, pp. 1–4 (2015)

Simulation of Web Service Selection from a Web Service Repository

Yann-Ling Yeap, R. Kanesaraj Ramasamy and Chin-Kuan Ho

Abstract This paper studies the dynamic Web service selection methods and simulates it with the same dataset to show the performance of each method. Quality of service (QoS) in Web services composition has various non-functional issues such as performance, dependability, and others. As more and more Web services become available, QoS capability is becoming a crucial factor in selecting the services. We observe the performance of each method and also the recommended hybrid methods by the selected authors. As a result of observation, we have proposed a new hybrid method which can produce better performance (better processing time) in Web service selection. The new approach is proposed based on particle swarm optimization.

Keywords Web service · Web service selection · QoS attributes

1 Introduction

Currently, nowadays, Web services play a vital role in the world of network such as the Internet. The responsibility of Web services is handling the execution of the business process. With the quick growth of developing Web services, there exist similar functional attributes of Web services. Although the functional attributes of Web services may be the same, the quality of Web services may be different.

Intending to select the best Web services among the Web services registry, Web service selection is a process used to choose the best Web services based on the quality of service (QoS). In the Web service selection process, different techniques or methods can be found from the Web service repository applied in the selection process.

Y.-L. Yeap · R. K. Ramasamy (✉) · C.-K. Ho
Multimedia University, Persiaran Multimedia, 63100 Cyberjaya, Selangnor, Malaysia
e-mail: r.kanesaraj@mmu.edu.my

© Springer Nature Singapore Pte Ltd. 2020
P. K. Mallick et al. (eds.), *Cognitive Informatics and Soft Computing*,
Advances in Intelligent Systems and Computing 1040,
https://doi.org/10.1007/978-981-15-1451-7_29

2 Literature Review

Web services are modular, self-aware, and self-describing applications [1]. Web services identify what functions it can execute and what kinds of inputs it needs to generate the outputs, and be able to report this to potential users and other WS.

Web services can be described as building blocks for making distributed applications that can be published and accessed through a network. Developers are allowing to implement distributed applications to produce corporate applications that combine available existing software modules from the system in differing organizational departments or from a different company. Besides, Web services can be combined and work together from distributed applications to perform substantially any business-related transaction.

Intending to select a service among the Web services registry, Web service selection is a process used to select the best Web service based on the quality of service (QoS). In the Web service selection process, different techniques or methods can be found from the Web service repository applied in the selection process. Different selections have their consideration matter, but how many of them from the selection methods concentrate on the selection process performed in term of computation time. In this paper, we simulated three existing Web service selections which authors make mention of those selection show lesser computation time.

2.1 Service-Oriented Architecture

The concept of SOA is enabling services to compose to application integration. Service-oriented architecture (SOA) is the construction of service that provides communication crosswire over various platforms and language by providing implementation loose coupling system so these existing services can be reused and compostable [2].

2.2 Functional and Non-functional Properties

Web services are described in term of description language. There are two major components in description language: functional and non-functional features. Non-functional properties define the service quality attributes and describe how services behave in terms of quality such as performance, availability, integrity, reliability, and so on [1].

Table 1 QoS definition used

Availability	The quality aspect of determining how likely the Web service is available to use when it is needed. A higher value of availability reflects on the longer time in use of the Web service [5]
Accessibility	The degree of capable of serving a Web service request. It may be expressed as the probability of success rate to achieve the Web service [5]
Integrity	The degree used to describe the accuracy of Web service sustains the exactness of the communication regarding the source. Aims to this, the Web service transactions need to be proper execution to deliver the correctness of interaction [5]
Performance	Two elements are used to measure Web service performance: throughput and latency [5]

2.3 Quality of Service

Quality of service (QoS) is known as non-functional properties. The importance of QoS properties is used to differentiate among the same functionally of Web services [3]. Papazoglou [4] has stated that "QoS is an important criterion that decides service usability and utility." Table 1 shows the QoS attributes which is being used in this paper (Table 2).

2.4 Web Service Selection Methods

In this section, we have studied a couple of Web service selection methods and reviewed those methods on its advantage and disadvantage.

2.5 Method Analysis

2.5.1 Analytic Hierarchy Process (AHP)

AHP is a multi-criteria decision making-method to solve complex decision problems [7]. AHP uses a multi-level hierarchical structure. The top of the hierarchy is defining the goal, and following criteria and the result are alternatives ways. The computation is making to find out the weightage and rank the alternatives. AHP is a matrix-based structure. Moreover, AHP dealing with integer and fractional numbers because of each value of matrix are diagonal and mathematical inverses of each other.

Table 2 Web service selection methods

Selection method	Explanation
TOPSIS algorithm	TOPSIS algorithm is a method that accounts user's desire and the non-functional attributes of services. In favor of delivering the best service for customers, the first thing required to determine the desires of users and then to use every possible QoS metrics of Web services to choose the best available service [1]
Dealing with user constraints in MCDM	In this approach, value constraints will be imposed into QoS parameters when it comes to the Web service selection. The resulting space might be insignificant when WS that do not fulfill user's constraints is eliminated from the search space. However, the eliminated Web services can provide an exciting solution to the end user in terms of performance [6]
An efficient approach	A user may find out the Web service that fulfills their desired functional requirements. Aims to this, QoS properties or known as non-functional properties will be used in the selection process, to choose the best candidate for Web service [2]
PROMETHEE	This approach provides a solution in the end user satisfaction by inclusive the end user request during service evaluation and handling the different QoS parameters. Further, the use of a classification-based profiler layer advances the performance of WS selection and cut down overall of selection period [3]

2.5.2   VIKOR

VIKOR method used to solve discrete decision issue with non-commensurable (different units) and different criteria [6]. VIKOR concentrates on finalize ranking and choosing the solution from a variety of alternatives. This method allows compromise for conflict resolution since the decision maker desires a solution that is nearest to the ideal.

2.5.3   Reference Ideal Method (RIM)

RIM is introduced to select the ideal solution as a value that can be within the maximum value and minimum value, without being strictly the extreme values of criteria. This method concept comes from TOPSIS and VIKOR method which these calculated values are absolute values, and this will lead to the rank reversal problem. Rank reversal problem will occur when inserted or deleted one or more alternatives [8]. But the rank reversal problem does not happen on the RIM method.

2.6 Theoretical Framework

We have used three methods which consist of the hybrid methods of above.

2.7 Research Contribution

A set of Web service was prepared with the variable which was used for this simulation for all three methods as shown in Table 3, selected which contains 200 services, and each service consists of five QoS parameters, namely reliability, accessibility, integrity, performance, and security. Each service includes a list of random value on QoS attributes. The end of result is to calculate the total time taken used to complete the selection process. The measurement of time used is second (s).

Three (3) separate models are created for each model used to implement a selection. The first selection approach is "Large-Scale Transactional" which consists of skyline and ACO. The second selection approach is "An Efficient Approach" which is a combination of the skyline, AHP, and VIKOR. The third selection approach is PSO.

After executing the simulation, we had observed the variation from the selection method A as shown from Table 3, it is shown the skyline will go through dot-by-dot starting from the bottom right toward the top left. The processing time of skyline can be very long if the number of services is enormous. Furthermore, the ACO shown the poor execution time due to ants randomly move around and search service. Besides, the simulation of selection method B is mathematical methods which means the solving process involves calculation. The simulation result was showing in the numerical digit (Fig. 1).

The result of service selected after executed selection three is not good as previous selection, but processing time shows it was better than selection method A. After fewer time implements the selection method C, the minimum number of services been found by particles is 1. The experiment is conducted five times for each selection, and Table 4 shows the average time taken for each selection methods.

Table 3 Hybrid methods used for simulation	Selection method	Types of method
	A	Skyline and ant colony optimization (ACO)
	B	Skyline, analytic hierarchy process (AHP), and VIKOR
	C	Particle swarm optimization (PSO)

Fig. 1 Selection method A; test 1

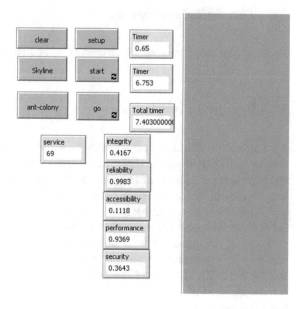

Table 4 Time execution of selections

Timer (ms)	Selection method A	Selection method B	Selection method C
Test 1	7.403	1.165	4.854
Test 2	6.387	0.979	4.854
Test 3	6.784	1.302	5.056
Test 4	5.922	1.437	4.945
Test 5	6.388	1.159	4.856
Average time	6.577	1.208	4.913

2.8 Proposed Solution

Particle swarm optimization is a suitable selection method to solve a selection problem with concentrated on time execution. But it is better to select a service including user constraint to focus on which QoS attributes user desires. Use of threshold value to control the selection service has the better desired QoS value. Equation 1 shows the calculation of the threshold value.

$$\text{Threshold Value} = \frac{\text{Sum of Desired QoS attributes in the WS class}}{\text{Number of service on the WS class}} \quad (1)$$

In the beginning, the user required to select desired QoS attributes from the chooser. Next, calculate threshold value based on the formula shown in Fig. 2. After computing the threshold value, apply particle swarm optimization. First, randomly

Fig. 2 Flowchart of particle swarm optimization (PSO)

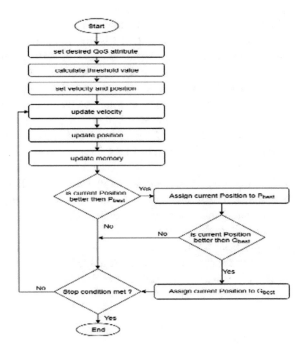

set velocity and the position for the swarm within a range. Second, update velocity for each swarm, this is used when the personal best or global best had been found, and the swarm will move to global best, and keep searching for any better position. Third, update the position if there found any personal best or global best is better than the current position. Fourth, upgrade the memory to all swarm. At this stage, every swarm will check whether the current position is better than personal best if the current position is better than personal best then allocate the current position to personal best. If current personal best had updated to the current position, then check also whether the current position is better than a global position. Repeat the second step to the fourth step until certain conditions are met such as iteration was reached in maximum value.

Table 5 shows the five times the execution of the proposed solution.

Table 5 Time execution of proposed solution

Timer (ms)	Proposed solution
Test 1	2.127
Test 2	1.217
Test 3	0.110
Test 4	0.812
Test 5	1.112

After executing five tests as shown in Table 5 of the proposed solution, the result of every test indicates time taken is not an approach to each other. Based on the observation of experimenting, any considered services are near to any particle swarms from the initial random setup will be comfortable and quickly selected. The result of Test 3 is an example of regarded service and is close to a particle swarm. Moreover, the result of Test 1 indicates the longest time taken used to WS selection.

There is a difference between PSO and the proposed solution. The process of PSO is a looping function, and a stop condition is needed to terminate the looping and to report the obtained result. In the simulation of selection 3, each particle has 50 times of movement to searching possible service among the search space. In other words, the looping function will stop once the task has executed for 50 times. A case may have happened when the loop has reached the limitation, and none of any swarm has found a service. PSO does not guarantee the selected service is the best service. While the proposed solution introduces the use of the threshold to choose more potential service, compare to selection 3. The looping function will stop once any of the swarms found the desired QoS attributes are more significant than the threshold value, and the service is selected as a result. Figure 3 shows the initial setup of proposed solution.

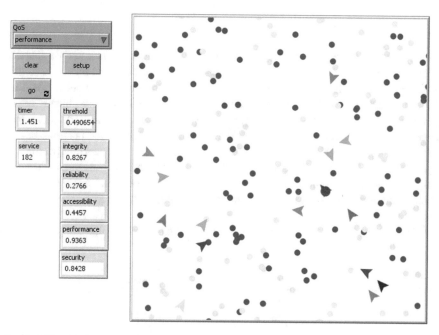

Fig. 3 PSO with threshold value interface

3 Conclusion

Web service selection is the process of selecting service from Web service discovery. The process will be proceeding when service requester is invoking the desired service on the Web.

Three approaches were selected to simulate and test the processing time for the selections, then rank those approaches based on the result of processing time. After simulating, investigate and rank those selection approaches based on processing time. The result shows the first ranking is an efficient approach, second ranking is PSO approach, and third ranking is large-scale transactional approach.

After the simulation, I proposed a solution which the selection is based on PSO with including user constraint to focus on which QoS attributes user desire. The processing time of the proposed solution is better than three simulated selections.

References

1. Belouaar, H., Kazar, O., Rezeg, K.: Web service selection based on TOPSIS algorithm. In: 2017 International Conference on Mathematics and Information Technology (ICMIT), pp. 177–182 (2017)
2. Serrai, W., Abdelli, A., Mokdad, L., Hammal, Y.: An efficient approach for Web service selection. In: 2016 IEEE Symposium on Computers and Communication (ISCC), pp. 167–172 (2016)
3. Purohit, L., Kumar, S.: A classification based web service selection approach. IEEE Trans. Serv. Comput. (2018)
4. Papazoglou, M.: Web Services: Principles and Technology. Pearson Education (2008)
5. Anbazhagan Mani, A.N.: Understanding quality of service for web services. https://www.ibm.com/developerworks/library/ws-quality/index.html. Accessed 01 Jan 2018
6. Serrai, W., Abdelli, A., Mokdad, L., Serrai, A.: Dealing with user constraints in MCDM based web service selection. In: 2017 IEEE Symposium on Computers and Communications (ISCC), pp. 158–163 (2017)
7. Saaty, T.L.: A scaling method for priorities in hierarchical structures. J. Math. Psychol. 15(3), 234–281 (1977)
8. Cables, E., Lamata, M.T., Verdegay, J.L.: RIM-reference ideal method in multicriteria decision making. Inf. Sci. 337 (2016)

Optimized Algorithm for Restricted Governor Mode of Operation of Large Turbo Generators for Enhanced Power System Stability

A. Nalini, E. Sheeba Percis, K. Shanmuganathan, T. Jenish and J. Jayarajan

Abstract The Central Electricity Regulatory Commission, India has guided that every Coal/lignite-based thermal generating units with more than 200 MW capacity should adopt restricted governor mode of operation (RGMO) for effective compensation of sudden frequency changes in the power grid. Presently, restricted governor mode of operation algorithm, developed by various boiler turbine generator suppliers, is used to meet the expectations of CERC. In this paper, it is discussed about an optimized algorithm for the effective implementation of RGMO for enhanced power system stability of large turbo generators of 600 MW or more capacity, fitted with electronic governors.

Keywords RGMO · RFGMO · FGMO · Frequency regulation · Speed droop

1 Introduction

The operational discipline of Indian power grid is getting improved in a sustained manner by the combined efforts taken by the state and central electricity statutory bodies. As an advancement measure, the CERC has introduced dual-band governor mode of operation with RGMO as inner band and FGMO as outer band for enhanced power grid system stability. This consistently restricted band of frequency regulation concept is the brain child of honorable CERC which does not exist anywhere else in the world. The advocated RGMO is implemented using distributed control system software of the BTG supplier in modern Indian power plants [1]. In this paper, an optimized algorithm for the realization of RGMO is presented which can be implemented by using the contemporary DCS platforms along with electro hydraulic turbine controller (EHTC) [2].

A. Nalini (✉) · E. Sheeba Percis · T. Jenish · J. Jayarajan
EEE Department, Dr. M.G.R. Educational & Research Institute, Maduravoyal, Chennai, India
e-mail: nalinitosiva@gmail.com

K. Shanmuganathan
AEE/Transmission, Tamil Nadu Transmission Corporation Ltd, Chennai, India

© Springer Nature Singapore Pte Ltd. 2020
P. K. Mallick et al. (eds.), *Cognitive Informatics and Soft Computing*,
Advances in Intelligent Systems and Computing 1040,
https://doi.org/10.1007/978-981-15-1451-7_30

273

2 Restricted Governor Mode of Operation

2.1 Need for the RGMO

Power systems in real time require maintaining the requisite instantaneous and continuous balance between total generation and load. The grid frequency drops from the nominal value when demand exceeds the generation and rises when generation exceeds demand [3, 4]. Fluctuations in frequency will affect the health of the equipment and may lead to power system collapse. Hence, frequency has to be tightly regulated through the combined efforts of generators and system operators. The frequency regulation services are achieved in three levels to maintain the generation and demand balance such as primary control, secondary control and tertiary control.

In the recent past, the frequency profiles of Indian power grid have significantly improved due to the introduction of availability-based tariff (ABT) and frequency remains within 49.0–50.5 Hz [5]. Hence, CERC has made RGMO as a statutory requirement for the thermal generating units of 200 MW and above.

2.2 Speed Droop Characteristics of Governor

The characteristic speed response of turbines for a change in applied load, plotted with load in horizontal axis and frequency in vertical axis is termed as the speed droop characteristics which determines the sensitivity of the governor [6]. The governor with less droop will respond more for a change in turbine speed as shown in Fig. 1.

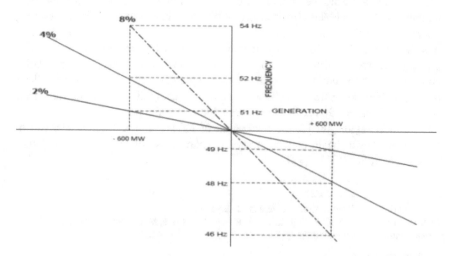

Fig. 1 Speed droop characteristics

Increasing the speed setting will not cause the change in speed of the turbine but it will augment the load supplied by the machine while it is on bar [7]. The speed droop of the steam turbines will be 4–5% and for hydro turbines will be around 2–3%.

2.3 Control Logics of EHTC for RGMO

Reference frequency is a floating frequency which depends upon previous operating frequency between 49.0 and 50.05 Hz. If the frequency changes from the operating frequency, then load changes are made. These changes are restricted to ±5% of maximum continuous rating and maintained approximately for 2 min (Hold Time) [8]. If there is still a frequency variation in this period, hold time becomes zero and count starts again. After this period, the load value changes to the original value following a ramp curve. Small changes in frequency up to ±0.03 Hz is ignored for load variations.

For changing the load, boiler fuel firing has to be changed by giving a signal to boiler master control (BMC) or coordinated master control (CMC). As per the CERC regulations, the ramped fall back to earlier operating load is at the rate of 1% per second. The RGMO output is blocked when frequency is raising from 49 to 50 Hz and frequency is reduced to 50 Hz from 50.3 Hz.

3 General Guidelines of Algorithm for RGMO

The actual speed signal is used to generate a floating frequency set point. The time delayed set point is generated by the actual value of speed [9]. While frequency change happens, the set point also changes to a value equal to the new frequency value in a ramp which will take about 2 min as set time.

If the change of frequency is more than ±0.03 Hz, then a hold signal is generated with a time duration around 2 min which is in direct relation to the change of frequency. This generated signal maintains the floating set point to the previous set value. This is required for effecting the changes in boiler parameters with sufficient duration. A frequency influence signal is generated with a proportional gain corresponding to droop setting by the proportional controller by comparing the set value and actual value. This proportional component will be at full magnitude (+5% maximum) at the beginning and starts reducing as deviation reduce on account of frequency set point slowly reaching the new actual value.

The frequency influence (+5% maximum) is summed to load set point for the formation of the load set point in final of the load controller to increase the load signal. This signal is in addition to the already available pressure dependant load correction. The hold signal is removed once the frequency fluctuations stabilize within ±0.03 Hz [10]. The set point will settle to a new frequency after the removal of hold signal. As the deviation reduces, the proportional frequency influence also starts coming down.

Now, system becomes ready to face next frequency change. If the system frequency is beyond RGMO band (<49 and >50.5 Hz) limits, then FGMO droop characteristics will be effected. The EHTC's pressure controller which is in pressure control mode of operation assures that the boiler header pressure is maintained by reducing the load, if there is a sudden pressure drop (10 kg/cm^2) than the set limit [11].

4 Optimized Algorithm for RGMO

4.1 Derivation of Speed Follow-up Signal

A time lag function is used to compute the speed follow-up signal (Inb) from the actual speed (Ina) of the turbine for a time lag of 60 s for 0–3600 RPM.

4.2 Derivation of Differential and Ripple Components

The actual speed (Ina) is subtracted from the speed follow-up signal (Inb) to compute the speed deviation (Ind) analog value. If the absolute value of the speed deviation (Ind) is greater than the ripple factor, then a digital output (Inc) is set as 'True.'

4.3 Calculation of Dead Band Range

The digital output (Z4) is set as 'True' if the turbine actual speed (Ina) is not in the dead band (i.e. between 2994 and 3003 RPM) and if ripple check factor digital component (Inc) is 'True' as shown in Fig. 2.

Fig. 2 Calculation of dead band

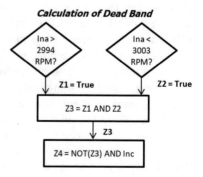

Fig. 3 Calculation of MW response

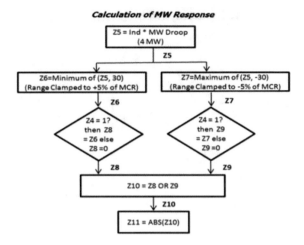

4.4 Calculation of MW Response

By multiplying the speed deviation value (Ind) with the MW value corresponding to droop setting of the turbine, a factor 'Z5' is calculated. The minimum between 'Z5' or maximum positive range response of MW is set as 'Z6,' for clamping the output response to +5% of maximum continuous rating (MCR), and calculated as 'Z6.' The maximum between 'Z5' or maximum negative range response of MW is set as 'Z7,' for clamping the output response to −5% of maximum continuous rating (MCR), and calculated as 'Z7.' The output 'Z10' will be set either 'Z6' or 'Z7' and if turbine actual speed (Ina) is not in dead band range (Z4). The absolute value of 'Z10' is assigned as 'Z11' as shown in Fig. 3.

5 Derivation of RGMO Final Output Response

It is checked, whether 'Z11' is greater than 1 MW and remains 'True' for continuously 5 s. If the above step (a) is 'True,' then the response 'Z10' is released to electro hydraulic turbine controller instantly through a time lag function which will be applied for 300 s in a 1% per second ramp fall down fashion. If the above step (a) is 'False,' then the previous RGMO response will continue through a time lag function which will be applied for 300 s in a 1% per second ramp fall down fashion.

Thus, we are getting machines primary response according to the grid's rate of change of frequency by varying the generators power output through the regulation of entrapped steam input to the turbines. Troubles like hunting, flat response and opposite response are also avoided using this algorithm. The results are depicted in Fig. 4 and Fig. 5, respectively.

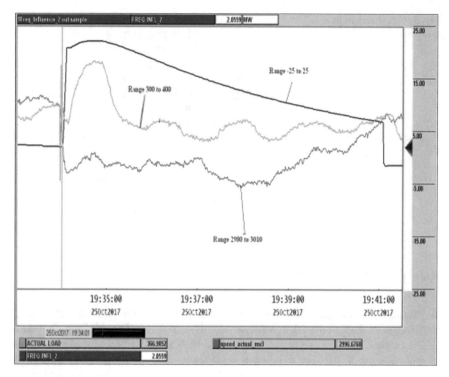

Fig. 4 Result-1

6 Conclusion

To impose grid discipline and for better regulation of the Indian power grid frequency, the CERC has introduced this novel idea of implementing RGMO concept at all significant power generators. The application of RGMO has also shown good impact in the all India average frequency. It is seen as a good symptom since the maximum and minimum frequency fluctuation band has narrowed down drastically in the recent years. The RGMO implemented along with supplementary controls has helped to enhance the overall Indian grid stability by this constant frequency control concept. Suitable retrofits, even though expensive, are suggested for the modification of mechanical hydraulic governors into electro hydraulic governors for the realization of RGMO in old machines. In this paper, an optimized algorithm has been discussed as an implementation tool of RGMO in distributed control systems (DCS) vendor platforms like max DNA©, Invensys Foxboro©, Plant Scape© of large thermal power stations. This optimized RGMO algorithm will be an efficient tool to maintain the frequency range within the desired bandwidth fulfilling the CERC statutory requirement which got tested in modern DCS software environment and the results are presented in this paper.

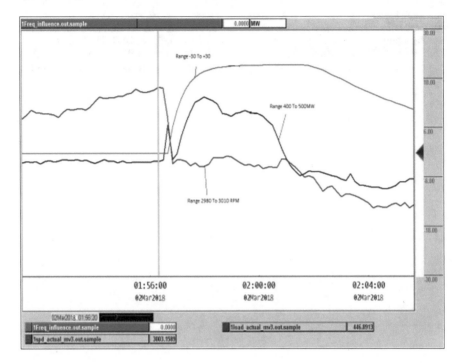

Fig. 5 Result-2

References

1. Sheikh, M.R.I., Haque, M.M., Hossain, A.: Performance of governor system on minimising frequency fluctuations with wind power generation. Int. J. Electr. Comput. Eng. **2**(1), 46–56 (2012)
2. Parida, S.K., Singh, S.N., Srivastava, S.C., Chanda, P., Shukla, A.K.: Pros and cons of existing frequency regulation mechanism in Indian power industry, 1st edn. (2008). IEEE 978-1-4244-1762-9/08
3. Xavier, P.N.V.K., Muthukumar, S.: Frequency regulation by free governor mode of operation in power stations. In: IEEE International Conference on Computational Intelligence and Computing Research (2010)
4. Central Electricity Regulatory: Guidelines of Central Electricity Regulatory a Commission Regulations and Amendments. www.cercind.gov.in
5. WRLDC: Guidelines for Restricted Governor Mode of Operation (RGMO) Test in WR Grid. www.wrldc.in (2010)
6. SRLDC: Details of Restricted Governor Mode of Operation, SRLDC. www.srldc.org
7. Geetha, T., Mala, E., Jayashankar, V., Jagadeesh Kumar, V., Sankaran P.: Coordinated measurements for governor operations in the southern Indian grid. In: 16th IMEKO TC4 (2008)
8. Soonee, S., Saxena, S.: Frequency response characteristics of an interconnected power system—a case study of regional grids in India. In: International R&D Conference on Sustainable Development of Water and Energy Resources—Needs and Challenges (2007)
9. Nalini, S.M., Sheeba Percis, E.: Intelligent identification for wide area monitoring in power system. ARPN J. Eng. Appl. Sci. **10**(20), 9401–9407 (2015)

10. Sheeba Percis, E., Arunachalam, P., Nalini, A.: Modeling and intelligent control of hybrid microgrid in a wide area system. Int. J. Pure Appl. Math. **120**(6), 11437–11446 (2018)
11. Janani, R., Santhiya, K., Swetha, S.R., Anusha, S.: Embedded based control for steam turbine to maintain frequency using RGMO. Int. J. Sci. Res. Sci. Eng. Technol. **2**(2), 161–166 (2016)

Implementing and Evaluating the Performance Metrics Using Energy Consumption Protocols in MANETs Using Multipath Routing-Fitness Function

P. Monika, P. Venkateswara Rao, B. C. Premkumar
and Pradeep Kumar Mallick

Abstract The energy consumption plays a key role in mobile ad hoc networks in a day-to-day life. Mobile ad hoc network (MANET) structure is a temporary network organized dynamically with a possible family of wireless mobiles independent of any extra infrastructural facilities and central administration requirements. In addition, it provides the solution to overcome the minimal energy consumption issues. Nodes are battery-operated temporarily does not operated on permanent batteries, so energy consumed by a battery depends on the lifetime of the battery and its energy utilization dynamically decreases as the nodes change their position in MANETs. Multipath routing algorithm in MANETs provides the best optimal solution to transmit the information in multiple paths to minimize the end-to-end delay, increases energy efficiency and moderately enhances the lifetime of a network. The research mainly focused on minimum energy consumption techniques in MANET is of a great challenge in industries. In this paper, the author highlights a novel algorithmic approach ad hoc on-demand multipath distance vector (AOMDV) routing protocol to increases the energy efficiency in MANET by incorporating the demand multipath distance and fitness function. The ad hoc on-demand multipath distance vector-fitness function (AOMDV-FF) routing protocol short out minimum distance path that consumes minimum energy and the simulation performance is evaluated

P. Monika
Department of CSE, EWIT, Bangalore, India
e-mail: pmounika9596@gmail.com

P. V. Rao
Department of ECE, VBIT Hyderabad, Hyderabad, TS, India
e-mail: raopachara@gmail.com

B. C. Premkumar
Department of ECE, NNRG Hyderabad, Hyderabad, TS, India
e-mail: pramukhatech15@gmail.com

P. K. Mallick (✉)
School of Computer Engineering, Kalinga Institute of Industrial Technology (KIIT) Deemed to be University, Bhubaneswar, Odisha, India
e-mail: pradeepmallick84@gmail.com

© Springer Nature Singapore Pte Ltd. 2020   281
P. K. Mallick et al. (eds.), *Cognitive Informatics and Soft Computing*,
Advances in Intelligent Systems and Computing 1040,
https://doi.org/10.1007/978-981-15-1451-7_31

using network simulator-2 (NS2) tool. Two protocols are proposed in this work AOMDV and AOMDV-FF and compared some of the performance parameters, like energy efficiency, network lifetime and routing overhead in terms of data transfer rate, data packet size and simulation time, etc. The overall simulation results of the proposed AOMDV-FF method is to be considered as a network with 49 nodes and the network performance factor end-to-end delay 14.4358 ms, energy consumption 18.3673 joules, packet delivery ratio 0.9911 and routing overhead ratio 4.68 are evaluated. The results show an enriched performance as compared to AOMDV and AOMR-LM methods.

Keywords Energy efficiency · MANET · Multipath routing · Fitness function · Packet delivery ratio · Throughput

1 Introduction

As the scaling down technology is progressing tremendously in a day-to-day life, in turn, the growth of the computer performance, as well as the scope and applications of wireless technologies, is also drastically increasing. With the advent of new technologies, the research has been mainly focused on minimum power consumption protocol developments and its cost-effective hardware architecture implementation. The awareness of power consumption, investigation on performance, effective routing protocols and more efficient path selection algorithms are major concern to improve the performance of the metrics and also to acquire maximum energy efficiency in MANETs. The performance factors of MANET nodes are having limitations in terms of speed and power, battery, storage and transport mechanisms. Most critical and challenging issues in MANETs are development of routing techniques and also development of routing algorithms. Multipath routing protocols most reasonably and efficiently find the shortest path among the many identified paths to establish the link to transmit the packets information source nodes to destination nodes. It is not necessary that each time the source will relay on only shortest path available to improve the lifetime of the network performance but also other factors, like power and battery life may also influence the performance of the networks. So the saving the power consumed at the source is limited by the number of dependant mobile nodes, the researchers made attempts to provide the minimal power consumption at these nodes which in turn enhances the lifetime of MANETS. There are several critical issues and facts exist in multipath routing protocols developing techniques. The researches focused on novel methodologies and algorithmic approaches that could able to identify the shortest path between the source nodes and the destination nodes. As the number of modes increases, this identity issue becomes much more complicated in MANETs to transfer the data/information in the form of packets. In most of the cases, partial amount of energy is being spent on identifying the optimum path and this being one of the drawbacks or critical issues is trying either to minimize or eliminate during transmission of data. So the part of energy being wasted during identifying the shortest path is given major priority to resolve of this paper. The more energy is being wasted during the data transmission. In this paper, the author identifies the approach to incorporate some of these major issues and to develop an innovative idea ad hoc On-demand multipath distance vector (AOMDV) routing protocol to minimize energy consumption algorithm in MANET with add on functions the demand multipath distance and fitness function too. Ad hoc on-demand multipath distance vector-fitness function (AOMDV-FF) an efficient multipath routing protocol is able to detect the possibility of conceiving nominal power with minimum path distance. The AOMDV-FF uses the fitness function to obtain an optimized solution by considering six parameters to choose the best possible shortest route with minimum energy level of the route and increasing the network lifetime of the path to transfer the packets from the source to the destination more efficiently. From the obtained simulation results as an overall performance, AOMDV-FF routing protocol shows the better performance factors like: throughput, packet delivery ratio, end-to-end delay, energy consumption, network lifetime and routing overhead ratio as compared to both AOMDV and ad hoc on-demand multipath routing with life maximization (AOMR-LM) routing protocols.

The work of this paper is structured as follows: Sect. 2 elaborates the related literature survey of AOMDV with fitness function as major concern and critical issues involved in the current work; Sect. 3 explores with a novel AOMDV-FF method; Sect. 4 highlights some of key factors based on the obtained results and evaluation process adopted; Sect. 5 performs the conclusion based on the obtained results and explores insights to further enhance as a future scope.

2 Background and Related Work

The literature survey explores the new developments and their tremendous utilization methods of Internet protocols (IPs) in mobile ad hoc networks and wireless communications applications mainly depend on their bandwidth limitations.

Taha et al. [1] and Carson and Macker [11] proposed that the networks that utilized in MANETs mainly depend on routing protocols and topologies like multi-hop, dynamic and random types. Recent developments in many routing protocols have been made efforts by researchers to enhance the lifetime of all routes and in turn made to select the best path to increase the network lifetime. Iswarya et al. [2] and Uddin et al. [3] both proposed similar multipath routing protocols to provide best solutions to select the shortest path to transfer information (data) packets efficiently between the source nodes and destination nodes. The same is preserved and used as a backup data whenever the route failures occur and also to provide the better performance without degrading the performance factors.

Due to lack of non-availability of permanent power supplies in mobile energy consumption plays a vital role in the research area as how to minimize power consumption during the transmission of data in MANETs. Ramya et al. [4] proposed a model on how to minimize energy consumption with the routing protocol AOMDV using fitness function and that establishes the shortest route to transfer the data more efficiently between the source nodes and destination nodes. With the help of the performance parameters that are used to calculate the metrics using NS_2 tool and also compared with protocols AOMDV AOMR-LM for better performance. The protocol AOMDV-FF was able to establish an optimal link among the nodes in a network and also provides less energy consumption in multipath routing.

Aye and Aung [5] highlights the major issue in mobile ad hoc networks in energy consumption and provides a solution by which this issue is minimized through more efficient multipath routing protocol that consumes less energy during the transmission of power and residual energy is used to maximize the network life time.

Tekaya et al. [6] and Gatani et al. [7] elaborated what are the possible ways to increase the network performance from the literature survey and explains the two types of the routing protocols implemented for ad hoc networks: Table driven (proactive) based on consistency of updated information and on demand reactive [8, 9]. At every instant of time, these routing protocols are shown more consistent and provide updated routing information efficiently at each mobile host. These protocols explain that each host mobile maintains with one or more tables regarding the routing information of all other host mobiles in the network. However, the network topology changes at any instant of time, the mobile hosts transmit the relevant messages to be updated to all other nodes in the network to increase the performance efficiency in establishing the routing information.

Montazeri et al. [15] AOMDV protocol is an extension of AODV protocol in locating all possible routes or multipath solution and to identify the shortest path among the identified one between the source and destination node in a given network frame. Hiremath and Joshi [14] AOMDV's ideal goal is to identify multiple routes. The implementation of AOMDV is inbuilt to overcome the critical issues, like the link failure rate and route/path discontinuities in MANETS and WSNs that degrades the performance of the network [10, 12, 14] (Fig. 1).

The disadvantage of AOMDV is requires additional requests; extra route requests (RREPs) and route errors (RERRs) in a multiple path detection process, and to maintain with additional fields; route replies (RREPs) [14].

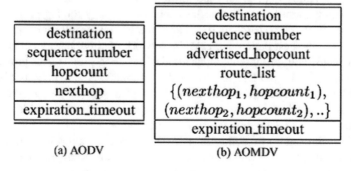

(a) AODV   (b) AOMDV

Fig. 1 Structures of routing table

Identification of route and its route maintenance is an integral part of the route discovery process. To establish a possible set of routes between source node and destination node, first these multipath protocols need to identify the link disjoints, node disjoints, or non-disjoint routes [7, 16]. The link disjoint routes does not possess a common path, these nodes are referred to be in common path. Node-disjoint routes totally disjoint routes does not possess either common nodes or links. Non-disjoint routes possess both nodes and links that are in common [16]. AOMDV major role is to evaluate multiple additional routes during a route discovery process. The development of AOMDV routing protocol is focused to establish highly dynamic ad hoc networks that have more frequent of link failures occurrences as well as route breaks. A novel approach has been incorporated to identify the route path which is applicable in this event process from all paths to the destination end.

3 Proposed Method

In this paper, the author incorporates an innovative idea by using fitness function (FF) to increase the efficiency of energy in multipath routing protocol is known as ad hoc on-demand multipath distance vector using the fitness function (AOMDV-FF) is of major concern. This system uses the fitness function to provide an optimized solution by highlighting one parameter out of two parameters as major concern in order to select the optimum shortest route/path; first is maximum energy absorbed at the route level and second is the route/path distance while transferring the data from source to the destination more efficiently with minimal energy consumption and by prolonging the network lifetime. To improve the performance of MAC layers by adopting some of the parameters, like average energy consumption and network lifetime and also the maximum level of elimination of congestion by adopting variable contention window sizes.

This work focus on energy consumption in MANET's by applying the fitness function technique to optimize the energy consumption in ad hoc on-demand multipath distance vector (AOMDV) routing protocol. The modified protocol AOMDV is with the fitness function as major concern is AOMDV-FF. The main purpose of fitness function is to identify an optimal solution among the multipath routing between source and destination nodes with minimum energy consumption. To evaluate the performance of the proposed AOMDV using FF protocol is being carried out using network simulator version 2 tools, simultaneously the results of it made comparison with the results of AOMDV protocol and ad hoc on demand multipath routing with life maximization (AOMR-LM) protocols. From the simulated results of the following metrics, like energy consumption, throughput, packet delivery ratio, end-to-end delay, network lifetime and routing overhead ratio performance metrics, varying the node speed, packet size and simulation time is being compared to acquire an accurate solution for MANETs.

A fitness function is mainly an objective function to measure the metrics of a design with optimal solution in selecting a path among many paths. Fitness functions are incorporated in genetic programming as well as genetic algorithms to provide a suitable optimal solution for a particular design in MANET's. In particular, in the fields of genetic programming and genetic algorithms, each design solution is commonly represented as a string of numbers (referred to as a chromosome). In recursive testing operations using NS_2 tool, fitness function provides a best possible round of testing among number of worst design simulated solutions. Each solution/round it provides closest proximity that meets the required minimal solution with the help of fitness function. The fitness functions mainly classified into two based on:

i.   Stationary fitness function is a set of possible test solutions.
ii. Mutable fitness function based on niche differentiation with a set of test solutions.

In MANET's, the research mainly focuses on the fitness factor depending on the factors energy, distance, delay, bandwidth, etc. This overall performance factors have shown an opportunity to design a novel routing protocol to enrich that network performance and it resources. From the literature survey, an integral part of fitness function is incorporated in particles swarm optimization (PSO) algorithms [13]. It was used to find an optimized solution for the path/route if the primary path fails/route. Some of the factors that affect to select the best possible optimized route depend on:

- Utilization of energy functional properties at each node.
- The path distance identity function of all links connectivity of all adjacent nodes.
- Evaluation of energy efficiency at each node.
- Measure path delays between the adjacent nodes.

Here, the author proposed a novel method by bringing the modifications required in AOMDV routing protocol and incorporating fitness function as major concerned. The message packets are able to transmit through these shortest routes depending upon the accuracy and quality that are needed with the help of request handshake signals by the node to highlight the principle of increasing energy efficiency factor and network life time. The major concerns are to be made clear at the source node before broadcasting and receiving the data packets in terms of shortest optimized route, maximum energy efficiency. To avail this minimum requirement of the nodes includes:

- Complete information of each node.
- The distances between the nodes.
- All possible energy consumptions of all the paths.

At each source node the above two parameters are calculated by permitting highest energy levels at source end and at receiving end the energy consumption is measured by route discovery recursive process. The advantage of route discovery process is used when all the routes at the destination end are failed. In such case, the source node will not select an alternate approach from the existing routing table by satisfying all parameters. Thus, the optimal shortest route with lower energy utilization is identified by Eq. (1).

$$\text{Optimum route} = \frac{\sum v(n) \in r\text{ene}(v(n))}{\sum v \in V\text{ene}(v)} \tag{1}$$

v No. of nodes in the optimum route 'r'
V Maximum no. of nodes in a given network

From the obtained simulation results for the parametric of MANET using NS2 tool. The comparison is made for the shortest route and optimized, maximum energy efficiency. The alternative path is chosen based on its distance parameter. The AOMDV protocol effectively evaluates the route with the lowest hop count. The modified AOMDV-FF implements the same principles of AOMDV to evaluate the route with peak energy level, minimum energy consumption during transmitting and shortest possible route. The evaluation of shortest route of AOMDV-FF technique is evaluated by Eq. (2):

$$\text{Optimum route} = \frac{\sum e(n) \in r\text{dist}(e(n))}{\sum e\varepsilon E} \tag{2}$$

where 'e' indicates the edges/links formed in the optimum route 'r'.
$E = $ sum of all the edges/links in a network.

A. **Algorithm Implementation**:

An algorithmic code for adapted fitness function is generated is mentioned in the following steps:

- Choose the source node and the destination node.
- Set the parameters for the source node.
- Initialize the control message path request and broadcasting the data packets and updating the routing table.
- Update all possible routes information in terms of energy levels and keep sending updated information to the routing table.
- Check the best possible route by route discovery process to evaluate the metrics.

B. **Simulation Measuring Metrics**:

The following simulation measuring metrics of AODMV-FF routing protocol is mentioned below:

1. **Packet Delivery Ratio (PDR)**: It is measurement of delivered packets to destination against to packets that were formed at the source [1].

$$\text{PDR} = \frac{\text{No of packets received}}{\text{No of packets sent}} * 100 \tag{3}$$

2. **Throughput**: The data size (bits) received by destination node successfully and is represented in kilobits per sec.

$$\text{Throughput} = \frac{\text{No of Bytes} * 8}{\text{Simulation time}} * 1000 \tag{4}$$

3. **End-to-End Delay**: The mean time (ms) observed by packets during the transmission between source-destination pair in a given network [1].

$$\text{End to End delay} = \sum_{i=1}^{n} \frac{(Ri - Si)}{n} \tag{5}$$

4. **Energy Consumption**: The total amount of energy (Joules) consumed by nodes in a network within the simulation time.

$$\text{Energy Consumption} = \sum_{i=1}^{n} (\text{ini}(i) - \text{ene}(i)) \tag{6}$$

5. **Network Lifetime**: The total time (ms) required for diminishing the battery life of n nodes and is calculated by Eq. (7)

$$\text{Network lifetime} = \sum_{i=1}^{n} (\text{ini}(i) - 0) \tag{7}$$

6. **Routing Overhead Ratio**: It is a ratio factor defined as no. of routing packets divided by the total number of data packets that were delivered.

$$\text{Routing overhead} = \frac{\text{No of routing packets}}{\text{No of routing} + \text{sent data packets}} \tag{8}$$

4   Analysis of Simulation Results

In this paper, the author is implemented a novel approach an energy-efficient multipath routing protocol ad hoc on demand multipath distance vector using fitness function to measure the performance metrics. Here, some of the typical parameters for AOMDV-FF is considered during the simulation operation are shown in Table 1.

It provides an optimized solution to measure the important factors like highest energy level, network lifetime and shortest path, etc., to transfer the data packets among the source and the destination nodes more efficiently to increase lifetime of the network and with less energy conserved. The simulations are performed by number of iterative operations in the discovery process. The mentioned parameters are analyzed for the best performance results and compared with previous existing methods from the literature survey.

Figure 2 shows the performance of throughput metric for 45 nodes in MANET. The throughput is measured in terms of joules and the graph shows for 30 ms simulation time.

Table 1 Simulation metrics of AOMDV-FF

Parameter-units	Value
No of simulation runs	5
Number of nodes	49
Node speed m/s	3020
Queue size	50
Simulation area-m^2	1200 × 1100
Packet size—bytes	64
Transmission range-m	250
Initial energy value J	100
Power consumption at Transmitter end J	10
Power consumption at Receiving end J	1.0

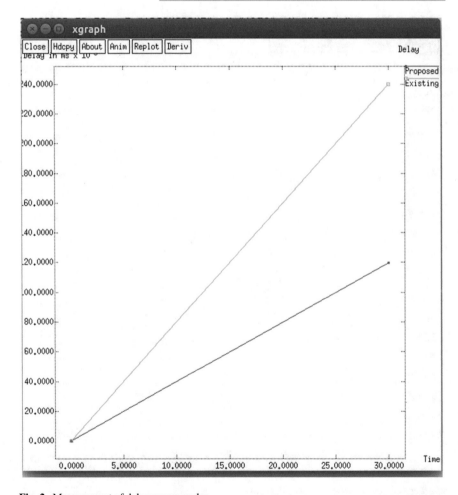

Fig. 2 Measurement of delay versus nodes

The proposed method shows the throughput obtained is 4 joules as compared to existing method 5.6 joules. The comparison of the simulation results shows that the throughput of this work is improved by 28.5%.

The scenario of network overhead ratio is mainly focuses on the sum of routing packets, delivery of packets over a single data packet. This measurement able to explore the relation between the overhead and additional bandwidth that is required in high-density data traffic, and the energy consumed is of main concern, and an effort is made to bring the extra bandwidth to be reduced due to overhead. The routing overhead can able to lower the bandwidth and power utilization of the nodes. The simulation results are delivered in Fig. 3.

Figure 4 shows the simulation results of energy consumption for 30 milliseconds. The energy consumption drastically decreased by approximately 50% as compared

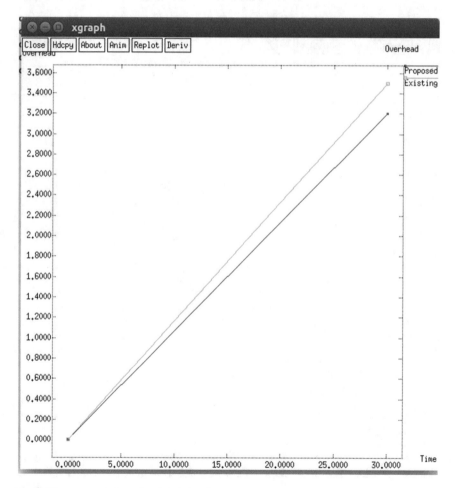

Fig. 3 Measurement of overhead versus nodes

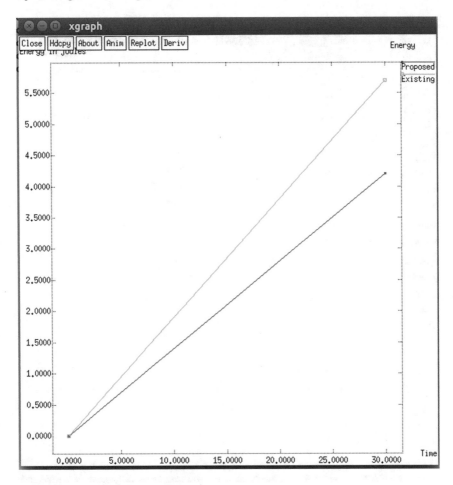

Fig. 4 Measurement of energy versus nodes

to existing method. The existing method consumes energy nearly 240 joules as compared to 120 joules by the proposed method. Figure 5 shows the simulation results of throughput of a network among 49 nodes.

The proposed AOMVD-FF method and existing methods remain approximately stationary values form the obtained results. The PDR value is estimated for 30 ms and shows a slight variation which is negligible. This is considering as one of the trade-off factors in this technique. The network lifetime is also improved by considerable amount of time.

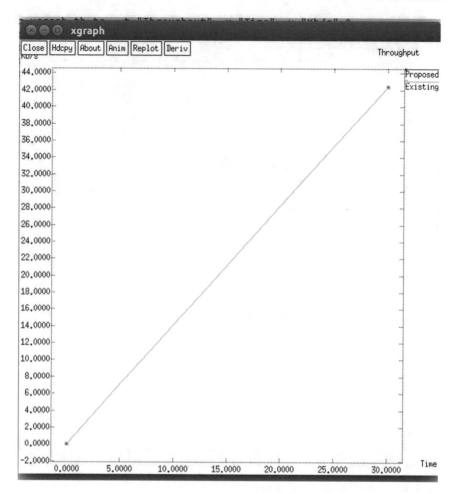

Fig. 5 Measurement of throughput versus nodes

5   Conclusion

In this research work, the author implemented a novel approach energy-efficient multipath routing algorithm using AOMDV-FF to evaluate four parameters. The estimated values run for 30 milliseconds with variable recursive operations 1, 5, 10. The performance factors throughput, overhead, energy consumption and PDR whose performance is estimated from the obtained results using NS-2 tool. The results show much improvement in the parameters throughput, overhead, energy consumption where as PDR remains stationary. End-to-end delay 14.4358 ms, energy consumption 18.3673 joules, packet delivery ratio 0.9911 and routing overhead ratio 4.68 are evaluated.

Additional parameters are to be considered to evaluate the performance of AOMDV-FF algorithm by incorporating highest energies at source level and also increasing the capacity of the nodes along with the iterations performed to evaluate the network life time.

Acknowledgements The authors express their sincere thanks to FIST-DST R&D Dept & Management of Vignana Bharathi Institute of Technology, Hyderabad, T. S. for their continuous and concurrent support for the usage of laboratory facility during this research work.

References

1. Taha, A., Alsaqour, R., Uddin, M., Abdelhaq, M., Saba, T.: Energy efficient multipath routing protocol for mobile ad-hoc network using the fitness function. IEEE Access **5**, 10369–10381 (2017). https://doi.org/10.1109/access.2017.2707537
2. Iswarya, J., Sureka, R., Santhalakshmi, S.R.: Energy efficiency multipath routing protocol for mobile ad-hoc network using the fitness function. Int. J. Inn. Res. Comput. Sci. Eng. (IJIRCSE) **3**(3), 11–15 (2018). ISSN: 2394-6364, www.ioirp.com
3. Uddin, M.,Taha, A.,Alsaqour, R., Saba, T.: Energy efficient multipath routing protocol for mobile ad-hoc network using the fitness function. IEEE Access https://doi.org/10.1109/access.2017.2707537
4. Ramya, V., Sangeetha, M., Rajeswari, A.: Energy efficient multipath routing protocol for manet using fitness function. IOSR J. Eng. (IOSRJEN) 59–63. ISSN (e): 2250-3021, ISSN (p): 2278-8719
5. Aye, M.C., Aung, A.M.: Energy efficient multipath routing for mobile ad hoc networks. Int. J. Info. Technol. Model. Comput. **2**(3) (2014)
6. Tekaya, M., Tabbane, N., Tabbane, S.: 'Multipath routing mechanism with load balancing in ad hoc network. Int. Conf. Comput. Eng. Syst (ICCES) 67–72 (2010)
7. Gatani, L., Re, G.L., Gaglio, S: Notice of violation of IEEE publication principles an adaptive routing protocol for ad hoc peer-to-peer networks. In: Sixth IEEE International Symposium on A World Of Wireless Mobile And Multimedia Networks, pp. 44–50. IEEE
8. Chaba, Y., Patel, R.B., Gargi, R.: Issues and challenges involved in multipath routing with DYMO protocol. Int. J. Inf. Technol. Knowl. Manage **5**(1), 21–25 (2012)
9. Mueller, S., Tsang, R.P., Ghosal, D.: Multipath routing in mobile ad hoc networks: issues and challenges. In:Performance Tools and Applications to Networked Systems, pp. 209–234. Springer, Berlin, Germany (2004)
10. Balaji, V., Duraisamy, V.: Varying overhead ad hoc on demand vector routing in highly mobile ad hoc network. Jrn. Comput. Sci. **7**(5), 678–682 (2011)
11. Carson, S., Macker, J.: Mobile Ad Hoc Networking (MANET): Routing Protocol Performance Issues and Evolution Considerations. RFC Editor (1999)
12. Poonam, M., Preeti, D.: Packet Forwarding using AOMDV algorithm in WSN. Int. J. Appl Innov Eng. Manage (IJAIEM) **3**(5), 456–459 (2014)
13. Hu, Y.-F., Ding, Y.-S., Ren, L.-H., Hao, K.-R., Han, H.: An endocrine cooperative particle swarm optimization algorithm for routing recovery problem of wireless sensor networks with multiple mobile sinks. Inf. Sci. **300**, 100–113 (2015)
14. Hiremath, P.S, Joshi, S.M.: Energy efficient routing protocol with adaptive fuzzy threshold energy for MANETs. Int. J. Comput. Netw. Wireless Commu. (IJCNWC) **2** (2012). ISSN: 2250–3501

15. Montazeri, A., Poshtan, J., Yousefi-Koma, A.: The use of 'particle swarm' to optimize the control system in a PZT laminated plate. Smart Mater. Struct. **17**(4), 045027 (2008)
16. Mueller, S., Tsang, R.P., Ghosal, D.: Multipath routing in mobile ad hoc networks: issues and challenges. In: International Workshop on Modeling, Analysis, and Simulation of Computer and Telecommunication Systems, pp. 209–234. Springer, Berlin, Heidelberg (2004)

Ms. Monika Received her B.E degree (CSE), 2016 from SJBIT, Bangalore, Visvesvaraya Technological University, Belagavi. She Received M.Tech (CSE), 2019 from East West Institute of Technology, Bangalore, Visvesvaraya Technological University, Belagavi. Her Research area interests are Wireless Sensor Networks, Data mining, IOT

Dr. P. Venkateswara Rao Currently working as Professor and R&D Coordinator, Dept. of ECE, Vignana Bharathi Institute of Technology, Hyderabad. Telangana State, India. He pursued his B.Sc. Electronics degree in 1986, B.E in Electronics Engg., in 1991 from Marathwada University, Aurangabad, University, Dharwad, and Ph.D. in Low Power VLSI Signal Processing, in 2011 from DR. MGR University, Chennai. His research areas of interest are Low Power VLSI, Signal and Image Processing, Biomedical Image Processing and Nano-Sensors. 3 Research scholars awarded Ph.D. VTU, Karnataka and Six Research scholars from Jain University, Karnataka under his supervision and currently 3 Research scholars perusing Ph.D from VTU, Belagavi and one from Jain University, Bengaluru. He Published 47 International Journals and 35 papers presented in International/National conferences. He is Life member of professional bodies FIETE, MISTE, ITES and IMAPs and also Reviewer and Editor for Peered Journals Springer, Teklanika, EURASIP, IJCT BORJ, ARJECE and Inder Science Publications.

Dr. Pradeep Kumar Mallick Currently working as Associate Professor in the School of Computer Engineering, Kalinga Institute of Industrial technology (KIIT) Deemed to be University, Odisha, India and Post Doctoral Fellow (PDF) in Kongju National University South Korea, Ph.D. from Siksha 'O' Anusandhan University, M. Tech. (CSE) from Biju Patnaik University of Technology (BPUT), and MCA from Fakir Mohan University Balasore, India. Besides academics, he is also involved various administrative activities, Member of Board of Studies, Member of Doctoral Research Evaluation Committee, Admission Committee etc. His area of research includes Algorithm Design and Analysis, and Data Mining, Image Processing, Soft Computing, and Machine Learning. He has published 5 books and more than 55 research papers in National and international journals and conference proceedings.

Performance Evaluation of Vehicular Ad Hoc Networks in Case of Road Side Unit Failure

C. M. Raut and S. R. Devane

Abstract Vehicular ad hoc networks (VANETs) are one of the important entities for implementing smart city projects. Traffic monitoring and controlling of the city during the emergency conditions such as vehicular congestion due to accident and vehicle failure are the important challenges of advance intelligent transport system (ITS). The performance of ITS in smart cities is based on the QoS of VANET, where mobility and dynamic changing topology degrade the performance. VANET primarily depends on the type of underlying routing protocol such as proactive (DSDV), reactive (AODV) and clustering-based routing protocols (CAODV). Network failure conditions in VANET caused by either node or RSU failure due to natural disasters, electrical faults or accidents, etc., degrade the overall QoS in VANET. This paper carries out the deep analysis of the effect of such failure by simulating failure condition against the normal VANET using all the above routing protocols.

Keywords Road side unit · Routing protocols · Vehicular networks · QoS · Throughput

1 Introduction

Vehicular ad hoc network systems (VANETs) are extremely encouraging that assumes an essential part of intelligent transportation system (ITS). VANETs assist vehicular drivers in conveying to keep away from various parts of the basic driving circumstances. VANETs support the assortment of security applications, for example, cooperative traffic monitoring, control of traffic streams, dazzle crossing, the avoidance of impacts, close-by data administrations and ongoing detour route calculation through engaging vehicle-to-vehicle (V2V) and vehicle-to-foundation (V2F) correspondences. VANETs comprise two components: access point, called road side

C. M. Raut (✉) · S. R. Devane
DMCE, University of Mumbai, Navi Mumbai, India
e-mail: Cmraut75@gmail.com

S. R. Devane
e-mail: srdevne@yahoo.com

© Springer Nature Singapore Pte Ltd. 2020
P. K. Mallick et al. (eds.), *Cognitive Informatics and Soft Computing*,
Advances in Intelligent Systems and Computing 1040,
https://doi.org/10.1007/978-981-15-1451-7_32

units (RSSUs), and vehicles called board units (OBUs). RSUs are settled and can go about as a dispersion point for vehicle networks.

VANETs allow web network to vehicles while progressing, so travelers can download music, send messages, book restaurants and play games. Due to vehicles varying speed, vehicular frameworks are depicted by fast topology changes. The last makes outlining a productive steering convention for vehicular condition extremely troublesome. Planning versatile steering conventions to such rapidly changing framework topologies is uncommonly fundamental to various vehicular prosperity applications as neglecting to routing influence evasion messages to their fundamental vehicles can render these messages to be useless.

The major entities of the VANET are vehicles, road side units (RSUs) and onboard units (OBUs), where they transmit information by using the vehicle-to-vehicle correspondence (V2V), infrastructure-to-infrastructure (I2I) and vehicle-to-foundation (V2F) interchanges shown in Fig. 1. These correspondences are supported by the dedicated short-range communication (DSRC) [1]. This vehicular communication system supports the wireless access in vehicular environments (WAVE) [1] using IEEE 802.11p. The IEEE 802.11p defines the link layer that supports Internet protocol and the WAVE short message protocol (WSMP). The WAVE standard is used for the purpose of minimizing the critical situations such as prevention or identification of the occurrence of accidents. The intelligent transport systems (ITS) use the WAVE protocol to broadcast information such as weather conditions, roadways maintenance and road traffic conditions.

This paper is to investigate the impact of the failure in network infrastructure like RSU in the overall QoS of VANET. The RSU failure condition has been implemented with the reactive, proactive and clustering routing protocols and by varying mobility scenarios. The mobility pattern is generated with various conditions and evaluating the various routing protocols with respect to QoS parameters. Possible scenarios are considered which may occur in the practice such as

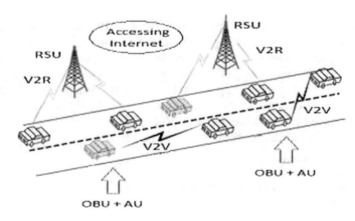

Fig. 1 Example of VANET topology and communications

1. VANET network using RSU.
2. VANET network using RSU and clustering.
3. VANET network as per case 1 and 2 where RSU is failed.

Section 2 states the study of recent works reported on routing protocols for VANET followed by the study on protocols investigated is presented along with the freeway mobility model. In Sect. 6, simulation results, its analysis and finding are discussed. Finally, the conclusion and future work are discussed in Sect. 7.

2 Related Work

The impacts of versatility, street topology and network applications have on the execution evaluation of VANET routing protocols. They assessed the execution of built-up VANET routing protocols by utilizing sensible portability from an extensive scale urban topology and forcing network load via an exemplary VANET-based, traffic query and application [2].

Chowdhury et al. presented the comparative investigation of responsive routing protocols named DSR, AODV and especially ad hoc on request multipath separation vector in VANET utilizing distinctive adaptability models gave in VanetMobiSim structure. By varying the moveability, the showcases were assessed, and the number of sources and focus point speed, while packet movement part, the end to end put off and institutionalized routing burden, were examined [3].

Mohammed et al. evaluated the execution of AODV, DSR and DSDV routing similarly as bundle conveyance proportion, average end to end put off, latency and throughput. The objective of this examination is to locate the best routing protocol overall conditions. In view of their approved outcomes, AODV plays out the best among all assessed protocols [4].

Arshad and Javaid, R presented the approach for node circulation with respect to density, network connectivity and conversation time. They assessed and broke down the execution of three directing conventions: AODV, DSR and FSR both in MANETs (IEEE 802.11) and VANETs (IEEE 802.11p). They enhanced these conventions by changing their steering data trade interims: MOD AODV, MOD DSR and MOD FSR [5].

Kamlesh Chandra Purohit and Sushil Chandra evaluated the flexibility of existing MANET routing protocols for VANET. Creators broke down the effect of the vehicle thickness and speed on the bundle conveyance proportion, institutionalized routing trouble, typical end to end delay, average throughput, average way length and average misfortune rate [6].

Shakshuki et al. reported another recent work on performance investigation of routing protocols for VANET. They investigated the execution of topology-based MANET routing protocols AODV, DSDV and DSR in a VANET expressway plan using NCTUNS 6.0 simulator. They considered the diverse parameters vacillate including speed, hub thickness, and propagation difficult model, blurring impacts, data rate and payload.

Work presented in this paper is different from the above works in below points:

– Focused on high mobility variations for the investigation of routing protocols using the real-time traffic.
– Generated the mobility model using the freeway mobility model.
– Investigated reactive, proactive and clustering-based routing protocols.

Investigated the impact of RSU failures on VANET performances.

3 Methodology

Reactive protocols such as (AODV), proactive protocol (DSDV) and clustering-based routing protocol (CAODV) are used to implement VANET, and the methodology is stated below.

3.1 AODV

This is the well-known reactive routing protocol. This routing protocol utilizes the on request approach for discovering routes; the route is built up as instructed by a hotspot and utilizes an objective progression numeral to identify current path. The home center point and the central of the road hubs store the accompanying trust information comparing with each follow in data bundle transmission. At the point when a route requires a course to an objective, it surges the framework with a route request (RREQ) bundle. On its way through the network, and when it accomplishes the objective, a route reaction (RREP) packet is unicast back along a comparable route on which the RREQ bundle was transmitted. A mobile node send message to the neighbouring mobile node [7].

3.2 DSDV

This is the proactive routing protocol and is a table-driven routing arrangement for especially ad hoc appointed network uses concept of Bellman–Ford algorithm. DSDV utilizes the briefest way routing algorithm to pick a singular path to an objective, to go without routing circles. Complete dumps and incremental updates are scratch among center points to ensure that routing information is correct. DSDV is one of the principal endeavors to adjust a built-up routing system to work with MANET. Each routing table records all goals with their present hop check and arrangement number, routing data is communicated or multicast, and DSDV is not suitable with the large network.

3.3 CAODV

This is the modified AODV routing protocol with the introduction of clustering. Amid the route discovery procedure, AODV floods the whole network with the substantial number of control packets, and thus, it finds numerous unused routes between the source and destination. This turns into a noteworthy downside to AODV since it causes routing overhead, expending data transfer capacity and hub strength. Further, the optimization strategy utilizes clustering the nodes of the network using overseeing routing by cluster heads and gateway nodes. Routing utilizing clusters viably diminish the control informed overwhelmed amid the route discovery process by supplanting broadcasting of RREQ packets by sending of RREQ packets to cluster heads. The clustering method is working in two main phases such as cluster formation and cluster routing.

3.4 Cluster Formation

The neighbor table is basically used for supporting the process of cluster formation. In the neighbor table, the data about its neighbor hubs is put away, for example, their IDs, their part in the cluster (cluster head called CH) and the status of the connection to that hub (uni-/bi-directional). The periodic broadcasting of HELLO messages is used for maintaining the neighbor table. The information about the one node state, its cluster adjacency table, its neighbor table, etc., is maintained by the HELLO message.

3.5 Cluster Routing

For the clustering routing, there are two data structures used such as two-hop topology database and cluster adjacency table (CAT). The information about neighboring clusters is maintained by CAT. It stores the information like whether they are connected either bi-directionally or uni-directionally.

- Routing updates: Because of node movement, it may happen that many nodes fail due to which route repair functionality must be carried out by routing protocol. In cluster-based AODV, we have added two such functions such as route maintenance and route shortening.

- Route maintenance: If a connection among two nodes fizzles, the HTR can repair the route. Consequently, one of the going with hubs of the route must be in the two-hop topology database of the hub.
- Route shortening: As a less than dependable rule, a hub may discover an association among itself and another succeeding center point of the route, which is not its quick successor or an association between two after center points, independently. This ought to be conceivable by analyzing the information set away in the two-hop topology database. Given this is valid, it abbreviates the route by notwithstanding the repetitive nodes from the route.

4   VANET Mobility Model for RSU Failure Impact

Freeway is one of the Manhattan models proposed to the development behavior of mobile hubs on a freeway the freeway map utilized as a part of our simulations has appeared in Fig. 2 which utilizes interchanging traffic status among the vehicles or tracking a vehicle on a freeway with path in both directions.

The NAM visualization result using parameters as per Table 1 is shown in Fig. 3, and 20 vehicles are moving according to the freeway model. There are five fixed RSU nodes and showing few vehicles are selected CHs.

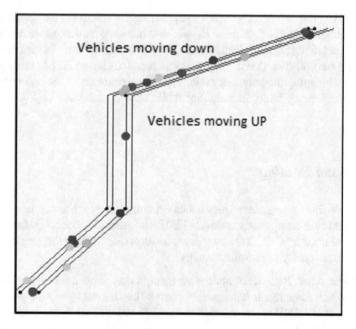

Fig. 2 Example of freeway mobility model

Table 1 Simulation parameters

Number of vehicles	20
Number of RSUs	5
Simulation time	200 s
Mobility (Km/s)	10–35
Routing protocols	AODV, DSDV and CAODV
MAC	802.11Ext (IEEE 802.11p)
Propagation model	Two-ray ground
Mobility	Freeway
Antenna	Omni antenna
Performance Metrics	Average throughput, average delay, packet delivery rate (PDR) and loosed packets

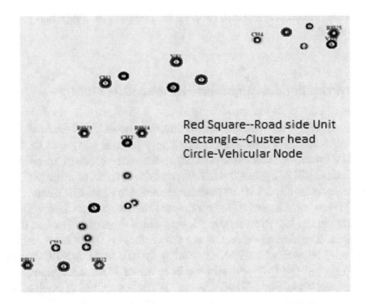

Fig. 3 NAM visualization of VANET

5 Parameter Consideration for Model

Mobility pattern is generated which composed of total 20 vehicles with varying speed from 10/second to 35 km/second for evolution purpose. The available 802.11Ext MAC protocol, which is the standard form of 802.11p protocol, is considered for the VANET networks simulation. The parameter of varying speed is considered for getting the variety of the mobility patterns.

Fig. 4 Average throughput
performance

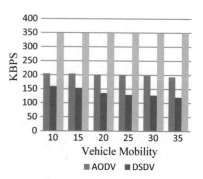

6   Simulation Result with and Without RSU Failure

As per the mentioned performance metrics, the comparative analysis is presented in section A and B.

6.1   Performance Investigation Without RSU Failure

Average throughput performances for all three protocols are given in Fig. 4 which clearly show that DSDV protocol performs very poor as compared to the AODV and CAODV protocols. There is a significant difference between the performances of AODV and CAODV against the DSDV protocol in terms of average throughput, packet delivery rate and packet loss performances due to table-driven nature of DSDV protocol. The construction of table takes more time for the data forwarding process which results in the loss of excessive packets which degrades PDR and throughput performances. The simulation results are shown that DSDV is not suitable for the high dynamics networks like VANET. Among AODV and CAODV, the clustering protocol improved the QoS performance in terms of PDR, throughput, delay and packet loss rate. The delay of DSDV is very less as compared to reactive protocols as there is minimum data transmission performed (Figs. 5, 6 and 7).

The CAODV routing protocol shows the improved routing performance for VANET as compared to AODV and DSDV protocol in terms of performance parameters, i.e., throughput, PDR, delay and packet loss as shown in Table 2.

6.2   Performance Evaluation with RSU Failure

For the simulation of RSU failure condition, single RSU failure condition is explicitly introduced in the network in order to check its impact of routing performance using AODV and CAODV protocols. The results discussed in section A shows that DSDV is

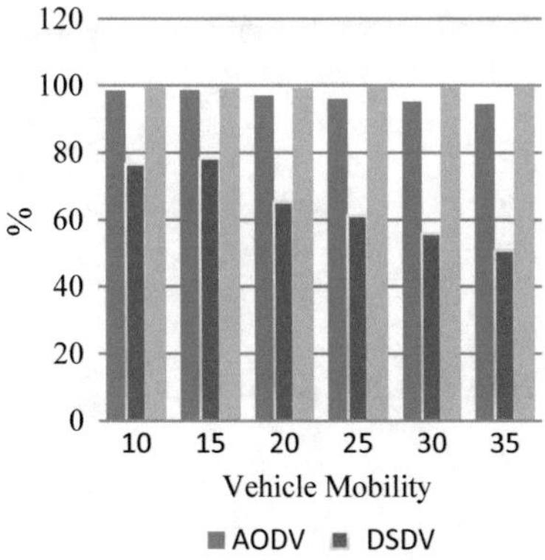

Fig. 5 Packet delivery ratio performance

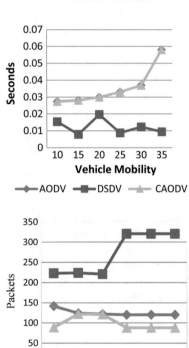

Fig. 6 Average delay performance analysis

Fig. 7 Packet dropped performance analysis

Table 2 Analysis of average performances

Performance parameter	AODV	DSDV	CAODV
Avg. throughput	199.88	137.57	348.83
Avg. delay	0.0355	0.012	0.0356
PDR	96.66	64.67	99.36
Packet loss	124	271	99

Table 3 AODV performance

Performance parameter	AODV-normal	AODV-RSU failure
Avg. throughput	204.69	163.4
Avg. delay	0.0274	0.0652
PDR	98.51	78.47
Packet loss	142	247

Table 4 CAODV performance

Performance parameter	CAODV-normal	CAODV-RSU failure
Avg. throughput	348.73	307.44
Avg. delay	0.00371	0.04143
PDR	99.43	79.47
Packet loss	89	226

no longer suitable for the high dynamic-based networks. Therefore, DSDV protocol is omitted to evaluate the impact of RSU failure condition. In the simulation process, at some interval of time, one of the RSU is failed, and this information is passed to the routing protocol. The result is shown in Tables 3 and 4 for RSU failure condition using AODV and CAODV protocol for 20 vehicular nodes and 5 RSU nodes with mobility 10.

6.3 Analysis and Findings

From the above simulation results, it is showing that as the mobility increases, the QoS performance of all protocols decreases. The mobility increases the packet drops. The simulation results are showing that DSDV is not suitable for the high dynamics networks like VANET. Among AODV and CAODV, the clustering protocol improved the QoS performance in terms of PDR, throughput, delay and packet loss rate. The delay of DSDV is very less as compared to both reactive protocols as there is minimum data transmission performed. The impact of RSU failure is significant on QoS performance. The throughput is decreased; PDR rate is decreased approximately by 20%, and the number of packet dropped increased by approximately three times than normal VANET network. In such RSU failure situation, when packet coming to that failure node that packet will not be processed which results in loss of packet and retransmission is carried out through other route. The RSU failure condition leads to the severe impact on QoS in VANET which increases data loss and delay in network and decreases the PDR. Additionally, the emergency conditions in network such as accidents and traffic jam may get distracted due to RSU failure.

Therefore, the RSU failure is one of the major research problems which are not addressed by previous works. The routing protocol should be self-adaptive to recover from such disasters in VANET.

7 Conclusion and Future Work

In this paper, the issues related to proper management of high mobility which is a main research challenge for VANET is discussed. The communications in VANET are mainly affected by the choice of routing protocol. Therefore, in this paper, the investigation of three categories of routing protocols with respect to varying mobility of vehicle nodes such as reactive, proactive and clustering-based routing protocols is evaluated. Further, the very important challenge of advance ITS is the RSU failure conditions and its impact on VANET QoS performance is discussed. The simulation results are presented and claimed the impact of mobility and RSU failures on routing performances. For future, it is suggested to work on RSU failure management in order to mitigate the impact of RSU failure on VANET.

References

1. Sven, J., Marc, B., Lars, W.: Evaluation of routing protocols for vehicular ad hoc networks in city traffic scenarios. In: Proceedings of the 11th EUNICE Open European Summer School on Networked Applications, pp. 584–602. Colmenarejo (2005).https://sourceforge.net/projects/mobisim/
2. Chowdhury, S.I., Lee, W.I., Choi, Y.S., Kee, G.Y., Pyun, J.Y.: Performance evaluation of reactive routing protocols in VANET. In: The 17th Asia Pacific Conference on Communications
3. Mohammed, F., Mohamed, O., Abedelhalim, H., Abdellah, E.: Efficiency evaluation of routing protocols for Vanet. In: 2014 Third IEEE International Colloquium in Information Science and Technology (CIST) (2014)
4. Arshad, W., Javaid, N., Khan, R.D., Ilahi, M., Qasim, U., Khan, Z.A.: Modeling and simulating network connectivity in routing protocols for MANETs and VANETs. J. Basic Appl. Sci. Res. 3(2), 1312–1318 (2013)
5. Purohit, K.C., Dimri, S.C., Jasola, S.: Performance evaluation of various MANET routing protocols for adaptability in VANET environment. Int. J. Syst. Assurance Eng. Manag. 8, (2[7]) 2017
6. Shakshuki, E.M., Maratha, B.P., Sheltami, T.R.: Performance evaluation of topology based routing protocols in a VANET highway scenario. Int. J. Distrib. Syst. Technol. 8(1) (2017)
7. Loulloudes, N., Pallis, G., Dikaiakos, M.D.: On the performance evaluation of VANET routing protocols in large-scale urban environments (Poster). In: 2012 IEEE Vehicular Networking Conference (VNC) (2012)

Issues of Bot Network Detection and Protection

Surjya Prasad Majhi, Santosh Kumar Swain and Prasant Kumar Pattnaik

Abstract The paper studies the various aspects of botnet detection. It focuses on the different methods available for detection of the bot, C&C and botherder. There is also the elaboration of different botnet protection methods that can be utilized by systems users to protect their systems before bot infection and also after bot infection.

Keywords Bots · Botnet · Bot network · Botherder · C&C channel · Botnet detection · Bot infection · System-level · Network-level

1 Introduction

A botnet is a collection of computers connected with each other via Internet communicating with similar machines whose actions are remotely controlled through a hidden command and control (C&C) channel by a botmaster or botherder. Botnet word consists of Bot and Net which comes from Robot and Network. Botnet performs malicious activities on demand after receiving orders from botherder anytime from anywhere. The performance after command makes botnet unpredictable, extremely flexible and their reaction to external events dynamic, which is indeed dangerous. Botnet detection is an important step for building a secure network for systems. As the botnets are random in nature, the botnet detection sometimes becomes difficult. Once the bot detection is done, it is easier to protect the system from bot infection.

Botnet detection is possible in different ways. They can be classified according to the botnet detection mechanisms available till date. There are various aspects in which the detection process can be classified further. Figure 1 describes the broad classification of the various directions of botnet detection.

S. P. Majhi (✉) · S. K. Swain · P. K. Pattnaik
School of Computer Engineering, KIIT Deemed to be University, Bhubaneswar, India
e-mail: surjyainepfo@gmail.com

S. K. Swain
e-mail: sswainfcs@kiit.ac.in

P. K. Pattnaik
e-mail: patnaikprasant@gmail.com

© Springer Nature Singapore Pte Ltd. 2020
P. K. Mallick et al. (eds.), *Cognitive Informatics and Soft Computing*,
Advances in Intelligent Systems and Computing 1040,
https://doi.org/10.1007/978-981-15-1451-7_34

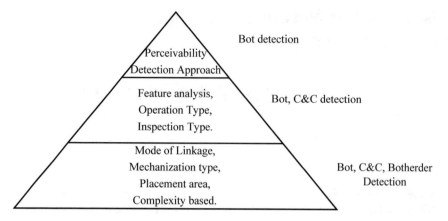

Fig. 1 Aspects of botnet detection

2 Bot Detection

Bot detection is generally done by system users and network managers who have the primary intention to safeguard their systems and networks from attacks. The users and the network managers are usually uninterested of the bot network. So the bot detection can be done considering or without considering the bot network. Bot detection clearly indicates that the system and network are vulnerable to bot network attack. The bot detection must be followed by counteractive measures for recovering from the bot attack and also prevent itself from being attacked again. There are some aspects which are used only for bot detection not for any C&C or botherder detection like:

Perceivability: Perceivability is the capability of the detection technique to discriminate among bot networks. The systems present in any bot network have less knowledge about other bots in the network. The researches that study the behaviour of bots and their networks need the information gathered by the trails of the bot network where the bot is present. The trails can be further studied to have more efficiency in detection methods. Perceivability can be broadly classified as:

Distinct methods: Distinct methods are termed for those methods which detect the bot-infected system and also provide information about the bot network in which the infected system is present. Bot networks can be detected here by combining the systems with similar behavioural traits or communication traits [1]. The use of library routine and system request sometimes disclose the bot network where the bot or infected system belongs to.

Indistinct methods: Indistinct methods are the methods which detect the system that is affected by bot attack but they are incapable of providing information about the bot network they belong to. As a result bot, networks cannot be distinguished from each other in these methods. BotHunter detects the infected systems but cannot provide any details or data about the bot network [2].

Detection Approach: The detection approach can also be an aspect of bot detection. This dimension explains the direction in which the bot detection is carried out. Some detection techniques follow a sequence and some are just a random method to find the infected system. While following a sequence, detection may be starting from the system going to the network or vice versa. They can be divided into two types:

Sequential approach: This approach explains about following a particular order or sequence for detection of bot and bot families. Some analysis methods may start with a bot and then trace the connection upwards to the bot network. For example, observing network modules such as spams [3, 4], differences in DDNS [5] and traffic behaviour of DNS [6]. Some methods analyse the behaviour of bots in the top network and identify the bot or bot families by organizing them according to their similarities in actions. For example, BotMiner [1] analyses the behavioural pattern and traits and combines the systems that follow the similar characteristics resulting in the identification of the bot family.

Random approach: Random approach does not follow any kind of order or sequence to trace down the bot or bot families. It is arbitrary in nature as it has no classified order in the botnet modules. The analyses can start from any bot and after observing the behavioural traits and list of systems connected to that particular bot, other bots or systems can be detected. The random approach is mostly associated with p2p bot networks.

Bot, C&C Detection: The bots are connected to the C&C server either directly or indirectly to receive the orders from the botherder for execution. The C&C server acts as an interface to transfer the information between bots and botherder. Analysing the C&C server helps to trace down the bots connected to it. Detection methods that can observe the C&C communication channel can easily detect the bots associated with it. There are some directions in which one can focus to detect the C&C depending on its need to analyse.

Feature analysis: Feature analysis describes the method of observing the occurrence of features in any bot network that helps in the detection process. Some detection methods focus only on some particular methods while some methods observe the general behaviour of the bot network. They can be broadly classified into two types:

Intensive analysis: Intensive analysis defines the detection methods that concentrate only on the detection of specific kind of bot and bot network. They are user defined to notice the behaviour and detect only a particular type of botnet. But they cannot trace the behaviour and detect other than the defined bot network. For example, IRC bot networks are detected by Rishi [7]. But it is inefficient to detect any other botnet.

Extensive analysis: Extensive analysis studies the general behavioural traits or characteristics of the bot network. They do not concentrate on particular botnets [1, 8]. Analysis of the general characteristics helps in detecting botnets with a different organization. As a result, they can track down a different type of botnets rather than sticking to a specific one.

Operation Type: Operation type refers to the type of action being carried when C&C server is in an active state. When the communication between bots and botherder is continuing, the C&C server can be detected by directly involving in the process. The network process has to be manipulated online to gather information about the communication channel and to detect the C&C server and the bots associated with it. They can be classified into the following types:

Adding type: Adding type operation adds packets of information into the communication network channel. If the reply from the network flow is similar to the representative bot's reaction, then one can consider the network to be the part of C&C. Addition can be implemented by either understanding the C&C conventions, or by simply reiterating the arriving packets in the suspicious network communication channel with or without any small changes. Hereby, the C&C server can be detected. For example, BotProbe can be used to detect botnet that uses chat-like C&C communication channel [9].

Blocking type: Blocking type operation described the detection method where the arriving or leaving packet of information is blocked for the generation of known reaction from any of the arriving/leaving end of the C&C network channel. Suppose a suspected system sends a request to C&C server for updating and at the receiving end, the request is dropped knowingly. After trying for some more times, the system will initiate its back-up procedure that is predefined. If one knows the actions that are initiated as part of the back-up procedure, then the C&C communication network can be detected. For example, SQUEEZE blocks the communication to the C&C server which the bots try to connect which later on initiates the back-up mechanism resulting in detection of C&C [10].

Inspection type: Inspection type denotes the property type that is used for inspection for the detection of C&C server and bots. The participation in the communication in the network sometimes gets difficult because of the policy limitations. Moreover, the participation is done when the communication is online or going on which may not be possible at all the times. So if the inspection for detecting can be done outside the network, it would be more efficient and beneficial for detection purpose. Here, the inspection is done by analysing the network flow for signs of C&C propagation. Some of the properties that can be inspected for the detection purpose are:

Signature-based: The signature-based inspection type involves the generation of signature models of C&C network flow. The network is observed carefully and the signatures are obtained by the repeatedly occurring string pattern or sequences in the C&C communication channel. Snort [11] is a detection structure that observes the C&C network and traces the cues of intrusion. The detection systems are already constructed with the set of signatures to analyse the traffic. Constructing of manual signature seems less appropriate and more time taking. So many automated models or signature preparation are recently been proposed [12–14].

Behaviour-based: The behaviour-based inspection type describes the detection which is an effective way to monitor and take down the botnet. It looks for different patterns followed in the network traffic to track the botnet rather that tracking the information being propagated [15]. Behaviour detection can be used to detect unknown threats of botnet [16]. Both attack and operational behaviour can be monitored for detection of a botnet. Behaviour-based detection focuses on the analysis of the individual C&C network flow or system. It does not compare the behaviour with other systems or network flow.

3 Bot, C&C and Botherder Detection

Botherder is the most difficult to detect. There are very fewer botnet techniques that target botherder and detect it. The bot network can face serious consequences after the detection of botherder. The botherder can face legal trials and heavy fines imposed on itself after being detected. The information given by the botherder may lead the entire bot network to be demobilized and failing it completely. As a result, the botherder tries to make themselves the most secured part of a botnet so that they are protected from detection mechanism. The botherder gives some orders to C&C server that is being communicated among the bots by the communication server. The fewer commands generate very less traffic and it can be encrypted too making the traffic analysis also difficult. Hence, the botherder detection is the most difficult one to pursue but once it is detected the whole bot network along with the C&C server can be traced down. There are many aspects which help in detecting botherder and the overall bot network along with the C&C server. Some of them are:

Mode of Linkage: Mode of linkage refers to the detection methods that are related to the network in different mode or type. Detection methods can function in either online mode or offline mode.

Online mode: Online mode describes the detection methods which operate on real-time basis and detects the botnets at the time when the examining bot or channel is operating. For example, botherder is detected live with the use of watermarks [17]. The live detection of botherder also depends on the deep packet inspection (DPI). A method that does complete DPI will weaken in a huge and random traffic while operating in a live situation.

Offline mode: Offline mode describes the detection methods that give accurate results when operated on offline components like log files, communication channel dumps, etc. These methods help the researchers for network analysis for the further study. For example, BotHunter operates in both online and offline mode [2]. Log files and other evidence can also be analysed to inspect the occurrences related to bot activities [18, 19].

Mechanization Type: Mechanization type defines the type of detection methods that use some human help or are fully mechanical in nature (no involvement of human required). They can be broadly classified into:

Semi-mechanized Type: Semi-mechanized type is the detection methods that require human involvement in the process. The methods require human acquisition for development of signatures that are manually fed to the detection system made by customization. Even a small change in bot's functionality also involves manual interference. The change requires manual preparation of signature and integrated into the detection system. Mostly, the detection methods for p2p botnet are manual

[20, 21]. Whereas some mechanized or automatic system is used for development of signature which requires very less human involvement [22].

Fully mechanized Type: Fully mechanized type is the detection methods that do not require any kind of human involvement after the preliminary development of the detection system. Any method used for detection should be preferably universal and automated in nature. It should not be dependent on a human for any kind of interference in the middle of the process. Detection systems generated mechanically [14] that are dependent on the behavioural property of bot networks [8] can be processed in a fully mechanized way.

Placement Area: Placement area defines the location where the detection method can be placed or applied so that the bot networks or bot themselves can be detected. The methods can be placed either in the system or in the network for analysis or detection. The methods are not restricted to the area of placement. A distributed detection process can use both the placement areas for a broad detection of a bot network.

System-based: System-based are the detection methods that are placed in the infected system to analyse the behaviour of the system and trace if there is any bot infection in the system. These detection methods are capable of spotting infection on the individual system. But they cannot provide the evidence of the presence of other bots present in the same family of the bot. For example, detection methods like BotHunter [23] and BotSwat [24] are placed in the system itself for identifying the bot infection.

Channel-based: Channel-based are the detection methods are placed in the communication channel to examine the communication channel's traffic. It can be placed in any place in the hierarchy level of the network depending on the opportunity to examine the network. Placement can be done in a proxy server, Internet service providers, etc. Botnets can also be detected on a huge Tier 1 Internet service provider network [25].

Complexity-based: The methods can be studied on the basis of the complexity analysis they perform. Some methods do the analysis by observing the data that are easily reachable or can be obtained easily. They involve less complexity whereas some methods observe deeply the data to detect botherder or the bot network. They can be broadly studied in two types:

High-Complexity: High-complexity detection methods are the methods that execute detailed investigation on data. Exposure of suspicious payloads [2] by signature method or communication [7, 26] uses deep packet check to find the bot network or botherder completely. These methods involve high functional and computational complexity when there is huge traffic on the network.

Low-Complexity: Low-complexity methods analyse only the data involved in the flow traffic and provide minimum complexity. These methods are efficient for large area networks as they involve less complexity. The methods can also compare the flow data records of doubtful systems with the network traffic models to recognize the botherder or the bot network complete activity.

4 Botnet Protection Methods

The botnet detection shows us a pathway to have certain protection methods against botnet infection. Botnet protection methods can be broadly classified into awareness methods, proactive methods and reactive methods. Awareness methods are methods used for protecting the systems before infection of the bots. Proactive methods are the methods that try to recover the system and networks by removing the infection after bot attack. Reactive methods are those who attack the complete bot network as a result of which no further infection is possible by the bot networks. It is very necessary to have protection methods relating to botnet because it can save from huge data and financial losses of users. But before that the common users should be made aware of the botnet who have no technical knowledge about them. As a result, they fall prey of the bot network. Figure 2 gives the broad classification of the protection models that can be used to safeguard the systems from bot infection or to remove the infection from the infected systems.

4.1 Awareness Methods

Awareness methods try to avoid the infection made by the bot network. These methods can be used in the systems or the network itself to shield from the possibility of being attacked by the botnet. Awareness methods are carried out prior to the attack by the botherder through the bot network. There are different awareness methods that can be used by the user to protect the systems. They can be defined as:

Actions: There are some actions that can be carried out by the network or system is infected by a bot and can be saved from being attacked in future. These methods depend on the system to be carried out and are connected to the network to be operated. Actions can be taken at system-level and network-level.

System-level Actions: Bot infection is spread in the network by targeting the susceptibility present in the individual system. There exist many actions that should be

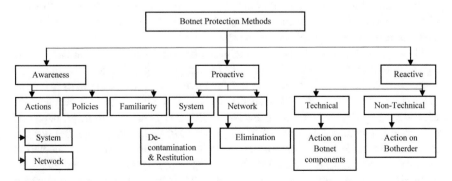

Fig. 2 Classification of botnet protection methods

carried out to protect the system from getting infected from the bot attack. Actions should be taken to keep the systems clean from any kind of chances that accelerate the infection from an unknown entity. A strong software or program must be present in the OS that will automatically upgrade it and other applications to the latest feature.

Network-level Actions: Network-level actions are the awareness methods that are carried out in network-level. The above system-level actions describe the awareness methods used in the individual system to remain safe from bot attack. These methods also give advantage to the network to protect systems. There are also some procedures that can be applied in network-level that decreases the threat of bot attack. Firewalls also help in defending from bots in the network. The gateways of the network can also have a proxy website through which the materials should be sent to check for a trace of malware, spyware, bots or any such malicious activity. Depending on the network specifications, honeypots or darknets can be installed that may trace the possibilities of definite attacks and help in identifying infected systems in the network. There is always a need for developing and implementing a complete network secure rule.

Policies: Awareness method also includes policies that need to be followed in order to protect the network from being attacked by a botnet. There is a necessity of enforcing a strong law along with the defensive methods against the bot attack. The motivation of botherder for the spreading of the botnet is mostly financial gain. The attraction towards spam business, thefts by clicks, malware propagation, and other such suspicious activities are being suggested by the researchers [27, 28]. To forbid the further expansion of botnet, legal actions should be carried out against the botherder. As there are no strict legal rules against botherder, they use to escape on the Internet. As a result, any kind of defensive actions or methods are not sufficient enough to stop the bot attack in other systems present in the network. Nowadays, the security government is more anxious to have a complete administration that will try to reduce the bot attacks [29]. The bot Rustock's putdown by Microsoft also supports in this respect [30]. The C&C communication server is present all over the world in different countries having different laws against cybercrime. So damaging of C&C server in any of the countries will not help in the international scenario. China, Japan, USA and European countries have very strict legal actions against crimes in the cyber world [31]. Still, a survey conducted by Damballa, a security organization, states that maximum C&C servers are introduced in France, USA and Germany [32]. This proves that stopping C&C server to propagate in one country does not takedown the whole bot network.

Familiarity: Familiarity is one of the major awareness methods that can help to a great extent to decrease the bot network from spreading. It is necessary that users should be familiar with the concept of bot and bot network. All the methods to protect the system from bot attack will be successful only if the user applies it. But usually, user do not pay attention to security measures till their system is running properly and not showing any problem. The lack of security becomes the reason why the systems become vulnerable for the bot attack. Most of the personal systems connected to the Internet do not have any security which makes them prone to the bot attack by botherder. Any technical security method may fail if the system user does not have the interest to apply it. The main reason for not applying the security is unawareness of the user about the consequences of the bot network. The user needs to be educated about the disruptive effects of bot attack and the methods they can use to protect their systems. This will motivate the user to secure their systems with the help of defensive methods. There exist some legal rules in some countries which make the user maintain security standards and impose fines if the standards are not followed [33].

4.2 Proactive Methods

Proactive methods are the methods used to remove the bot infection from the system or the network. They help in partial or complete removal of the infection depending on the techniques used. They focus in the retrieval of a system after the bot infection. As the name suggests, they are active methods do not trace the behaviour or activities. They attempt to solve the infection after the bot attack. Like awareness methods described above, the proactive methods can be applied in system-level or network-level.

System-level Methods: System-level methods help in recovering the system that is affected by bot attack. Detection techniques only detect the presence of botnet in the network and awareness methods make the system safe from bot attack before the attack occurs but they carry no actions to make a system infection free if it is already attacked by a bot. Some methods stop the command and control server but they bring no gain for the already infected systems. As the C&C channel provides the commands to the bots for execution of the task, the infected system or bots are still vulnerable to danger even after the take down of C&C server. It is because the bots execute the commands given by the botherder and keep performing malicious functions assigned until they have received commands to stop for the pre-issued start instruction through C&C server. For these reasons, it is necessary to have system-based methods that bring the infected system to a condition free from bot infection. There are some steps to be followed to clean the infection. They can be defined as:

Decontamination and Restitution: Decontamination is the first process to be followed. If decontamination is successful in removing bot infection, there is no need for restitution process. It involves in removing infection technically with the help of programs. There are some standard programs that emphasize on decontamination of bot infection from individual system. Considering their efficiency in removing bot infection, it cannot be completely reliable. Registry keys, drives and files are not the only places to look after to have a completely clean system. Sometimes the rootkit is also infected by bots. The contamination in Master Boot Record is the most difficult to clean and makes it necessary for the system user to restore operating system.

Restitution is the next process to be carried on after decontamination fails to clean the system to infection-free stage. In a maximum attack by bots, it is impossible to remove the infection in the system. So it is recommended to the system user to restore to a previous uninfected image. If such image is not found then the complete restitution of operating system is performed to bring the infected system to a normal state.

Network-level Methods: Network-level methods are the methods that are applied in the network to protect it from the already infected systems in the network. As soon as the network administrators knew about any bot attack in the systems or in any part of the network, immediately steps are taken to remove the infected system or part from the network. The most important process is to remove the infection so that the other systems connected to the network stay safe. After the infection is removed from them, it is again connected to the network. Elimination is the easiest and safest method in the network to have a control over the already infected machines.

Elimination: Elimination is a process where network administrators separate the systems being affected by bots or the part of network infected by a bot. All the separated systems are stored in a separate place, treated and again connected to the network back. Internet service providers (ISP) isolate the systems which are found to be under bot attack from the outside network so that no further harm is made to the system itself or to the network it is connected. These systems are prohibited to access any network facility except a group of domains that help in recovering the system back to clean stage. As per the standards of ISP rule, if a system passes then it is reconnected to the network. This concept is described as three-phase methods that are the detection of bot affected system, notifying about the system in the network and finally, restitution of those systems [34]. If the C&C channel is identified, it can be blocked so that no further attack is attempted by the channel. The URLs and IP address can be blocked that are involved in the communication channel. But in a simpler process, the communication of network traffic through suspicious ports should be stopped. Traffic inspection is also necessary to know about the presence of bot.

4.3   Reactive Methods

Reactive methods are used to terminate the bot operation or to significantly reduce the operation by attacking the botnet infrastructure and destroying it. As a result, the suspicious activities being carried out are either stopped as the botnet infrastructure is hampered. They are destructive method that targets directly to attack and stop the ongoing propagation of bots in the network. There can be two types of reactive methods:

Technical Methods: Technical methods are the methods that involve the botnet directly by technical means. Here, the technical constituents of bot network are directly attacked or targeted to stop the bot operation and propagation.

Action on bot components: Action on bot components involve technical target on the constituents of a botnet like the infected system, C&C channel and sometimes the server itself. The bot binary program and the C&C contain bugs just like different software programs [35, 36]. The remedial team can utilise these loopholes, gain access to the systems and can remotely access C&C to provide commands to stop the bots or infected systems. But the removal of bot infection without the system user's knowledge is seen as privacy intrusion and involves legal and ethical concerns. An unwanted command can be injected into the C&C channel. In a peer-to-peer botnet, some commands can be passed along with the command keys used by bots so that the real message never reaches the infected systems [21]. Some of the known attacks are Sybil attack and Eclipse attack. Sybil attack is against peer-to-peer botnet where the Sybil are placed so well that possess control over a part of the network and disrupt it. They can also change the destination to another peer rather than the actually infected systems [37]. Eclipse attack also has similar technique but it has control over a small part of the network. As the remedial team has an interpretation of the C&C channel, it can change and introduce commands according to their choice.

Non-technical methods: Non-technical methods are the methods that are not related to systems directly. These methods try to lessen the importance of bot to the botherder. Understanding the objective of the botherder, the methods try to futile the goal, thereby discouraging the botherder to spread bots over the network. The main focus of the method is to reduce the financial gain that the botherder has by attacking the business model running for it. The methods take mostly action on botherder.

Action on Botherder: Action on botherder is carried out so as to attack the business and the major objective of spreading bots by botherder. Multihost Adware Revenue Killer is a wide spread network of systems enabled to decrease the false advertisement, click per rates and unwanted software installation in the system [38]. Bots are also capable of gathering personal sensitive data [39, 40]. These data are being updated in a drop area in the network from where it is sold to the third party. Injecting fake data in this drop area will hamper botherder financial gain which will reduce the interest of botherder to spread bots again in the network. Traceable data can also be introduced into the network so that the botherder can be identified and penalised under law for such cyber offense.

5   Conclusion

Botnets have evolved as one of the most serious threats for communication over networks. As most of the day to day activities rely on networks, it has become vulnerable to bot infection by botherder. The insufficient knowledge about botnets, unawareness of users about bot propagation, detection and protection from them act as a catalyst for the cybercrime. In this paper, we describe various aspects of botnet detection and the methods involved in them. The detection methods help the user to know the protection against the bot infection and can utilise those methods to be safe from any unwanted attack. Future scope of work must include a complete protection method for the network and the systems associated with it so that user can safeguard their systems from bot infection.

References

1. Gu G., Perdisci R., Zhang J., and Lee W.: Botminer: clustering analysis of network traffic for protocol- and structure-independent botnet detection. In: Usenix Security Symposium, vol. 5, No. 2, pp. 139–154 (2008)
2. Gu, G., Porras, P., Yegneswaran, V., Fong, M., Lee, W.: Bothunter: detecting malware infection through ids-driven dialog correlation. In: Usenix Security Symposium, vol. 7, pp. 1–16 (2007)
3. Zhuang, L., Dunagan, J., et al.: Characterizing botnets from email spam records. In: USENIX Workshop on Large-Scale Exploits and Emergent Threats, vol. 8, pp. 1–9 (2008)
4. Rajab, M.A., Zarfoss, J., Monrose, F., Terzis, A.: A multifaceted approach to understanding the botnet phenomenon. In: Proceedings of the 6th ACM SIGCOMM Conference on Internet Measurement, pp. 41–52 (2006)
5. Villamarín-Salomón, R., Villamarín-Salomón, J.: Identifying botnets using anomaly detection techniques applied to DNS traffic. In: Proceedings of the 5th IEEE Consumer Communications and Networking Conference, pp. 476–481 (2008)
6. Choi, H., Lee, H., Kim, H.: Botnet detection by monitoring group activities in DNS traffic. In: Proceedings of the 7th IEEE International Conference on Computer and Information Technology, pp. 715–720 (2007)
7. Goebel, J., Holz, T.: Rishi: identify bot contaminated hosts by IRC nickname evaluation. In: Usenix Workshop on Hot Topics in Understanding Botnets, vol. 7, p. 8 (2007)
8. Strayer, W.T., et al.: Botnet detection based on network behavior. In: Botnet Detection, vol. 36, pp. 1–24. Springer, US (2008)
9. Holz, T., Gorecki, C., Rieck, K., Freiling, F.: Detection and mitigation of fast-flux service networks. In: Proceedings of the 15th Annual Network and Distributed System Security Symposium (2008)
10. Gu, G., Yegneswaran, V., Porras, P., Stoll, J., Lee, W.: Active botnet probing to identify obscure command and control channels. In: Computer Security Applications Conference IEEE, pp. 241–253 (2009)
11. Snort IDS web page. http://www.snort.org, March (2006)
12. Rossow, C., Dietrich C.J.: Provex: detecting botnets with encrypted command and control channels. In: International Conference on Detection of Intrusions and Malware, and Vulnerability Assessment, pp. 21–40. Springer (2013)
13. Perdisci, R., Lee, W., Feamster, N.: Behavioral clustering of HTTP-based malware and signature generation using malicious network traces. In: Usenix Symposium on Networked Systems Design & Implementation, pp. 391–404 (2010)
14. Wurzinger, P., et al.: Automatically generating models for botnet detection. In: European Symposium on Research in Computer Security, pp. 232–249. Springer (2009)
15. Rehak, M., Pechoucek, M., et al.: Adaptive multiagent system for network traffic monitoring. IEEE Intell. Syst. 3(24), 16–25 (2009)
16. Caglayan, A., Toothaker, M., et al.: Behavioral analysis of botnets for threat intelligence. Inf. Syst. E-Bus. Manag. 10(4), 491–519 (2012). (Springer)
17. Ramsbrock, D., Wang, X., Jiang, X.: A first step towards live botmaster traceback. In: Proceedings of the 11th International Symposium on Recent Advances in Intrusion Detection, pp. 59–77. Springer (2008)
18. FireEye: Next generation threat protection. FireEye Inc. (2011)
19. Damballa,: Damballa::homepage (2011)
20. Grizzard, J.B., Johns, T.: Peer-to-peer botnets: overview and case study. In: Usenix Workshop on Hot Topics in Understanding Botnets (2007)
21. Holz, T., Steiner, M., Dahl, F., Biersack, E., Freilling, F.: Measurements and mitigation of peer-to-peer-based botnets: a case study on storm worm. In: Proceedings of the 1st Usenix Workshop on Large-Scale Exploits and Emergent Threats, vol. 8, pp. 1–9 (2008)
22. Caballero, J., Poosankam, P., Kreibich, C., Song, D.: Dispatcher: enabling active botnet infiltration using automatic protocol reverse engineering. In: Proceedings of the 16th ACM Conference on Computer and Communications Security, pp. 621–634 (2009)

23. Liu, L., Chen, S., Yan, G., Zhang, Z.: Bottracer: execution-based bot-like malware detection. In: International Conference on Information Security, pp. 97–113. Springer (2008)
24. Stinson, E., Mitchell, J.C.: Characterizing bots' remote control behaviour. In: International Conference on Detection of Intrusions & Malware and Vulnerability Assessment, pp. 89–108. Springer (2007)
25. Karasaridis, A., Rexroad, B., Hoeflin, D.: Wide-scale botnet detection and characterization. In: Proceedings of the First Conference on First Workshop on Hot Topics in Understanding Botnets, p. 7 (2007)
26. Gu, G., Zhang, J., Lee, W.: Botsniffer: detecting botnet command and control channels in network traffic. In: Network and Distributed System Security Symposium (2008)
27. Holz, T., Engelberth, M., Freiling, F.: Learning more about the underground economy: a case-study of key loggers and dropzones. In: European Symposium on Research in Computer Security, pp. 1–18. Springer (2009)
28. Kanich, C., Kreibich, C., et al.: Spamalytics: an empirical analysis of spam marketing conversion. In: Proceedings of the 15th ACM Conference on Computer and Communications Security, pp. 3–14 (2008)
29. Furfie, B.: Laws must change to combat botnets Kaspersky. Feb (2011)
30. Bright, P.: How Operation b107 decapitated the Rustock botnet (2011)
31. A.P.E.C, AEC: Guide on Policy and Technical Approaches against Botnet. Dec (2008)
32. Leyden, J.: Botnet-harbouring survey fails to accounts for sinkholes (2011)
33. Orgill, G.L., Romney, G.W., et al.: The urgency for effective user privacy-education to counter social engineering attacks on secure computer systems. In: Proceedings of the 5th Conference on Information Technology Education, pp. 177–181. ACM (2004)
34. Mody, N., O'Reirdan, M., Masiello, S., Zebek, J.: Common best practices for mitigating large scale bot infections in residential networks, July (2009)
35. Li, P., Salour, M., Su, X.: A survey of internet worm detection and containment. IEEE Commun. Surv. Tutorials 10(1), 20–35 (2008)
36. Cho, C.Y., Caballero, J.: Botnet infiltration: finding bugs in botnet command and control (2011)
37. Dinger, J., Hartenstein, H.: Defending the sybil attack in p2p networks: taxonomy, challenges, and a proposal for self-registration. In: First International Conference on Availability, Reliability and Security, p. 8. IEEE (2006)
38. Ford, R., Gordon, S.: Cent, five cent, ten cent, dollar: hitting botnets where it really hurts. In: Proceedings of the 2006 Workshop on New Security Paradigms, p. 310. ACM (2006)
39. IEEE 802.11ah. 2018: Accessed 23 Feb 2018. Retrieved from https://en.wikipedia.org/wiki/IEEE_802.11ah
40. Lee, A., Atkison, T.: A comparison of fuzzy hashes: evaluation, guidelines, and future suggestions (2017)

An Experimental Study on Decision Tree Classifier Using Discrete and Continuous Data

Monalisa Jena and Satchidananda Dehuri

Abstract Classification is one of the fundamental tasks of pattern recognition, data mining, and big data analysis. It spans across the domain for classifying novel instances whose class labels are unknown prior to the development of model. Decision trees like ID3, C4.5, and other variants for the task of classification have been widely studied in pattern recognition and data mining. The reason is that decision tree classifier is simple to understand, and its performance has been comparable with many promising classifiers. Therefore, in this work, we have developed a two-phase method of decision tree classifier for classifying continuous and discrete data effectively. In phase one, our method examines the database, whether it is a continuous-valued or discrete-valued database. If it is a continuous-valued database, then the database is discretized in this phase. In the second phase, the classifier is built and then classifies an unknown instance. To measure the performance of these two phases, we have experimented on a few datasets from the University of California, Irvine (UCI) Machine Learning repository and one artificially created dataset. The experimental evidence shows that this two-phase method of constructing a decision tree to classify an unknown instance is effective in both continuous and discrete cases.

Keywords Decision tree · Classification · Discretization · Data mining

1 Introduction

Classification is a way of fitting objects to a category, which best suits its characteristics. There are number of classification techniques in pattern recognition and data mining [1]. Decision tree (DT) induction is a technique of supervised learning in which a knowledge-based expert system is built by inductive inference from examples [2]. Inductive learning systems are helpful in extracting useful knowledge from

M. Jena (✉) · S. Dehuri
Department of I & CT, Fakir Mohan University, 756019 Balasore, Odisha, India
e-mail: bmonalisa.26@gmail.com

S. Dehuri
e-mail: satchi.lapa@gmail.com

© Springer Nature Singapore Pte Ltd. 2020
P. K. Mallick et al. (eds.), *Cognitive Informatics and Soft Computing*,
Advances in Intelligent Systems and Computing 1040,
https://doi.org/10.1007/978-981-15-1451-7_35

patterns. The main objective is to build a model, which predicts the class label of an unknown instance by observing the behavior of several input instances. A DT is a tree-like structure in which leaf nodes represent the class values, each intermediate node corresponds to one of the attributes, and each branch represents attribute values. In general, DT divides a big dataset into smaller ones until it obtains efficiency up to the mark, which helps users to predict the class labels of unknown data samples. More use of DT in the field of classification is due to its noise handling capacity, low computational cost, and capability of handling redundant attributes. The main challenge in the DT is to identify which attributes to consider as the root node, intermediate node, and each level of division. This is achieved by attribute selection measures. There exist different types of attribute selection measures like information gain, gain ratio, gini index, and many more [3]. In this work, we use the information gain-based method, i.e., Iterative Dichotomiser (ID3) as an attribute selection measure [4]. ID3 is suitable for discrete-valued attributes but in real world, datasets do not contain discrete values only, they also contain continuous or mixed-mode data values. To handle continuous data, various methods have been adopted earlier. In this work, we use discretization to convert continuous to discrete data. Discretization is a method of converting continuous-valued attributes into discrete ones by assigning categorical values to intervals. Various discretization methods like equal width partitioning, equal frequency partitioning, multisplitting, fuzzy discretization and many more exist [5] of which, we use equal width partitioning method in this work as it is effective and easy to understand.

The rest of the sections are set out as follows: Sect. 2 includes related work; in Sect. 3, some preliminaries are discussed. Section 4 represents our contribution, i.e., it describes the proposed two-phase framework for classification and prediction of class label of the unknown sample. Numerical results are shown in Sect. 5, and lastly, in Sect. 6, the paper is concluded along with future works.

2   Related Work

A decision tree is one of the popular data mining algorithms used for both classification and regression. Various methods and algorithms have been proposed earlier for inducing DTs. During the 1980s, Quinlan, a pioneer in machine learning, developed the ID3 algorithm. ID3 uses entropy-based attribute selection measure [4]. He then proposed C4.5, an extended version of ID3 and uses 'gain ratio' as the default splitting criterion. C4.5 is used for both discrete and continuous attributes. For samples containing continuous attributes, Quinlan used a threshold-based splitting criterion for building the DT classifier and predicting the class label of unknown samples [6]. Breiman et al. reported the initiation of binary DTs in the form of classification and regression trees (CART) [7]. ID3, C4.5, and CART follow a non-backtracking approach in which DTs are constructed in a top-down recursive divide-and-conquer manner [8]. Jearanaitanakij proposed the modified ID3 algorithm, which divides the continuous features into intervals and performed the classification within those intervals [9].

Discretization helps in number of applications in machine learning models for classification. One of the applications is label ranking where data are ranked based on the features given and the constraint required by the end users. Sa et al. have carried out multiple works based on entropy-based discretization on ranking data [10]. They used minimum description length principle (MDLP) to measure information gain by comparing entropy value before and after the splitting. Ching et al. have proposed an elegant algorithm for class-dependent discretization for both continuous and mixed data [11]. Unlike the above work, Liu et al. have provided a global optimum algorithm for discretization for inductive learning [12]. They have used the normalized mutual information to quantify the inter-dependency between the class label and continuous variables. They have utilized the concept of fractional programming for obtaining the global optimum value.

Uther et al. have presented a decision tree model for discretization for pruning irrelevant states for state space reinforcement learning [13]. Chen et al. have proposed an elegant approach for discretization environment data for modeling Bayesian network classifier and logistic regression. In their work, they have used both balanced and unbalanced data for optimizing discretization [14].

3 Preliminaries

While inducing a decision tree, it is very important to choose the heuristic for selecting the splitting criterion. ID3 uses entropy-based attribute selection measure. It decides ordering of attributes in the nodes of DT. The expected information or entropy measure is based on the probability of belongingness (P_i) of any random tuple of a dataset D to a particular class [8]. For 'm' classes, entropy can be defined as follows:

$$\text{En}(D) = \sum_{i=1}^{m} P_i \log_2 P_i \tag{1}$$

where, $Pi = |Si|/|S|$. $|Si|$ is the number of samples of S in class C_i, and $|S|$ is the set containing total number of samples.

For k distinct values $\{a_1, a_2, a_3, \ldots, a_k\}$ of each attribute A, the entropy of the partition with respect to A is calculated as follows:

$$\text{En}_A(D) = \sum_{i=1}^{n} \frac{|D_k|}{|D|} \times \text{En}(D_k) \tag{2}$$

where n refers to the number of partitions, and En (D_k) is the entropy of the partition with respect to the values of an attribute A. D_k consists of those tuples in D that have outcome a_k of A. This is needed to obtain an exact classification of the instances.

The information gain, i.e., $G(A)$ is the difference between entropy of the dataset and that of the partition.

$$G(A) = \text{En}(D) - \text{En}_A(D) \tag{3}$$

The attribute with the highest gain value is chosen as the splitting attribute. For better understanding of the readers, 'lenses' dataset from UCI machine learning repository [15] is taken here. Attributes of lenses dataset are discrete-valued. The dataset is having three class labels such as patient with hard contact lenses, patient with soft contact lenses, and patient without contact lenses. The dataset is having four attributes: age of the patient: (1) young, (2) pre-presbyopic, and (3) presbyopic; spectacle prescription (SP): (1) myope and (2) hypermetrope; astigmatic (ast): (1) no, (2) yes; tear production rate (TPR): (1) reduced and (2) normal. The entropy of the dataset can be calculated using Eq. (1) as follows:

$$\text{En}(D) = -\frac{15}{24}\log_2\left(\frac{15}{24}\right) - \frac{5}{24}\log_2\left(\frac{5}{24}\right) - \frac{4}{24}\log_2\left(\frac{4}{24}\right) = 1.326$$

The entropy of the partition with respect to attribute age is

$$\begin{aligned}
\text{En}_{\text{age}}(D) = \frac{8}{24} &\times \left(-\frac{2}{8}\log_2\left(\frac{2}{8}\right) - \frac{2}{8}\log_2\left(\frac{2}{8}\right) - \frac{4}{8}\log_2\left(\frac{4}{8}\right)\right) + \frac{8}{24} \\
&\times \left(-\frac{1}{8}\log_2\left(\frac{1}{8}\right) - \frac{2}{8}\log_2\left(\frac{2}{8}\right) - \frac{5}{8}\log_2\left(\frac{5}{8}\right)\right) + \frac{8}{24} \\
&\times \left(-\frac{1}{8}\log_2\left(\frac{1}{8}\right) - \frac{1}{8}\log_2\left(\frac{1}{8}\right) - \frac{6}{8}\log_2\left(\frac{5}{24}\right)\right) = 1.287
\end{aligned}$$

The information gain value on the partition with respect to age using Eq. (3) is $G(\text{age}) = 1.326 - 1.287 = 0.039$. Similarly, $G(\text{Ast}) = 0.377$; $G(\text{SP}) = 0.039$; $G(\text{TPR}) = 0.549$. Hence, from this, it is observed that the tear production rate attribute is having maximum gain. So, it is used as the splitting criterion. Hence, attribute TPR is taken as root node, and its attribute values are its branches.

Table 1 shows the lenses dataset, and Fig. 1 depicts the partitioned dataset. It is realized that in real-life situations, the database may not always contain discrete values. It may contain continuous values. For approaches, which can directly handle continuous data, learning is often less effective and efficient. Hence, discretization has been playing an active role in data mining and pattern recognition.

4 Our Contribution

Our proposed two-phase approach is depicted in Fig. 2. The dataset is supplied to phase I, and then it is checked whether it is discrete or continuous. If the given

Table 1 Lenses dataset

S. No.	Age	Spectacle prescription(SP)	Astigmatic	Tear production rate (TPR)	Class label
1	1	1	1	1	3
2	1	1	1	2	2
3	1	1	2	1	3
4	1	1	2	2	1
5	1	2	1	1	3
6	1	2	1	2	2
7	1	2	2	1	3
8	1	2	2	2	1
9	2	1	1	1	3
10	2	1	1	2	2
11	2	1	2	1	3
12	2	1	2	2	1
13	2	2	1	1	3
14	2	2	1	2	2
15	2	2	2	1	3
16	2	2	2	2	3
17	3	1	1	1	3
18	3	1	1	2	3
19	3	1	2	1	3
20	3	1	2	2	1
21	3	2	1	1	3
22	3	2	1	2	2
23	3	2	2	1	3
24	3	2	2	2	3

Fig. 1 Tuples obtained after partition for lenses dataset

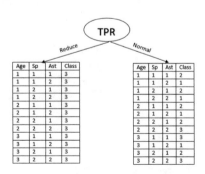

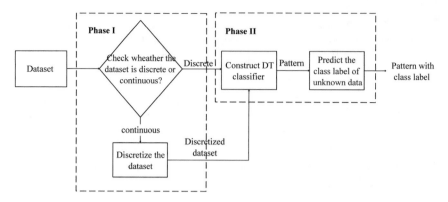

Fig. 2 Structure of two-phase framework

dataset contains discrete values, then the decision tree classifier is built, and the class label of unknown instance is predicted. If it contains continuous values, then our discretization module will transform it into discrete intervals. Through discretization, the continuous variables are divided into equal width intervals. Then, a fixed number of intervals divide the range equally. In phase II, the classifier is built using the training samples. Then, the classifier is provided with testing samples to predict the class label of unknown instances. The algorithm of our two-phase approach is as follows:

Algorithm for Classification using Two-Phase Approach

> **Input:** Dataset
> **Output:** Pattern with class label
> **Step 1:** Read the dataset
> **Step 2:** Check the dataset
>   – if it is discrete then go to step 3
>   – if it is continuous then
> • discretize the dataset, and
> • go to step 3
> **Step 3:** Build the classifier
> – draw decisions
> – find efficiency
> **Step 4:** Assign class label to unknown instance

5 Numerical Results

The proposed two-phase classifier has been implemented using MATLAB 8.1 (R2013a Sun). All the experimental works have been carried out on a system with i7 processor having 2.90 GHz clock speed. It is having 64-bit OS, 1 TB hard disk and 8 GB RAM configuration.

In order to validate the proposed framework, we have taken two datasets tic-tac-toe and lenses dataset from UCI repository, the details of lenses dataset are already explained in Sect. 2, and a brief description of the other one is given here. Besides these two, another dataset is artificially created containing continuous-valued attributes. The artificially created dataset is depicted in Table 2.

i. **Tic-Tac-Toe**: This dataset consists of all possible arrangements of 'X' and '0' on a board of 3×3 square at end of the game. The first player is said to have symbol 'X.' Out of all arrangement, the first player is said to win if any three horizontals, vertical, or diagonal square have 'X' on it. The target output is the chance to win (positive) or lose (negative) for the first player. However, the same can be considered for the second player also when the end game arrangements have three zeros in horizontal, vertical, or diagonal square. This dataset consists of nine features, which are the coordinates of all nine squares on the board. Total number of instances is 958.

ii. **Synthetic dataset**: It consists of five attributes including the class attribute and 37 instances. It is a bi-class problem (1 and 2). In contrast to other datasets taken, this one contains continuous-valued attributes. Through discretization, it is converted into discrete intervals as illustrated in Table 2. This discretized dataset is given as the input to the second phase of the two-phase model for the creation of DT model to classify an unknown instance.

Each of the above-mentioned datasets is partitioned into training and testing sets in 7:3 ratio. The performance of the model has been measured by considering the testing sets of the datasets, which are presented in the respective confusion matrices as shown in Table 3. The performance of the proposed two-phase model has been evaluated based on the values obtained from the confusion matrices. Confusion matrix is the way of capturing the class belongingness of different instances of the dataset [8]. The performance measures accuracy, precision, recall, and F-measure have been obtained from the confusion matrix generated through the experiments. The obtained results are depicted in Table 3.

6 Conclusions and Future Work

In this work, we have developed a two-phase method of decision tree for both discrete and continuous data. Two real-world datasets having discrete-valued attributes and one artificially created dataset having continuous-valued attributes have been considered for evaluation of the framework. From the experimental results, it is observed that our method is very suitable for handling both categories of data effectively and efficiently. Our future work includes the study of different types of measure for attribute selection, automatic construction by using evolutionary algorithms, handling of mixed-mode data and its applications in many domains of interest.

Table 2 Sample dataset and the discretized version of it

S. No.	Dataset before discretization					Dataset after discretization				
	A1	A2	A3	A4	Class	A1	A2	A3	A4	Class
1	1	2	3	5	1	30	32	32	34	1
2	2	3	56	4	2	30	32	61	34	2
3	2	4	62	56	1	30	62	90	64	1
4	2	56	45	22	1	30	32	61	34	1
5	2	2	5	5	2	30	32	32	34	2
6	6	8	25	58	2	30	32	32	64	2
7	2	5	5	205	2	30	32	32	34	2
8	5	25	2	22	1	30	32	32	34	1
9	22	5	5	5	2	30	32	32	34	2
10	2	22	5	5	2	30	32	32	34	1
11	5	5	22	22	1	30	32	32	34	2
12	2	22	5	14	2	30	32	32	34	1
13	5	21	1	11	1	30	32	32	34	2
14	1	11	4	1	1	30	32	32	34	1
15	14	41	1	1	2	30	62	32	34	1
16	1	25	2	25	2	30	32	32	34	2
17	41	68	2	5	1	30	92	32	34	2
18	25	5	1	44	1	59	32	32	34	1
19	58	78	9	75	2	30	92	32	64	1
20	5	55	8	25	2	59	62	32	94	2

(continued)

Table 2 (continued)

S. No.	Dataset before discretization					Dataset after discretization				
	A1	A2	A3	A4	Class	A1	A2	A3	A4	Class
21	9	51	5	89	2	30	62	32	34	2
22	5	7	1	5	1	30	32	32	94	2
23	15	8	42	4	1	30	32	32	34	1
24	5	84	1	1	1	30	92	61	34	1
25	14	5	85	5	2	30	32	32	34	1
26	5	22	5	22	1	30	32	90	34	2
27	78	89	4	8	2	30	92	32	34	1
28	4	22	5	2	2	88	32	32	34	2
29	5	5	2	54	1	30	32	32	34	2
30	2	2	5	4	1	30	32	32	64	1
31	86	4	58	7	2	30	32	32	34	1
32	5	11	25	84	1	88	32	61	64	2
33	5	25	5	7	2	30	32	32	94	1
34	2	1	25	5	1	30	32	32	34	2
35	51	1	25	55	2	30	32	32	34	1
36	21	2	2	1	1	59	32	32	64	2
37	12	2	22	1	1	30	32	32	34	1

Table 3 Performance metrics of the datasets

Dataset		Confusion matrix					Evaluation parameter			
			Predicted				Precision	Recall	F-measure	Accuracy
			Class 1	Class 2	Class 3					
Lenses (three class)	Actual	Class 1	2	0	0		0.93	0.83	0.88	0.75
		Class 2	0	1	1					
		Class 3	0	0	4					
			Class 1	Class 2	Class 2					
Tic-tac-toe (two class)	Actual	Class 1	151	9			0.94	0.95	0.94	0.94
		Class 2	8	119						
			Class 1	Class 2						
Synthetic dataset (two class)	Actual	Class 1	5	1			1	0.83	0.91	0.91
		Class 2	0	5						

Acknowledgements Thanks to Mr. Sagar Muduli, MCA student, Dept. of I & CT, F. M. University, Balasore, Odisha, for his notable contribution in this work.

References

1. Phyu, TN.: Survey of classification techniques in data mining. In: Proceedings of the International Multi Conference of Engineers and Computer Scientists, vol. 1, pp. 18–20 (2009)
2. Wang, R., Kwong, S., Wang, X.Z., Jiang, Q.: Segment based decision tree induction with continuous valued attributes. IEEE Trans. Cybern. **45**(7), 1262–1275 (2015)
3. Loh, W.Y.: Fifty years of classification and regression trees. Int. Stat. Rev. **82**(3), 329–348 (2014)
4. Quinlan, J.R.: Decision trees and decision-making. IEEE Trans. Syst. Man Cybern. **20**(2), 339–346 (1990)
5. Garcia, S., Luengo, J., Saez, J.A., Lopez, V., Herrera, F.: A survey of discretization techniques: taxonomy and empirical analysis in supervised learning. IEEE Trans. Knowl. Data Eng. **25**(4), 734–750 (2013)
6. Quinlan, J.R.: Improved use of continuous attributes in c4.5. J. Artif. Intell. Res. **4**, 77–90 (1996)
7. Breiman, L.: Classification and regression trees. Routledge (2017)
8. Han, J., Pei, J., Kamber, M.: Data mining: concepts and techniques. Elsevier (2011)
9. Jearanaitanakij, K.: Classifying continuous data set by id3 algorithm. In: Information, Communications and Signal Processing, 2005 Fifth International Conference, pp. 1048–1051. IEEE (2005)
10. De Sa, C.R., Soares, C., Knobbe, A.: Entropy-based discretization methods for ranking data. Inf. Sci. **329**, 921–936 (2016)
11. Ching, J.Y., Wong, A.K.C., Chan, K.C.C.: Class-dependent discretization for inductive learning from continuous and mixed-mode data. IEEE Trans. Pattern Anal. Mach. Intell. **17**(7), 641–651 (1995)
12. Liu, L., Wong, A.K.C., Wang, Y.: A global optimal algorithm for class-dependent discretization of continuous data. Intell. Data Anal. **8**(2), 151–170 (2004)
13. Uther, W.T., Veloso, M.M.: Tree based discretization for continuous state space reinforcement learning. In: Aaai/iaai, pp. 769–774 (1998)
14. Chen, Y.C., Wheeler, T.A., Kochenderfer, M.J.: Learning discrete bayesian networks from continuous data. J. Artif. Intell. Res. **59**, 103–132 (2017)
15. Dheeru, D., Taniskidou, E.K.: UCI machine learning repository (2017)

Analysis of Speech Emotions Using Dynamics of Prosodic Parameters

Hemanta Kumar Palo and Mihir N. Mohanty

Abstract In this paper, an attempt is made to explore the dynamics of speech prosody to characterize and classify emotional states in a speech signal. The local or fine variations describing the prosodic dynamics are combined with the static prosodic parameters for a possible enhancement in the emotional speech recognition (ESR) accuracy. The efficient vector quantization (VQ) clustering algorithm has been applied to compress the static and dynamic parameters before further processing in a radial basis neural network (RBFNN) platform. Results reveal an improvement in ESR accuracy of 86.05% by involving both static and dynamic prosodic features as compared to 84.92% accuracy when the combination of static prosodic feature simulated alone.

Keywords Speech emotion · Speech prosody · Feature extraction · Radial basis function neural network · Classification

1 Introduction

Emotion perception is almost evidenced in every sphere of human life and is manifested from facial expression, vocal conversion, and cultural artifacts such as pictures or music. Among these modalities of expressive emotions, speech remains a natural medium and is most convenient, particularly when the conversation takes place via phone. The task of recognizing speech emotions remains difficult as the performance of identification system models depends heavily on many [1]. These are (a) suitable and reliable features that contain rich affective information, (b) an authenticated database balanced with genders, emotions, samples, etc., and (c) a suitable machine learning algorithms for efficient classifications of the intended emotions.

H. K. Palo · M. N. Mohanty (✉)
Department of Electronics and Communication Engineering, ITER, Siksha 'O' Anusandhan University, Bhubaneswar, Odisha, India
e-mail: mihirmohanty@soa.ac.in

H. K. Palo
e-mail: hemantapalo@soa.ac.in

© Springer Nature Singapore Pte Ltd. 2020
P. K. Mallick et al. (eds.), *Cognitive Informatics and Soft Computing*,
Advances in Intelligent Systems and Computing 1040,
https://doi.org/10.1007/978-981-15-1451-7_36

333

Among mostly discussed features, the spectral features are extracted at frame-level, whereas the prosodic features are extracted at utterance level. The existing works in this field mostly focus on speech prosody involving statistical parameters that represent a speech sample as a whole [2–4]. However, these descriptions are static in nature and do not consider the dynamism or variation as time elapses during a spoken emotional outburst. The fine or local variations of the prosodic information with time can provide crucial inputs in describing emotional attributes and need to be explored in the current scenario. This lays the foundation for this work and motivates the authors to characterize and classify the speech emotions using both static and dynamics speech prosody for better speech emotion portrayal.

The novelty of this work is formalized as follows:

i.   Initially, the prosodic features such as the fundamental frequency or pitch, zero-crossing rate (ZCR), autocorrelation coefficient (ACF), and energy are extracted from each emotional sample at a frame level.
ii. The extracted features of each frame are then clustered into a single feature using the VQ compression technique.
iii. Extraction of derivative (delta) VQ-prosodic features is carried out for the characterization of the desired speech emotions.
iv. The ESR classification accuracy has been studied by using both static and dynamic prosodic features in a RBFNN platform.

The rest of the paper is organized as briefed below. The materials and methods used for the proposed work are explained in Sect. 2. Section 3 of this work analyzes the findings of this work graphically as well as in tabular form. Section 4 provides the conclusion.

2 Materials and Method

The chosen emotional speech database and the feature extraction technique used in this work have been explained in this section.

2.1 Feature Extraction Techniques

A few of the frame-level prosodic features used in this work are the ZCR, ACF, Pitch (f_0), and the energy. A 30 ms frame size with 10 ms frame overlapping using a Hamming window has been employed to obtain the windowed signal of a speech sample. The VQ feature compression technique applies to each set of framed feature of a sample for the extraction of the static and dynamic VQ-prosodic features.

2.1.1 Autocorrelation Coefficient (ACF)

The ACF provides the correlation or similarity between two adjacent frames, features, or characteristics of a signal with respect to time. It can play an important role, as it gives the complementary energy information in a signal [5]. For a speech signal, $s(n)$ defined for all 'n', it can be represented by

$$R_s(\tau) = \lim_{M \to \infty} \frac{1}{2M+1} \sum_{n=-M}^{M} s(n)s(n+\tau) \tag{1}$$

where M is the size or the frame or the length of the section to be analyzed and 'τ' is the delay or shift parameter.

2.1.2 Pitch or Fundamental Frequency (F_0)

The fundamental frequency (F_0) or pitch of a signal varies with the sub-glottal air pressure, tension, and vibration of the vocal folds [6]. The pitch period can be estimated as

$$\text{pitch period}(m) = \arg_m \max R_m(\tau), \quad m = 1, 2, \ldots, M \tag{2}$$

The pitch is computed from Eq. 2 as

$$F_0 = \frac{f_s}{\text{pitch period}(\tau)} \tag{3}$$

where f_s denotes the sampling frequency of the signal under consideration.

2.1.3 Zero-Crossing Rate (ZCR)

When a signal crosses the zero axes number of times, it indicates the presence of higher frequency contents in this signal, i.e., the signal fluctuates more rapidly. Since speech emotions are frequency dependent, hence, the ZCR can provide the desired emotional information of a speech signal. The short-time average ZCR can be expressed as.

$$z(m) = \sum_{n=-\infty}^{\infty} |\text{sgn}(s(n)) - \text{sgn}(s(n-1)| w(m-n) \tag{4}$$

where

$$\text{sgn}(s(m)) = \begin{array}{l} 1 \text{ for } s(m) \geq 0 \\ -1 \text{ for } s(m) < 0 \end{array} \tag{5}$$

and $w(m)$ is the windowed speech signal.

2.1.4 Short-Time Energy (STE)

The energy of a speech signal provides the variation in its amplitude and reveals the signal loudness. As human emotions have different arousal levels, the signal possesses different level of amplitude and energy during an emotional outburst. For a signal $s(n)$, the STE is given by

$$E_{ST}(m) = \sum_{n=-\infty}^{\infty} |s(n)w(m-n)|^2 \tag{6}$$

where $w(m)$ is the analyzing window.

2.1.5 Delta Prosodic Features

The delta prosodic ZCR features (Z_Δ) of an emotional sample are derived as per the equation [7–9].

$$Z_\Delta = \varphi \times \sum_{p=1}^{2} n \times [Z(q+p) - \text{MFCC}(q-p)], \quad q = 1, 2, \ldots Q \tag{7}$$

Where, Z_Δ denote the delta ZCR features with a frequency scaling of $\varphi = 2$ and Q indicates the number of Z_Δ features corresponding to an emotional sample. Similarly, the other prosodic feature dynamics are extracted.

The proposed model is shown in Fig. 1.

2.2 The Chosen Database

This work uses the Surrey Audio-Visual Expressed Emotion (SAVEE) emotional speech database as it is in English language [10]. Further, much of the work has already been done with the authenticated results with this dataset that makes it a potential candidate to make a comparable standard for this work. It comprises emotional utterances simulated by four British speakers and involves seven emotional states. There are four hundred eighty emotional samples having sixty utterances for each category of emotions. However, the neutral state has one hundred twenty

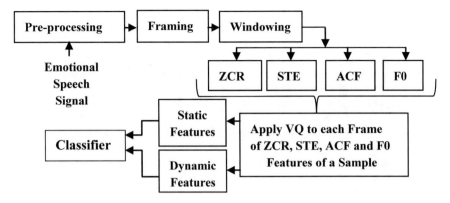

Fig. 1 Proposed model

utterances. The recordings are carried out with a sampling rate of 44.1 kHz and the average classification accuracy with this dataset has been 61%.

3 Results and Discussion

The characterization of emotional states such as angry, happy and sad state using the dynamic prosodic feature sets is graphically shown in Fig. 2 through Fig. 5. The variation in dynamic prosodic features of the f_0_VQ, STE_VQ, ZCR_VQ and ACF_VQ

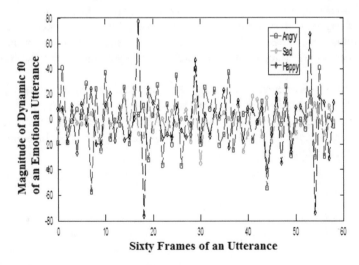

Fig. 2 Characterization of angry, sad, and happy SAVEE speech emotion using dynamic F_0_VQ features

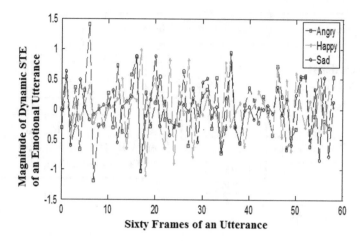

Fig. 3 Characterization of angry, sad, and happy SAVEE speech emotion using dynamic STE_VQ features

for SAVEE database has been shown in Figs. 2, 3, 4 and 5. The graphs are drawn for sixty frames of a single utterance of these three emotional states. In all these figures, angry emotion has shown higher magnitude compared to all other classes of emotion followed by happy emotion. This is due to the presence of higher frequency components, higher energy contents in these emotions compared to sad emotion. The sad emotion has the largest variety in terms of autocorrelation coefficients although adjacent values are quite closely correlated as observed from these figures. This reveals its irregularity between adjacent frames. Further, the sad state is described by a low level of arousal with low-frequency components as observed from these figures. A detailed analysis of these figures reveals a wide variation among frames of the emotional sample under consideration. It indicates the dynamic prosodic information can lead to a better description of the emotional signal by providing more discriminating features.

The RBFNN classification accuracy with the static, dynamic and the combination of static and dynamic prosodic features are shown in Table 1. It can be inferred that the use of prosodic dynamism with static information leads to a better recognition accuracy as compared to either the static or the dynamic prosodic combination simulated separately by the classifier.

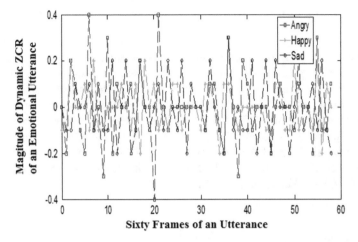

Fig. 4 Characterization of angry, sad, and happy SAVEE speech emotion using dynamic ZCR_VQ features

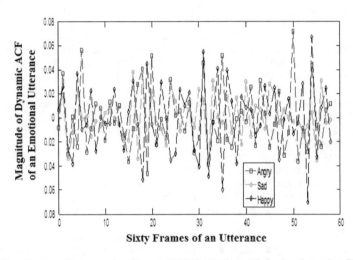

Fig. 5 Classification of angry, sad, and happy SAVEE speech emotion using dynamic ACF features

Table 1 Accuracy with different features for emotions

Features	RBFN accuracy percentage			
	Angry (%)	Sad (%)	Happy (%)	Average (%)
Static	87.20	82.11	85.45	84.92
Dynamic	84.51	79.83	81.60	81.98
Static + dynamic	88.22	83.74	86.18	86.05

4 Conclusions

Variation in speech emotions with respect to time is essential to estimate an accurate affective state of a person. As features describing human emotions vary widely in a sentence during expressive emotions, the local or variation among adjacent samples is more informative as observed from our results. In a way, it helps to characterize the low arousal state from the high arousal state more efficiently. Hence, the combination of the static and dynamic prosodic parameters is expected to provide an enhanced SER accuracy as observed in our results. Further, such combinational framework may provide a new direction to this field due to availability of more emotional relevant information. It has been kept as our future work.

References

1. Palo, H.K., Mohanty, M.N.: Compartive analysis of neural networks for speech emotion recognition. Int. J. Eng. Technol. 7(4), 111–126 (2018)
2. Rao, K.S., Reddy, R., Maity, S., Koolagudi, S. G.: Characterization of emotions using the dynamics of prosodic features. In: Speech Prosody 2010-Fifth International Conference (2010)
3. Mannepalli, K., Maloji, S., Sastry, P.N., Danthala, S., Mannepalli, D.P.: Text independent emotion recognition for Telugu speech by using prosodic features. Int. J. Eng. Technol. 7(4), 111–126; 7(2), 594–596 (2018)
4. Cao, H., Verma, R., Nenkova, A.: Speaker-sensitive emotion recognition via ranking: studies on acted and spontaneous speech. Comput. Speech Lang. 29(1), 186–202 (2015)
5. Palo, H.K., Mohanty, M.N.: Modified-VQ features for speech emotion recognition. J. Appl. Sci. 16(9), 406–418 (2016)
6. Ramakrishnan, S.: Recognition of emotion from speech: a review. In: Speech Enhancement, Modeling and Recognition-Algorithms and Applications. InTech (2012)
7. Mishra, A.N., Chandra, M., Biswas, A., Sharan, S.N.: Robust features for connected Hindi digits recognition. Int. J. Sign. Process. Image Process. Pattern Recogn. 4(2), 79–90 (2011)
8. Kwon, O.W., Chan, K., Hao, J., Lee, T-W.: Emotion recognition by speech signals. In: Interspeech (2003)
9. Palo, H.K., Chandra, M., Mohanty, M.N.: Recognition of human speech emotion using variants of Mel-Frequency cepstral coefficients. In: Advances in Systems, Control and Automation, pp. 491–498. Springer, Singapore (2018)
10. Jackson, P., Haq, S.: Surrey audio-visual expressed emotion (SAVEE) database, pp. 398–423. University of Surrey, Guildford, UK (2014)

Energy Harvesting Using the Concept of Piezoelectric Tile Using Motors and Wooden Plank

C. Subramani, Anshu Jha, Sreyanko Mittra, Adesh Shrivastav, Durga Menon and Nilanjan Sardul

Abstract Energy is the source that drives the entire earth. In this era of technological development, the main concern is the need to develop new energy sources so as to meet the needs and demands of the consumers. The most viable and feasible fields after all these years of testing and research have led us to transducers, devices that convert one form of energy to the other. The purpose of this "Energy Harvesting Using the Concept of Piezoelectric Tiles using Motors and Wooden Plank" project is to develop a low-voltage electrical energy based on the concept of green technology. Here using the same concept of tile mechanical vibration energy that results when people walk across the tile (wooden plank) is received by the motors which will generate a small electric energy, this electric energy will be connected to a circuit having a diode as a rectifier and convert an alternating voltage to a direct current voltage. When there is a direct current voltage, this voltage can be stored in the bank of capacitor connected in parallel circuit which acts as a collector of electric energy to be used further.

Keywords Piezoelectric tile · Materials · Motor and wooden plank

C. Subramani (✉) · A. Jha · S. Mittra · A. Shrivastav · D. Menon · N. Sardul
Department of Electrical and Electronics Engineering, SRM Institute of Science and Technology, Kattankulathur, Chennai, India
e-mail: csmsrm@gmail.com

A. Jha
e-mail: Jhaanshu.79@gmail.com

S. Mittra
e-mail: deep28mittra@gmail.com

A. Shrivastav
e-mail: adeshshrivastav0@gmail.com

D. Menon
e-mail: drgmnn@gmail.com

© Springer Nature Singapore Pte Ltd. 2020
P. K. Mallick et al. (eds.), *Cognitive Informatics and Soft Computing*,
Advances in Intelligent Systems and Computing 1040,
https://doi.org/10.1007/978-981-15-1451-7_37

1 Introduction

In this era of world, the majority and most difficult task is to innovate and generate environmental friendly and economically efficient electricity such that there is no shortage of energy across the world. The need to generate energy efficiently is the requirement for today's world and most importantly the need to develop a clean and green source of energy [1]. The deprecate is a continuing trend in the electronics industry. The electronics devices usages of power has been decreased but their ability and way for dealing with huge amounts of power are simply a boon for the industries, thanks to power electronics. The innovation of capturing or generating lost energy from a vibrating source or a system from the environment is inescapable and entrenched in the society of today's world. This kind of energy harvesting is mostly from engines of different kind of automobile equipment like rotating rotor and the body of human being, in all cases this energy produced due to vibration convert into electrical energy due to the deformation of a piezoelectric tile material [2, 3]. Thus, this generating of this kind of energy harvesting is one of the most advanced and promising ways for generating more energy and adequate power it provides.

Crystal of piezoelectric tiles such as quartz exhibits different such as electromechanical resonance, which shows different characteristics with time and temperature. They depend on the different factors and when exposed to vibration, they produced electrical energy. These crystals are used for both large and small scale. But some material is not used for higher productivity, so they are doped accordingly for the user-based applications. Some materials have higher frequency and are expensive [4, 5]. The combination of different organic and inorganic compounds produced different piezoelectric crystals for producing pollutant-free energy.

2 Types of Piezoelectric Materials

There are mainly two types of materials, natural and synthetic materials. Some of the material is made of domestic and agricultural waste and some of them are made from doping of the several materials together [6].

A. Advantages of Piezoelectric tiles
    The tile is available in different desired shape. It is of tough and rugged construction. It is smaller and portable. It has higher frequency response with almost no shift in phases.
B. Disadvantages of Piezoelectric tiles
    They are temperature sensitive, and the values change due to humidity and moisture and energy produced with any proper advancement in structure is very low.

2.1 Comparison with Recentness for Wireless Energy Harvesting

A. Advantages

   The lifetime of the rectennas is almost unlimited. It does not need replacement (unlike batteries).
B. Disadvantages

   The process is expensive and is still in the research stage. It is slower than many processes.

2.2 Proposed Model

The proposed model is shown in Fig. 2 that yields high amount of volts. To improve the efficiency and to increase the output received, we are using motors and concept of piezoelectric tiles. The working is explained as follows

When force is applied on the wooden plank, which is shown in Fig. 2, acting as a plate by putting pressure on the plate, the compression force is made by the spring (made of wooden plank and rubber band). Thus springs on the bolt get compressed. The recoil spring attached to the pulley rotates the motor in clockwise direction (Table 1).

When compression is taken off, the plate moves upward, pulling the pulley in anticlockwise direction due to the recoil of the spring. Motor thus rotates in both directions in one strike [7]. The output is fetched through a series of electronic circuits to fetch the desirable and efficient dc output to run a load (Fig. 1).

The final working hardware model is shown in Fig. 2.

Table 1 Piezoelectric material

Natural piezoelectric material	Synthetic piezoelectric material	Natural piezoelectric material	Synthetic piezoelectric material
Quartz crystal (mostly used)	Lead zirconate titanate (PZT)	Sucrose	Gallium orthophosphate ($GaPO_4$)
Rochelle salt crystal	Zinc oxide (ZnO)	Enamel	Lithium tantalite ($LiTaO_3$)

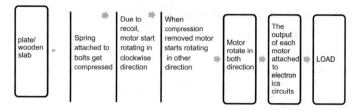

Fig. 1 Basic block diagram of model

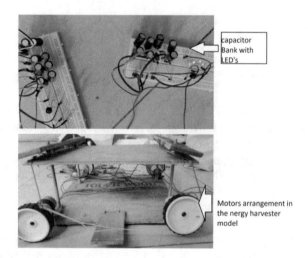

Fig. 2 Final model of hardware

3 Result and Analysis

3.1 *Mathematical Calculation Motor Part Calculation*

DC motor voltage = 12 V, Current = 3 A

Output of motor when driven $V * I = 36$ W

When turned mechanically, 1/6th of amps produced. Thus producing around 2 W by each motor

Total no. of motor = 4

Thus total power = $4 * 2 = 8$ W Electronics circuit calculation

Capacitors capacitance for 1 = $(2000 * 10^{-6})$ F Capacitors

Capacitance for 8 = $(8 * 2000 * 10^{-6})$ F = 0.016 F

Voltage at full charge = 15 V

Current at full charge = 1.2 A Charge stored by 8 capacitors $Q = C * V = (2000 * 10^{-6} * 15 * 8)$ coulomb = 0.24 C

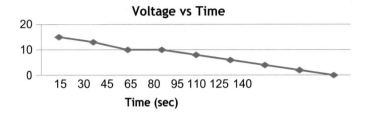

Fig. 3 Voltage characteristics

Table 2 Parameters of proposed method

S. No.	Name	Total No.	Specifications
1	DC motor	4	12 V, 3 A
2	Nut bolts	4	–
3	Rectifier	4	W10
4	In4007 diode	4	
5	Capacitor	16	$2200 * 10^{-6}$ F
6	Breadboard	2	–
7	Wires	–	–

Energy stored, $E = (\frac{1}{2} * C * V^2)$ Joule Energy stored in 8 capacitors $= (\frac{1}{2} * 0.016 * (15 * 15)) = 1.8$ J Power $= V * I = (15 * 1.2) = 18$ W

Total no. of LEDS for each motor set has 4 LED's. For first 15 s voltage is 15 then after each 15 s voltage drop by 2 V when it reached 10 V it remains constant for 20 s. After this, gain voltage starts dropping by 2 and reach to 0.

In Fig. 3, the LED starts glowing as capacitor gets fully charged and voltage at 15 V and till 10 V for next 20 s it remains constant and LED start turning off for next half i.e., from 80th second.

Thus total time LED is on = 2 min 20 s

Time during which LED has maximum glow, $T_{on} = 65$ s (Table 2).

In Fig. 4, the graphical representation shows that discharging and charging respect to LED turn on and turn off.

Advantages

One of the advantages of this over the previous piezoelectric crystal model is that, here when damage occurs, it can be repaired as the setup is built mechanically. Whereas in piezoelectric crystal model, repair and rework is not possible as the piezoelectric crystal should be completely replaced.

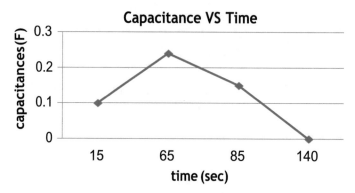

Fig. 4 Capacitance and time characteristics

3.2 Applications

The power generated from foot stepping can be utilized in different parts of sectors like for industrial sectors, domestic appliances, farming industries, street lamp and lightening. It can be also used in emergency during failure of power supplies. Time during which LED start turning off, $T_{off} = 75$ s.

4 Conclusion

Thus, we can conclude from the paper that the power or energy generated from each object can be utilized so that there is maximum utilization of the resource and creating a less pollution to the environment. Thus, our project also produced power from energy produced by human step or it can be also maximum pressure of water falling on it. This is the kind of energy does not require any supply of power from the mains. This module and concept is very efficient and useful in populated areas or country like India where population and consumption is higher than usual.

This module has a high scope for producing environmental friendly energy. We can build a road or path where population is much higher as people each stepping on wooden plank acting as a plate will produce ample amount of energy to light up street lights. This can be used in railway station as India population more than 80% uses train as transportation so it can produce enough energy efficiently.

References

1. Soderkvist, J.: Dynamic behavior of a piezoelectric beam. J. Acoust. Soc. Am **90**, 686–691 (1991)
2. LTC-3588-1 Data sheet. http://cds.linear.com/docs/en/datasheet/35881fb.pdf

3. Paul, C.: Fundamentals of Electric Circuit Analysis. Wiley. ISBN 0-471-37195-5 (2000)
4. The Piezoelectric Effect, PZT application manual
5. Xu, X., Wei, X., Gao, H., Zhu, J. et al.: Research on Charging Strategy of Lithium-ion Battery. SAE Technical Paper 2015-01-1192 (2015)
6. Laughton, M.A.: Power semiconductor devices. In: Electrical Engineer's Reference Book, pp. 25–27. Newness (2001)
7. Erickson, R.W.: Fundamentals of Power Electronics. Chapman and Hall, New York (1997)

Automatic Room Light Intensity Control Using Soft Controller

C. Subramani, Varun Sah, G. R. Vishal, Abhimanyu Sharma,
Indranil Gupta, Suraj Das and Shubham

Abstract LEDs for lighting purpose have become vastly popular as it has high lumens yield with less watt consumption. Lumens differ for different tasks and locations. Our system will efficiently control the intensity of light and provide the required amount of intensity for user and will automatically switch off in absence of the user. The analysis of energy consumption with and without our cost-effective smart system is discussed. Our system uses MOSFET to control voltages, which proportionally control the light intensity.

Keywords Illuminance · Lux · Power consumption · Utilization factor

C. Subramani (✉) · V. Sah · G. R. Vishal · A. Sharma · I. Gupta · S. Das · Shubham
Department of Electrical and Electronic Engineering, SRM Institute of Science and Technology,
Chennai, India
e-mail: csmsrm@gmail.com

V. Sah
e-mail: varunsah1997@gmail.com

G. R. Vishal
e-mail: g.r.vishall@gmail.com

A. Sharma
e-mail: abhishinu.2010@gmail.com

I. Gupta
e-mail: guptaindranil111@gmail.com

S. Das
e-mail: sohaldas10@gmail.com

Shubham
e-mail: shubham13402@gmail.com

© Springer Nature Singapore Pte Ltd. 2020
P. K. Mallick et al. (eds.), *Cognitive Informatics and Soft Computing*,
Advances in Intelligent Systems and Computing 1040,
https://doi.org/10.1007/978-981-15-1451-7_38

1 Introduction

Lighting system is one of the most important and common electrical systems. It is used in every sector. Even reducing few units individually by increasing the efficiency will save vast power in grid system and that power can be used for other purpose. The user has great advantage by this system, it will help in reduction of unit proportionally reduce cost of electricity bill and also decrease the strain on the eye [1].

The system monitors the intensity of light on worktable and occupants in the room using microcontroller. Using this parameter, we change the voltage and isolate the system with relay.

As lighting system, our system also varies depending upon the size of the room. It also varies depending upon number of windows. At present, the lighting systems which are used is mostly not automatically controlled, and in most of the places, fluorescent lamps are used as illumination which is less efficient and consume more power as compare to LED [2].

This paper proposed automatic lumens control of the room which is affected by the surrounding lumens. In Sect. 2, calculation of lumens required and home lighting is analyzed. In Sect. 2.1, proposed system architecture and block diagram are presented; in Sect. 2.2, software and hardware are presented. In Sect. 6, result and conclusion are presented [3].

2 Lumen Calculation

The room with length of 3.26 m, breadth of 3.82 m and height of 2.54 m is given which has one window facing north having a length of 1.53 m and breadth of 1.53 m and one door facing south having a length of 1.93 m and breadth of 0.9 m. For calculating the lumen required on working plane, LUMEN FLUX method has been used. This method provides uniform illumination over the working plane of the room. Basically, it is a method to calculate the required lux in the room. And also, the utilization factor has also been calculated for the room; Room Index of the room is also calculated. Direct lamping method is used in which 90% of light fall in work plane [4].

Area of the room $= 10.6928$ m^2
Area of window—2.7999 m^2
For calculating the Room Index

$$\text{Room Index} = (a \cdot b)/(h \cdot (a + b))$$

where

a = Length of the room = 3.26 m
b = breadth of the room = 3.27 m
Worktable length—0.75 m
Height of the table—1.79 m
Room Index—$\frac{10.6928}{6.54*1.79} = \frac{10.7}{11.7}$
Room Index = 0.9

Calculation lumen required in the working plane

$$N = \frac{E * A}{F * Mf * Uf}$$

$N = 2$ where N is number of lamps
$E = 250$ where E is no of lumens required in work plane
$F = X$ (Lumens per lamps)
$Mf = 0.8$
$Uf = 0.46$

Ivory white 76.00% reflection

$$2 = \frac{250 * 10.69}{x * 0.8 * 0.46}$$
$$2 = \left(\frac{726.22}{x}\right)$$
$$X = 3631.114$$

Energy consumption for the full day

$$22\,W * 24 = 528\,W$$

Energy consumption per hour = 22 W.

2.1 Maintaining the Integrity of the Specifications

The template is used to format your paper and style the text. All margins, column widths, line spaces and text fonts are prescribed; please do not alter them. You may note peculiarities. For example, the head margin in this template measures proportionately more than is customary. This measurement and others are deliberate, using specifications that anticipate your paper as one part of the entire proceedings, and not as an independent document. Please do not revise any of the current designations.

2.2   Description About the Block Diagram

Figure 1 is the overview block diagram of the hardware. Basically, the hardware is consisting of four major components Arduino mega, relay, voltage-controlled MOSFET and load which can vary as per the requirement. This block diagram has two different kinds of sensors each of two-two quantity. The sensors used are photo sensor and occupancy sensor which is nothing but an IR module. Photo sensor will sense the intensity of the surrounding light in lumen and how much it will affect the system. The second sensor which is an IR sensor works as an occupancy sensor for the room; it will calculate the number of entries in the room. By the requirement, it will switch on or off the supply. The block diagram is shown in Fig. 1.

Supply is given to the relay module which is connected to the IR module. It notifies the relay module when to pass the current which basically depends upon the entry or occupancy in the room. IR module is further connected to the voltage-controlled MOSFET which will pass the required voltage to the load. Before that photo sensor is connected to the MOSFET via Arduino mega. Arduino mega is the brain of the system; all the peripherals are connected to it and it will give all the signals to the sensors and other peripherals. Photo sensor senses the lumen of the surrounding and

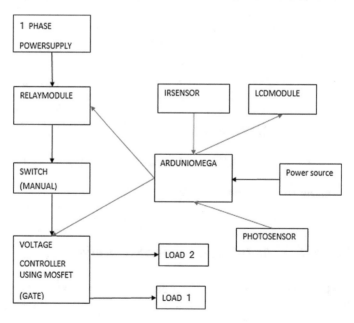

Fig. 1 Block diagram

by the value of the lumens present MOSFET will pass the voltage to the load which is connected. Required lumen is provided by the LED which is using as a load.

3 Analysis and Simulation

Simulation for the intensity control circuit was performed using MATLAB 2017 Simulink. The load is a resistance which is equivalent to LED bulb; this circuit diagram can be split into four simple processes as shown in Fig. 2.

(1) Step down process
(2) Rectification and filtering process
(3) MOSFET control
(4) Voltage across load

Circuit diagram is consisting of a transformer, rectifier and load. Many sensors were used, such as voltage sensor and current sensor for calculating the power.

Full bridge rectifier is used to convert AC into DC, before rectification a transformer is used for stepping down the voltage to 12 VAC from 220 VAC and after operation of rectification 10–12 V DC is achieved. That rectified supply pass through a capacitor and it works as a filter. Though that a MOSFET is connected in series with the current sensor (1) Here MOSFET is used because it can control the power output of the load by varying the R_{ds}. And though that a current sensor (2) is connected with the load in series and a voltage sensor is connected in parallel to the load. Voltage sensor and current sensor are used to measure the voltage and current, respectively, in the respected line.

The operating range for the LED which is integrated with photo sensor for the lux needed in the room to get better illuminance is calculated below.

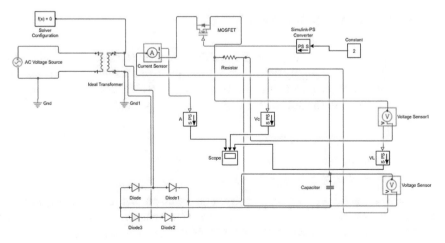

Fig. 2 Circuit diagram

Formula for calculating various parameters

$$P(\text{power } o/p) = V_{\text{load}} * I_{\text{load}}$$
$$P(\text{power } o/p) = I^2 R$$

Voltage and current division rule

$$V_{\text{load}} = \frac{V_{\text{source}} * I_{\text{source}}}{R_{\text{load}} + R_{\text{mosfet}}}$$
$$V_{\text{load}} = \frac{V_{\text{source}}}{R_{\text{load}} + R_{\text{mosfet}}}$$

where

V_{load} = load voltage
I_{load} = load current
V_{source} = source voltage (rectified)

Parameters to calculate

R_{mosfet} = resistance b/w drain and source
R_{load} = load resistance

By the above calculation, we calculated the R_{mosfet} and R_{load} as follow for reducing the current in the load that will also decrease the intensity of the LED because intensity of the light is directly proportional to the current needed by the load.

Operating voltage for LED is $6-10$ V dc.
Operating current -1.2 to 5 A.

4 Output of the Simulation

After supplying the AC supply to the rectifier, rectifier rectified the supply into pulsating DC rectified supply that sends to the capacitive filter which decreases the ripple present in the supply that is connected to the MOSFET, and by varying the V_{gs} through R_{ds} of the MOSFET, the desired output shown in Fig. 3.

5 Methodology

The MOSFET gate is connected to the Arduino digital pin and sensors are connected to Arduino. 24VDC supply from step down transformer and rectifier is connected to the load that is LEDs via MOSFET. When TEMP6000 analog value is zero, the duty cycle is 100% and hence the maximum power is drawn, the illumination is maximum

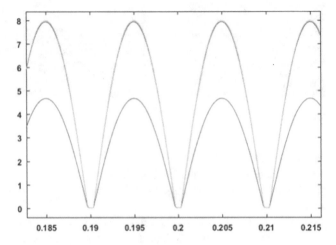

Fig. 3 Simulation output

as per lumen calculation. When the lumen is different due to surrounding light, the TEMP6000 value is some analogy value.

As per calibration and analogy value, the particular duty cycle is given to gate of MOSFET and power drawn is reduced to the load; hence, the illumination is controlled. For example, when analogy value is 250 then duty cycle given is 50% and hence the power drawn by load is reduced as compare to maximum power drawn.

Another sensor IR sensor which keeps the count of the occupant when count is zero it make the circuit open and prevent wastage of power.

6   Conclusions

Illumination was controlled as per surrounding lumen of room and occupant control was successful. Energy was conserved with this proposed system which shown in Fig. 4.

Fig. 4 Hardware setup

Some cases are given below

Case 1—when there is no external light or illumination.

This system will be consuming same power as the old system which does not have this system.

Case 2—when occupant fails to switch off, he/she leaves the room.

This system will automatically switch off and therefore save the energy. Without this system the energy will be wasted.

Case 3—when illumination varies with respect time.

In this case, we can consider three different period for given period (1 h) first period when there is no external illumination (30 min), in second period external illumination is maximum (SUPPLY is ON) (15 min) and last some illumination is in the room (15 min).

Old system for 1 h energy consumed is 22 W.

This system for 1 h energy consumed is $11.5 + 0 + (3 * 4 * 0.25) + 0.1^2 * 4 = 14.54$ W.

Above cases show that this system successfully save energy with illumination control and also, this article tells us about the smart, cheap and efficient way to save energy by including several components like Arduino, light sensor, temperature sensor, IR sensor.

References

1. Bhimra, P.S.: Power electronics, 3rd edn. Khanna Publishers, Delhi (2004)
2. Smart Street Lighting System, with Light Intensity Control Using Power Electronics. Visvesvaraya National Institute of Technology, Nagpur, India
3. Luo, Y., Li, Y., Zhan, X.: Intelligent control system for energy saving of street lamp. In: International Conference on Transportation, Mechanical, and Electrical Engineering, pp. 1024–1027 (2011)
4. The design of a street lighting monitoring and control system. In: International Conference and Exposition on Electrical and Power Engineering (EPE 2012), pp. 25–27 October, Iasi, Romania (2012)

IoT-Based Smart Irrigation System

C. Subramani, S. Usha, Vaibhav Patil, Debanksh Mohanty, Prateek Gupta, Aman Kumar Srivastava and Yash Dashetwar

Abstract In today's world due to scarcity of non-renewable energy and increasing demand for energy, we require effective and clean energy for this harnessing energy from solar power which is the best way as it is cost effective and can fulfill all our energy needs. Solar-powered smart irrigation systems are the new technology of advance farming method to the Indian farmer. This system consists of solar-powered water pump along with an automatic water flow control using a moisture sensor and mobile device using the concept of IoT. It is the proposed solution for the present energy crisis for the Indian farmers. This system conserves electricity by reducing the load of grid power and saving water by reducing water wastages.

Keywords Energy · IoT · Sensor

C. Subramani (✉) · S. Usha · V. Patil · D. Mohanty · P. Gupta · A. K. Srivastava · Y. Dashetwar
Department of Electrical and Electronics Engineering, SRM Institute of Science and Technology, Chennai, India
e-mail: csmsrm@gmail.com

S. Usha
e-mail: ushakarthick@gmail.com

V. Patil
e-mail: vp1022752@gmail.com

D. Mohanty
e-mail: debankshmohanty@gmail.com

P. Gupta
e-mail: prateekgupta_jaikishan@srmuniv.edu.in

A. K. Srivastava
e-mail: amankumar_srivastava@srmuniv.edu.in

Y. Dashetwar
e-mail: yashd13698@gmail.com

© Springer Nature Singapore Pte Ltd. 2020
P. K. Mallick et al. (eds.), *Cognitive Informatics and Soft Computing*,
Advances in Intelligent Systems and Computing 1040,
https://doi.org/10.1007/978-981-15-1451-7_39

1 Introduction

Solar energy is the present in a lot of quantity abundantly. Solar power with few other renewable energy resources is not only a parallel option to today's energy crisis but also an eco-friendly form of energy which does not harm nature. The cost of solar panels has been constantly decreasing which persuades its usage in various sectors domestic as well as commercial [1].

One of the important usages of this technology is in irrigation and agricultural systems by farmers. Solar-powered smart irrigation system using Arduino can be a suitable option for farmers across the world in the current days of energy crisis in India as well as globally. Farming is the basic source of living of people in India. In past years, it is noted that there is not much crop enhancement in agriculture sector. Food prices are constantly increasing because crop the rate of crop production is decreasing [2]. It has got over 40 million farmers into poverty since 2010. There are various factors which affect this; it may be because of wastage of water, low soil fertility, more use of fertilizer, climatic changes, or any other such factors. It is very essential to make strict action or technological up-gradation in agricultural field, and the solution is information of technology in association with solar energy [3]. This is a greenway or non-harmful way of producing energy that gives free energy on an initial investment.

In this research paper, we all proposed a smart irrigation system using solar power and Arduino which makes water pumps to pump the water from underground well to a tank and the outlet of tank is self-regulated using microcontroller and moisture sensing devices and temperature sensors to control the rate of water flowing from the tank to the farm field which make a more effective way of use of water [4].

2 Literature Survey and Background Study

India has got a great resource of fertile land and abundant water source, to use this resource precisely, and in a modern and utilized way, government is planning new techniques like smart ways of irrigation, construction of pumped storage dams, channels, loans for small farmers which are done. After all, in the end to have proper framing we need electrical power supply to run pumps, motors, lawnmowers, jet sprays. Thus, there is a growing demand of power consumption day by day, and seeing on the other side, energy cannot be fulfilled with the present power generation; capacity of India for this farmers has moved to new renewable sources of energy [5]. Like solar energy, India consumed around 51.84 GW (21.83%) of the total electricity production in agriculture in the year 2003/04 and day by day it is increasing at high pace; only generating power is not the only goal [6]. This smart irrigation system is helps to reduce the water consumption and wastage of water. Recycling of water should be taken care around 45% water in the normal irrigation system is wasted because of overuse, these can be reduced and in water scarcity area irrigation can be done efficiently. Everything can be done in a smart way if we use new and current resources into account; our demand increases day by day with it; our resources are limited so by implementing solar power around 5–10% of burden on the grid can be reduced [7].

3 The Proposed Solution

In the solution, the water pump is driven by the help of solar energy which is obtained by the sunlight. This energy is converted into electrical energy with the help of rated solar panel. Our system uses single-stage energy consumption system where water is pumped direct to the field from the storage tank. This saves a significant amount of energy and time and efficient use of renewable energy. The pump is controlled with the use of intelligent and simple algorithm in which the flow of water is being regulated according to the requirement of the soil. In this system, a conventional soil sensor with two electrode plates is used to determine the moisture level with the help of calculation using voltage supply. These voltage changes will trigger the Arduino Uno to start the pump when the moisture levels goes below a certain point. It will operate till the water is completely passed through the field. A LCD display will tell us the current position of the moisture inside the soil and the temperature of the soil. The over all block diagram of the work and hardware implementation are shown in Fig. 1 and Fig. 2 respectively.

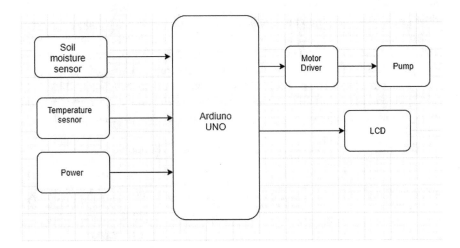

Fig. 1 Block diagram

Fig. 2 Hardware setup

3.1 Cost Analysis

In India, farmers use over nine hundred thousand tube wells across every state. For pumping water in irrigation system, an estimate of Rs. 18 million cost of energy is used as water pumping to the field. This is a huge amount of money, and it can be saved by the use of solar energy (five million KWh). It saves forty percent of the total energy in electrical supply system which is of cost twenty-seven million per annum in India.

At the initial stage, it is a very huge investment; the cost of solar panel is very high, but as the year passes the amount of energy saved is large enough to earn back the initial investment for minimum time span (an amount of 4.8 KWh energy is saved).

The Ministry of New and Renewable Energy (MNRE) is testing universal solar pump controllers (USPC). These controllers help the utility of solar array system (PV system). These controllers run at least four equipment like apple grading and grain (flour) machine. The power range of equipment is between 3 and 10 HP.

PV System rating for USPC/Motor Pump (3–10 HP) is shown in Table 1 and the corresponding controller eficiency is given in Table 2. The cost analysis is noted in Table 3.

USPC efficiency for operation is above 80% rated STC power.

Table 1 Rating of the proposed method

Motor pump set capacity (HP)	PV array rating (STC) (W)
3	2,800–3,100
5	4,800–5,000
7.5	6,750–7,200
10	9,000–9,500

Table 2 Efficiency rating of the proposed method

Motor pump set capacity (HP)	Controller efficiency (%)
3	93
5	93
7.5	94
10	95

Table 3 Cost estimation

Materials	Unit cost	Quantity	Total cost (Rs.)
Solar panel (1.4 m^2)	24,000	4	96,000
Converter circuit	400	1	400
Battery (24100 A)	8250	1	8250
Pump (5 HP)	80,700	1	80,700
Total			185,350

4 Working

The working of the system as follows: Initially, the soil will be dry. This is due to low amount of water present inside it. This will indicate the microcontroller to switch on the pump with the help of soil moisture sensor. The moisture sensor will tell the amount of water level present inside the soil. The working of moisture sensor is as follows.

Two plates are present there in moisture sensor: One plate will act as ground, and second plate will be connected to 5 V supply through Arduino Uno along with 10 K ohms resistor. It simply acts as voltage divider network, and output is taken from the second pin of the sensor. The output will range in 0–5 V, in proportion with change in contents of water in the soil. Ideally, when the moisture of the soil is 0, it will act as open circuit; the sensor will have infinite resistance. How this condition we get 5 V as output. If the level is below certain value, the pump will me switch on. The value of moisture of the soil is mapped on some predefined data which was taken by soil testing.

Similarly, the temperature is also calculated by change in voltage supply. The formula for calculating the temperature is taken as:

Voltage at pin in mV $=$ (reading from ADC) $*$ $(5000/1024)$

Centigrade temperature $= \big[(\text{analog voltage in mV}) - 500\big]/10$

The unit of measuring temperature is in °C. The pump will be ON till the moisture level is achieved inside the soil. The predefined data is been stored inside the microcontroller (Arduino Uno). That memory will be used to control the switching ON/OFF of the pump.

5 Future Scope

Using the solar-powered smart irrigation technique will not only ease the burden on the grid but also help to reduce the pollution in the environment. This technique is maximizing the output from the resources that are presently available; in this technique, water used in the irrigation is reduced by almost 40% because of this with less amount of water available farmers can get the yield.

As it is a self-driven system, farmers with less income source will face difficulty to implement it in the initial period, but once the method is implemented, the output from the system can be taken back in almost in one and half year; it has one-time investment of around 1–1.5 lac (for approx. 2 ha field), but the return can be assured with guaranteed results. The electric energy consumption is reduced from 18% to 15% with help of smart irrigation system. Because of which burden will be less on the power plants and if the power output is more from the solar panels, they can send it to grid through net metering which is a further scope in this project.

6 Conclusions

With every new method, there are some positive and negative points; starting with the positive points, this system has got low operating cost and low maintenance and works in hot and dry climates, flexibility, and controllable as it is fully automated, but this makes to case that the person who is operating should have bit prior knowledge about the electronics component; if we want to increase the output, more panels are required which will increase the cost so it increase the initial investment. But overcoming all the points, it is a new sustainable and modern technique for the new-generation irrigation system.

References

1. Garg, H.P.: Advances in solar energy technology, vol. 3. Reidel Publishing, Boston, MA (1987)
2. Halcrow, S.W., Partners.: Small-scale solar powered irrigation pumping systems: technical and economic review. UNDP Project GLO/78/004. Intermediate Technology Power, London, UK (1981)
3. Harmim, A., et al.: Mathematical modeling of a box-type solar cooker employing an asymmetric compound parabolic concentrator. Solar Energy **86**, 1673–1682 (2012)
4. Tse, K.K., Ho, M.T., Chung, H.S.-H., Hui, S.Y.: A novel maximum power point tracker for PV panels using switching frequency modulation. IEEE Trans. Power Electron. **17**(6), 980–989 (2002)
5. Haley, M., Dukes, M. D.: Evaluation of sensor-based residential irrigation water application. ASABE 2007 Annual International Meeting, Minneapolis, Minnesota, ASABE Paper No. 072251 (2007)

6. Persada, P., Sangsterb, N., Cumberbatchc, E., Ramkhalawand, A., Maharajh, A.: Investigating the feasibility of solar powered irrigation for food crop production: a caroni case. J. Assoc. Prof. Eng. Trinidad Tobago **40**(2), 61–65 (2011)
7. Priyadharshini, S.G., Subramani, C., Preetha Roselyn, J.: An IOT based smart metering development for energy management system. Int. J. Electr. Comput. Eng. **9**(4), 3041–3050

Multi-antenna Techniques Utilized in Favor of Radar System: A Review

Subhankar Shome, Rabindranath Bera, Bansibadan Maji,
Akash Kumar Bhoi and Pradeep Kumar Mallick

Abstract What is new in a radar system, what advancement is going on, and what may be the road map for future radar system development, this is the question for which authors tried to find out the solution throughout this review. Radar transformation is described in two parts. Like every communication system, analog radar baseband is transformed into digital baseband which is adding the advantage of digital signal processing (DSP) of the transmitted and received signal which helps to improve target parameter characterization. Software-defined radio (SDR) is one of the powerful tools which are heavily used to develop digital radar baseband in recent days. In the second part, single antenna-based old radar system is transformed into multi-antenna-based modern radar which helping to improve signal-to-noise ratio (SNR) in the radar receiver. This front end antenna part is still more or less analog, but several multi-antennas techniques are adding different advantages to the total system. A good amount of signal reception using multi-input multi-output (MIMO) system is helping in target characterization in the digital baseband section in which advance signal processing is working. These days few other multi-antenna techniques

S. Shome (✉) · R. Bera
Department of Electronics and Communication Engineering, Sikkim Manipal Institute of
Technology, Sikkim Manipal University, Sikkim, India
e-mail: subho.ddj@gmail.com

R. Bera
e-mail: rbera50@gmail.com

B. Maji
National Institute of Technology Durgapur, Mahatma Gandhi Rd, A-Zone, Durgapur 713209,
West Bengal, India

A. K. Bhoi
Department of Electrical and Electronics Engineering, Sikkim Manipal Institute of Technology,
Sikkim Manipal University, Sikkim, India
e-mail: akash730@gmail.com; akash.b@smit.smu.edu.in

P. K. Mallick
School of Computer Engineering, Kalinga Institute of Industrial Technology (KIIT) University,
Bhubaneswar, India
e-mail: pradeepmallick84@gmail.com

© Springer Nature Singapore Pte Ltd. 2020
P. K. Mallick et al. (eds.), *Cognitive Informatics and Soft Computing*,
Advances in Intelligent Systems and Computing 1040,
https://doi.org/10.1007/978-981-15-1451-7_40

365

like Array antenna, Phased MIMO antenna are becoming popular for improving SNR of the system. In this article all multi-antenna techniques are reviewed to find out the best one in favor of Radar.

Keywords DSP · SDR · MIMO · Array · Phased MIMO

1 Introduction

Analog radar systems used in world war are faced several changes in recent days. Almost like every electronic system, analog radar became digital which is a revolutionary change. The primary job of radar is the detection of the desired target, which means it will recognize a reflected return of the transmitted signal from several environmental objects. Characterization of the desired target using the received signal reflected from the target is the main challenge for radar. Target characterization depends on a few important radar parameters like range resolution, doppler resolution, high processing gain, etc. Use of different code in radar baseband enhances the security and introduces the anti-jamming feature in radar system like communication systems. Transmission of code also increases the processing gain at the radar receiver using match filtering, which improves target radar detectibly. Modern radar is using digital pulse modulation technique which produces large time-bandwidth product waveforms which are helping to improve the above radar perimeters. This pulsed compression technique [1] also helps to improve the processing gain which makes a good effect in target detection [2]. Now, the implementation of digital baseband needs digital hardware architecture for which software-defined radio (SDR) is one of the popular choices these days. A main advantage of SDR hardware is they are reconfigurable [3]. An FPGA processor is working as a heart of these devices, which can be reprogrammed every time as per our needs. Another advantage is the reconfigureability of the radio boards. The SDR board is a built-in radio transmitter and receiver radio cards, which provide us a platform to design a complete hardware system. Reconfigurable radio cards are flexible enough to change every parameter of the radio at any point in time.

Digital radar baseband is now a part of modern radars, but RF frontend improvement is also an important research area these days. Researches are applying all the advance antenna technique in radar. MIMO radar, array radar, phased array radar are outcomes of this research [4]. In literature survey, the author has a review of every recent application in radar research and tried to find out the advantages out of it.

2   Review on MIMO Radar

In the above section, digital radar is reviewed which is basically a single antenna-based system. In this type of conventional radar, the target range and RCS are measured out of the amplitude, and the time delay of the echoed signal, respectively, for moving target the doppler frequency shift of the received signal is processed for velocity extraction of the moving target [5]. Instead of a single antenna, multiple antennas enable a number of signal paths to carry the data. MIMO is mainly used in two different formats. One called special diversity, and another called special multiplexing. Special diversity means sending the same data across multiple propagation paths, which increase the reliability against various forms of fading. Finally, this diversity method is improving the signal strength in the receiver side resulting in better SNR at the receiver. If the data rate is the main concern, then special multiplexing is the better option in the MIMO system. In special multiplexing, each propagation channel carries independent information, which basically increases the traffic over the channel resulting high throughput of the system. In single-antenna system, capacity is defined by Eq. (1)

$$\text{Capacity} = \text{BW} \log_2(1 + \text{SNR}) \tag{1}$$

where C is the channel capacity in bits per second, BW is the bandwidth in Hertz, and SNR is signal-to-noise ratio. But in the case of MIMO system, the capacity is determined by Eq. (2)

$$\text{Capacity} = N \, \text{BW} \log_2(1 + \text{SNR}) \tag{2}$$

where N is the number of the antenna.

Equation 2 is clearly mentioning how the number of the antenna is helping to increase the system capacity [6].

Application of radar is different from the communication system. In radar, we are more concern about target detection and characterization, where the reliability of the detection is the main aim. Special multiplexing is very advantageous for radar as multiple antennas give us an opportunity to transmit orthogonal waveform using different propagation channels. This provides huge benefits such as the diversity in the paths [7], virtual aperture extension [8], beam pattern improvement [8], and a higher probability of detection (Fig. 1).

However, the benefits offering by the MIMO radar come at the price of sacrificing coherent processing gain at the transmitting array offered by the phased array radar. Hence, the MIMO radar with collocated antennas may suffer from beam-shape loss which leads to performance degradation in the presence of a target's RCS fading.

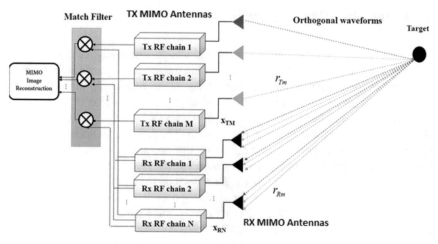

Fig. 1 MIMO radar concept [9]

3 Review on Phased Array Radar

Every country needs surveillance radar in border areas, even this surveillance radar is an integral part of airports. To surveil over a large area, radar beam steering is the main criterion, most of the radar mechanically rotating the radar antenna to spread the beam over a large area. Array radar comes into the picture when people start thinking of electrical beam steering instead of mechanical antenna rotation. Lincoln Laboratory, USA, starts working on it and came up with a solution of a large planar array of UHF elements which is capable of electrical beam steering over a large geographical area [10, 11] (Fig. 2).

An array is a combination of smaller identical antenna elements which get active by applying a complex voltage or current and generate a radiation pattern in the desired direction. In most application, arrays are arranged in matrix formation, and it also can be divided into a small elemental group called subarray. A linear array formation is shown in Fig. 3 for ready reference [13, 14]. This array formation will add few advantages as follows:

(i)   Without any mechanical movement which means any antenna parts physically moving, the beamforming allows electrical steering of the radar antenna beam to scan over a large area. It can also produce a pattern of multiple beams, each covering a cross section of the 360° plane.

(ii) The array's gain within the beam is very high, which leads to improved target detection.

(iii) In microsecond order, control unit of the antenna array permits very fast jumps from one target to another which allows tracking of multiple targets.

(iv) Good Coherency gain.

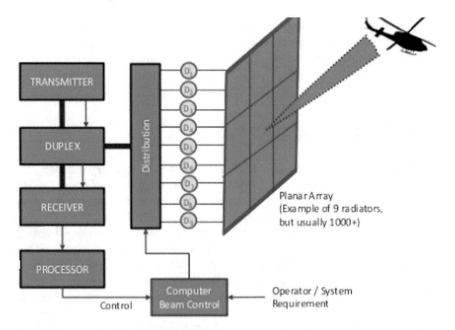

Fig. 2 Concept of phased array radar [12]

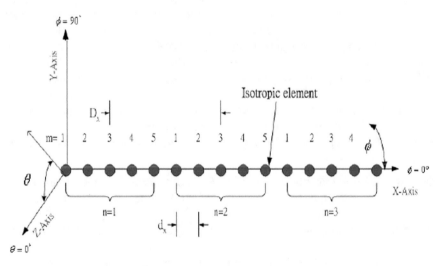

Fig. 3 Linear array formation [17]

But this phased array radar is suffering from minimal diversity gain. Here, the requirement of a new technique comes into the picture. A system needs to be designed which can provide minimal diversity gain and coherent processing gain into a single system [15, 16].

4 Review on Phased MIMO Radar

In Sects. 2 and 3, authors reviewed the advantages of MIMO and array radar [18]. But both the systems are having a few disadvantages. To nullify this weakness of MIMO radar and array radar, author found out a new technique which integrates the advantages of the MIMO radar which is waveform diversity [19, 20], with the advantages of the phased array radar [21, 22], which is coherent processing in a single system. In MIMO radar, every antenna is creating a beam over the space, so it is enabling multiple propagation paths for data, but in array radar, several antenna elements are creating a single beam in a specific direction in space, which is creating a beam-forming effect toward target [23]. To achieve an integrated system, the total antenna array is divided into a number of the subarray. Now, one array basically becomes multiple subarray [24] which will create multiple beams over space to create MIMO effect using an array antenna system. Each subarray consists of multiple number of elements which able to create beam in a certain direction over the space, number of beam in a certain direction, created by different subarrays will enable multiple propagation paths for the signal. In this way, MIMO and phased array technique integrated into a single system called phased MIMO system. Multiple subarrays are helpful to coherently transmit orthogonal waveform. Weight vector can be designed for every subarray to achieve coherent processing gain, and using weight vector beamforming is possible. Parallel subarrays jointly form multiple beams resulting MIMO effect which provides higher resolution capabilities. The new advanced technique is enjoying the below advantages by exploring two multi-antenna techniques [25]:

(a) Angular resolution improvement, multi-target detection, parameter identifiability improvement, these all advantages of MIMO radar are now part of the new technique.
(b) Existing beamforming techniques is a part of array radar, now a part of the new system
(c) A tradeoff between resolution and robustness against beam-shape loss.
(d) Improved robustness against strong interference (Fig. 4).

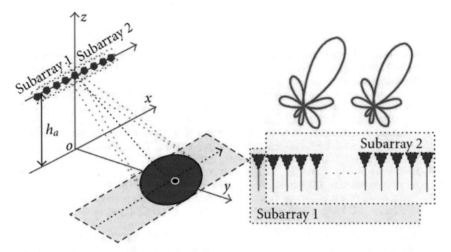

Fig. 4 Concept of phased MIMO radar [26]

5 Conclusion

In this article, the author tried to investigate a road map for the modern radar system. Digital radar baseband is already in action which is mostly implemented using reconfigurable hardware, and SDR is one of the popular choices for this application. The researcher is also working in the field of radar front end development which is still mostly analog. In this area, research is mostly focused on multi-antenna-based system development for radar. In case of multi-antenna-based system, few well-established technologies like MIMO, phased array are already working fine in several communication systems, and the same is incorporated for radar also. MIMO radar and phased MIMO radar are already in the market from last few years, they have several advantages in their class which is already discussed in the article, but both have few disadvantages also. To overcome these disadvantages, researchers are trying to integrate advantages of both the system in a single system. Modern radar research is targeted to integrate both the system in a single system which is called phased MIMO system. This is having both the advantages of MIMO and array system and nullifying all the disadvantages. After reviewing all working radar system, authors find a road map for the future radar research which is the integration MIMO and array techniques and develop a new technique called phased MIMO system, which will perform in every critical condition.

References

1. Muralidhara, N., Rajesh, B.R.C., Biradar and Jayaramaiah, G.V.: Designing polyphase code for digital pulse compression for surveillance radar. In: 2nd International Conference on Computing and Communications Technologies (ICCCT), Chennai (2017)
2. Lewis, B.L., Kretschmer, F.F.: A new class of polyphase pulse compression codes and techniques. In: IEEE transactions on aerospace and electronic systems AES-17, pp. 364–372. https://doi.org/10.1109/taes.1981.309063 (1981)
3. Costanzo, S., Spadafora, F., Borgia, A., Moreno, H.O., Costanzo, A., Di Massa, G.: High resolution software defined radar system for target detection. J. Electr. Comput. Eng. **2013**, 7. Article ID 573217 (2013). https://doi.org/10.1155/2013/573217
4. El-Din Ismail, N., Mahmoud, S.H., Hafez, A.S., Reda, T.: A new phased MIMO radar partitioning schemes. IEEE Aerospace Conference, Big Sky, MT (2014)
5. Patole, S.M., Torlak, M., Wang, D., Ali, M.: Automotive radars: a review of signal processing techniques. IEEE Signal Processing Magazine **34**(2), 22–35 (2017). https://doi.org/10.1109/msp.2016.2628914
6. Study Paper on Multiple-Input Multiple-Output (MIMO) Technology. Source http://tec.gov.in/pdf/Studypaper/Test%20Procedure%20EM%20Fields%20From%20BTS%20Antennae.pdf
7. Fishler, E., Haimovich, A., Blum, R., Cimini, L.J., Chizhik, D., Valenzuela, R.A.: Spatial diversity in radars: models and detection performance. IEEE Trans. Sign. Process. **54**(3), 823–838 (2006)
8. Bekkerman, I., Tabrikian, J.: Target detection and localization using MIMO radars and sonars. IEEE Trans. Sign. Process. **54**(10), 3873–3883 (2006)
9. Kpre, E.L., Decrozel, C., Fromenteze, T.: MIMO radar pseudo-orthogonal waveform generation by a passive 1 × M mode-mixing microwave cavity. Int. J. Microw. Wirel. Technol. **9**(7), 1357–1363 (2017). https://doi.org/10.1017/s175907871700023x
10. Fenn, A., Temme, D.H., Delaney, W.P., Courtney, W.: The development of phased-array radar technology (2000)
11. Butler, J., Lowe, R.: Beam-forming matrix simplifies design of electronically scanned antennas. Electron. Des. **9**, 170–173 (1961)
12. Satyanarayana, S.: Multi-function phased array radar. Source https://www.slideshare.net/mistral_solutions/multifunction-phased-array-radar
13. Vera-Dimas, J.G., Tecpoyotl-Torres, M., Vargas-Chable, P., Damián-Morales, J.A., Escobedo-Alatorre, J., Koshevaya, S.: Individual patch antenna and antenna patch array for wi-fi communication. Center for Research of Engineering and Applied Sciences (CIICAp), Autonomous University of Morelos State (UAEM), 62209, Av. Universidad No. 1001, Col Chamilpa, Cuernavaca, Morelos, México (2010)
14. Ghosh, C.K., Parui, S.K.: Design, analysis and optimization of a slotted microstrip patch antenna array at frequency 5.25 GHz for WLAN-SDMA system. Int J Electr Eng Inform **2**(2), 106 (2010)
15. Hassanien, A., Vorobyov, S.A.: Why the phased-MIMO radar outperforms the phased-array and MIMO radars. In: 2010 18th European Signal Processing Conference, pp. 1234–1238. Aalborg (2010). Keywords (array signal processing; MIMO radar; phased array radar; radar signal processing; phased array; multiple input multiple output; signal to noise ratio; SNR analysis; phased MIMO radar beam pattern; processing gain; transmit beamforming; MIMO radar; Signal to noise ratio; Radar antennas; Arrays; MIMO; Array signal processing). http://ieeexplore.ieee.org/stamp/stamp.jsp?tp=&arnumber=7096493&isnumber=7065143
16. Fuhrmann, D.R., Browning, J.P., Rangaswamy, M.: Signaling strategies for the hybrid MIMO phased-array radar. IEEE J. Select. Topics Sign. Process. **4**(1), 66–78 (2010)
17. Monterey, California.: Distributed subarray antennas for multifunction phased-array radar. Master of Science in System Engineering, Naval Postgraduate School September 2003
18. Fu, H, Fang, H, Cao, Lu, S.M.: Study on the comparison between MIMO and phased array antenna. In: IEEE Symposium on Electrical and Electronics Engineering (EEESYM), Kuala Lumpur (2012)

19. Fuhrmann, D., Antonio, G.: Transmit beamforming for MIMO radar systems using signal cross-correlation. IEEE Trans. Aerosp. Electron. Syst. **44**, 171–186 (2008)
20. Stoica, P., Li, J., Xie, Y.: On probing signal design for MIMO radar. IEEE Trans. Sign. Process. **55**, 4151–4161 (2007)
21. Haykin, S., Litva, J., Shepherd, T.J.: Radar Array Processing. Springer, New York (1993)
22. Van Trees, H.L.: Optimum Array Processing. Wiley-Interscience, New York (2002)
23. Hassanien, A., Vorobyov, S.A.: Transmit/receive beamforming for MIMO radar with colocated antennas. In: 2009 IEEE International Conference on Speech, Signal Processing (ICASSP'09), pp. 2089–2092. Taipei, Taiwan, Apr 2009
24. Ismail, N., Hanafy, Sherif & Alieldin, Ahmed & Hafez, Alaa. (2015). Design and analysis of a phased-MIMIO array antenna with frequency diversity, pp. 1745–1750
25. Mucci, R.: A comparison of efficient beamforming algorithms. IEEE Trans. Acoust. Speech Sign. Process. **32**(3), 548 (1984)
26. A flexible phased-MIMO array antenna with transmit beamforming—scientific figure on research gate. Available from https://www.researchgate.net/figure/Illustration-of-the-flexible-phased-MIMO-antenna-array_fig3_258385610. Accessed 18 Jun 2019

Home Automation Using IoT-Based Controller

C. Subramani, S. Usha, Maulik Tiwari, Devashish Vashishtha, Abuzar Jafri, Varun Bharadwaj and Sunny Kumar

Abstract At present, the technology has been developing day by day. In this era, the home automation has become the most prominent and developing design. This is consist of an automated system that can provide luxurious and at the same time for conserving energy. Basically, this is the intersection of various devices to perform a single task. This paper is depicted about how to perform switching operations and at the same time to control the intensity of light as well as motor. This paper has limited applications, and this is very useful in old age homes and in hospitals to add up enhancement to the luxurious and helpful to old people. The goal of this paper is to create a fully controlled home automated system using smart phone that is wirelessly connected to Arduino via Bluetooth module. With the help of application, we can control the appliances and at the same time control it.

Keywords Arduino Nano · Bluetooth module · Home automation · ULN2003 · Arduino · Internet of Thing

C. Subramani (✉) · S. Usha · M. Tiwari · D. Vashishtha · A. Jafri · V. Bharadwaj · S. Kumar
Department of Electrical and Electronics Engineering, SRM Institute of Science and Technology, Kattankulathur, Chennai, India
e-mail: csmsrm@gmail.com

S. Usha
e-mail: ushakarthick@gmail.com

M. Tiwari
e-mail: mauliktiwari196@gmail.com

D. Vashishtha
e-mail: devashish5198@gmail.com

A. Jafri
e-mail: abuzarjafri1@gmail.com

V. Bharadwaj
e-mail: varun.fpl.3@gmail.com

S. Kumar
e-mail: sunnysingh20467@gmail.com

© Springer Nature Singapore Pte Ltd. 2020
P. K. Mallick et al. (eds.), *Cognitive Informatics and Soft Computing*,
Advances in Intelligent Systems and Computing 1040,
https://doi.org/10.1007/978-981-15-1451-7_41

1 Introduction

The home automation is a very common word nowadays spelt by many engineers, and there are many applications of this automatic design. The world is now moving towards fully automated system to reduce the man power [1]. The electronics field is running towards the compact design and controlling things. The home automation can be controlled either by manual or smart switches. The sense of using smart is to get control over these switches via smart phone application. If something went wrong, it will prevent us from electrical shock as we are controlling it from smart phone [2, 3].

This paper is about controlling light, fan, switchboard by utilizing Arduino boards and Bluetooth modules by the principle of Internet of Things.

2 Literature Review

In many papers, the authors had designed automated systems utilizing GSM technology. That suffers many disadvantages someone has to make a call or throw a message to get control over appliances. This exhibits the delay [4]. In this paper, the Bluetooth module is used as the communicating medium between smart phone and the microcontroller. In addition, we are keeping eye over controlling the intensity as well as speed of light and fan [5].

3 Realization

A general block diagram is given in Fig. 1.

Module 1:

Figure 1 is for switching the appliances. In the given block diagram, we are making use of Arduino Uno board. It is used to operate relay. Basically, Arduino is programmed such that it can provide voltage to its pin. In particular, pin relay is planted. Relay is basically an electromagnetic switch that is utilized for high-power switching operations [6]. Area of uses of relays is in starters that are used to start a machine.

The supply is needed for Arduino board 9 V supply either from battery or some another source. Aim is basically to influence the whole appliances by our smart phone. For that, Bluetooth module is planted that provides the connectivity to the smart phone via Bluetooth, and with the help of android application, we can influence whole appliances through any smart phone or any android device [7]. For this module, we are utilizing four-channel relay. An electric lamp and a charger are connected to two of its relay. Atmost we can control four appliances with the help of the proposed diagram shown in Fig. 1.

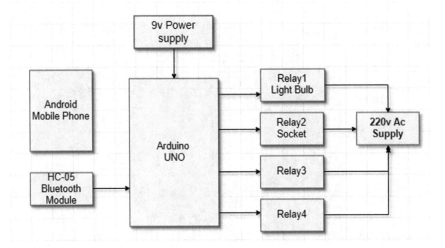

Fig. 1 Block diagram

4 Speed Control of Fan and Intensity of Light

Module 2:

As per Fig. 2 demonstrated by the square, the batteries supply ability to the Arduino nano and rest of the sections are available in this module. ULN2003 is adjusted as a controller which helps in the control of the speed of fan and power of light [8]. Flexible application is used to control the piles.

Steps related to controlling of the module by methods for Bluetooth:

1. Partner the Bluetooth to the phone.
2. Select the BT04-A Bluetooth in the application and partner with it.
3. By and by we can swipe the bar to change the component of power required.

Arduino nano is associated with the Bluetooth module just as it is associated with ULN2003 IC in the event that we associate it by means of versatile application through Bluetooth. It can control the driver circuit that is ULN2003. Consequently, by varying according to picture appeared beneath, we can control the speed of fan just as in the meantime it can control the power of light. The connections as per the circuit diagram shown in Figs. 10 and 11.

4.1 Arduino Nano

Arduino Nano is a microcontroller board arranged by Arduino.cc. The microcontroller used in the Arduino Nano is Atmega328, a comparable one as used in Arduino

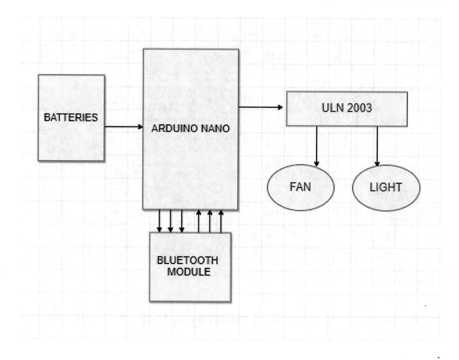

Fig. 2 Speed control of fan

Uno. It has a wide extent of employments and is a critical microcontroller board because of its little size and flexibility. The system circuit is shown in Fig. 3.

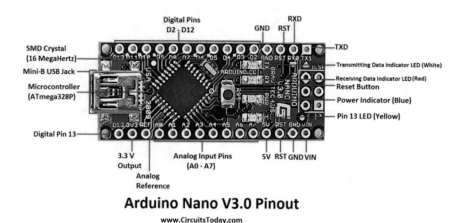

Fig. 3 Arduino Nano

Fig. 4 Microcontroller board

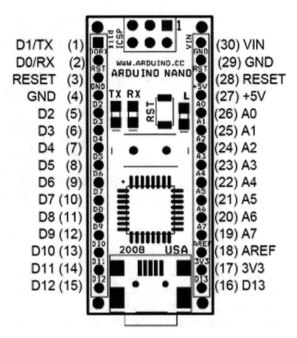

Features are:

- It has 22 input/yield sticks through and through.
- 14 of these pins are propelled pins as shown in Fig. 4.
- Arduino Nano has eight straightforward pins.
- It has 6 PWM pins among the automated pins.
- It has a valuable stone oscillator of 16 MHz.
- Its working voltage changes from 5 to 12 V.

4.2 Arduino Uno

It is a microcontroller (ATmega 328). Performs its operation at 5 voltage. It can in take 7–12 voltage. It has 14 digital pins following by 6 analog pins. It can uphold 40 mA. In addition, for 3.3 V pin can uphold 50 A in this Uno is the most modern board that come in the existence massive clock speed of 16 MHz. As shown in Fig. 5.

Fig. 5 Arduino Uno

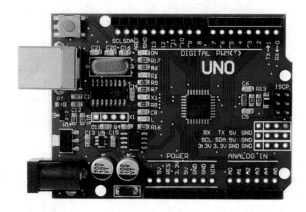

4.3 Relay

The relay is basically an electromagnetic switch that performs the switching operations based upon the output of the microcontroller. Here, we are applying Arduino board for controlling the operations of relay. The appliances that a relay can control is based upon the number channels. Here, operations have been executed by four-channel relay means we control four appliances. Figure 6 is shown the figure of four-channel relay.

ULN2003:

As shown in Fig. 7, ULN2003 is a driver circuit. Basically, it consists of two transistor pairs termed as Darlington pairs. As shown in Fig. 8, the emitter of one transistor is connected to the base of another transistor and as a result, double amplification

Fig. 6 Chip details

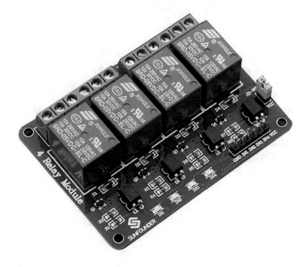

Fig. 7 Driver circuit ULN
2003

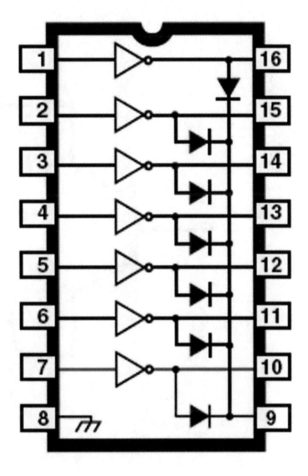

takes place; this amplification is used to regulate the supply for influencing the appliances which shown in Fig. 9.

4.4 Bluetooth Module

It is used to perform the SPP. It can be for master and slave config. It is an asset to wireless communication. From the beginning, it acts as a slave. It is used to get the connections from another Bluetooth module such as smart phones. Here, we are using it from smart phone to influence over appliances.

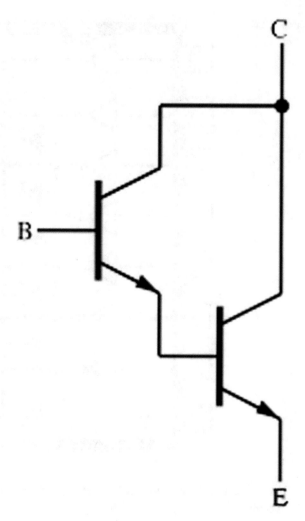

Fig. 8 Darlington pairs

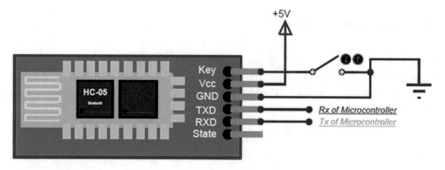

Fig. 9 Bluetooth module

5 Results

As per our target, the appliances can be controlled by the proposed circuits. In Fig. 10, the picture of the application is used for controlling switches. Followed by Fig. 11, the picture of the application is used to control the light intensity along with the speed of the motor.

Arduino is associated with Bluetooth module just as hand-off the Bluetooth module is associated with portable application. The Arduino can be control and can work the hand-off by the giving it voltage high or low. In Fig. 9, picture is shown. This output we get from Module 1 is we can switch the 230 V appliances easily via Bluetooth. Here, two channels of relay are occupied simply for switching the bulb and one plug as shown in Fig. 11.

In this module, we are influencing the speed of the fan as well as intensity of the light. Particularly, in a single application options for both motor speed and led brightness is given by varying the nob it will affect the appliances. While connecting led and a simple motor to bluetooth to get influence over them and vary the output as per our need. In Fig. 12, picture of both the module is in a single stream.

Fig. 10 Results

Fig. 11 Switching the bulb
and one plug

Fig. 12 Hardware set-up

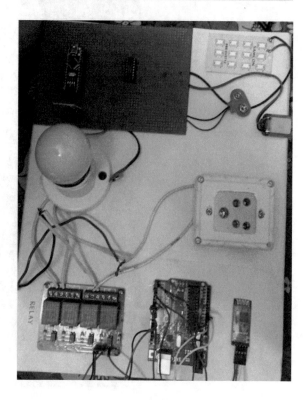

6 Conclusion

The following paper explored the possibility of controlling home appliances by using IoT and has shown successfully the ways to do so. The performance of the influence over light intensity and control the fan speed, operated and controlled the circuit by ULN2003 and Arduino. The applications of this experiment are varied and the implications can be a boon for people in all walks of life. For instance, the physical and emotional well-being of the senile in old age homes can be taken care of by making the appliances around them to be more accessible and more safe thus empowering them while taking care of their safety. The applications are limitless, and hence, IoT itself has limitless potential.

References

1. Atukorala, K., Wijekoon, D., Tharugasini, M., Perera, I., Silva, C.: Smart eye integrated solution to home automation, security and monitoring through mobile phones. In: IEEE Third International Conference on Next Generation Mobile Applications, Services and Technologies, pp. 64–69 (2001)
2. Zhai, Y., Cheng, X.: Design of smart home remote monitoring system based on embedded system. In: IEEE 2nd International Conference Control and Industrial Engineering, pp. 41–44 (2011)
3. Yamazaki, T.: Beyond the smart home. In: IEEE International Conference on Hybrid Information Technology, vol. 2, pp. 350–355 (2006)
4. Ogawa, M., Tamura, T., Yoda, M., Togawa, T.: Fully automated bio signal acquisition system for home health monitoring. In: IEEE Proceedings of the 19th Annual International Conference on Engineering in Medicine and Biology Society, vol. 6, pp. 2403–2405 (1997)
5. Easambattu, T., Reddy, P., Kumar, A., Ramaiah, G.N.K.: Controlling home appliances through GSM modem and internet. Int. J. Electr. Eng. Res. 1–7 (2013)
6. https://www.instructables.com/id/Arduino-controlled-light-dimmer-The-circuit/
7. https://en.wikipedia.org/wiki/Internet_of_Things
8. http://developer.android.comltools/studio/index.html

Bidirectional Converter with ANN-Based Digital Control and State Transitions

B. Vinothkumar, P. Kanakaraj, C. Balaji and Jeswin George

Abstract Bidirectional converters are used where bidirectional power flow is required. It reduces the space occupied drastically as compared to the individual converters utilized otherwise. Usually, the system contains a push–pull conversion stage on the lower voltage side (LV) and a full-bridge phase shift conversion stage on the higher voltage side (HV). The transitioning between buck and boost stages in a bidirectional converter is not subject to high currents and voltages which increases the life of the system and its dependability. Bidirectional converters that are isolated in circuit are normally utilized in electric vehicles as well as data storage applications where between the higher voltage side and lower voltage side, power flow is observed. The system execution incorporates the power flow between the higher voltage side and lower voltage side by using a current sustained push–pull conversion stage on the lower voltage side and a PSFB stage that synchronously rectifies the power flow to the lower voltage side. The limitations faced during the implementation of this system are: (1) quick and consistent transition from buck to boost mode and vice versa; (2) during the reverse power in the push–pull converter, the rectification of the high-voltage bridge.

Keywords Boost converter · Buck converter · Bidirectional DC–DC converter · Artificial neural network

B. Vinothkumar (✉) · P. Kanakaraj · C. Balaji · J. George
Department of Electrical and Electronics Engineering, SRM Institute of Science and Technology, Kattankulathur, Tamilnadu, India
e-mail: bpvinoth@gmail.com

P. Kanakaraj
e-mail: pkrajpse@gmail.com

C. Balaji
e-mail: balaji.c@ktr.srmuniv.ac.in

J. George
e-mail: jeswin.geo4@gmail.com

© Springer Nature Singapore Pte Ltd. 2020
P. K. Mallick et al. (eds.), *Cognitive Informatics and Soft Computing*,
Advances in Intelligent Systems and Computing 1040,
https://doi.org/10.1007/978-981-15-1451-7_42

1 Introduction

The bidirectional DC–DC converters (BDCs) are prevalently utilized in numerous mechanical applications. As of late, the utilization of such converters has expanded in the electric vehicles (EVs), half and half electric vehicles (HEVs), and energy unit vehicles (FEVs) applications because of the fuel cost and their ascent of worldwide emanations. In such applications, the energy storage systems (ESSs) typically require the batteries or supercapacitors. The BDCs must almost certainly give the ability to the engines from the ESSs amid speeding up and revive the ESSs amid the braked recovery. In the writing, the BDCs are grouped into two sorts: disconnected and non-detached converters, contingent upon the particular applications. The benefit of utilizing secluded BDCs is the way that any electrical breakdowns in both the low-voltage (LV) and high-voltage (HV) sides will be contained inside the comparing side, accordingly evading harm of the ESS framework or the unpredictable engines. BDCs can lessen the span of the change frameworks by a factor of up to 40% because of its decreased impression.

The upside of utilizing PWM conspires in the control and change of the DC–DC voltage is by the reality of delicate exchanging which can lessen pointless losses that come because of hard exchanging. The proposed framework has a push–pull arrange on the low-voltage (LV) side. The push–pull organize is constrained by PWM exchanging [1]. The best possible phase width and switching frequency will guarantee the ideal yield voltage in the AC structure which is corrected by the full-bridge arrangement on the high-voltage (HV) side. The best possible switching grouping joined with the switching frequency and the phase width of the door beats in the full extension framework on the HV side will bring down the voltage to the ideal lower voltage on the LV side.

The consistent exchanging among buck and boost mode should be possible with the assistance of a microcontroller. At the point when the LV side normally associated with the ESS goes about as the source amid the boost mode where the engine goes about as the yield because of motoring activity. Amid regenerative breaking, the engine goes about as a generator just as a source currently creating energy, this vitality to be put away over into the ESS is changed over to a lower voltage, and it helps in energizing the ESS keeping up steady power exchange subject to exchange losses.

2 Literature Survey

Hua Bai had led an investigation on bidirectional converter in a hybrid electric vehicle. This converter interfaces a high-voltage battery at a lesser voltage with a higher voltage transport and is considered to be a very powerful converter. Usually, the battery pack is at a voltage of 300–400 V. The ideal voltage for an inverter and an engine to work is 600 V. Hence, this can be utilized to coordinate between voltages of battery framework and engine framework. A portion of different elements of this DC–DC converter incorporates decreasing swell current in the battery, advancing the task of the power train framework, and keeping up DC connect voltage, henceforth, high-power activity of the power train.

R. Goutham Govind Raju had defined a zero-voltage exchanging (ZVS) bidirectional segregated DC–DC converter which is utilized in high-power applications particularly for power supply in energy component vehicles, electric vehicle driving framework, and age of intensity where high-power thickness is required. This specific method has different points of interest like minimal effort, lightweight, and it is a high unwavering quality power converter where the power semiconductor gadgets (MOSFET, IGBT, and so on) and bundling of the individual units, and the framework combination assumes a critical job in disconnected DC–DC converter crossover/energy component vehicles.

Young-Joo Lee had detailed a book coordinated bidirectional charger, and DC–DC converter (hence, the incorporated converter) for PHEVs and half and half/module cross-breed transformations were proposed. This bidirectional converter can exchange power between battery and the higher voltage bus and can work as AC–DC battery.

Lisheng Shi displayed the fundamental necessities and determinations for HEV air-conditioning DC converter plans. There are commonly two kinds of topologies utilized for HEVs: an autonomous topology and a mix which uses the engine's inverter. The duties of the two topologies are broken down in full, and the mix topology investigation is favored in light of the fact that it is increasingly worthwhile in HEVs, in regard to reserve funds in cost, volume, and weight.

B. Tanmay had proposed a multi-control port topology equipped for taking care of numerous power sources and as yet keeping up effortlessness and certain different highlights like acquiring lower yield current swell, high increase, wide burden varieties, and capacity of parallel battery vitality because of the secluded structure. This plan joins a transformer winding strategy in which the spillage of the inductance of the inductor is reduced definitely.

João Silvestre planned a bidirectional DC–DC converter for a little electric vehicle. This DC–DC converter was planned and tried; it was discovered that it is equipped for raising the voltage from the battery pack (96 V ostensible) to 600 V important to encourage the variable-frequency drive which controls the acceptance engine. This converter circuit is additionally fit for working the other way (600–96 V); this is done to catch the vitality from regenerative braking and downhill driving.

Hyun-Wook Seong had structured a non-disconnected high advance up DC–DC converter utilizing zero-voltage exchanging (ZVS) alongside lift joining strategy (BIT) and their light-load recurrence regulation (LLFM) control. The proposed ZVS BIT converter coordinates a bidirectional lift converter alongside an arrangement yield module as a parallel information and arrangement yield (PISO) setup.

Zhe Zhang et al. structured a bidirectional disconnected converter constrained by stage move and obligation for the power device half and half vitality framework. The suggested framework topology diminishes the quantity of the switches and its related door driver segments fundamentally by using two transformers of high frequency.

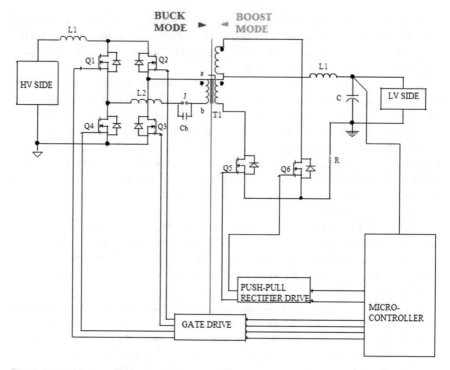

Fig. 1 Block diagram of the proposed bidirectional converter

which join a full-bridge circuit and a half-bridge circuit on the primary side. The flyback converter has an added advantage of simple structure the straightforward control; however, it experiences hard switching of the MOSFET and high-voltage switching voltage losses with the frequency of switching increments with lessening the load. In substantial load condition, the pinnacle current of the power MOSFET and major current path are extremely high, and it will at last reason higher conduction losses and leads to poor productivity. The block diagram of the proposed bidirectional converter is shown in Fig. 1.

3   Methodology and Operation

The bidirectional converter is basically a buck–boost converter and is also sometimes called a step-up/step-down power stage. In this converter, the output voltage is inverted and its magnitude can be varied to be higher or lower than the magnitude of the input voltage [2]. This converter consists of two different non-isolated topologies. Each topology consists of switches, output capacitor, and an output inductor. The two topologies can be explained separately as two different modes, i.e., boost mode and buck mode. The circuit has a full-bridge converter which consists of four MOS-FETs, and a push–pull converter which consists of two MOSFETs, a center-tapped transformer, inductors, and capacitors. The input is a DC current source.

3.1 Buck Mode

It is also called a step-down mode. The circuit consists of an inductor, a capacitor, and two switches. In this mode, the input voltage is fed through a push–pull converter and the output voltage has a magnitude less than the magnitude of the input voltage. The magnitude can be controlled by using switches to control the duty cycle. These switches also control the current in the inductor. The inductor–capacitor combination works as filter to eradicate ripples from the output waveform. When the switches are closed, then the current changes in the circuits. This changing current brings about an opposing voltage across the inductor. The inductor then continues to oppose the voltage from the source. This reduces the voltage drop across the load. The inductor is energized by the magnetic field [3]. When the switch is opened, there will be a change in current which will ensure that there is a voltage across the inductor. Hence, voltage drop across the load tends to reduce. The energy is stored in the inductor and tends to aid the flow of current through the load. This increased current covers up for the reduction in voltage and maintains the magnitude of power across the load.

3.2 Boost Mode

This mode is also called a step-up converter. In this mode, the input voltage is fed through a full-bridge converter which is made of MOSFETs. The final output voltage is observed to be greater than the voltage across the input. At the point when switch is in the off state, then inductor stores power. A field is generated which is generated in nature [4]. The current in the open state experiences high resistance, and hence, the current reduces. Therefore, the magnetic field is destroyed to maintain the help of the flow of the current. When the switching is at very fast pace, the inductor does not discharge completely, and therefore, the final voltage across the output side will be higher than the voltage in the input side. Capacitor connected in parallel to the load charges to the new higher voltage. So, as the switch is brought to the off state, then capacitor is able to supply the power to load (Fig. 2).

The converter in this state functions as the current fed push–pull stage, charges up, and increases the voltage to a high-voltage AC. Upon transferring this higher voltage AC by the transformer to the secondary side, the full-bridge switching devices are turned off and their internal diodes function to rectify the AC to DC voltage. The inductor–capacitor combination works to reduce the ripples as well as the settling time of the output voltage to a stable DC level [5, 6]. The switches of the push–pull

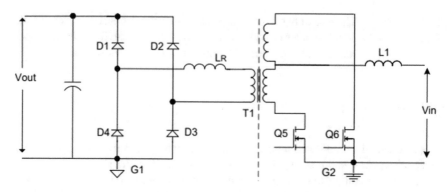

Fig. 2 Operation of boost mode

conversion stage, Q_5, and Q_6 are given PWM pulses which are more than half of $180°$ out of phase with the corresponding duty cycles.

T_0–$T_1 \rightarrow Q_6$ and Q_5 are both switched ON. Inductive vitality in L_1 raises and the current is also raised.

T_1–$T_2 \rightarrow$ At time T_1, the switching device Q_6 is switched off. Due to this, the inductive energy stored begins to discharge and gives power to the higher voltage side (HV side) through the switching devices diode's Q_3 and Q_1.

T_2–$T_3 \rightarrow$ Operation amid T_2–T_3 is actually equivalent to T_0–T_1.

T_3–$T_4 \rightarrow$ Similarly, this operates like T1–T_3, and then again, actually at T_3, Q_5 is turned off and diodes of Q_4 and Q_2 give capacity to HV side.

The complex nature of the PWM switching scheme poses a challenge to the controller to generate a waveform in the buck mode. Advanced computerized controllers can generate the waveforms that are required. Furthermore, the transition between the boost and buck mode should be seamless and whenever required. The input and output voltages should not affect the smooth and quick transition between the modes and thus increase reliability and reduce the stress on the components. To ensure that the voltages and current on both the lower voltage and higher voltage side are within safety limits, the initial condition of the compensator as well as its gains must be tuned accurately.

The wide ranges of voltages on both the buses along with different operating states pose a monumental challenge to overcome by using observational approaches. The main challenges of the bidirectional converter are solved by this approach. Initially, different execution enhancements were presented, while using PWMs with 50% of the duty cycle with the control of the phase shift for the rectification of HV FET in the reverse flow of power in the bidirectional converter [7].

Secondly, to achieve seamless switching between boost and buck modes under a wide spectrum of output and input current and voltages, a technique was proposed. The simulation of a 300 W bidirectional converter system confirmed the accuracy of this method. What this paper suggests is that the usage of the command value of the duty cycle in per unit to be used as the initial input in the equation to switch seamlessly between the buck mode and the boost mode. The phase command value in per unit should be applied as the initial input in the equation when switching from the boost mode to the buck mode.

The VLV and VHV values in the equations compare to the values of voltage on the buses before the mode changes.

4 Controller Techniques

The main theme of the ANN is processing information like how the brain, process information through neurons. Using ANN can process numerous highly interconnected variables to solve specific problems. The ANN has different algorithms such as feed-forward (FF), support vector machine (SVM), and self-organizing map (SOM).These algorithms are categorized based on supervising and un-supervising data. In this paper, multi-feedback scheme was introduced. Because it have many advantages over SVM, SOM, such as supervised learning algorithm, no cycles or loops, the information flows in one direction. Mainly supervised algorithm can compare the old values with new values for produce the optimum result. But unsupervised algorithms process the way of trial and error method and do not consider old values for process the new values. During the leaning process, the supervised algorithm process the data within the short period compared to unsupervised algorithms. ANN is suitable for real-time operation control and fault tolerance of redundant information coding.

In DC–DC bidirectional converter, ANN is used to achieve the optimal switching performance. The controller of the converter requires the pre-trained data set for implementing real-time operation and control. This is can be obtained through the multi-feedback data set. Initially, the controller was designed for linear load operation. Because to obtain the control variable data set, the controller was trained for the nonlinear load of operation. In learning process, the neural net identifies the pattern that will be processed to the layer of the network. This process will continue until it reaches the output layer. The multi-feedback control scheme is introduced to manipulate the current control between the source 1 and source 2, the multi feedback networks is shown in Fig. 3.

For the simulation, data set consists of 150 instances of input current and input voltage which are used. With this algorithm, fast estimation of the nonlinear load data was obtained. Finally, the effective control of bidirectional DC–DC converter was simulated in MATLAB.

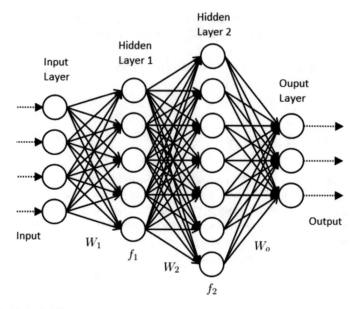

Fig. 3 Multi feedback networks

5 SIMULATIONS

5.1 Boost Mode

See Figs. 4, 5, 6 and 7.

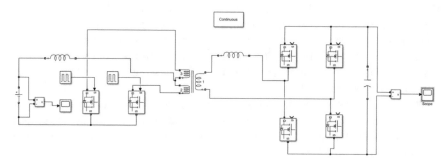

Fig. 4 Boost mode simulation

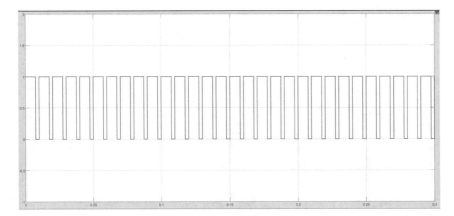

Fig. 5 Gate pulse for boost mode

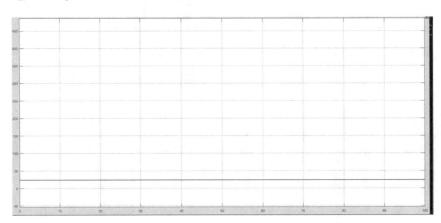

Fig. 6 Input voltage waveform

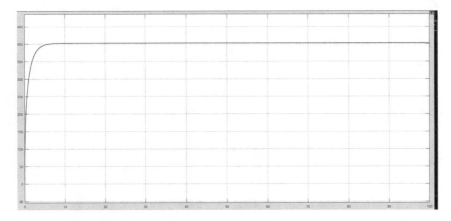

Fig. 7 Output voltage waveform

5.2 *Buck Mode*

See Figs. 8, 9, 10, and 11.

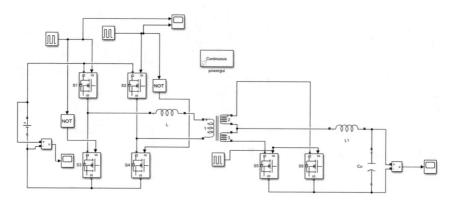

Fig. 8 Buck mode simulation

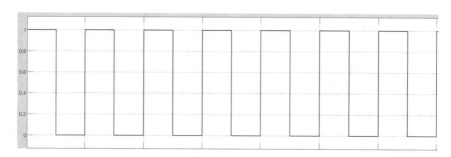

Fig. 9 Gate pulse for buck operation

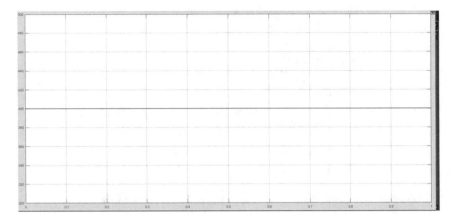

Fig. 10 Input voltage waveform

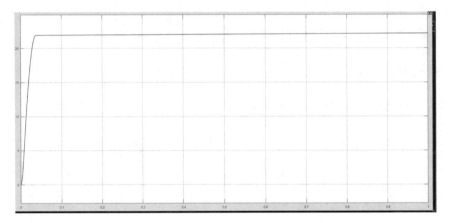

Fig. 11 Output voltage waveform

6 Design Calculations

1. Maximum duty cycle, $D = \dfrac{+V_o + V_F}{V_o + V_F + V_i(\min)} = \dfrac{400 + 0.5 \text{ V}}{400 + 0.5 + 24} = 0.943$

2. DC converter ratio, $= \dfrac{V_o}{V_i} = \dfrac{-D}{1 - D} = \dfrac{-0.943}{1 - 0.943} = -16.6875$

3. Maximum peak switch current

$$= I_{\text{swm}}$$
$$= (I_{\text{om}} \frac{(V_i(\min) + V_F - V_o)}{V_i(\min)}) + \frac{I_{L1}}{2}$$

$$= 0.1 \, \text{A} \times \frac{24 \, \text{V} + 0.5 \, \text{V} + 400}{24} + \frac{365}{2} = 1.951 \, \text{A}$$

4. Maximum output current

$$= I_{\text{OM}}$$

$$= \left(I_{\text{sw(min)}} + = \frac{I_L(\text{pp})}{2} \right) \times \left(\frac{V_i(\text{min})}{V_i(\text{min}) + V_F - V_o} \right)$$

$$= \left(1.951 + \frac{0.365}{2} \right) \times \left(\frac{24}{24 + 0.5 + 400} \right) = 0.75 \text{A}$$

5. Maximum peak switch voltage, $V_{\text{om}} = V_{\text{im}} + V_F - V_0 = 440 \, \text{V}$

6. Inductance $L_i = \dfrac{V_i D}{I_{\text{LPP}} f_{\text{sw}}} = $ Data sheet recommendation $L_i = 4.7 \, \mu\text{H}$

7. Average Induction Current $L_{\text{iav}} = \dfrac{I_o}{1 - D} = \dfrac{1.95}{1 - 0.94} = 32.516 \, \text{A}$

8. Maximum Diode DC reverse voltage, $V_{\text{rm}} = V_o - V_{\text{im}}$

9. Maximum Input Capacitance

$$= \frac{I_{\text{Liav}} \times D}{f_{\text{sw}} \times \left(V_{\text{ipp}} - \left(I_{\text{Lipp}} \times \text{ESR}_{\text{CI}} \right) \right)}$$

$$= C_i = 2.3 \, \mu\text{F}$$

10. Maximum output capacitance

$$= \frac{I_o D}{f_{\text{sw}} \times \left[\left(V_{\text{opp}} - \left(\frac{I_o}{1-D} \times \frac{I_{\text{cpp}}}{2} \right) \right) \right] \times \text{ESR}_{\text{co}}}$$

$$= 9.6 \, \mu\text{F}$$

7 Conclusion

A novel and universal solution for the bidirectional DC–DC converter with higher effectiveness and stable yield voltage, unwavering quality, and simple and improved advanced control has been proposed. The significant difficulties of the bidirectional converter, which utilizes computerized control, are settled. To start with the correction in the inverted power flow of a bidirectional converter displayed different execution upgrades which utilized half-duty cycle PWM with phase shift control of HV FET.

Secondly to achieve a smooth transition between buck and boost modes, with a wide range of input and output voltages, a different method was proposed. The above system was simulated and yielded expected results.

References

1. Sabaté, J.A., Vlatkovic, V., Ridley, R.B., Lee, F.C., Cho, B.H.: Design contemplations for high-control full-connect ZVS-PWM converter. In: Proceedings of IEEE APEC90, pp. 275284 (1990)
2. Kim, K.-M., Park, S.-H., Lee, J.-H., Jung, C.-H., Won, C.-Y.: Mode Change Method of bi-directional DC/DC Converter for Electric Vehicle. In: Power Electronics and ECCE Asia (ICPE & ECCE), 2011 IEEE 8th International Conference, pp. 2687–2693
3. Mweene, L., Wright, C., Schlecht, M.F.: A 1 kW, 500 kHz based front-end converter for a conveyed power supply framework. In: Proceedings of IEEE APEC89, pp. 423432 (1989)
4. Vinothkumar, K.B., Love, K.M.: Interleaved boost converter with high efficient switch control for PV systems. J. Adv. Res. Dyn. Control Syst. **10**(07), 1197
5. Ayyanar, R., Mohan, N.: A tale delicate exchanging DC–DC Converter with wide ZVS-extend and decreased channel necessity. In: Proceedings of IEEE PESC99, pp. 433438 (1999)
6. Full-load-go ZVS half breed DC–DC Converter with two full-spans for high-control battery charging. In: Proceedings of INTELEC99 (2018)
7. Ngo, K.D.T., Kuo, M.H., A. Fisher, R.: A 500 kHz, 250 W bidirectional converter which has numerous yields constrained by stage move PWM and attractive speakers. In: Proceedings of High Frequency Power Conversion Conference, pp. 100110 (1998)

Public Opinion Mining Based Intelligent Governance for Next-Generation Technologies

Akshi Kumar and Abhilasha Sharma

Abstract The term technology signifies the applied scientific knowledge utilized to manipulate and transform the human environment. The successful implementation of such knowledge simplifies the practical aims of human life that makes a substantial impact over socio-economic condition of a nation. With the changing lifestyle of society and to cater their growing demands, next-generation technologies pave the way for a continuous process of global development. They may be viewed as a solution space for global challenges of upcoming century. Various application areas/domains are open with a strong urge to embed these technologies for better productivity and performance such as health, agriculture, transport, energy utilization, building and infrastructure, mobility and so on. Government is also taking different measures to incorporate these technologies as a practical solution towards the routine problems of various domains faced by citizens. The accelerating pace of technological innovation brings numerous challenges. It is not easy to synchronize day-to-day service delivery with the quick change in new technology where citizen's interest is at stake. So, the contemplation of public perception is a critical step in the entire process of growth and expansion in intelligent governance system. This paper propounds an opinion prediction model using machine learning algorithms for the next-generation technology in exploring universe. A recent space mission, Gaganyaan, or orbital vehicle that is nation's first manned space flight introduced by Indian government has been chosen for this conduct. In this paper, an effort has been made to examine and evaluate the public perception for this space operation by exploiting social big data in order to realize the positive and negative inclination of Indian citizen over upcoming technologies. Various social networking tools are available to collect and extract public opinion or sentiments. Amongst them, Twitter has been used as a social media tool for data acquisition concerning the availability of real-time data for this space technology.

A. Kumar · A. Sharma (✉)
Department of Computer Science & Engineering, Delhi Technological University, Delhi 42, India
e-mail: abhilasha_sharma@dce.ac.in

A. Kumar
e-mail: akshikumar@dce.ac.in

© Springer Nature Singapore Pte Ltd. 2020   401
P. K. Mallick et al. (eds.), *Cognitive Informatics and Soft Computing*,
Advances in Intelligent Systems and Computing 1040,
https://doi.org/10.1007/978-981-15-1451-7_43

Keywords Opinion mining · Government intelligence · Machine learning ·
Space · Twitter

1   Introduction

Technology has transformed the shape of various sectors of life across world. Even
the most trivial of our day-to-day tasks involve the use of some or the other form of
technology. From making a phone call using a mobile phone to heating food in the
microwave, from driving to the office in a car to sleeping in an air-conditioned room,
technological inventions have been providing solutions to our problems and have
now become a part of our everyday lives. Technology can be embedded in absolutely
any sphere of life to ensure a convenient and more efficient functioning. Figure 1
represents various domains [1] where technology is being applied currently and able
to make progressive developments with the use of next-generation technologies for
their competitive advantage.

The rapid speed of technological changes contributes towards improving various routine life activities. Better communication technology, improved mobility
and remote connectivity, ease in information storage and access, reduced human
effort/automation, better tools portability, cost-efficient systems, improved learning
systems, artificial intelligence are some of the examples [2, 3] showcasing technology as a boon. Future technologies, also commonly referred to as "next-generation
technologies", have long been an issue of interest and demand the formation/creation
of new, innovative and technology-based models in different real-time application
areas. Outer space technology is one such upcoming area that is gaining attention
amongst people and has captured the attention of the government as well. Moreover,

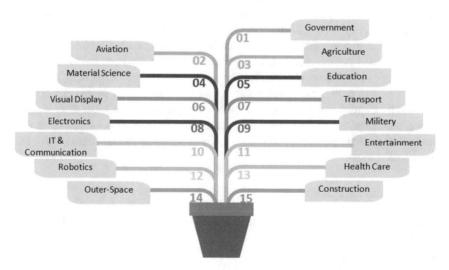

Fig. 1   Open application areas for adoption of next-generation technologies [1]

government is constantly trying to advance itself by incorporating science and technology in its processes and practices [4] for the public welfare. However, if decisions are being made for the public, the opinion of the public on the decision must be taken into consideration whilst evaluating the outcome of the decision. Hence, an effective process of opinion mining [5] helps in proper assessment of any governmental initiative, in accordance with the views of the public.

Therefore, the objective of this paper is to make an intelligent and sustainable governance able to embed next-generation technologies for boosting the living standards of a society by utilizing the concept of public opinion mining. This research involves the analysis of public opinion over an initiative taken by government of India that is *Gaganyaan*, with space technology as the chosen application area of study. Gaganyaan is an Indian crewed orbital spacecraft intended to be the basis of the Indian Human Spaceflight Programme [6]. To be launched in 2022, this project is expected to promote research and development in India. On the contrary, assigning a budget of Rs. 10,000 crores [7] for a future project (that may or may not be successful) involves high risk and investment. Consequently, it is crucial to know the inclination of the public towards such projects that are introduced with the intention of improving the lives of the public. This paper entails the use of ten machine learning algorithms, namely Naive Bayes, Support Vector Machine, Decision Trees, k-Nearest Neighbour, Multi-Layer Perceptron, Random Forest, Simple Logistic, K-Star, Bagging and Adaboost in order to figure out the best classifier for the classification of opinion polarity of the tweets captured on the subject *Gaganyaan*. Twitter, being a social networking channel and the single available platform for sharing and posting views for governmental practices and proceedings, has been used for dataset compilation.

The remaining paper is lined up as follows: Sect. 2 discusses the proposal of *Gaganyaan* and provides an overview of its progress graph till date. Section 3 evaluates the public inclination towards the initiative that includes collection of real-time data (tweets), various forms of statistical representation of data and opinion classification process by measuring Precision, Recall and Efficiency. Section 4 discusses the results and findings along with their comparative graphical representations. At last, Sect. 5 concludes the paper.

2 Gaganyaan—Target to Turn Inside Out of Universe

Gaganyaan [8] is the name of a crewed spacecraft that is planned to be launched in 2022 by Indian Space Research Organization (ISRO). It is the first step towards ISRO's Indian Human Spaceflight Programme [9] that aims to carry astronauts into lower earth orbit (Fig. 2).

The initial work for Gaganyaan started in 2006, under the generic name "Orbital Vehicle". It was supposed to have an endurance of a week in space with a capacity of two astronauts. It was finalized in March 2008 and then sent for funding. The funding for the Indian Human Spaceflight Programme was given a positive response in February 2009. In 2010, announcements [11] were made that Gaganyaan would

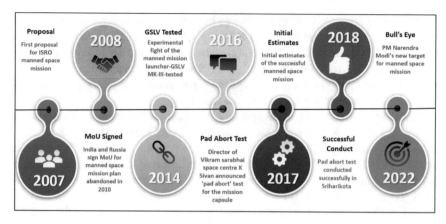

Fig. 2 Timeline-based work progress for Gaganyaan project [10]

be launched in 2016. However, due to issues with the funding for the project, it was put off ISRO's priority list. In early 2014, the project was reconsidered and benefitted from the increase in budget [12] announced in 2017, February. Prior to this in January 2014, the ISRO Chairman revealed [13] that the orbital vehicle would be tested in flight in 2014. ISRO's pad abort test was set for March 2017, but it was delayed until 5 July 2018. It was a success and [14] conducted in Sriharikota.

Gaganyaan [13, 14] weighs 3.7 tones and is meant to carry three astronauts. It has endurance of a few orbits or seven days in space. It is shaped like Soyuz reentry spacecraft. Two liquid-propellant engines power the spacecraft [14]. The spacesuit is orange, developed and designed over the past two years at Vikram Sarabhai Space Centre, Thiruvananthapuram. It can hold one oxygen cylinder, breathing capacity for 60 min. Currently, two have been made, and more suit is yet to bee finished. All three astronauts are planned to travel in capsule for five to seven days. It is planned for 400 km orbit from the earth's surface. The capsule is equipped with a thermal shield since during the reentry into earth's surface, the capsule will get extremely heated and catch flames. The shield is to ensure that the temperature stays below 25°. Flames will be visible to the astronauts from the capsule windows. The spacecraft is set to rotate around [14, 15] the earth every 9 min. India can be seen from it, and the astronauts will be conducting experiments on microgravity. It will take 16 min to reach the 400 km lower earth orbit, and the return will take 36 h. It is to land in the Arabian Sea where the coast guard and the navy will be present for assistance. ISRO displayed the Gaganyaan crew model [15] and the spacesuits at the Bengaluru Space Exposition, 6th edition on 6 September 2018. Along with the suits, the crew model and escape model were also displayed. Success of this spacecraft mission will make India the fourth country to have done so, right after USA, Russia and China.

3 Opinion Mining of Gaganyaan

The research and development in the field of outer space are set to establish a nation as a nation trying to make a mark in the world. The proposal of manned spacecraft is the next milestone that India plans to achieve in its outer space exploration. The announcement of this mission is a big news because of the amount of resources and funds that go into a project with a huge size and also because such projects put the nation at a spotlight in front of the world, leaving no room for error. Contrarily, the resources spent in these fields might be considered by many as a focus in the wrong direction. These resources could be spent in a number of basic amenities that a country needs work in. This brings forth a lot of public opinion, ideas, suggestions or sentiments to be considered where social media provides a convenient platform to post and share them. Therefore, opinion mining [16] is done in order to analyse the public opinion over a project that requires loads of funding and resources. Twitter, the social networking site, has chosen for the extraction of data. Figure 3 represents sequential block diagram for the process of opinion classification of tweets.

3.1 Data Collection and Pre-processing

Twitter [17] has been used for the collection of data related to the Gaganyaan space mission. It is a social networking service through which users post and interact with messages known as "tweets". It is a platform which helps in determining the opinion of millions of people over a specific topic or a scheme.

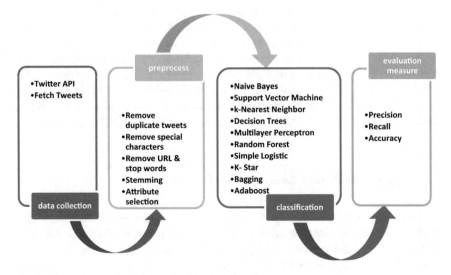

Fig. 3 Block diagram for opinion classification of tweets

Table 1 Stats of tweets collected on daily basis

Week-I		Week-II		Week-III		Week-IV	
Date	Tweets count	Date	Tweets count	Date	Tweets count	Date	Tweets count
15/8/2018	372	22/8/2018	1	29/8/2018	50	05/9/2018	3
16/8/2018	279	23/8/2018	5	30/8/2018	23	06/9/2018	50
17/8/2018	155	24/8/2018	26	31/8/2018	16	07/9/2018	87
18/8/2018	11	25/8/2018	6	01/9/2018	23	08/9/2018	19
19/8/2018	51	26/8/2018	17	02/9/2018	0	09/9/2018	4
20/8/2018	7	27/8/2018	43	03/9/2018	63	10/9/2018	12
21/8/2018	5	28/8/2018	64	04/9/2018	6	11/9/2018	43
						12/9/2018	8
						13/9/2018	12
						14/9/2018	43

Twitter search application package interface (API) has been used having specific keywords in them. The various hash tags used for the mission are #gaganyaan, #manspacemission and #orbitalvehical. The data collection processing was carried out for a duration of 4 weeks giving a total of 1504 tweets. The count of number of tweets collected on daily basis has been listed in Table 1.

To carry out opinion mining, the data should be well organized, uniform and standardized, and therefore, cleaning of data was done, i.e. unnecessary words were removed from the collected data set such that the resultant data contains only required information for the analysis. Tweets obtained after this process were then classified by using various machine learning algorithms. The data pre-processing technique involves various steps such as elimination of duplicate tweets followed by removal of numbers and special characters such as @, #, URL links and stop words like is and are. After this, removal of inflectional and derivational endings has been done to reduce words in their root form. Finally, the resultant data is pre-processed for feature selection and classified into three categories based on opinion polarity, namely positive, negative and neutral. Some examples of the attribute selected are Lighting, Shedding, Connectivity, Families, Homes, Poor, Scheme, Bulb, Crackdown, etc.

3.2 Applied Machine Learning Techniques

The process of sentiment classification can be carried out using various approaches. Machine learning and lexicon-based are more popular as per the literature. Machine learning approaches use statistical analysis [18] to make predictions using input data. In this paper, we empirically analyse ten standard machine learning algorithms, [18, 19] namely Naive Bayes, Support Vector Machine, Decision Trees, k-Nearest Neighbour, Multi-Layer Perceptron, Random Forest, Simple Logistic, K-Star, Bagging and

Adaboost, and compare them over three key performance parameters, i.e. Precision, Recall and Accuracy. Table 2 illustrates the brief summary of machine learning techniques used, and description of standard efficacy measures to evaluate opinion classification process is listed in Table 3.

3.3 Opinion Classification of Tweets

Opinion mining involves building a model in order to collect and categorize opinions about any subject, topic, processes, practices, project or a product. In this paper, machine learning-based opinion mining has been implemented to classify tweets polarity and qualitatively analyse the public opinion inclination towards Indian first manned space project Gaganyaan. Table 4 represents the classification of tweets for all the four weeks of data set into three categories: positive (P), negative (N) and neutral (Nu).

Figure 4 represents the weekly distribution of positive, negative and neutral tweets. The statistics of first week show that around 880 tweets collected, 28.9% of the tweets were neutral, 67.04% were positive, and 3.9% of the tweets were negative in polarity. The second-week statistics reflect mixed public opinions, and still, the percentage of positive tweets (58.6%) were higher than negative tweets (8.6%). Second week captures 32.7% neutral tweets. From the third week figures, it is evident that there is similarity in number of positive tweets with 51.9% as compared to 40.8% of neutral tweets and 7.1% of negative tweets. Similar scenario has been reported for the fourth week with total tweets being 281 out of which 38.7% positive, 53.7% neutral and 7.4% negative tweets. The percentage fall in number of positive tweets as we move from week 1 to week 4 clearly states that most of the people expressed their views during the initial proposal of the project, and there is a rise in number of neutral tweets due to consistent and continuous posting of informational tweets regarding the mission by government, civic or media.

4 Results and Findings

A count of 1504 tweets of Gaganyaan project has been collected for the classification process of tweets opinion polarity. The performance of ten machine learning classifiers has been computed and analysed to figure out the best classifier for the process of tweet classification. Results have been compiled based on three performance parameters, i.e. Precision, Recall and Accuracy for every classifier and are listed in Table 5.

Results determine that Adaboost performed with minimum accuracy of 73.8%, whereas SVM classifier achieved the best amongst all with 92.7% accuracy followed by MLP with 92.2%. kNN results in 91.5% that is almost equivalent to DT with an accuracy of 91.4%. The remaining classifiers accuracy fall in the following range as

Table 2 Brief description of machine learning algorithms

Algorithm(s) used	Description
Naive Bayes	This model belongs to the family of "probabilistic classifiers". It is based on conditional probability and assumes that features are statistically independent. It requires a number of parameters linear in the number of variables (features/predictors) in a learning problem
Support Vector Machine	It is based on supervised learning model which can be used for both classification and regression tasks. Its objective is to find an optimal hyperplane in an N-dimensional space (N—the number of features) that distinctly classifies the data points
Multi-Layer Perceptron	It is a deep, feed-forwarding artificial neural network consisting of more than one perceptron. It comprises an input layer to receive the signal, an output layer that makes a decision or prediction about the input, and in between those two, an arbitrary number of hidden layers that are involved in computation of the input
k-Nearest Neighbours	This algorithm is one of the most fundamental and simplest classification algorithms with a non-parametric feature, i.e. does not make any underlying assumptions about data distribution. Object classification is based on the majority vote of its neighbours, with the object being assigned to the class most common amongst its k-Nearest Neighbours
Simple Logistic	It is a form of predictive modelling technique which investigates the relationship between a dependent and independent variable
Decision Tree	It is a non-parametric supervised learning-based algorithm where the data set is continuously split (i.e. decision is taken) according to a certain parameter. Its major goal is to predict the value of a target variable by making simple decision rules inferred from the data features
Random Forest	Random decision forest is a method that operates by constructing multiple decision trees during training phase. The decision of the majority of the trees is chosen by the random forest as the final decision
K-Star	It is based on instance of the classifier. In this, the test instance class is based on the training instance class which is alike to it, and as resolved by some homogeneous function, it differs from other learner in that it uses entropy function
Bagging	In bagging, we take multiple samples from training data and combine the result to individual trees to create a single prediction model; this method runs in parallel to each bootstrap sample is not dependent on others
Adaboost	It is a simple classifier that combines various weak classifiers into a strong one. A base algorithm has been selected and equal weights are assigned to all training classes. Decision stump splits the data into two regions, increasing the weight of incorrect classified examples. Iterate it for n times, with updated weights over base classifier. Resultant model is the weighted sum of n learners.

Table 3 Standard performance measures description

Standard measures	Definition
Precision	It refers to the degree of correctness or quality of classifier, how precisely the classifier measures. Quantitatively, it is the ratio of true positives to predicted positives
Recall	It describes the sensitivity of the classifier, its ability to find all the relevant cases within a dataset. It is the ratio of true positives to actual positives
Accuracy	The degree of closeness of a measurement to the true value is termed as *Accuracy*. It is the ratio of number of correct predictions to the total number of input samples

Table 4 Daily status of tweets classified based on opinion polarity

Date	N	Nu	P	Total	Date	N	Nu	P	Total
Week-I					Week-II				
15/8/2018	16	43	313	372	22/8/2018	0	0	1	1
16/8/2018	10	104	165	279	23/8/2018	0	4	1	5
17/8/2018	04	65	86	155	24/8/2018	4	12	10	26
18/8/2018	0	8	3	11	25/8/2018	1	1	4	6
19/8/2018	3	26	22	51	26/8/2018	1	2	14	17
20/8/2018	1	6	00	7	27/8/2018	3	8	32	43
21/8/2018	1	3	1	5	28/8/2018	5	26	33	64
Total	35	255	590	880	Total	14	53	95	162
Week-III					Week-IV				
29/8/2018	4	22	24	50	05/9/2018	0	1	2	3
30/8/2018	2	11	10	23	06/9/2018	0	34	16	50
31/8/2018	2	4	10	16	07/9/2018	3	23	61	87
01/9/2018	1	1	21	23	08/9/2018	0	13	6	19
02/9/2018	0	0	0	0	09/9/2018	1	1	2	4
03/9/2018	2	34	27	63	10/9/2018	1	8	3	12
04/9/2018	2	2	2	6	11/9/2018	8	28	7	43
Total	13	74	94	181	12/9/2018	2	4	2	8
					13/9/2018	2	8	2	12
					14/9/2018	04	31	8	43
					Total	21	151	109	281

RF with 90.8%, Simple Logistic with 89.1%, k-star with 87.3%, Bagging with 85.2% and NB with 78.7%. Analysing the resulting figures, we found that Support Vector Machine outperforms over the remaining classifiers. The graphical stats analysis for opinion classification of the ten machine learning classifiers used for Gaganyaan project is represented in Fig. 5.

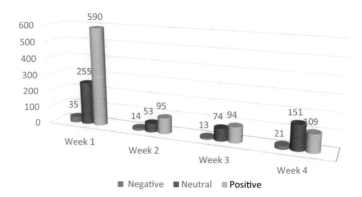

Fig. 4 Weekly distribution of positive, negative and neutral tweets for Gaganyaan

Table 5 Standard measure of evaluation results for machine learning classifiers

Machine learning techniques	Accuracy	Precision	Recall
Support Vector Machine	91.2	90.7	92.7
Decision Trees	90.1	92.1	91.4
Multi-Layer Perceptron	90.4	92.8	92.2
k-Nearest Neighbour	90.3	90.2	91.5
Random Forest	86.9	84.96	90.8
Simple Logistic	84.2	86.7	89.1
K-Star	82.7	81.5	87.3
Bagging	80.1	78.7	85.2
Naive Bayes	79.3	73.9	78.7
Adaboost	67.8	69.2	73.8

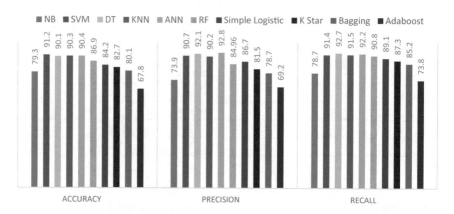

Fig. 5 Standard measure of evaluation for machine learning techniques

5 Conclusion

Gaganyaan, the first manned space project of India introduces different new dimensions of progress for the nation at international forum. It tends to enhance the scope of science and technology, shape up the current industrial growth and to provide the vision for the usage of upcoming technologies for social benefits. Along with the appealing benefits, unpromising limitations and challenges are also associated that either boost the human living standards or may impact the lifestyle of any society. Hence, the purpose of this paper is to evaluate the rapid increase in the usage of next-generation technologies for social welfare considering the recent advancement in Indian space technology, i.e. Gaganyaan. The results reveal that approximately 59% of the citizen are favouring the concept of this space mission, whereas 5.5% people are not agreed with the concept. The remaining 35.6% of the tweets are neutral, or basically, informational tweets contain the information updates of latest processing of the mission. There is a high probability of increase in positive percentage of tweets after the successful launch of this mission.

Opinion mining has been performed by applying total ten machine learning classifiers used for computation of Precision, Recall and Accuracy. SVM attained the best accuracy, and Adaboost scored the lowest one. The future scope of this research work may be enhanced in the direction of optimizing opinion classification process by the use of hybrid classifiers. More computing techniques such as deep learning or swarm-based may be used in combination in order to improve the overall accuracy of classifiers.

References

1. Applications: https://www.data.gov/applications
2. Top 8 Areas that Define Modern Applications: http://www.surroundtech.com/top-8-areas-that-define-modern-applications
3. Applications of Information Technology: https://sites.google.com/site/viveklpm/information-technology-in-veterinary-science/applications-of-information-technology
4. Kumar, A., Sharma, A.: Paradigm shifts from e-governance to s-governance. In: The Human Element of Big Data: Issues, Analytics, and Performance, vol. 213 (2016)
5. Kumar, A., Sharma, A.: Systematic literature review on opinion mining of big data for government intelligence. Webology 14(2), 6–47 (2017)
6. Gaganyaan: https://en.m.wikipedia.org/wiki/Gaganyaan
7. Mission Gaganyaan: https://www.google.com/amp/s/www.indiatoday.in/amp/science/story/mission-gaganyaan-cabinet-approves-10-000-crore-budget-send-indians-to-space-2022-1418941-2018-12-28
8. Indian Manned Spacecraft: http://www.astronautix.com/i/indianmannedspacecraft.html
9. Rs. 171 Crore Boost to Manned Space Project: http://timesofindia.indiatimes.com/business/india-business/Rs-171-crore-boost-to-manned-space-project/articleshow/30591353.cms
10. Explained: How to send an Indian into space? https://www.civilsdaily.com/news/explained-how-to-send-an-indian-into-space/
11. Indian Human Spaceflight Programme: https://en.wikipedia.org/wiki/Indian_Human_Spaceflight_Programme

12. Budget 2018 Allocates Rs. 89.6 bn to Department of Space for Satellite Launches: https://www.business-standard.com/budget/article/budget-2018-allocates-rs-89-6-bn-to-dept-of-space-for-satellite-launches-118020200311_1.html
13. ISRO Setting up Launch Pad for Gaganyaan Mission: https://www.thehindu.com/news/national/isro-setting-up-launch-pad-for-gaganyaan-mission/article25010147.ece
14. Third Launch Pad to be Set up at Sriharikota for Gaganyaan Mission: https://iasscore.in/current-affairs/prelims/third-launch-pad-to-be-set-up-at-sriharikota-for-gaganyaan-mission
15. Gaganyaan: Astronauts on mission likely to be pilots, crew module design to be finalised soon. https://www.indiatoday.in/science/story/gaganyaan-mission-isro-pilots-crew-module-1435145-2019-01-20
16. Pang, B., Lee, L.: Opinion mining and sentiment analysis. Found. Trends® Inf. Retrieval 2(1–2), 1–135 (2008)
17. Kumar, A., Sharma, A.: Opinion mining of Saubhagya Yojna for Digital India. In: International Conference on Innovative Computing and Communications, pp. 375–386. Springer, Singapore (2019)
18. Kumar, A., Sharma, A.: Socio-Sentic framework for sustainable agricultural governance. In: Sustainable Computing: Informatics and Systems (2018)
19. Kumar, A., Jaiswal, A.: Empirical study of twitter and tumblr for sentiment analysis using soft computing techniques. In: Proceedings of the world congress on engineering and computer science, vol. 1, pp. 1–5 (2017)

The Multifaceted Concept of Context in Sentiment Analysis

Akshi Kumar⑩ **and Geetanjali Garg**⑩

Abstract The contemporary web is about communication, collaboration, participation, and sharing. Currently, the sharing of content on the web ranges from sharing of ideas and information which includes text, photos, videos, audios, and memes to even gifs. Moreover, the language and linguistic tone of user-generated content are informal and indistinct. Analyzing explicit and clear sentiment is challenging owing to language constructs which may intensify or flip the polarity within the posts. Context-based sentiment analysis is the domain of study which deals with comprehending cues which can enhance the prediction accuracy of the generic sentiment analysis as well as facilitate fine grain analysis of varied linguistic constructs such as sarcasm, humor, or irony. This work is preliminary to understand the what, how and why of using the context in sentiment analysis. The concept of 'context in use' is described by exemplifying the types of context. A strength–weakness–opportunity–threat (SWOT) matrix is made to demonstrate the effectiveness of context-based sentiment analysis.

Keywords Sentiment analysis · Context · Social media · SWOT

1 Introduction

Undeniably, the cross-platform, cross-lingual, multimodal social web is omnipresent. Popular sites such as Twitter, Facebook, and Instagram have become a critical part of our daily lives to share content, stay connected, and gain insights. The expansive user base on such social networking sites generates voluminous data which can be intelligently filtered and analyzed for building a real-time knowledge discovery framework. The user-generated data can be used by organizations to analyze the trends of market, opinion for the elections, recommendation of the products, and

A. Kumar · G. Garg (✉)
Department of Computer Science and Engineering, Delhi Technological University, Delhi, India
e-mail: geetanjali.garg@dtu.ac.in

A. Kumar
e-mail: akshikumar@dce.ac.in

© Springer Nature Singapore Pte Ltd. 2020
P. K. Mallick et al. (eds.), *Cognitive Informatics and Soft Computing*,
Advances in Intelligent Systems and Computing 1040,
https://doi.org/10.1007/978-981-15-1451-7_44

services to the users [1, 2]. Sentiment analysis is one such popular natural language processing task to mine the web content. This classification task determines the opinion polarity of the post to comprehend the sentiment and/or emotion specified explicitly. Multiple studies have been conducted to analyze the sentiment of the posts, but the language used by users on the web is a mixture of formal and informal language [3, 4]. Detecting the accurate sentiment of the post is not an easy task, and the presence of constructs like sarcasm, irony, and humor makes it exigent even for humans [5]. For example in the tweet, *'Unlike Twitter, LinkedIn is full of positivity. People whom I have never worked with are endorsing me for the skills that I don't possess'* conveys a jest which is difficult to understand without cues. Thus, it is imperative to comprehend supplementary cues from users linguistic input that is aware of 'context' and aids right interpretation.

Context is a set of facts or circumstances that surround a situation or event. Understanding context is one of the most difficult aspects of content moderation. Contextual assistance has been studied across pertinent literature, and its effectiveness in sentiment analysis has been validated. As sentiment reflects more latent information in text, the meanings that sentiment words contain are often context-sensitive. Context-based sentiment analysis is thus a well-recognized task-based solution to improve the conventional sentiment analysis.

Studies reveal that context is a multifaceted concept with no standard categorizations. This paper intends to formalize the concept of context in sentiment analysis by defining types of contextual cues which may assist fine grain sentiment analysis, emotion analysis, sarcasm detection, irony detection, humor detection, among others. Finally, a strength–weakness–opportunity–threat (SWOT) matrix is proffered, which determines the effectiveness of context-based sentiment analysis. The paper is organized as follows: The second section elaborates the context-based sentiment analysis followed by section three which discusses the past relevant studies conducted on the context-based sentiment analysis. The fourth section puts forward the multiple facets of context-based sentiment analysis followed by the SWOT matrix in section five. Finally, the sixth section illustrates the conclusion.

2 Context-Based Sentiment Analysis

Sentiment analysis has been thriving to facilitate knowledge extraction for decision making within the omnipresent social web setting. As a generic text classification task, it indispensably relies on the understanding of the human language and emotions expressed via textual or non-textual content. Detecting fine grain emotions, sarcastic tone, cutting expression, remarks, humor, and taunts in natural language is tricky even for humans, making its automated detection more arduous. Moreover, the expansive uses of emblematic language markers such as punctuations, emojis, and slangs are some commonly seen phenomenon which increase the complexity of computational linguistics to analyze the social media content, thus making sentiment analysis a non-trivial challenge with a lower prediction accuracy.

Contextual clues can help detect fine grain sentiment from text by resolving the ambiguity of meaning and improve the generic polarity classification of voluminous user-generated social textual data. Context in sentiment analysis is defined as any complementary source of evidence which can either intensify or flip the polarity of content. The accuracy of polarity classification will thus depend on a context vector, and the learning model will ensure that the overall decision making (classification) is more reliable. The next section discusses the types of context which can be used in sentiment analysis of social data.

3 Related Work

Social web is the platform for extracting and analyzing the sentiments from the huge volume of user-generated opinion data. Twitter, Epinion, Facebook, Instagram, etc., are some of the popular sources for finding sentiments. In 2018, Han et al. [6] use neural network and local contextual features for the improving sentiment classification task. In the same year, different works have been done based on feature fusion approach, context modeling for multimodal sentiment analysis, extracting most influential sentiment for topics, SentiCircle, etc. [7–9].

In 2017, Deng et al. [10] extracted contextual information by using dependency features. In [11, 12] user information, behavior on social network and user's network was used as context by making graphs between user and content. Muhammad et al. [13] in 2016 proposed a SA model using local and global contextual information. In [14], the authors used contextual semantics to improve polarity and strength of sentiments. Meire et al. [15] used pre-post and after-post information for SA. Nakov et al. [16] took context around the target term to improve upon the results of SA. In [17], authors used social context and topical contexts individually as well as in combination sentiment classification. In 2015, authors in [18] proposed a method to improve classification accuracy by taking 'popularity at a location' and 'gender information' into account. Other works using factors like fuzzy logic, opinion holder, topics, and situations have also been carried out in the same year.

From papers published in the years 2009 to 2014, various factors like dependency-based rules, topic, user profile, social influence, product ontologies, lexical knowledge, discourse knowledge, textual, dialogical, appraisal expressions (e.g., attitude, engagement, graduation, etc.), opinion homophily, temporal changes in the data, number of followers, and domain knowledge have been used for finding context to improve the classification accuracy.

4 Multifaceted Context in Sentiment Analysis

Contextualization of words is imperative to bridge the gap between what you have experienced and what you are trying to say. Basically, context creates meaning by providing precise and useful information. The polarity shift and other contextual clues can help detect sarcasm, irony, satire, emotion, or humor from text and improve the generic sentiment classification of voluminous user-generated social textual data. Figure 1 depicts the various broad category types of 'context' defined in the literature to complement sentiment analysis.

These types of context are defined as follows:

- **Social graph-based**: Any cue that associates a message or user directly or indirectly with other messages or with members of the underlying social network is termed as social context. It can be linked users, connection between messages of the same author, data about likes, Retweet, etc.
- **Temporal**: Any cue giving time related information is termed as temporal context. It can be origination time of post, passage of time, etc.
- **Content-based**: Any cue about the lexical knowledge, i.e., interaction of terms with their neighborhood (modifiers like negations and discourse structure like capitalization), domain, topic, sequence of dialog, and semantics is termed as content-related context.
- **User profile-based**: Any cue that is associated with background information of the user is termed as user profile context. It can be personal information, interests, online activity, etc.
- **Modality-based**: Any cue from different modalities (text supporting images or images assisting text, emojis assisting text) used in data can be associated with intensification or diminishing the polarity strength.

The recent systematic literature review conducted within the domain of context-based sentiment analysis [1] explicates that the maximum amount of research within the domain is done using content-based context (textual, global, local, semantic, topical and semantics); whereas, the modality of user-generated content that is visual, typographic, infographic, emojis and acoustic is least explored as context (Fig. 2).

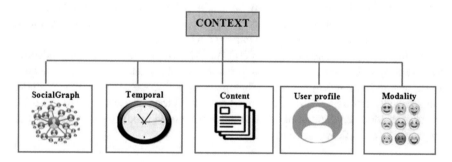

Fig. 1 Types of *'context'*

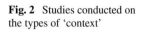

Fig. 2 Studies conducted on the types of 'context'

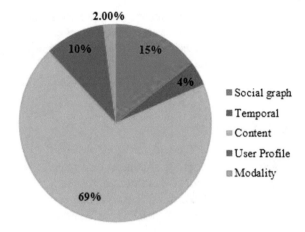

5   SWOT of Context-Based Sentiment Analysis

In this section, we discuss the strengths, weaknesses, opportunities, and threats (SWOT) of using context in sentiment analysis.

5.1   Strengths

Context provides additional information about the text which can be used to improve the accuracy of sentiment classification. Various categories of context, viz. language-based, spatial, temporal, social, user profile, etc., can be applied to achieve better results.

- Context-based sentiment analysis considers metadata (non-textual features) which is able to contribute greatly to the performance of sentiment analysis algorithms.
- Context-based sentiment analysis is effective both for regular texts and texts with a high degree of noise which do not follow the grammatical rules.
- It is able to detect the shift in polarity of a certain term considering the scenario related to it. Moreover, context gives hint in the background which is able to effectively identify the strength of polarity.
- The concept of context in sentiment analysis is able to handle issue of polysemy.
- Context is imperative as sentiment polarities are context-dependent. Change in context can lead to variation in polarity.
- Context provides the knowledge of domain which is useful in cases when only the text does not provide enough information about the hidden sentiment.
- Shifting of neutral polarity to positive or negative is possible with the addition of context as context is domain-dependent.
- Temporal context if considered is able to identify change in sentiment of a user for a topic with the change in time.

- It enables the integration of information from multiple sources, such as the metadata to identify the polarity of text. The metadata includes identification of influential users, biased users, rumor mongers, the time of creation, the use of emoticons, the length of the text, etc.

5.2 Weakness

- Although the concept of context when applied to sentiment analysis is able to detect the difficult task of sarcasm, it is intricate to deal with language constructs showing orientation toward multiple figures of speech like humor, irony, and satire. These are figures of speech with a thin line of demarcation between them.
  - Humor is something that is funny and comical. It is aimed at inducing amusement and laughter. Commonly, casual jokes fall in the category of humor.
  - Irony is aimed at highlighting a contrasting situation or an outcome or behavior that is completely opposite to what is expected.
  - Satire, as opposed to the other two, is aimed at ridiculing the weakness or shortcomings of someone or someone's work.

  For example, a post '*Anushka Sharma was named the hottest Vegan by PETA, while Virat runs a restaurant that serves many non-vegetarian delicacies*' can fall in either of the irony or satire categories.
- Another weakness of context-based sentiment analysis is dealing with constructs which cannot be solved by a single type of context. For example, a post '*summers are so wonderful, the blistering heat, the hot winds and sweaty clothes make a pleasant environment ☺(confounded face)*,' cannot be analyzed accurately exclusively on the basis of text. The accompanying emoji also needs to be taken into account.

5.3 Opportunities

- Context-based sentiment analysis considers context of terms and does not rely on grammatical structures, and hence, it is capable effectively of handling noisy text.
- Slangs, non-standard abbreviations, misspelt words, and colloquial words which are beyond the reach of conventional sentiment analysis can be handled by the context-based sentiment analysis.
- Addition of context with sentiment analysis is able to deal with multilingual content.
- Context-based sentiment analysis is able to detect sarcasm, irony, and humor into some extent if proper context is known.

- Knowledge of context in sentiment analysis is able to handle ambiguity in data. Topical context or discourse information has the capability to resolve the ambiguity.
- Many a times the meaning of words varies as per the context in which they are used. This is called polysemy. Polysemy may lead to mis-classification of terms using conventional sentiment analysis technique. This can be solved by using context-based sentiment analysis.
- Sentiment analysis in automatically transcribed text is challenging due to the fact that spoken language tends to be less structured when compared to written language. Focusing on non-textual aspects of the call such as loudness intonation and rhythm may help in improved sentiment classification.

5.4 Threats

Some classic problems of NLP pose a threat to context-based sentiment analysis. These include

- **Co-reference Resolution**: It is the task of resolving a mention in a sentence refers to which entity. For example a post *'iPhone's camera quality is much better than Nokia Lumina but it is more reliable.'*
- **Negation Handling**: A negative sentiment in a sentence is negated when used in combination with another negative word. Such sentences need to be handled carefully to avoid ending up with sentiments opposite to the writer's intention. For example a post: *'Anita is not a cruel teacher. She did not punish the class for no reason. Students do not have any hard feelings for her.'*
- **Ellipsis Resolution**: To make a sentence compact, certain words and phrases are omitted to avoid repetition. For example a post, *'Brothers Billy and John are lucky to have got such intelligent wives who make up for their foolishness. Although Billy recognizes his foolishness, John does not, even though his wife does.'* Such a post would be more comprehensible even to humans if it was rephrased as Brothers Billy and John are lucky to have got such intelligent wives who make up for their foolishness. Although Billy recognizes Billy's foolishness, John does not realize John's foolishness, even though John's wife realized John's foolishness. Even the context of the sentence does not help clarify who is being referred to here.
- **Slangs and Abbreviations**: Social media posts are abundant with slangs and non-conventional abbreviations that are a challenge to any NLP task. For example: *'The weather makes me feel soooohappyyyyy!! I LUUVV it!!'*

Figure 3 depicts the SWOT matrix.

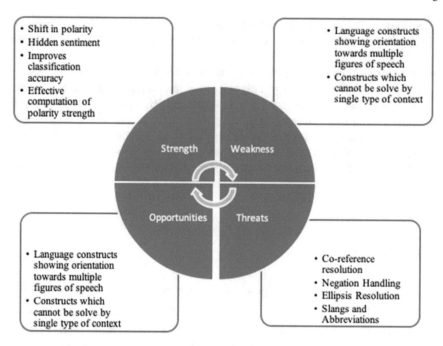

Fig. 3 SWOT matrix for context-based sentiment analysis

6 Conclusion

Social web users generate a voluminous amount of unstructured data which can be mined to extract sentiments for market, business, and government intelligence. The array of language constructs and usage styles increase the complexity of mining task and call for approaches which can leverage auxiliary add-on information. The use of context in sentiment analysis is one such practical approach which can find use cases to determine sarcasm, irony, and humor in real-time user posts. The paper discussed the importance of 'context' of the situation, the specific topic, and the environment in analyzing the sentiment by describing the various types of contextual cues which can be put in use, and finally a SWOT matrix demonstrated the research dynamics within the area of context-based sentiment analysis.

References

1. Kumar, A., Garg, G.: Systematic literature review on context-based sentiment analysis in social multimedia. Multimedia Tools Appl., 1–32 (2019)
2. Kumar, A., Garg, G.: Sentiment analysis of multimodal twitter data. Multimedia Tools Appl., 1–17 (2019)

3. Bouazizi, M., Ohtsuki, T.O.: A pattern-based approach for sarcasm detection on Twitter. IEEE Access **4**, 5477–5488 (2016)
4. Bamman, D., Smith, N.A.: Contextualized sarcasm detection on twitter. In: Ninth International AAAI Conference on Web and Social Media (2015)
5. Wang, Z., Wu, Z., Wang, R., Ren, Y.: Twitter sarcasm detection exploiting a context-based model. In: International Conference on Web Information Systems Engineering, pp. 77–91. Springer, Cham (2015)
6. Han, H., Bai, X., Li, P.: Neural Comput. Appl. (2018). https://doi.org/10.1007/s00521-018-3698-4
7. Majumder, N., Hazarika, D., Gelbukh, A., Cambria, E., Poria, S.: Multimodal sentiment analysis using hierarchical fusion with context modeling. Knowl. Based Syst. **161**, 124–133 (2018)
8. Sheik, R., Philip, S.S., Sajeev, A., Sreenivasan, S., Jose, G.: Entity level contextual sentiment detection of topic sensitive influential Twitterers using SentiCircles. Data Eng. Intell. Comput. 207–216. Springer, Singapore (2018)
9. Feng, S., Wang, Y., Liu, L., et al.: World Wide Web (2018). https://doi.org/10.1007/s11280-018-0529-6
10. Deng, S., Sinha, A.P., Zhao, H.: Resolving ambiguity in sentiment classification: the role of dependency features. ACM Trans. Manage. Inf. Syst. (TMIS) **8**(2–3), 4 (2017)
11. Jiménez-Zafra, S.M., Montejo-Ráez, A., Martin, M., Lopez, L.A.U.: SINAI at SemEval-2017 Task 4: user based classification.. In: Proceedings of the 11th International Workshop on Semantic Evaluation (SemEval-2017), pp. 634–639 (2017)
12. Fersini, E., Pozzi, P.A., Messina, E.: Approval network: a novel approach for sentiment analysis in social networks. World Wide Web **20**(4), 831–854 (2017)
13. Muhammad, A., Wiratunga, N., Lothian, R.: Contextual sentiment analysis for social media genres. Knowl.Based Syst. **108**, 92–101 (2016)
14. Saif, H., He, Y., Fernandez, M., Alani, H.: Contextual semantics for sentiment analysis of Twitter. Inf. Process. Manage. **52**(1), 5–19 (2016)
15. Meire, M., Ballings, M., Van den Poel, D.: The added value of auxiliary data in sentiment analysis of Facebook posts. Decis. Support Syst. **89**, 98–112 (2016)
16. Nakov, P., Rosenthal, S., Kiritchenko, S., Mohammad, S.M., Kozareva, Z., Ritter, A., Stoyanov, V., Zhu, X.: Developing a successful SemEval task in sentiment analysis of Twitter and other social media texts. Lang. Resour. Eval. **50**(1), 35–65 (2016)
17. Wu, F., Huang, Y., Song, Y.: Structured microblog sentiment classification via social context regularization. Neurocomputing **175**, 599–609 (2016)
18. Hridoy, S.A.A., Ekram, M.T., Islam, M.S., Ahmed, F., Rahman, R.M.: Localized twitter opinion mining using sentiment analysis. Decis. Anal. **2**(1), 8 (2015)

An Improved Machine Learning Model for Stress Categorization

Rojalina Priyadarshini, Mohit Ranjan Panda, Pradeep Kumar Mallick and Rabindra Kumar Barik

Abstract Stress and depression are now a global problem. 25% of world's population is facing this problem. With the growing use of sensor equipped intelligent wearables, physiological parameters can be easily extracted and effectively analyzed to predict it at an early stage. Stress management is a complex problem, as to predict stress; there exist several parameters to be considered. Choosing the right parameter is a challenging task, to predict the stress more accurately. In this work, to select the most efficient parameters the unsupervised algorithm, *K*-Means is used; and after getting the right parameters, a radial basis function based neural network is utilized to group the captured data to be stressed or non-stressed. The model also identifies the type of the stress. The work is validated in Python based environment and gives a promising result, in terms of accuracy.

Keywords Stress · Classification · Prediction · Machine learning

R. Priyadarshini (✉)
Department of Computer Science and Information Technology, C. V. Raman College of
Engineering, Bhubaneswar, Odisha, India
e-mail: priyadarshini.rojalina@gmail.com

M. R. Panda
Department of Computer Science and Engineering, C. V. Raman College of Engineering,
Bhubaneswar, Odisha, India
e-mail: mohit1146@gmail.com

P. K. Mallick
School of Computer Engineering, Kalinga Institute of Industrial Technology (KIIT) Deemed to be
University, Bhubaneswar, Odisha, India
e-mail: pradeepmallick84@gmail.com

R. K. Barik
School of Computer Academy, Kalinga Institute of Industrial Technology (KIIT) Deemed to be
University, Bhubaneswar, Odisha, India
e-mail: rabindra.mnnit@gmail.com

© Springer Nature Singapore Pte Ltd. 2020
P. K. Mallick et al. (eds.), *Cognitive Informatics and Soft Computing*,
Advances in Intelligent Systems and Computing 1040,
https://doi.org/10.1007/978-981-15-1451-7_45

1 Introduction

Management of stress has been becoming a challenging task. According to World Health Organisation (WHO), 25% of people are suffering from this mental health ailment [1]. In this fast and busy life, stress becomes an unwelcomed part of our daily life. If stress continues for a long duration it turns into depression, which may further lead to severe consequences like suicidal tendency. Some of the existing methods for stress detection are evolved around, getting the physiological data and to analyze the behaviour and correlation among these data and the stress. But this task is a complex due to several factors like (i) subjective nature: The reason or stimuli which may cause stress in one may not generate in case of another person. (ii) Due to its subjectivity, it is very difficult to find out the root cause and its duration. (iii) There are multiple parameters to be taken into account when measuring the stress [2]. So choosing the right parameters are one of the important components, while dealing with stress prediction problem. There is a tremendous growth seen in the last years in intelligent wearable products. The use of sensors have increased, which are responsible for collecting data of different physical parameters like body temperature, heart rate, amount of sweat secretion etc. These could be used to predict the stress in advance, so that precautions could be taken in advance to mitigate the consequences generated by severe stress [3, 4]. In this work, a hybrid machine leaning model is used which can effectively detect the stress. This model is tried not only to detect the stress, but also identify the category of the stress. It is better to know the category of the stress then just knowing for its presence. The physiological data such as (1) electro dermal activity (EDA); (2) heart rate (HR); (3) arterial oxygen saturation (SPO_2) and (4) temperature are used to find out whether stress is there, if yes what type of stress it is [5]. The main contributions of this work are mentioned as below.

- To build a machine learning model, that can select the appropriate parameters to detect the stress.
- To categorize the stress whether it is a physical or cognitive or emotional stress, this can further help the physician to choose proper medication for the same.

The contents of the remaining of the paper are arranged as follows. Section 2 tells about the significant work in the same field by different individual researchers. Section 3 illustrates the proposed model and Sect. 4 represents the implementation details and the obtained results. Section 5 will conclude the paper with describing briefly about the future work.

2 Related Works

S. Sriramprakash et al. made use of sensors to extract data features like Electro-cardiogram (ECG), Galvanic Skin Response (GSR). These data are then processed by supervised machine learning techniques like Support vector machine (SVM) and

K-Nearest neighbourhood (KNN) on SWELL-KW dataset [6]. They exploited different kernel functions for SVM to get higher accuracy, also with giving emphasis on correct choice of features. S. Casale et al. also made use of SVM with sequential minimum optimization to detect stress component from speech data. But their main objective is to make out the presence of emotional stress in the speech of a person [7]. SVM is also used in [8] to obtain the stress level by using the EEG signals. A. Ghaderi et al. extracted biological data signals like respiration, heart rate etc. Then they applied KNN and SVM to classify the stresses into three categories like high, low and medium [9]. They collected 78 data features and their goal is to choose the best feature which can give maximum accuracy. But they have used the wrapper method to do feature selection which may take longer time if the dimension of data is more. Maria Viqueira et al. proposed a hardware solution to predict the mental stress by building a sensor based on GSR [10]. ECG signals are also used by David Liu et al. for predicting stress [11]. They utilized a linear SVM to do the task, but ECG signals are good measures for stress prediction when it is a physical stress. Along with ECG, GSR and accelerometer data were captured and used by Feng-Tso Sun et al. to detect mental stress [12]. They used different related sensors to collect this type of data. X. Li et al. used a statistical method, hidden Markov model to predict the mental stress. They have got an accuracy of 96% but they have not used any feature extraction in their work [13]. Many works are done by using the machine learning techniques to process the data, captured through sensors [14] and then processing them to infer some results. But still there is always a scope to improvise the model by developing them with different parameters and then doing the fine tuning of those parameters.

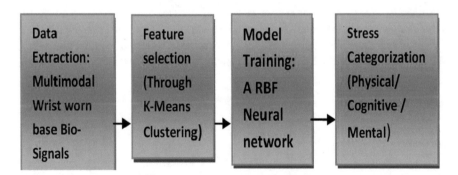

Fig. 1 Overall framework of proposed model: data extraction-feature selection-model training-stress categorization

3   Proposed Model and Implementation

The steps for the proposed model are shown in Fig. 1. The data can be extracted from the sensors. Then the required features are selected from the data. In this work, K-Means clustering algorithm is used for extracting the features. The features which are clearly classifying all the three types of stresses are selected, all other features are discarded. Then the selected features are passed through a radial basis neural network for doing the training. It can determine whether the data belong to a stressed category or non-stressed category. The types of stress considered are of three types. They are (1) Cognitive (2) Mental (3) Physical. The data used to validate the model is collected from www.utdallas.edu [15]. The data was collected using a couple of wrist worn devices. These are the data that carry the bio-signals of 20 students, and their readings were recorded of. The readings of data were captured when passing during four neurological conditions. The states were as the types of stress mentioned in the above paragraph with one more state i.e relaxed state. Here all the four states are taken into account. Pre-processing has been done on the data to classify the equal state data. The data were labeled to four types of numerical labels. After labels are assigned to each individual instance, they are segregated into two subgroups. 80% of data is applied as training input to the model and the remaining was employed as testing sample. Then the data are brought within the range from 0 to 1 through the process of normalization. Normalization is done through the expression given in Eq. 1.

$$\text{InpNorm} = \frac{\text{Inp} - \text{InpMin}}{\text{InpMax} - \text{InpMin}} \tag{1}$$

where Inp is the original input, InpMin and InpMax represents the lowest and highest values of total input data range and I InpNorm is the normalized output. Figures 2, 3, 4 and 5 presents the clustered values of instances over selected EDA, Temperature,

Fig. 2   Three clusters formed by K-means on electro dermal activity parameter

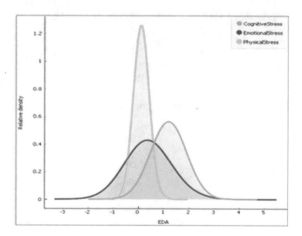

Fig. 3 Three clusters formed by *K*-means on temperature parameter

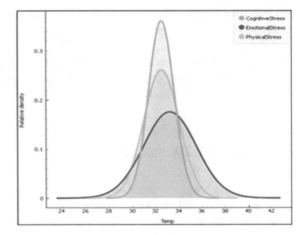

Fig. 4 Three clusters formed by *K*-means on heart-rate parameter

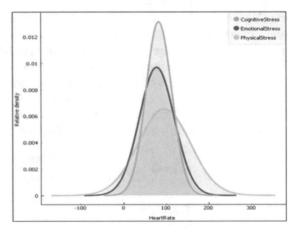

Fig. 5 Three clusters formed by *K*-means on SPO_2 parameter

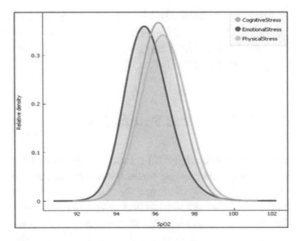

Heart-rate and SPO$_2$, which generates the overlapping clusters. But the features where minimum area is overlapped are considered as better features, and thus they are selected as the eligible features for providing training to the RBF neural network . The proposed framework is executed in an open source software environment. Python based Machine learning package, Scikit-learn is used along with Orange software to do the exploratory data analysis.

4   Results and Discussion

The model has been evaluated by taking four parameters namely Precision, True positive rate or Recall, number of correctly classified samples or Accuracy and F-Score. Accuracy is defined as the fraction of truly predicted instances to available number of instances. Precision is defined as the fraction of truly detected positive samples to the available positive instances. The value computed by truly classified positive instances to available samples in original class is the Recall. The quantity got over weighted averaging the value of precision and recall is said to be as F score. The expressions used to compute Precision and recall are presented in Eqs. 2 and 3. The shown results are taken from random 20 iteration and the average value is taken as the final performance. K-cross validation is implemented on the test data to validate the results more correctly. Here the value of K is taken as 10. Table 1 provides the results for the proposed model for 1000 and 6000 iteration.

$$\text{Recall} = \frac{(\text{TruePositive})}{(\text{TruePositive} + \text{False Negative})} \tag{2}$$

$$\text{Precesion} = \frac{(\text{TruePositive})}{(\text{TruePositive} + \text{False Positive})} \tag{3}$$

From the achieved results, it could be noticed that, after getting trained for 6000 iterations, the proposed model has obtained an accuracy of 92% on test instances of stress dataset. The present work is compared with some other related work. The comparison is given in Fig. 6.

5   Conclusion and Future Works

This work proposes a hybrid approach to detect stress at an early stage, so that precautions could be taken in advance. It made use of the K-means clustering algorithm for choosing the best features and then the selected features are being used by the radial basis neural network that has been trained to detect stress. The model not only detects the stress, but also identifies its category. It can correctly tell whether the stress is cognitive, mental or physical. Still the model is having limitation on the time taken

Table 1 Performance result on 4000 iterations and 6000 iterations

With 4000 iterations, learning rate parameter = 0.11				With 6000 iterations, learning rate parameter = 0.11			
Accuracy	Precision	Recall	F-score	Accuracy	Precision	Recall	F-score
0.8498	0.834	0.8273	0.768	0.8620	0.8353	0.82825	0.76701
0.8972	0.8232	0.8089	0.7215	0.9094	0.8245	0.80985	0.72051
0.887	0.8435	0.8596	0.7918	0.8992	0.8448	0.86055	0.79081
0.8461	0.8101	0.8000	0.7111	0.8583	0.8114	0.80095	0.71011
0.9055	0.8673	0.8874	0.8335	0.9177	0.8686	0.88835	0.83251
0.8758	0.8328	0.8423	0.7695	0.8880	0.8341	0.84325	0.76851
0.9678	0.9773	0.8965	0.7847	0.9801	0.9786	0.89745	0.78371
0.8907	0.8737	0.8513	0.7882	0.9029	0.875	0.85225	0.78721
0.9487	0.8155	0.8016	0.8115	0.9609	0.8168	0.80255	0.81051
0.9055	0.8673	0.8874	0.8335	0.9177	0.8686	0.88835	0.83251
0.8758	0.8332	0.8256	0.7397	0.8880	0.8345	0.82655	0.73871
0.9018	0.8555	0.8653	0.7925	0.9140	0.8568	0.86625	0.79151
0.9223	0.9214	0.7867	0.7762	0.9345	0.9227	0.78765	0.77521
0.9645	0.9401	0.7647	0.7615	0.9767	0.9414	0.76565	0.76051
0.9098	0.7668	0.7611	0.7315	0.9220	0.7681	0.76205	0.73051
0.8976	0.9068	0.8078	0.7216	0.9098	0.9081	0.80875	0.72061
0.9456	0.8801	0.8001	0.7247	0.9578	0.8814	0.80095	0.72371
0.9421	0.9125	0.7863	0.6647	0.9543	0.9138	0.78725	0.66371
0.9663	0.9512	0.7889	0.7738	0.9785	0.9525	0.78985	0.77281
0.959	0.9257	0.7851	0.8015	0.9712	0.927	0.78605	0.80051
Average result over 20 epochs							
0.912945	0.8719	0.82169	0.76505	0.925145	0.8732	0.82264	0.76406

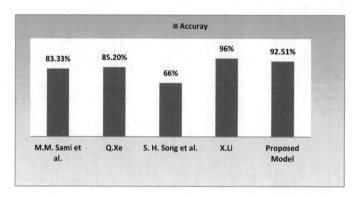

Fig. 6 Comparison of proposed work with some other works based on accuracy

for feature selection is more. But as it is not affecting on the new test instances, hence does not affect the overall performance. This model is validated against the dataset, which is extracted from a standardised bench mark dataset. It is giving an accuracy of 92% on the test data. A full-fledged automatic portable hardware device could be developed basing on the same idea, which could be made available as a ready-made app. Further more data size could be generated from a variety of population to make the model robust. More focus could be given on parameter tuning of the model.

References

1. Organization, W.H.: The World Health Report 2001: Mental health: new understanding, new hope. World Health Organization (2001)
2. Mukhopadhyay, S.: Measuring stress in humans: a practical guide for the field. In: Ice, G.H., James, G.D. (eds.), pp. 271. Cambridge university press, Cambridge (2006). ISBN 0-52184479-7; Hardback. J. Biosoc. Sci. **41**(1), 156–158 (2009)
3. Adams, Z.W., McClure, E.A., Gray, K.M., Danielson, C.K., Treiber, F.A., Ruggiero, K.J.: Mobile devices for the remote acquisition of physiological and behavioural biomarkers in psychiatric clinical research. J. Psychiatr. Res. **85**, 1–14 (2017)
4. Majumder, S., Mondal, T., Deen, M.: Wearable sensors for remote health monitoring. Sensors **17**(1), 130 (2017)
5. Priyadarshini, R., Barik, R., Dubey, H.: Deepfog: Fog computing-based deep neural architecture for prediction of stress types, diabetes and hypertension attacks. Computation **6**(4), 62 (2018)
6. Sriramprakash, S., Prasanna, V.D., Murthy, O.R.: Stress detection in working people. Procedia computer science **115**, 359–366 (2017)
7. Casale, S., Russo, A., Scebba, G., Serrano, S.: Speech emotion classification using machine learning algorithms. In: 2008 IEEE International Conference on Semantic Computing, pp. 158–165. IEEE (2008)
8. Sani, M., Norhazman, H., Omar, H., Zaini, N., Ghani, S.: Support vector machine for classification of stress subjects using eeg signals. In: 2014 IEEE Conference on Systems, Process and Control (ICSPC 2014), IEEE, pp. 127–131 (2014)
9. Ghaderi, A., Frounchi, J., Farnam, A.: Machine learning-based signal processing using physiological signals for stress detection. In: 2015 22nd Iranian Conference on Biomedical Engineering (ICBME), pp. 93–98 (2015). https://doi.org/10.1109/icbme.2015.7404123

10. Villarejo, M.V., Zapirain, B.G., Zorrilla, A.M.: A stress sensor based on galvanic skin response (gsr) controlled by zigbee. Sensors **12**(5), 6075–6101 (2012)
11. Liu, D., Ulrich, M.: Listen to your heart: stress prediction using consumer heart rate sensors. [Online]. Retrieved from the Internet (2014)
12. Sun, F.T., Kuo, C., Cheng, H.T., Buthpitiya, S., Collins, P., Griss, M.: Activity-aware mental stress detection using physiological sensors. In: International conference on Mobile computing, applications, and services, pp. 282–301. Springer (2010)
13. Li, X., Chen, Z., Liang, Q., Yang, Y.: Analysis of mental stress recognition and rating based on hidden Markov model. J. Comput. Inf. Syst. **10**(18), 7911–7919 (2014)
14. Gjoreski, M., Lutrek, M., Gams, M., Gjoreski, H.: Monitoring stress with a wrist device using context. J. Biomed. Inform. **73**, 159–170 (2017). https://doi.org/10.1016/j.jbi.2017.08.006, http://www.sciencedirect.com/science/article/pii/S1532046417301855
15. Birjandtalab, J., Cogan, D., Pouyan, M.B., Nourani, M.: A non-eeg biosignals dataset for assessment and visualization of neurological status. In: 2016 IEEE International Workshop on Signal Processing Systems (SiPS), IEEE, pp. 110–114 (2016)
16. Song, S.H., Kim, D.K.: Development of a stress classification model using deep belief networks for stress monitoring. Heal. Inform. Res. **23**, 285–292 (2017)

Password-Based Circuit Breaker for Fault Maintenance

Arijit Ghosh, Seden Bhutia and Bir Hang Limboo

Abstract It has been observed that in many power stations at the time of repairing the electrical lines due to lack of communication between the station and the maintenance staff, the number of accidents is increasing each day. This proposed circuit ensures lineman safety and makes it easier for them. With the help of 8051 microcontroller, a system has been developed which can ON/OFF the electrical lines after a fault is sensed by the relays or the circuit breaker and trips the faulty circuit and isolates them. By using passwords, the specific faulty line can be isolated by leaving the other lines working without any interference. After repairing it, the circuit is switched on and starts working. The main motivation of this circuit is to build a system which is safe for the staff working in the power stations.

Keywords 8051 microcontroller · Password · Fault maintenance · Relay

1 Introduction

It was observed that many linemen were injured because of lack of safety provided to them during their working hours in a substation. So to prevent the loss of human lives ahead, this proposed circuit was initiated not only for the protection purpose but also to make it easier for the staff to operate the system and upgrade the level of the substation just like how everything is evolving day by day. The concept of this system is considered an important matter in the field of electrical engineering. The electricians are always dealing with high current in their jobs, and even if they take caution, they always encounter accidents one way or the other. This circuit can be used in the substations where the electricians have to be alert day and night whenever a fault occurs in a line. So, the sole purpose of this system is to provide safety for the electricians and the staff working in the station. There has been a lot of research done on this topic, and it proves to be a successful operation. Another

A. Ghosh (✉) · S. Bhutia · B. H. Limboo
Department of Electrical & Electronics Engineering, Sikkim Manipal Institute of Technology,
Sikkim Manipal University, Gangtok, Sikkim, India
e-mail: arijit.g@smit.smu.edu.in

© Springer Nature Singapore Pte Ltd. 2020
P. K. Mallick et al. (eds.), *Cognitive Informatics and Soft Computing*,
Advances in Intelligent Systems and Computing 1040,
https://doi.org/10.1007/978-981-15-1451-7_46

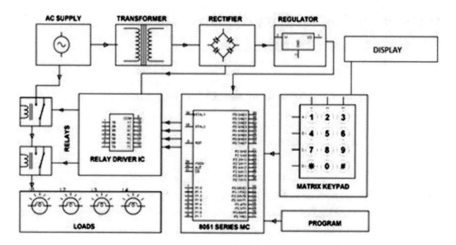

Fig. 1 Proposed block diagram of the scheme

research has been done where the circuit is much more advanced. The microcontroller is interfaced by a GSM module connected via a Max232 IC which transmits serial communication between them. Here, some of the commands like STOP, in case of any fault is occurred, are already programmed in the microcontroller. So, whenever a fault occurs, an SMS is sent to the microcontroller by the module to STOP the system in case the password entered by the operator is not valid or for any security reasons [1]. This proposed circuit can be implemented in the technical field which not only provides safety but also advances the technology by using innovative ideas of every engineer to make the future of technology evolved in every way possible.

2 Proposed Scheme

2.1 Proposed Block Diagram

See Fig. 1.

2.2 Working Principle

A 230 V AC supply is given directly to the loads via the relay switches. The AC supply is stepped down and rectified for the regulated DC supply given to the microcontroller and the relay driver IC. The keypad is connected to the controller using its ports which is used for entering the password. Another port is connected to the programming board connecting the controller IC. The relay driver IC is connected with the ports of the microcontroller and the relays used for the switching of the loads [2] (Fig. 2).

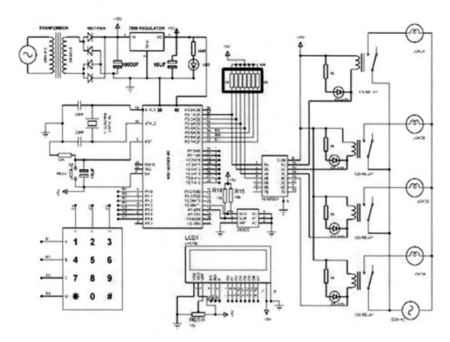

Fig. 2 Circuit diagram of the proposed scheme

2.3 Circuit Description

A 230 V AC supply is given to a 12 V step-down transformer. A rectifier is connected with the transformer which converts the AC signal to DC. Now, a 7850 voltage regulator is used to give a fixed 5 V to the microcontroller. A crystal oscillator is used for the microcontroller to provide a stable clock signal. A push button switch is used here to reset the circuit. A relay driver IC ULN2003A is used here to drive the four 12 V relay used to switch on or off the load in the given condition. This IC is interconnected with the pull-up resistor and the microcontroller which will both be controlling the operation of the IC. The EEPROM is used to store and set the password for the circuit breaker in the microcontroller which can be erased and reprogrammed. The output of the supply is 5 V. This is connected with pin number 40 of the microcontroller, and the ground is connected with pin number 20. Pin 0.1 to pin 0.7 of port 0 of microcontroller are connected to pull-up resistor. Pin 0.1 to pin 0.4 of port 0 are shorted and connected to the pins 1 to 4 of relay driver IC ULN2003. A 12 V supply is connected with ninth pin of ULN2003 which is a com pin. Pin number 16 to 13 of the relay driver IC are connected to the relays. Pin 2.0–2.7 of port 2 of microcontroller are connected to the data lines of LCD. Pin 0.5, 0.6 and 0.7 of port 0 of microcontroller are connected to RS, RW and EN. Write and enable pin of LCD pin 1.0–1.3 of port 1 of microcontroller are connected to rows of keypad, and pin 1.4–1.6 of port 1 of microcontroller are connected to columns of keypad [3] (Figs. 3 and 4).

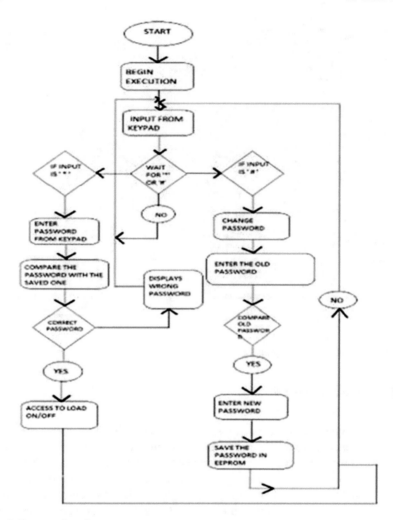

Fig. 3 Program flow chart

3   Conclusion

This proposed scheme is made specifically for the safety of the electricians since it is dangerous to be surrounded by an area of electrical conductors where high currents will be transmitted and received every minute. As an electrician, whenever there is a fault in these areas, it has to be recovered and maintained. And maintaining it is not easy always. So, if you know that the high currents are not travelling in the conductors and you are safe to operate in the specific area, it will be a lot easy and safe to work and know the condition of the fault section and hence.

Fig. 4 Final hardware circuit

References

1. Abdallah, F.E., Yassin, H.A., Hamed, M.H., Mahmoud, Y.O.: *Password Based Circuit Breaker with GSM Module* (Doctoral dissertation, Sudan University of Science and Technology)
2. Subrahmanyam, K.B.: Electric lineman protection using user changeable password based circuit breaker
3. https://www.electronicshub.org/password-based-circuit-breaker
4. http://www.essay.uk.com/essays/engineering/password-based-circuit-breaker-project-report

Performance Enhancement Using Novel Soft Computing AFLC Approach for PV Power System

Neeraj Priyadarshi, Akash Kumar Bhoi, Sudip Kumar Sahana, Pradeep Kumar Mallick and Prasun Chakrabarti

Abstract This work deals the comparison of perturbation and observation (P and O) as well as asymmetrical fuzzy logic controller (AFLC)-based soft computing approach as a maximum power point tracking (MPPT) means for photovoltaic (PV) network. The simulation of PV power system has been realized using both algorithms for different solar insolations. In this work, Cuk converter has been interfaced between PV array and load for smooth MPPT operations. The grid-tied PV system with unity power factor has estimated using MATLAB/SIMULINK approach. Simulation results explain the effectiveness and robustness of AFLC-based MPPT formulation.

Keywords Cuk converter · FLC · MPPT · PV · P and O

N. Priyadarshi (✉)
Department of Electrical Engineering, Birsa Institute of Technology (Trust), Ranchi 835217, India
e-mail: neerajrjd@gmail.com

A. K. Bhoi
Department of Electrical & Electronics Engineering, Sikkim Manipal Institute of Technology, Sikkim Manipal University, Gangtok, Sikkim, India
e-mail: akash730@gmail.com

S. K. Sahana
Department of Computer Science and Engineering, Birla Institute of Technology, Mesra, Ranchi 835215, India
e-mail: sudipsahana@bitmesra.ac.in

P. K. Mallick
School of Computer Engineering, Kalinga Institute of Industrial Technology (KIIT) University, Bhubaneswar, India
e-mail: pradeepmallick84@gmail.com

P. Chakrabarti
Department of Computer Science and Engineering, ITM University, Vadodara, Gujarat 391510, India
e-mail: drprasun.cse@gmail.com

© Springer Nature Singapore Pte Ltd. 2020   439
P. K. Mallick et al. (eds.), *Cognitive Informatics and Soft Computing*,
Advances in Intelligent Systems and Computing 1040,
https://doi.org/10.1007/978-981-15-1451-7_47

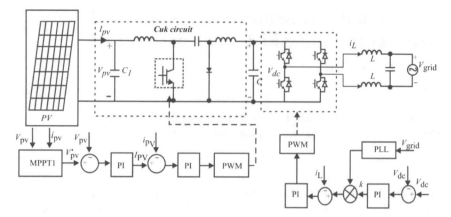

Fig. 1 Cuk converter employed PV power system

1 Introduction

Due to rapid deficiency of natural propellants, the requirements of renewable energy
sources are growing day by day [1–3]. Photovoltaic (PV) systems have considered
more accepted renewable technology because of its environmentally behavior as well
as lower maintenance cost. The PV systems provide the global requirement of energy
demand and fulfill the energy crisis issues [4–7] (Fig. 1).

Furthermore, the installation cost and tracking efficiencies of the PV systems are
the major issues of the recent PV systems. Maximum power point trackers (MPPT)
are major components which provide the solutions of the PV tracking efficiency
with varying solar insolation [8–11]. These MPPT systems are able to track peak
power from solar modules and transfer to load. Several MPPT approaches have been
discussed to optimize the tracking efficacy of PV power system. Sun insolation and
temperature are main factors on which PV power depends. Several MPPT methods
have been developed as PV power trackers. Perturbation and observation (P and
O) method detects PV parameters and provides perturbation in voltage to estimate
orientation in variation. In this work, fuzzy logic controller (FLC)-based MPPT is
employed because of simpler design and precise response. Figure 2a, b present PV
cell *P–V* and *I–V* characteristics are shown in Fig. 2 at different irradiance level.
Characteristics reveal that the maximum power point (MPP) is settled approaching
to each PV curve.

2 MPPT Algorithms: P and O/AFLC

Figure 3 presents the summary of operation of P and O-based MPPT technique. The
PV power is increasing with increase in PV voltage to achieve maximum power point

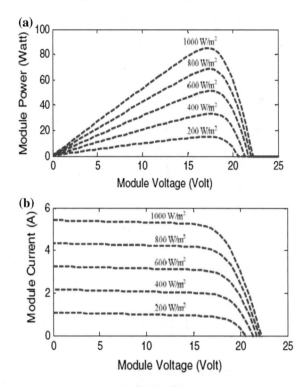

Fig. 2 *P–V* and *I–V* peculiarities of PV cell at different irradiance level

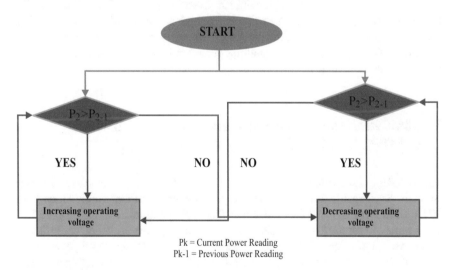

Pk = Current Power Reading
Pk-1 = Previous Power Reading

Fig. 3 Schema diagram of P and O-based MPPT controller

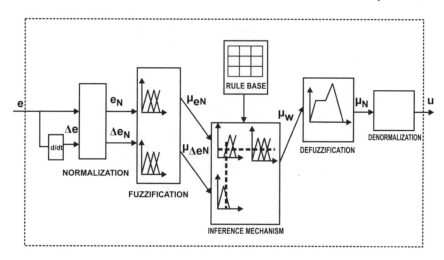

Fig. 4 Fuzzy logic controller design

(MPP) as per left side of PV curve. And MPP is achieved on right side of PV curve and PV voltage has been increased with decrement in PV power. It is noted that perturbation is applied in the command of increased power to achieve MPP region and vice versa.

Figure 4 describes the fuzzy logic controller design in which fuzzification, rule base and defuzzifications are the major steps. The crisp to fuzzy transformation is carried out using fuzzification with mamdani-based max–min composition approach to perform fuzzy agreement. And finally, numerical values of duty ratio of Cuk converter are achieved using defuzzification process with employment of centroid technique. The asymmetrical-assigned membership functions are also presented in Fig. 5 with error, change in error and Cuk converter duty cycle.

3 Simulation Comparison of P and O and FLC Methods as MPPT Trackers for Grid PV System

Figure 6 depicts the implementation of P and O MPPT tracker for PV grid power structure amidst Cuk converter. The transient sunlight irradiant value varies from 300 to 1000 W/m^2 at $t = 0.4$ s while environment temperature remains fixed (25 °C). The simulation responses of PV parameters are varying corresponding to variations in sun insolation values. The P and O algorithm-based MPPT approach provides PV power tracking with more oscillations near to MPP region under changing sun insolation level. The grid-tied PV system has unity power factor coefficient with injection of sine-shaped inverter current. The major drawback of this P and O algorithm-based MPPT approach is PV tracking power with less convergence speed and has more perturbation nearer to MPP region.

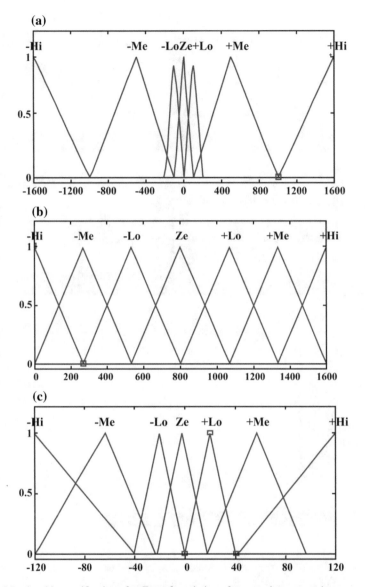

Fig. 5 Membership specification of. **a** Error, **b** variation of error, **c** duty proportion

The main target of this MPPT implementation for PV grid-tied PV system is to evaluate the performance in transient weather conditions. The system has been tested by providing 300–1000 W/m² sun irradiance at 0.4 s and corresponding PV responses are noticed in Figs. 6 and 7. The simulation results explain the effectiveness of the PV system implementation using P and O method as well as asymmetrical FLC-based MPPT. The proposed inverter controller gives sine-shaped current injection to utility

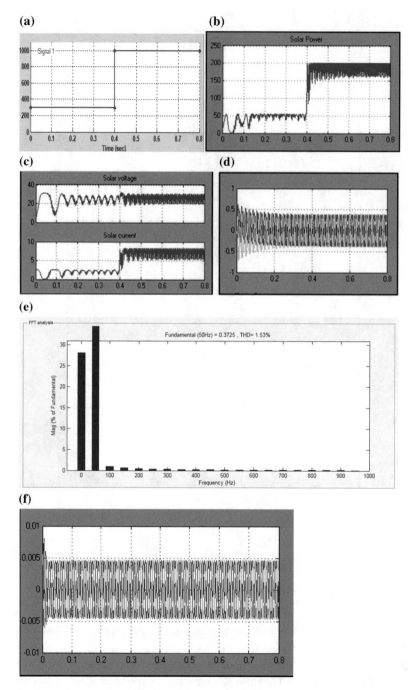

Fig. 6 **a** Variable irradiance level, **b** PV panel output power, **c** V_{PV}, I_{PV}, **d** inverter current **e** FFT inverter current, **f** inverter voltage, **g** grid voltage, **h** grid current

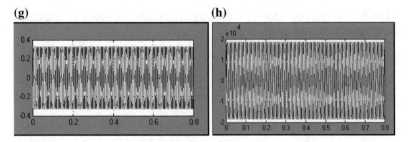

Fig. 6 (continued)

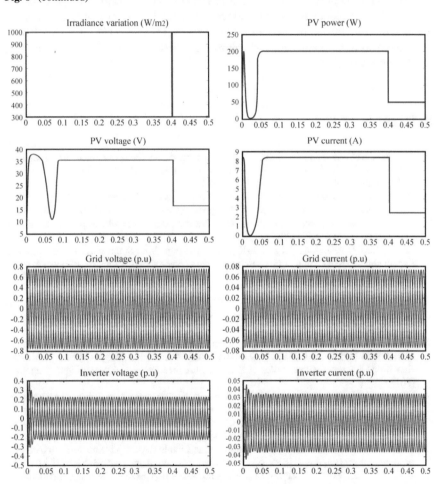

Fig. 7 Simulated responses during variable irradiance level and constant temperature. Irradiance variation, P_{PV}, V_{PV}, I_{PV}, V_{Grid}, I_{Grid}, inverter voltage, inverter current

grid. The current produced through inverter has overall harmonic distortion 1.53%, which satisfy the THD% requirements.

4 Conclusions

This paper deals the comparison of PV MPPT tracking using P and O and asymmetrical FLC methods. The soft computing-based FLC MPPT technique provides rapid PV tracking accurately under changing sun insolation with negligible perturbation nearer to MPP region, while classical P and O method has slow convergence of PV tracking ability with more oscillations nearer to MPP area. MATLAB simulation-based analysis has been presented to estimate the effectiveness of MPPT functioning of PV grid integration.

References

1. Priyadarshi, N., Padmanaban, S., Maroti, P.K., Sharma. A.: An Extensive practical investigation of FPSO-based MPPT for grid integrated PV system under variable operating conditions with anti-islanding protection. IEEE Syst. J. 1–11 (2018)
2. Priyadarshi, N., Padmanaban, S., Bhaskar, M.S., Blaabjerg, F., Sharma, A.: A fuzzy SVPWM based inverter control realization of grid integrated PV-wind system with FPSO MPPT algorithm for a grid-connected PV/wind power generation system: hardware implementation. IET Electric Power Appl. 1–12 (2018)
3. Priyadarshi, N., Anand, A., Sharma, A.K., Azam, F., Singh, V.K., Sinha, R.K.: An experimental implementation and testing of GA based maximum power point tracking for PV system under varying ambient conditions using dSPACE DS 1104 controller. Int. J. Renew. Energy Res. 7(1), 255–265 (2017)
4. Nishant, K., Ikhlaq, H., Bhim, S., Bijaya, K.P.: Self-adaptive incremental conductance algorithm for Swift and ripple free maximum power harvesting from PV array. IEEE Trans. Ind. Info. 14, 2031–2041 (2018)
5. Priyadarshi, N., Sharma, A.K., Priyam, S.: An experimental realization of grid-connected PV system with MPPT using dSPACE DS 1104 control board. In: Advances in Smart Grid and Renewable Energy. Lecture Notes in Electrical Engineering, vol. 435, Springer, Singapore (2018)
6. Priyadarshi, N., Sharma, A.K., Azam, F.: A hybrid firefly-asymmetrical fuzzy logic controller based MPPT for PV-wind-fuel grid integration. Int. J. Renew. Energy Res. 7(4) (2017)
7. Priyadarshi, N., Sharma, A.K., Priyam, S.: Practical realization of an improved photovoltaic grid integration with MPPT. Int. J. Renew. Energy Res. 7(4) (2017)
8. Bhim, S., Chinmay, J., Anmol, B.: An improved adjustable step adaptive neuron based control approach for grid supportive SPV system. IEEE Trans. Ind. Appl. 54, 563–570 (2018)
9. Abderazak, L., Dezso, S., Josep, M.G., Laszlo, M., Aissa, B.: Discrete model predictive control-based maximum power point tracking for PV systems: overview and evaluation. IEEE Trans. Power Elec. 33, 7273–7287 (2018)

10. Priyadarshi, N., Padmanaban, S., Mihet-Popa, L., Blaabjerg, F., Azam, F.: Maximum power point tracking for brushless DC motor-driven photovoltaic pumping systems using a hybrid ANFIS-FLOWER pollination optimization algorithm. MDPI Energies **11**(1), 1–16 (2018)
11. Sanjeevi, P., Neeraj, P., Jon, B. H.-N., Mahajan, B., Farooque, A. Amarjeet, K.S., Eklos, H.: A novel modified sine-cosine optimized MPPT algorithm for grid integrated PV system under real operating conditions. IEEE Access. **7**, 10467–10477 (2019)

A Novel Approach on Advancement of Blockchain Security Solution

Lukram Dhanachandra Singh and Preetisudha Meher

Abstract Blockchain technology is an emerging technology that has the potential to transform the way of sharing data and transfer of value. Blockchain platforms are currently being used across the world in several government processes and business domains, for their unique characteristics. Even though blockchain shows potentials in its ability to provide a limitless number of inventive commercial trading, payments, government, healthcare, military and other critical applications, they face a security weakness on their recent prominent breaches of exchanges. Hence, there remain major security issues that are essential to be overcome before blockchain adopts the mainstream. And currently, this technology relies on hardware security modules (HSM) for the management and protection of their digital keys. The HSM generates key pairs, has secure storage and can off-load cryptographic operations from the entire system. But recently, FPGAs are preferred for hardware realization of algorithms considering its flexibility, low cost and long-term maintenance. It also has an advantage of reconfigurable or reprogrammable hardware design whenever new security or adaptation of an algorithm is required to support higher security levels. The paper presents a review of hardware security modules and proposed to enhance the scalability and reliability of the HSM by implementing it with the silicon-based secure module. Integrating PUF technology into the chip for storing and securing encryption or private keys, its security level can also be improved.

Keywords Blockchain · Bitcoin · FPGA · Hardware security modules · Physical unclonable functions

L. D. Singh (✉) · P. Meher
ECE Department, NIT Arunachal Pradesh, Yupia, India
e-mail: dhana.lukram0@gmail.com

P. Meher
e-mail: preetisudha1@gmail.com

© Springer Nature Singapore Pte Ltd. 2020
P. K. Mallick et al. (eds.), *Cognitive Informatics and Soft Computing*,
Advances in Intelligent Systems and Computing 1040,
https://doi.org/10.1007/978-981-15-1451-7_48

1 Introduction

Blockchain technology is renovating the way of sharing data and transmission of the value. Blockchain is a decentralized, distributed ledger, which means that a ledger used for recording the transactions is spread whole over the network midst all peers in the network, and each peer holds a copy of the complete ledger. However, it got a weak spot while coming to the security part.

While blockchain is proving its limitless number of innovative applications in various business firms, government processes, etc., they face multiple security breaches that are solved by improving the hardware security modules. And various emerging research industries are working on HSMs to improve its scalability, efficiency, reliability and to provide highly secure blockchain solutions. An HSM is a cryptography-based processor that securely generates, safeguards and stores the keys, working with certifications like FIPS certification, Common Criteria Evaluation Assurance Levels (CC-EAL), an international standard providing sophisticated cryptography with tamper-resistant hardware.

Key generation, encryption, decryption, digital signature verification, time stamping, securing storage, protecting keys and intellectual property and protection from side-channel attacks are some key features of the HSMs used for a blockchain solution. While many researchers in hardware security have proposed or invented advanced security solutions for these features using physical unclonable functions (PUF) technology for securing keys, keys generation and authentication, some are using FPGAs for designing TRNG architecture, crypto-processor. Zetheron technology is on FPGA Mining. A VLSI hardware architecture was implemented using encryption algorithm for security system [1]. Even for blockchain, a caching system (hardware-based) on FPGA NIC was proposed and evaluated [2]. Others are also working for enhancing the reliability of these PUFs/FPGAs by machine learning [3].

The following parts of this paper will give details of blockchain technology, history, working principles and its past security issues and possible future outcomes are presented in Sect. 2. In Sect. 3, the key features of HSMs are discussed along with the requirements for the advanced security system and security certificates. Some role of VLSI technology in hardware security related to the blockchain is reviewed in Sect. 4, and finally, the conclusion is presented in Sect. 5.

2 Blockchain Technology

A blockchain is a chain formed by linking the blocks, which comprises with a cryptographic hash of the preceding block along with a timestamp, and these transaction data are encoded into Merkle tree which linked with each other using cryptography. Blockchain is used to record transactions across many computers with the aim of any involved record that cannot be changed ex post facto, without the modification of all consequent blocks [4].

Currently, blockchain networks can be classified into four types, public blockchain which has completely no access limitations, private blockchain, a permission one in which one cannot join without invitation from the network administrators, consortium blockchain is also a permission, but here multiple numbers of corporations might work with a node on such a network rather than one organization controls it, and the last one is hybrid blockchain which is a combination of others.

2.1 History of Blockchain

W. Scott, Stuart, Stornetta and Haber called the first work on a cryptographically secured chain of blocks in 1991 [5] implementing a system where tampering document timestamps was not possible. In 1992, they combined Merkle trees to the design improving its efficiency by collecting numerous document certificates into a single block [6]. It was mentioned that, as a way to track and verify the transactions for cryptocurrency bitcoin, blockchain was invented [4]. Cryptocurrencies are an online form of digital currency that depends on cryptography to control the construction of currency and authenticate transactions.

2.2 General Principles and Cryptographic Mechanics that Make Blockchain Works

The distributed nature and lack of authority as a central administrative to watch over and authenticate transactions of blockchain is a part of the attraction to the organization employing with cryptocurrencies. Within cryptocurrencies, blockchains are maintained by nodes inside the peer-to-peer blockchain network which use their hardware resources to execute a process of authenticating transactions and inserting them to the blockchain after encryption (Fig. 1).

Once the nodes accept transactions, they are gathered into blocks and linked to the chain. Every block is encrypted with a specific cryptographic hash that is distributed to the following blocks on the chain, ensuring that the blocks will remain linked. Blockchain not only relies on hashes to function but also on cryptography and public key infrastructure. All data goes into the chain after encryption, and each customer on the network owned their private key. Even though the blockchain is available to the public, it is safe as the process is encrypted thoroughly. As a node completes a new block, it is added to the chain, and the updated form of the blockchain is publicized till the last nodes.

Therefore, the properties of the blockchain technology which help it increase well-known compliment are

- Decentralization: The information is not kept by a single entity; everyone in the network possesses the information.

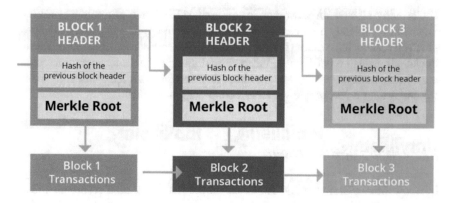

Fig. 1 Structure of blockchain

- Transparency: It secures the person's real identity, but anyone on the network can see all the transactions as they were done by their public address.
- Immutability: Once something has been entered into the blockchain, it is unalterable.

2.3   Securing Blockchain

Wallets or multi-signature wallets are the software which is currently used to keep digital assets and private keys safe by most of the people, but these solutions are determined more by accessibility than security. Hardware wallets, such as Keepkey or Trezor, were intended to suggest an advanced level of private key security, but even these are weak to many hacks, including fault injections attack [7, 8].

A fault injection attack is likely to alter the software execution by inserting an error in a computing device, and hence, it stops the execution of an instruction or corrupts the data the processor is operating with, which may lead to compromise the security of hardware wallets by leaking the private keys. And if the private key got stolen, then it will not be a matter that how much the blockchain is secure; anyone can monetize and exploit the asset, and any malevolent transfer of value is usually immediate and not reversible. In 2016, the Bitfinex hack occurred in which two private keys kept in a multi-signature wallet hosted by BitGo were compromised [9].

Whether gain access to sensors, miners, peer stakeholders or even third-party consumers, if endpoint entities are not genuine and safe, the blockchain and the whole transaction infrastructure can become polluted and unreliable, while the integrity of its data is still preserved. The hashing process of the Merkle tree used in the blockchain does not prevent this weakness. HSMs seal the gaps in secure digital storage, asset transaction security vigorous to the confidentiality of blockchain cryptographic keys and serious parameters that are essential to safeguard peer-to-peer, point-to-point or shared content.

The next generation of ultra-secure PCs [10] comes with an embedded HSM which has two factors of authentication, i.e., a key and a password to ensure that unofficial users cannot lay a hand to the device. An optimized blockchain solution for the IBM z14 as an application of cloud services is also presented [11] in 2017.

2.4 Future of Blockchain

The conception of quantum computers is one of the most important threats to blockchain technology in its current state. Current cryptographic algorithms are secured against common computers of these days but not in contradiction of quantum computers. Blockchain solutions should work HSMs with provision for quantum-safe cryptographic algorithms to be secure, to protect exclusive data and code.

3 Hardware Security Modules

An HSM is a physical device that protects and manages the digital keys, facilitating encryption, decryption, signing and authentication. It provides a safe environment to interact with a private key without which the key is at risk of being exposed. These modules usually come as an external device or a plug-in card that attaches right to a network server or computer.

The entire cryptographic key lifecycle from provisioning, handling and storing to arranging of or archiving the keys happens in the HSM. HSM can also capture digital signatures.

An HSM is made, so it cannot be tampered or damaged or may be placed in a physically protected area of a data center to avoid unauthorized exchange.

3.1 Important Key Features Which HSMs Look Forward to

- Key Generation
  - The HSM has a dual True Random Number Generator TRNG entropy source and NIST SP800-90 compliant RNG
  - Direct secure address generation (hash of the public key)
- Side-channel protection to prevent the extraction of keys without compromising the storage
- Hardware-based tamper response

- Segregated functions in hardware and firewalls

  - Prevents attack by the silicon vendor
  - Mitigate risk from compromised software

- Cryptographic function in hardware

  - Side-channel protection
  - Protection from specter/meltdown kind of attacks
  - Field upgradable FPGA implementation

- A role model with multi-factor authentication to mitigate risk associated with hacked admin accounts
- Integrated key access control
- Device clustering for HA redundancy and performance scalability with an integrated secure backup feature
- Smart key attributes like integrated multi-signature authentication scheme
- Support for various cryptocurrencies like ETH, BTC-based, Ripple, IOTA and many more.

3.2 Security Certifications of HSMs

HSMs have a certain level of regulatory assurance, such as the Federal Information Processing Standard (FIPS) certification and Common Criteria Evaluation Assurance Levels (CC-EAL), an international standard. With such other certifications, it ensures that each device reaches certain industrial-grade security control necessities (Tables 1 and 2).

Table 1 FIPS 140-2 requirements according to security levels

Level	Requirement
1	Basic security requirements
2	Tamper evidence, user authentication
3	Tamper detection/resistance, data zeroization, splitting user roles
4	Very high tamper detection/resistance, environmental protection

Table 2 Protection Profile (PP) for Target of Evaluation (ToE)

Target of Evaluation	Protection Profile
Key generation	CMCKG-PP
Signing operations	CMCSO-PP

4 Role of VLSI Technology in Hardware Security

Key features of the hardware security module for blockchain solution include key generation, encryption, decryption, digital signature verification, time stamping, securing storage, protecting keys and intellectual property and protection from side-channel attacks (SCA).

VLSI technology plays a very important role in security systems, and some of them will be discussed as follows. A VLSI hardware architecture was implemented using the AES algorithm for the security system [1]. FPGA platform was considered for the system to run the encryption and decryption of video through the JTAG communication port. Zetheron technology, as in their articles, is working on FPGA crypto mining.

A hardware-based caching system on FPGA NIC for blockchain was implemented on the NetFPGA-10G board, evaluated and found that the throughput was improved [2].

A Mux-based Physical Unclonable Functions was designed for the generation of an efficient and reliable secret key as a cryptographic application [12].

PUFs have been usually recognized as favorable hardware security basic and lately approved in many security applications like as hardware fingerprinting and authentication, intellectual property protection, side-channel analysis, secure key generation and secure storage for ciphers [13]. It has the property of non-clonability and unpredictability which gives advantages above the solutions to the security applications. To enhance the reliability of PUF, machine learning was implemented [14], and various defense mechanisms against machine learning attacks on strong PUF was reviewed [3].

5 Conclusion

Using these FPGA designs or PUF technology implemented with machine learning algorithm in HSMs will make blockchain impenetrable which gives security-conscious consumers and organizations guarantee that regardless the applications they used it will be safe, providing a far advanced security system to the blockchain, increasing its scalability and reliability and reducing its size and cost.

Improvements in blockchain security will make the technology progressively striking and functional for an extensive number of organizations and users.

References

1. Mate, A.Y., Khaire, B.R.M.: A VLSI hardware architecture implemented of security system using encryption algorithm. Int. J. Sci. Eng. Res. **7**(5) (May-2016)

2. Sakakibara, Y., Morishima, S., Nakamura K., Matsutani, H.: A hardware based caching system on FPGA NIC for blockchain. Inst. Electron., Inf. Commun. Eng. Trans. Inf. Syst. **E101-D**(5) (May 2018)
3. Noor, N.Q.M., Ahmad, N.A., Sa'at, N.I.M., Daud, S.M., Maroop N., Natasha, N.S.: Defense mechanisms against machine learning modeling attacks on strong physical unclonable functions for IoT authentication: a review. Int. J. Adv. Comput. Sci. Appl. **8**(10), (2017)
4. Blockchains: the great chain of being sure about things. The Econ. (31 October 2015)
5. Haber, S., Scott Stornetta, W.: How to time-stamp a digital document. J. Cryptol. **3**(2), 99–111 (1991)
6. Bayer, D., Haber, S., Scott Stornetta, W.: Improving the efficiency and reliability of digital time stamping. Sequences II, 329–334 (March 1992)
7. Tomshwom, Lessons from the Trezor Hack. Steemit, Aug 2017. Accessed Jan 2018. https://steemit.com/bitcoin/@tomshwom/lessonsfrom-the-trezor-hack
8. Redman, J. A Def Con 25 Demonstration Claims to Break Bitcoin Hardware Wallets. Bitcoin.com, 27 Jun 2017. Accessed Jan 2018
9. Redman, J. Small Ethereum Clones Getting Attacked by Mysterious '51 Crew. Bitcoin.com, 4 Sep 2016. Accessed Jan 2018
10. Calore, M. This ultra-secure PC self destructs if someone messes with it. Wired, 23 Jun 2017. Accessed Jan 2018
11. Nunez Mencias, A., et al.: An optimized blockchain solution for the IBM z14. IBM J. Res. Dev. **62**(2/3) (2018), paper 4
12. Pegu, R., Mudoi, R.: Design and analysis of Mux based physical unclonable functions. Int. J Eng. Res. Technol. **4**(05) (May 2015)
13. Wang, Q., Gao, M., Qu, G.: A machine learning attack resistant dual-mode PUF. In: GLSVLSI'18, May, Chicago, IL, USA
14. Wen, Y., Lao, Y.: Enhancing PUF reliability by machine learning. In: IEEE 2017

Design of a Traffic Density Management and Control System for Smart City Applications

Prashant Deshmukh, Devashish Gupta, Santos Kumar Das and Upendra Kumar Sahoo

Abstract Traffic congestion is a serious issue for urban cities. From city roads to highways, a lot of traffic problems occur everywhere in today's world, because of exponentially increase in the number of vehicles, the traffic management system and road capacity are not efficiently compatible with vehicles traveling on them. These frequent traffic problems like traffic jams have led to the need for an efficient traffic management system. This work focuses on the design of dynamic traffic control system based on real-time vehicle density present at the traffic post and highlights the experimental verification of outdoor density estimation system combined with traffic control unit. It provides good results under mixed traffic conditions and in adverse weather. Vehicle classification and counting are done using edge computing techniques and upload data to database; based on the vehicle count, density is estimated and green channel time is calculated for particular lane of traffic post.

Keywords IP camera · Local processing unit · Vehicle classification · Density estimation · Raspberry Pi · Time synchronization · Traffic management and control

1 Introduction

Various intelligent traffic management systems (ITMS) have been developed for efficient and smooth flow of traffic all over the world to solve some common and vital problems that exist. According to the Transportation Research and Injury Prevention

P. Deshmukh · D. Gupta · S. K. Das (✉) · U. K. Sahoo
Department of Electronics and Communication, National Institute of Technology Rourkela, Rourkela, India
e-mail: dassk@nitrkl.ac.in

P. Deshmukh
e-mail: 517ec6019@nitrkl.ac.in

D. Gupta
e-mail: 115ei0342@nitrkl.ac.in

U. K. Sahoo
e-mail: sahooupen@nitrkl.ac.in

© Springer Nature Singapore Pte Ltd. 2020
P. K. Mallick et al. (eds.), *Cognitive Informatics and Soft Computing*,
Advances in Intelligent Systems and Computing 1040,
https://doi.org/10.1007/978-981-15-1451-7_49

Programme (TRIPP), Indian Institute of Technology Delhi, Road Safety in India: Status Report 2016 dated 2017 [1], the number of cars and motorized two-wheelers (MTW) registered in the 2015 was 26.4 and 154.2 million, respectively. Inefficient traffic control system leads to congestion, accidents, unnecessary fuel consumption, pollution, and wastage of the commuter's time [2]. The inflow of traffic at signal changes with time of the day and situation and various other factors that cannot possibly be predicted. Therefore, the need of a traffic control module that accounts for all these factors in real time and uses the data collected from the sensors [3, 4] to optimize the flow of traffic, due to the trivial traffic system the crisis vehicles such as ambulance, fire rescue, and police vehicles face delay which may lead to loss of life and property damage. Most of ITMS uses a sensor network that consists of a device and entranceway nodes [5]. The duty of the device node is to watch traffic in associate allotted space, utilizing completely different devices that may live many physical traffic parameters like flow, density, volume, headway, waiting time, throughput, additionally as pollution [6]. The entranceway node then collects the traffic data from all the nodes and directs an equivalent to the base station [7]. In Indian traffic system, where the commuters do not tend to follow their respective lanes and cluster, the presence of unequally sized lanes, such a ITMS is not the most efficient solution. This paper discusses a traffic density management and control system density (TDMC) that uses visual data to prioritize lane and time allotment to each of them. It also notes down what future modifications that can be made to the module to optimize as well as increase the utility of the system.

2   System Model

The system model used is as shown in Fig. 1. It makes use of an IP camera and local processing unit (LPU) for each incoming road at the traffic post, cloud database, and traffic control unit for providing green channel time based on traffic density.

3   Block Diagram of TDMC System

According to the system model, dynamic traffic control system is designed for providing green channel based on the density of particular lane, and Fig. 2 shows the block diagram of TDMC system, where the camera is connected with LPU, it captures the traffic feed on the feed vehicle classification, and counting algorithm is applied. Further, the density is estimated based on number of count and area captured by camera. These density values are transmitted wirelessly to the MongoDB

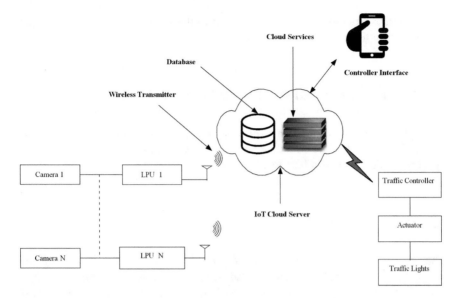

Fig. 1 System model of TDMC

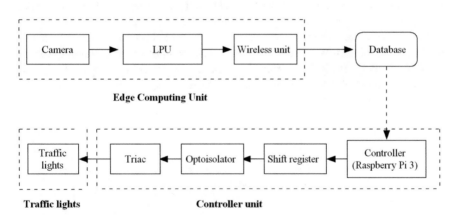

Fig. 2 Block diagram of TDMC

database and stored in respective lanes collection. In this proposed work, edge computing is performed for video analysis. Traffic control panel consists of Raspberry Pi as the main controller, a driver unit to drive the traffic lights. Raspberry Pi having capabilities of parallel processing, Wi-Fi enable, and clustering. Controller transmits data to shift register 74HC595 which takes data serially and fetches it parallel to 8 outputs, and two shift registers are employed in order to control the 16 lights. Output of the shift register drives FET/MMBTA16 and an optocoupler/MOC 3041. The optocoupler isolates the Raspberry Pi and shift registers from high voltages. The database density values retrieve, and green time is calculated for a particular lane based on time synchronization method and dive the signal for traffic lights through the driver unit mentioned in next sections.

4 Design and Fundamental Description of TDMC System

The work flow methods of the proposed TDMC system are divided into the following sections, e.g., vehicle classification, density estimation, time calculation, time synchronization, and error handling method.

4.1 Vehicle Classification

The vehicle classification procedure is depicting in Fig. 3 in the first step object detection model is loaded which is based on single-shot detector (SSD) algorithm [11]. Transfer learning freezes the first few layers and then changed the number of nodes in the final output layer according to the number of different classes. In the proposed system, model is trained on 6 classes (truck, car, bus, auto, motorbike, and bicycle). After changing the number of nodes in the final layer, the next step is to retrain the model. The final step is to train the model according to new training images with different class labels. After successful completion of the training, vehicle detection model is ready for testing. The next step is to process the video frame and get the number of objects with their classes which are required for the traffic density calculation.

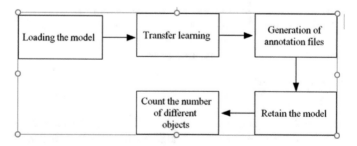

Fig. 3 Working flow of vehicle classification

4.2 Density Estimation

The algorithm of density estimation for a particular lane is as shown in Fig. 4 for vehicle detection and classification [8–10]. First trained the model based on single-shot detector (SSD) algorithm [11]. The number of nodes in the final output layer depends on the different number of classes, and the model is trained to detect objects in a video feed. The final output layer has the softmax activation function. The next step is to process the video frame and get the number of objects with their classes which are required for the traffic density calculation. In this proposed work, classifications are done for heavy motorized vehicles (bus, truck), light motorized vehicles (auto, car), and two-wheelers (motorbike, bicycle) [9, 12]. After finding the number of objects, assign a weight factor to different–different classes, calculate the density of each lane [13–15], and upload to the database. Traffic density (TD_i) of each i_{th} incoming lane is calculated in percentage, which is based on the number of classified vehicles with their respective weights and the area captured by camera (AC_i) of that lane. TD_i can be represented as

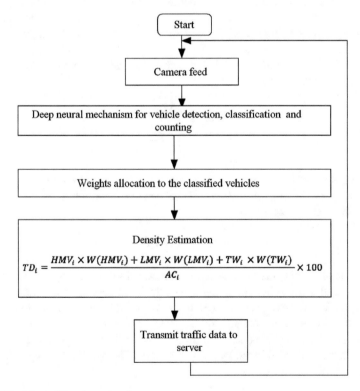

Fig. 4 Flowchart of density estimation

$$TD_i = \frac{HMV_i \times W(HMV_i) + LMVi \times W(LMV_i) + TW_i \times W(TW_i)}{AC_i} \times 100$$

$$(1)$$

where HMV_i is the number of heavy motorized vehicles in the i_{th} incoming lane; LMV_i is the light motorized vehicles in the i_{th} incoming lane; TW_i is the number of two-wheelers in the i_{th} incoming lane; $W(HMV_i)$, $W(LMV_i)$, $W(TW_i)$ are the respected weights and i is in range from 1 to N.

4.3 Signaling Time Calculation

The signaling time calculation script is running in the traffic controller side for the calculation of green time, and in this, controller is communicating with the cloud server with the correct credentials. After successful connection, the algorithm fetches the density data from the database. First step is to connect the system to the MongoDB database with the correct credentials. If any error occurs while connecting with database, the system provides an "error" output in a text file for the controller. Later on after a time interval of 10 s, it retries to connect to the database. After successful connection, the algorithm fetches the density data from the database and calculates the fractional density (DF_i) of each i_{th} incoming lane with respect to total density. DF_i can be represented as

$$DF_i = \frac{TD_i}{\sum_{j=1}^{N} TD_j}$$

$$(2)$$

where TD_i is the traffic density of the i_{th} incoming lane; i and j are in range from 1 to N. Now based on DF_i, green time (GT_i) of each i_{th} incoming lane is evaluated in seconds. GT_i can be expressed as

$$GT_i = CT \times DF_i$$

$$(3)$$

where CT is the total cycle time (seconds). Practically, CT is set by the traffic operator using traffic controller interface. This calculated data is then simultaneously uploaded back to the database for analytics purpose and also store as a list in a text file for internal processing; it also provides an error handling mechanism, e.g., if any error occurred in the running script, it informs the controller; so that it can work in static mode (parameters defined by traffic operator). The workflow of signaling time calculation is shown in Fig. 5.

Fig. 5 Flowchart of
signaling time calculation

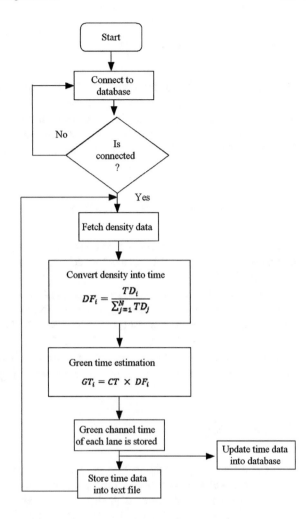

4.4 Time Synchronization

Time synchronization script provides synchronization between all the incoming lanes at the traffic junction. The script fetches the data stored in the text file by the signaling time calculation script. If there is no data, it kept on searching for the data. Once the data is received, it checks the data. If the data is "error," it informs the controller to operate in static mode sequence whose parameter is set by traffic operator using traffic controller interface; otherwise, it extracts the green time data from the text file and based on this data red time (RT_i) of each i_{th} incoming lane calculated. RT_i can be expressed as

$$RT_i = (N-1) \times YT_i + \left(\sum_{j=1}^{N} GT_j \right) - GT_i \qquad ((4))$$

where N is the number of incoming lanes at the traffic junction; YT_i is the yellow time of the i_{th} incoming lane; GT_i is the green time of each i_{th} incoming lane; i and j are in range from 1 to N. Practically, YT is set by traffic operator or it would be same for all lanes. Figure 6 depicts the flowchart of time synchronization. After calculation of timings, the control signals sent to shift register to display the lights accordingly.

5   Experimental Results and Discussion

The proposed system characterization has been verified using hardware, numerical, and software simulation. In this section, schematic of wireless traffic controller is shown in Fig. 7.

To verify the functionality of system configuration, experimental setup is installed at National Institute of Technology, Rourkela (NITR) as shown in Fig. 8. In the installed setup, network camera (Hikvision DS-2CD204RWD-I) and LPU (Jetson TX2) are used also verified on recorded video Fig. 9.

Traffic controller interface is designed for display the current status of density at the traffic post ant configuring the controller shown in Fig. 10.

6   Conclusion

Traffic density management and control system are designed for dynamic switching of green channel time based on traffic density. This design is tested and verified experimentally where density estimation is performed using LPU and the data is uploaded to database through wireless unit, according to these density values control the traffic lights of traffic junction. This module will be used as a backbone for emergency vehicle detection system, accident detection system, red-light violation system, etc.

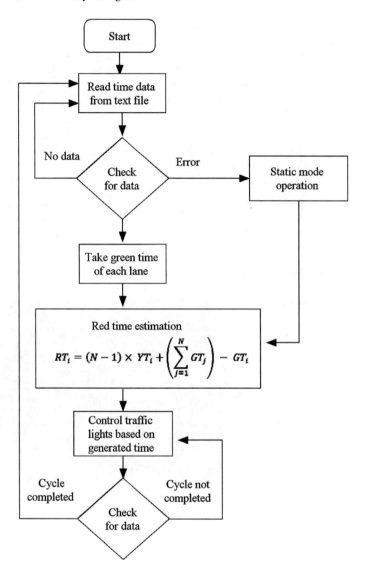

Fig. 6 Flowchart of time synchronization

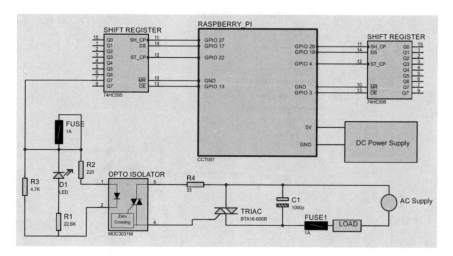

Fig. 7 Schematic of traffic controller

Fig. 8 Experimental setup

car:9
car:11
car:17
car:6
Type here to search

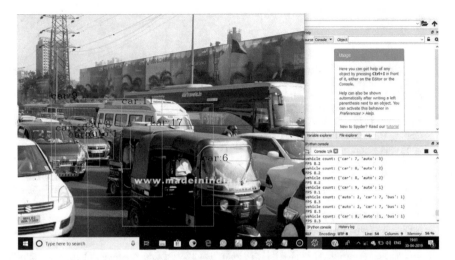

Fig. 9 Vehicle classification in heterogeneous traffic

Fig. 10 Traffic controller interface

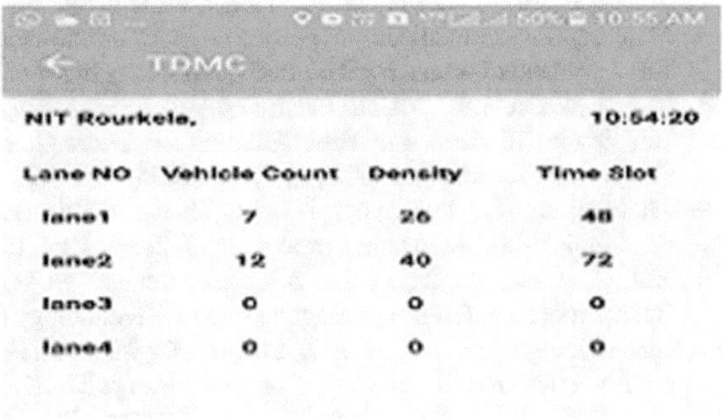

TDMC
NIT Rourkela,
10:54:20
Lane NO
Vehicle Count
Density
Time Slot
lane1 7 26 48
lane2 12 40 72
lane3 0 0 0
lane4 0 0 0

References

1. Bhalla, K., Khurana, N., Bose, D., Navaratne, K.V., Tiwari, G., Mohan, D.: Official government statistics of road traffic deaths in India under-represent pedestrians and motorised two wheeler riders. Inj. Prev. **23**(1), 1–7 (2017)
2. Singh, S.K.: Urban transport in India: issues, challenges, and the way forward. Eur. Transp./Trasporti Europei (52) (2012)
3. Kabrane, M., Krit, S., Elmaimouni, L., Bendaoud, K., Oudani, H., Elasikri, M., Karimi, K., El Bousty, H.: Smart cities: energy consumption in wireless sensor networks for road trafile modeling using simulator SUMO. In: International Conference on Engineering & MIS (ICEMIS), pp, 1–7 (2017)
4. Sundar, R., Hebbar, S., Golla, V.: Implementing intelligent traffic control system for congestion control, ambulance clearance, and stolen vehicle detection. IEEE Sens. J. **15**(2), 1109–1113 (2014)
5. Ali, S.S.M., George, B., Vanajakshi, L., Venkatraman, J.: A multiple inductive loop vehicle detection system for heterogeneous and lane-less traffic. IEEE Trans. Instrum. Meas. **61**(5), 1353–1360 (2011)
6. Jagadeesh, Y.M., Suba, G.M., Karthik, S., Yokesh, K.: Smart autonomous traffic light switching by traffic density measurement through sensors. In: International Conference on Computers, Communications, and Systems (ICCCS), pp. 123–126 (2015)
7. Masek, P., Masek, J., Frantik, P., Fujdiak, R., Ometov, A., Hosek, J., Andreev, S., Mlynek, P., Misurec, J.: A harmonized perspective on transportation management in smart cities: the novel IoT-driven environment for road traffic modeling. Sensors **16**(11), 1872 (2016)
8. Mei, X., Zhou, S.K., Wu, H.: Integrated detection, tracking and recognition for ir video-based vehicle classification. In: IEEE International Conference on Acoustics Speech and Signal Processing Proceedings, vol. 5, pp. V–V (2006)
9. Mallikarjuna, C., Phanindra, A., Rao, K.R.: Traffic data collection under mixed traffic conditions using video image processing. J. Trans. Eng. **135**(4), 174–182 (2009)
10. Kul, S., Eken, S., Sayar, A.: A concise review on vehicle detection and classification. International Conference on Engineering and Technology (ICET), IEEE, pp. 1–4 (2017)
11. Liu, W., Anguelov, D., Erhan, D., Szegedy, C., Reed, S., Fu, C.Y., Berg, A.C.: Ssd: single shot multibox detector. In Eur. Conf. Comput. Vis., pp. 21–37. Springer, Cham (2016)
12. De Oliveira, D.C., Wehrmeister, M.A.: Towards real-time people recognition on aerial imagery using convolutional neural networks. In: 2016 IEEE 19th International Symposium on Real-Time Distributed Computing (ISORC), IEEE, pp. 27–34(2016)
13. Ahn, J.W., Chang, T.W., Lee, S.H., Seo, Y.W.: Two-phase algorithm for optimal camera placement. Sci. Program. (2016)
14. Mallikarjuna, C., Rao, K.R.: Area occupancy characteristics of heterogeneous traffic. Transportmetrica **2**(3), 223–236 (2006)
15. Uddin, M.S., Das, A.K., Taleb, M.A.: Real-time area based traffic density estimation by image processing for traffic signal control system: Bangladesh perspective. In: International Conference on Electrical Engineering and Information Communication Technology (ICEEICT), IEEE, pp. 1–5 (2015)

IoT Aware Automatic Smart Parking System for Smart City

Manisha Sarangi, Shriyanka Mohapatra, Sri Vaishnavi Tirunagiri, Santos Kumar Das and Korra Sathya Babu

Abstract Unstructured parking and wrong positioned parking are illegal. Parking is a major issue in crowded and busy cities. It can be built with advanced sensors, IP cameras which provide smart services to the user in smart city. As there is increase of vehicles, it leads to traffic congestion. This problem can be effectively overcome by using smart parking system. Hence, we have proposed an IoT aware automatic smart parking system. The system is implemented by using sensor technologies for occupancy checking of the parking slots, computer vision algorithms for recognition of vehicle, and android application for providing user access. Through android application, user can get to know about the available slots and the amount can be deducted automatically.

Keywords Smart parking · Ultrasonic sensors · IoT · Camera

M. Sarangi · K. S. Babu
Department of Computer Science and Engineering, National Institute of Technology, Rourkela, India
e-mail: manishasarangi49@gmail.com

K. S. Babu
e-mail: ksathyababu@nitrkl.ac.in

S. Mohapatra · S. V. Tirunagiri · S. K. Das (✉)
Department of Electronics and Communication Engineering, National Institute of Technology, Rourkela, India
e-mail: dassk@nitrkl.ac.in

S. Mohapatra
e-mail: shriyankam@gmail.com

S. V. Tirunagiri
e-mail: srivaishu.1908@gmail.com

© Springer Nature Singapore Pte Ltd. 2020
P. K. Mallick et al. (eds.), *Cognitive Informatics and Soft Computing*,
Advances in Intelligent Systems and Computing 1040,
https://doi.org/10.1007/978-981-15-1451-7_50

469

1 Introduction

Parking of vehicles using smart technologies is one of the important pillars for the development of smart cities. Usually, people find problems for finding the space for the parking of their vehicles in crowded locations such as airports, universities, shopping malls, etc. Most of the times, the wrong parking of the vehicles unnecessarily leads to the traffic congestion and road blockages. This problem grows exponentially using the festive seasons, especially in the countries like India. With increase in number of vehicles, the problem of finding a suitable slot for parking becomes a difficult task for the drivers.

Various researchers have been trying to address this issue by developing smart parking systems (SPSs) using various state of art technologies. SPS gathers information from sensors about the parking spaces in a parking lot in real time to notify user about the availability of the slots. The SPS provides an integrated platform for fully automatic of parking services such as real-time status of availability of the slots, prebooking of slots, automated payment mechanism, etc. Simultaneously, it will provide a faithful communication between the user and system by taking real-time requests and providing real-time notifications. The SPS will help to reduce the overall time for the traveling by minimizing the wastage of time during parking and traffic congestion by avoiding the frequent occurrence of wrong parking.

Next section describes the related works of smart parking system.

2 Related Work

In [1], author stated that drivers have to search of available slot to park their vehicles after receiving ticket from the device without the need of manpower in the parking area. This system built on advanced sensing technique with time stamp sensing data from sensor network in parking lot to calculate total amount based on availability of vehicle parking in that lot and provides price information to drivers. Ticket vending machine is there to collect the ticket at the entry gate for billing which wasted lots of papers and it is not an eco-friendly solution.

In [2], Kianpisheh explained the multilevel parking system architecture which detects improper parking space, displays vacant parking spaces, and provides direction to vacant space through LED, payment facilities. Here, there are four different types of parking spaces that show different types of LEDs, i.e., vacant (green), occupied (red), reserved for VIP (blue), and handicapped (yellow). When user reaches to parking area, there is LED display board which shows available space. User can navigate to the available space through LED. When user occupied the slot, it shows red that means slot is now booked. There is Touch "n" Go module to pay the amount of parking.

In [3], author tries to minimize the parking waiting time in a large parking area. Here, user has to first register by using the mobile application then only user can get privileged to check whether there is any free slot available or not in that particular area and continuously updated in mobile application. Here, vehicle identification can be done by RFID tags where in every bike, there should be RFID tags and in parking area then RFID reader from which captures the RFID information. IR sensors detect whether slot is vacant or not. Through IR sensors and RFID tag, payment can be calculated. Payment can be done by using application through online banking. This paper also provides navigation through GPS. Here, prebooking of lots and security is not available.

In [4], paper provides nearest parking area to navigate the user from the remote location and also provides availability of parking slot in that a particular parking area which reduces the time and fuel consumption. This system is applicable for covered, open, and street-side parking where data is stored in cloud. All the information like availability, number of slots can be accessed through gateway. Here, parking area is continuously captured by camera to check the availability of slot. System also navigates the parking area from the current location of user. Cameras are mounted on ceiling in parking area where each and every slot can be detected. But if camera is mounted in closed-loop area, then more cameras are used and it is costly. Also, in this paper, smartphone is not used to display available space.

In [5], paper presents basic concept of using cloud and through application or web page, user can able to know the availability of slots in that area. Here, IR sensors are used to know whether there is empty slot available or not. After that, system will update the availability in app or web page and if parking area is filled, then message shows "PARKING FULL." Here, in this paper, problem is that number plate detection is not done in entry and exit gate and also not applicable for automatic billing system.

In [6], author mainly focuses on current demand of the customers and organization level. Here, in entry and exit point, camera is placed which detects and captures the number plate and sends it to the server. Server verifies the number plate of the car owner and if it matched, then it will provide information in mobile app. Once vehicle occupied slot, proximity sensors triggered and update in database that slot is occupied. At the exit, total amount is calculated on the basis of vehicle parking duration. The calculated amount will be deducted from end user wallet and the receipt will be sent to him/her through email. At the exit time, sensors triggered and timer stops and total amount is calculated on the basis of parking duration and deducts amount from user wallet, and receipt will be sent to user through mobile app.

In [7], author focuses on automatic license plate detection for Saudi Arabian license plate. There are three ways of acquisition image, i.e., using analog camera, digital camera, and frame grabber to select a frame. System first converts the image into gray scale. There are four steps to extract candidate region from a set of regions, i.e., vertical edge detection, si-and-shape filtering, vertical edge matching, B/W ratio, and license plate extraction.

In [8], author objective is to design authorized automatic vehicle number plate detection. Firstly, system will detect the vehicle and image is captured. Here, optical character recognition is used for recognizing the character. After that, resulting data is compared with the records and get the information of the vehicle like vehicle owner, place of registration, owner address, vehicle number, etc. Simulation is done in MATLAB. System provides robust detection of number plate but camera used in this project is very sensitive to vibration and fast changing targets. It can also improve the OCR recognition from different size and angles.

Researchers use various sensors for smart parking system. The sensor survey table is described in Table 1.

However, few specific areas of the SPS need further attention of the research community to improve its quality of service. Hence, in this research work, we have

Table 1 Sensor survey table

Sensor survey		
Title	Sensors used	Sensors work
Smart parking system (SPS) architecture using ultrasonic detector	Ultrasonic sensors, cameras	Ultrasonic sensors are used to detect vacant space in parking slot and camera is used for surveillance and check the occupancy in parking lot
IoT based smart parking system using RFID	RFID, IR sensors	IR sensors are used to detect slot detection and RFID is used for vehicle detection
Automatic smart parking system using internet of things	Camera	Camera is used to capture parking area continuously to check whether the slot is filled or vacant
Smart parking system for internet of things	Ultrasonic sensor, magnetic sensor	Ultrasonic sensors are used in indoor parking and magnetic sensors are used in outdoor parking. Both the sensors are used for detecting vehicle occupancy in parking lot
An approach to IoT based car parking and reservation system on cloud	PIR sensor, camera, proximity sensor	PIR sensors are used to detect the vehicle at the entry gate. Camera is used for number plate recognition. Proximity sensors are used to check the vehicle presence in parking slot
License plate recognition system	Camera	To take picture of vehicle for giving output as number plate
Advanced license plate recognition system for car parking	RFID camera	RFID sensors are used to detect number plate by RFID tags and camera also detects number plate by image processing

developed an SPS with enhanced features such real-time slot finding and allocation, automatic car recognition, automatic payment system, etc.

Next section describes the problem statement of smart parking system.

3 Problem Statement

The proposed SPS is developed to provide a completed integrated solution for the parking-related issues faced by the citizens of the smart city. The objectives of the problem statement can be outlined as follows.

- To reduce the wastage of time for searching the parking space.
- To develop a user-friendly android application for smooth functioning of the SPS.
- To make fully automatic parking management system by detecting the vehicle in entry and exit point, and updating the same to the database automatically.
- To validate the usage of correct parking slot and detect wrong parking in real time by entering auto-generated one-time password (OTP) through 4 * 4 matrix keypad.

Next section describes the development of the proposed work of smart parking system.

4 Development of the Proposed SPS

The detailed development of the proposed SPS can be explained as follows.

5 Working Principle

The working principle of the SPS can be explained by using its system architecture as shown in Fig. 1.

Here, we are using high-resolution IP camera, microcontroller, proximity sensor, i.e., ultrasonic sensors, matrix keypad, android application. IP camera will detect the vehicle and send the same to the database. Ultrasonic sensors are used to detect the availability of slot. Through android application user can search their preferable parking area and check the availability of parking slot. User can be notified about the availability of slot, OTP. After the user parked his vehicle in the allotted slot, he has to type OTP in 4 * 4 matrix keypad to confirm vehicle is in allotted slot.

In every slot, there will be an ultrasonic sensor which will check continuously whether the lot is empty or full by calculating the distance from the wall to underground and sends the data to the server and then update the system. If vehicle is already there and new vehicle is going to be placed in wrong slot, then buzzer will buzz.

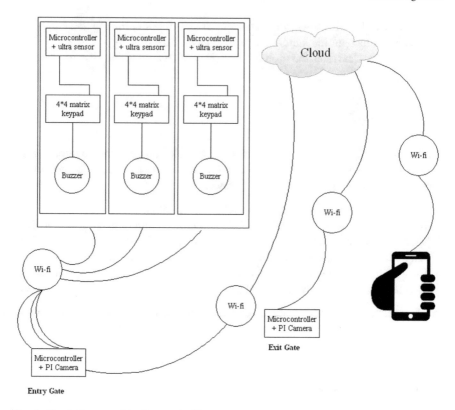

Fig. 1 Block diagram of the System Architecture

5.1 Algorithm

- **Step 1**: Detect the vehicle
- **Step 2**: If vehicle is detected then step 3, else repeat step 1
- **Step 3**: If slot available, then go to step 4, else regret
- **Step 4**: Allow access.

The flowchart for the providing entry access to the vehicle in current booking scenario is presented in Fig. 2. Once vehicle is in the gate, then camera detects vehicle. After the detection of the vehicle, it will be updated to the database. Then the system will check the availability slot. If there is any slot available, system will be assigned a slot to user and allow him/her to that particular slot. In current booking, when vehicle enters into the parking area, IP camera will detect the vehicle and send the data, i.e., vehicle number, entry time, entry date to the database. User will get notified about slot number, lane number, lot number, OTP in their android application. User has to park the car in the allotted slot and ultrasonic sensors will detect vehicle is present and send the data to server. After parking the car, user has to type four-digit OTP in 4 * 4 matrix keypad. User will get three times to type the correct OTP; if failed, then buzzer will buzz to indicate wrongly parked.

Fig. 2 Flowchart of current
booking working principle

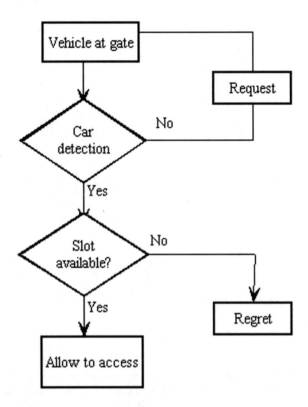

5.2 Hardware Implementation of Sensor Node

The hardware implementation at every parking slot is shown in Fig. 3. In order to ensure valid vehicle parking, each parking slot consists of a sensor node. The sensor node consists of a NodeMCU that is interfaced with Maxbotix MB1003 HRLV-MAX SONAR-EZ0 sensor, 4 * 4 matrix keypad, and a piezoelectric buzzer. The ultrasonic sensor first detects the presence of the vehicle once it has arrived in the parking slot. In case the slot is empty and a vehicle parks, the piezoelectric buzzer immediately sends an alarm. In case the slot is marked as filled, the driver has to enter the four-digit OTP accurately within 4 min using the 4×4 matrix keypad connected to 8 GPIO pins of NodeMCU. The inbuilt Wi-Fi module allows wireless communication with the cloud using MQTT protocol and MongoDB database.

5.3 Development of Mobile Application

An android application is developed for the SPS for the new user registration and login to the system. At the time of registration, app will ask to fill the user information

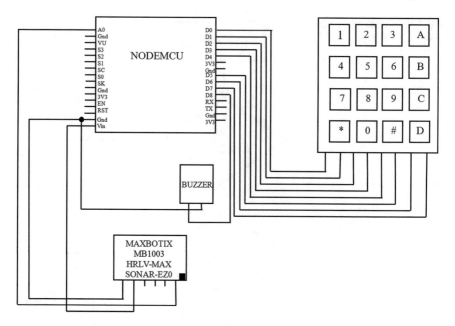

Fig. 3 Hardware design of sensor node

like user name, vehicle number, vehicle type, etc. There are two types of parking, Commercial Parking and Non-Commercial Parking. In application, when a user clicks on Commercial Parking, there are two types of services such as prebooking and current booking.

In current booking, when user is in front of parking area, app will provide that area long with nearby parking spot. When user entered into the parking area, then notification comes to user app that vehicle is entered with detailed information like vehicle number, allotted slot number, lot number, lane number, entry time, OTP, etc. At the time of exit, payment will be generated and notification comes to user with detailed information like entry time, exit time, total amount, etc. and directed to the third party for payment.

When user clicks on prebooking button, they can search areas they are interested to visit and application will provide full address of that parking area. As number of vehicles and their number are already registered, app will ask to choose vehicle type. System also asks to fill the entry and exit time of parking and submit the details. After that, system will provide the summary of user details with random slot number and total amount while counting the duration of parking.

Next section describes the results and discussions of smart parking system.

6 Results and Discussions

Figure 4 shows that when a vehicle enters into a parking lot, IP camera will detect the number plate and output shown in console i.e., in Fig. 5.

When a vehicle enters into a parking slot, the data is saved in database with entry time, entry date, and its vehicle number. When a vehicle exits from a parking slot, the data is saved in database with exit time and exit date by verifying the vehicle number.

Figure 6 shows that in current booking, when a vehicle entered into parking slot, then user gets notified that vehicle is entered into parking area and asks user wants to book this area or not. If user is interested in any other parking area, then we are providing option to park the vehicle in near by parking.

After booking, it shows all information with allotted slot, lot number, OTP which shown in Fig. 7. When a vehicle exits from parking slot, user gets notified that vehicle is exited and it shows all the details with entry time, exit time, user details, and amount to be paid which shown in Fig. 8.

Figure 9 shows that in prebooking, user can search and book their desirable parking area before hand to book the slot in that particular area. System will ask which type of vehicle, entry time, exit time, etc. while booking the slot.

Figure 10 shows the billing system of prebooking. After booking the slot, user will notified about the booking status with slot number, lot number, booking state, etc.

Fig. 4 Snapshot from input video

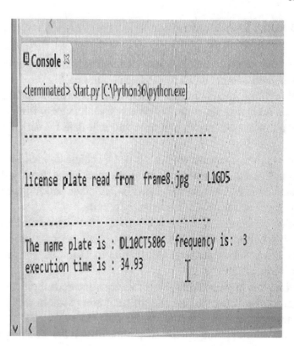

Fig. 5 Console results

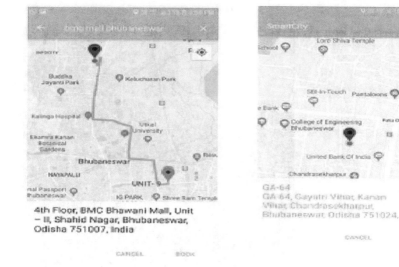

Fig. 6 Current booking

Fig. 7 Current booking entry notification

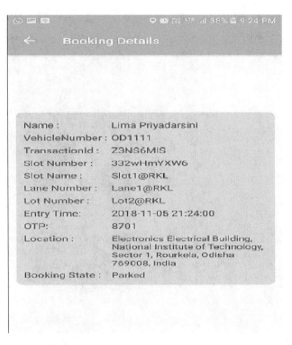

Fig. 8 Current booking exit notification

Fig. 9 Prebooking address
navigation

4th Floor, BMC Bhawani Mall, Unit
– II, Shahid Nagar, Bhubaneswar,
Odisha 751007, India

Fig. 10 Prebooking details

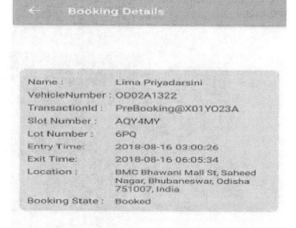

7 Conclusion

A proposed architecture of IoT aware smart parking system is presented in this paper. It provides efficient parking management solution for user to park their vehicle. The proposed system can easily detect vehicle while entering to the parking area through computer vision technique, presence of vehicle in the parking lot can be detected by ultrasonic sensors and confirmation of correct slot can be easily being identified by 4 * 4 matrix keypad by typing OTP. User can be notified about the booking status, OTP through user-friendly mobile application. System calculates the total amount of parking and notified to user through application.

Acknowledgements This work is a contribution of IMPRINT project sponsored by Ministry of Human Resource Development (MHRD) and Urban Development (UD).

References

1. El-Seoud, S.A. El-Sofany H., Taj-Eddine, I.: Towards the development of smart parking system using mobile and web technologies. In: 2016 International Conference on Interactive Mobile Communication, Technologies and Learning (IMCL), pp. 10–16, 2016
2. Kianpisheh, A., Mustaffa, N., Limtrairut, P., Keikhosrokiani, P.: Smart parking system (SPS) architecture using ultra-sonic detector. Int. J. Softw. Eng. Its Appl. **6**(3), 55–58 (2012)
3. Thorat, S.S., Ashwini, M., Kelshikar, A., Londhe, S., Choudhary, M.: IoT based smart parking system using rfid. Int. J. Comput. Eng. Res. Trends **4**, pp. 9–12 (2017)
4. Basavaraju, S.R.: Automatic smart parking system using internet of things. Int. J. Sci. Res. Publ. **5**(3), 629–632 (2015)
5. Lee, C., Han, Y., Jeon, S., Seo D., Jung, I.: Smart parking system for internet of things. In: 2016 IEEE International Conference on Consumer Electronics (ICCE), pp. 263–264, 2016
6. Hans V., Sethi P., Kinra, J.: An approach to IoT based car parking and reservation system on Cloud. In: 2015 International Conference on Green Computing and Internet of Things (ICGCIoT), pp. 352–354, 2015
7. Ahmed, M.J., Sarfraz, M., Zidouri, A., Al-Khatib, W.G.: License plate recognition system. In: 10th IEEE International Conference on Electronics, Circuits and Systems, 2003 ICECS 2003. Proceedings of the 2003, vol 2, pp. 898–901, 2003
8. Sudith Sen, E., Deepa Merlin Dixon, K., Anto, A., Anumary, M.V., Micheal, D., Jose, F., Jinesh, K.J.: Advanced license plate recognition system for car parking. In: 2014 International Conference on Embedded Systems (ICES), IEEE, pp. 162–165, 2014

Optimization and Control of Hybrid Renewable Energy Systems: A Review

Harpreet Kaur and Inderpreet Kaur

Abstract Sustainable power sources are sure to assume a key job later on energy production because of the quick exhaustion of ordinary wellsprings of energy. The major sustainable power sources like solar and wind can possibly meet the energy emergency somewhat. Though such sources when investigated autonomously are not reliable due to fickle character, their utilization as hybrid energy frameworks looks more reliable and practical, because of the corresponding character of these two assets. In this paper, an endeavor has been made to examine a logical evaluation linked with optimal control of hybrid sustainable power source frameworks.

Keywords Exhaustion · Sustainable · Power · Emergency · Hybrid · Reliable · Fickle

1 Introduction to Renewable Energy Sources

The requirement for world's power generation is expanding at a raising pace and cannot be satisfied totally by ordinary energy frameworks, because of their restricted supplies [1, 2]. So the utilization of hybrid systems for energy generation by sustainable power sources has drawn considerations overall [3–5]. A single source of renewable energy resources is not ready to give energy constantly to load; hence, hybrid energy systems turn into an essential solution [6]. Renewable power is gathered from inexhaustible assets, which are normally recharged on a human timescale, for example, sunlight, wind, rain, tides, waves, biomass, and geothermal warmth as presented in Fig. 1. Renewable energy frequently gives energy in four essential regions: power production, air and water heating/cooling, transportation, and rural (standalone) energy services [7–9]. Renewable power source assets exist over wide geological territories, as compared to other sources, that are amassed in narrow

H. Kaur (✉) · I. Kaur
Electrical Engineering Department, Chandigarh University, Gharuan, Mohali, India
e-mail: harpreet.ee@cumail.in

I. Kaur
e-mail: inder_preet74@yahoo.com

© Springer Nature Singapore Pte Ltd. 2020
P. K. Mallick et al. (eds.), *Cognitive Informatics and Soft Computing*,
Advances in Intelligent Systems and Computing 1040,
https://doi.org/10.1007/978-981-15-1451-7_51

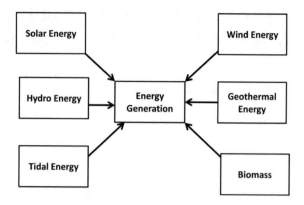

Fig. 1 Different types of renewable energy sources [16]

number of nations [10–13]. Fast organization of sustainable power source and energy efficiency is bringing about huge vitality security, environmental change relief, and financial advantages [14, 15].

2 Introduction to Microgrid

A microgrid is a small-scale control grid that can work autonomously or cooperatively with other small power networks. The act of utilizing microgrid is known as distributed, dispersed, decentralized, district, or embedded energy production [17]. Any small-scale limited power station that has its own generation and capacity assets and determinable limits can be considered a microgrid. If microgrid can be incorporated with the territory's principle power grid, it is frequently referred as hybrid microgrid [18]. Or on the other hand a microgrid is a limited gathering of power sources and loads that typically works associated with and synchronous with the wide traditional zone synchronous grid (macrogrid), however can likewise disconnect to "island mode"—and work self-governing as physical or monetary conditions dictate [19]. In this way, a microgrid can adequately incorporate different wellsprings of distributed generation (DG), particularly renewable energy sources (RES)—sustainable power, and can supply crisis control, changing among island and associated modes [18, 19].

An essential component is likewise to give different end-use needs as heating, cooling, and power in the meantime since this permits energy bearer substitution and expanded energy effectiveness because of waste heat use for heating, local high-temperature water, and cooling purposes (cross-sectoral vitality utilization) as appeared in Fig. 2 [20]. Microgrids are ordinarily supported by generators or renewable wind and solar energy-based assets and are frequently used to give strengthening power or supplement the fundamental power framework amid times of substantial interest. A microgrid procedure that coordinates nearby wind or solar resources can

Fig. 2 Basic block diagram of microgrid [18]

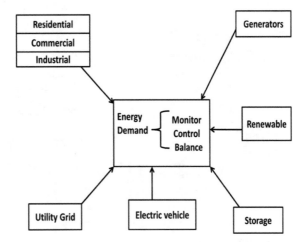

give repetition to fundamental administrations and make the main grid less helpless to confined failure [21].

3   Role of Hybrid Renewable Energy System

Energy requirements will keep on developing with its expanding population and urbanization. Environmental change, expanded carbon emissions, and the exhaustion of non-renewable energy assets have compelled to move to sustainable power source assets. Around one billion individuals, roughly fifteen percent of worldwide population, are still living without power as indicated by World Economic Outlook 2016 [22]. A large portion of this rate is in provincial zones of developing nations. People in these zones for the most part use diesel generator to satisfy their prerequisites, originating air contamination and thus getting worsen their health. Utilizing the new and suitable technologies and sustainable power sources in remote territories is exceptionally viable arrangement that lessens the utilization of conventional energy. The fundamental issue of standalone system is the variance of supply that can be abstained from utilizing hybrid sustainable power source assets [23].

Hybrid systems are utilized for providing energy in various regions to beat the discontinuity and changeability of sun and wind resources. It includes at least two sustainable power sources expanding the attainability of feasibility to load [24].

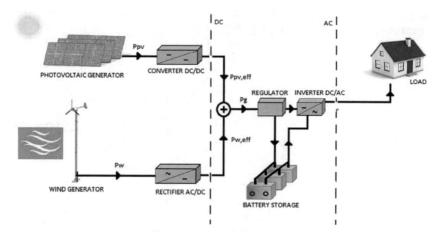

Fig. 3 PV-W-B system [25]

4 Different Hybrid Renewable Energy System

4.1 Solar Photovoltaic-Wind-Battery Hybrid Systems (PV-W-B)

PV-W-B comprises solar photovoltaic panels and small wind turbine generators for producing power as appeared in Fig. 3. For the most part, such systems are capable for small potential. The regular energy generation limits for PV-W-B systems are in the range from 1 to 10 kW [25].

4.2 Solar Photovoltaic-Wind-Diesel-Battery Hybrid System (PV-W-D-B)

This system incorporates solar panel, DG set, wind generator, and energy storage device as shown in Fig. 4 [26].

4.3 Solar Photovoltaic-Wind-Biodiesel-Battery Hybrid System (PV-W-BD-B)

The system comprises PV, wind (WT), energy storage device, and Bio-DG set, as appeared in Fig. 5. The supply need is to such an extent that the load is first fulfilled by

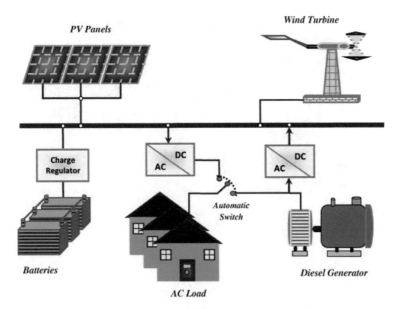

Fig. 4 PV-W-D-B system [26]

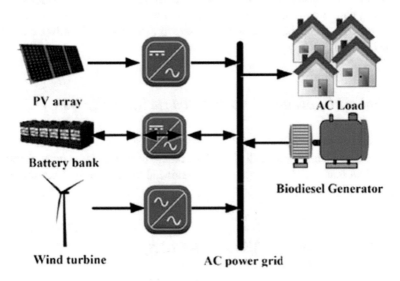

Fig. 5 PV-W-BD-B system [27]

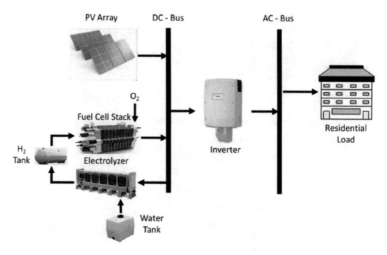

Fig. 6 PV-FC system [28]

the sustainable power source generators, and the battery operates when the sustainable power source generators' yield is not sufficient to meet the energy requirement, if working within its limits. If load requirement is more and the storage device is insufficient to produce the aggregate power requirement, Bio-DG fulfills the residual load [27].

4.4 Solar Photovoltaic-Fuel Cell Hybrid System (PV-FC)

PV and fuel cell energy frameworks synchronized with electrolyzer to produce hydrogen and phase charging-dispatch control procedure (energy unit works to fulfill the AC essential load, and the excess energy is utilized to operate the electrolyzer) as appeared in Fig. 6. [28]

4.5 Solar Photovoltaic-Wind-Fuel Cell System (PV-W-FC)

This framework comprises PV panels, wind power system, and fuel cell system. Electrolyzer is utilized to assimilate the quickly fluctuating output power with load and produce hydrogen [29] (Fig. 7).

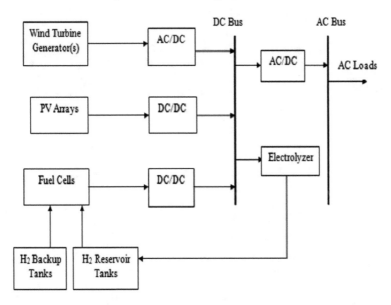

Fig. 7 PV-W-FC system [29]

5 Different Types of Solar Concentrators

Global energy requirement is expected to rise by 60% of current requirement by 2030. In this way, the capability of sustainable power source should be examined. Sustainable power source is the power obtained from natural and unnatural accessible structures consisting of biomass, solar, wind, and residual heat energy created through different manmade exercises. Solar power is an accessible and clean type of sustainable power source utilized as an option in contrast to non-renewable energy source in producing energy. On the other hand, the amount of thermal energy extracted from the sun is generally difficult [30]. The image shows the four main types of solar concentrators as shown in Fig. 8.

(a) Parabolic trough (line focus)
(b) Linear Fresnel (line focus)
(c) Heliostat field (point focus)
(d) Parabolic dish (point focus)

Table 1 shows some technical parameters of solar thermal power systems. Parabolic troughs and power towers are best suited for large power projects. Even though dish collectors are better for smaller power projects, they can be arranged in dish farms in order to supply larger demand. Table 2 shows the status of solar collector in India.

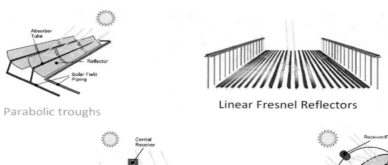

Fig. 8 Main types of solar concentrators [30]

Table 1 Technical parameters of solar thermal power systems [31]

Technology	Parabolic trough	Heliostat field	Parabolic dish
Unit size	30–200 MW	10–50 MW	5–25 kW
Operating temperature (°C)	390	565	750
Peak efficiency (%)	20	23	29
Commercial status	Commercially available	Scale-up demonstration	Prototype demonstration

Table 2 Status of solar collector in India [32]

Type of solar collector	Operational		Under construction		Planning	
	No. of plants	Total capacity (MW)	No. of plants	Total capacity (MW)	No. of plants	Total capacity (MW)
Parabolic trough	–	–	1	10	11	590
Linear Fresnel	–	–	2	250	–	–
Heliostat field	1	2.5	1	7.5	–	–
Parabolic dish	–	–	–	–	1	10

Table 3 Findings of different solar collectors used by different researchers in their study

Type of solar concentrator	Method/technology	Observations/findings
Parabolic trough solar collector	Monte Carlo method (MCM) in MATLAB	Optical effectiveness of 65% was acquired for collector segments with opening breadth of 0.6 m, edge of 100°, and recipient distance across of 0.025 m. Additionally, the optical proficiency of 61% was acquired for opening breadth of 0.7 m, edge of 90°, and diameter receiver of 0.025 m [33]
Parabolic trough collector	Geometry analysis	Solar collector adds to greater enhancement in thermal applications. The PTC can be delivered up to 400 °C temperature of heat [34]
Photovoltaic/thermal linear focus 20× solar concentrator	Experimental verification	A general effectiveness of solar energy change in both electric and thermal energy is as high as 61.5% [35]

6 Literature Review

Table 3 shows findings of different solar collectors used by different researchers in their study. The distinctive combinations of hybrid system utilized by different analysts for various sites for India and worldwide are presented in Tables 4 and 5.

7 Objectives of Proposed Research Work

1. Analyze the performance of different solar concentrators of different shapes having same ratings.
2. Implementation of optimal solar hybrid systems using artificial intelligence.

8 Methodology

The following steps will be used to achieve the proposed objectives.

1. Analyzing different solar concentrators of different shapes hing same ratings.
2. Assessment of load demand and resources available.

Table 4 Different hybrid systems in India

Hybrid energy system combinations	Method/technology	Site	System configuration	Observations
Solar wind-diesel-battery different combinations	HOMER software	Punjab, India 31.6340°N, 74.8723°E	Standalone	Mohammad Junaid Khan et al. described that solar wind-diesel-battery was the most feasible combination for maximum power generation in Amritsar with lowest cost of energy 0.164$/Kwh [36] 2017
Solar wind-battery	HOMER software	Jharkhand, India 22° 47′N, 86° 12′E	Standalone	Abhi Chatterjee and Ramesh Rayudu proposed framework that can be used to recover electricity infrastructure in electricity communities [37] 2017
Solar wind-battery-micro-hydropower-biogas-biomass	HOMER software	Uttarakhand, India 30.0668°N, 79.0193°E	Standalone	Anurag Chauhan and R. P. Saini explained that the system under study was sensitive toward the fluctuation of biomass prices ($15/ton–$50/ton) and the cost of energy varied from $0.092/kWh to $0.115/kWh [38] 2016

(continued)

Table 4 (continued)

Hybrid energy system combinations	Method/technology	Site	System configuration	Observations
Solar-micro-wind-battery	Microsoft excel software	Himachal Pradesh India 31.1048°N, 77.1734°E	Grid connected	Sunanada Sinha and S. S. Chandel explained that the state has good potential of power generation from hybrid systems with major solar (1034–1796 kWh/kWp/Yr) and minor wind generation (222–616.8 kWh/kWp/Yr) [39] 2015
Solar-hydro–wind-battery	HOMER software	Damla village, Haryana, India 30° 5′ 0″N 77° 13′ 0″E	Grid connected	Om Krishan and Sathans examined different combinations and found that Solar-hydro–wind-battery had least cost of energy and net present cost [40] 2018
Solar wind-hydro-biogas-biomass	HOMER software	Jharkhand, India 22° 47′N, 86° 12′E	Standalone	Aditya Kumar Neg and Shibayan Sarkar proposed that solar wind-hydrokinetic-bioenergy had lower cost of energy in the range $0.36/kWh–$0.68/kWh [41] 2017
Solar wind-biomass-diesel-battery different combinations	HOMER software and PSO	Barwani, India 22.0363°N, 74.9033°E	Standalone	Yashwant Sawle et al. described that solar wind-biomass-diesel-battery was the most feasible combination for maximum power generation in Barwani with lowest cost of energy 0.2899$/Kwh [42] 2017

(continued)

Table 4 (continued)

Hybrid energy system combinations	Method/technology	Site	System configuration	Observations
Solar-hydrogen-fuel cell	Fuzzy logic, HOMER Pro software	Bhopal, India 23° 12′ N 77° 24′ E	Standalone	Anand Singh et al. observed the technological and economic feasibility of renewable energy-based system for an academic research building. Net generation is 20,712.63 kWh/year and cost of energy is 0.2033415$/kWh [43] 2017
Solar wind integrating solar thermal unit and storage unit	LabVIEW	Northeast, India	Standalone	Subhadeep Bhattacharjee and Shantanu Acharya predicted the performance of solar wind system with annual average efficiency 65.52 and 24.21% [44] 2016

Table 5 Different hybrid systems worldwide

HES combinations	Method/technology	Site	System configuration	Observations
Solar wind-diesel-battery	iHOGA Software based on Genetic algorithm	Egypt 30° 40′ 0″N, 30° 4′ 0″E	Standalone	T. M. Tawfik et al. explained designing and simulation of HES for remote area for determining best sizing and control strategy for the system [45] 2018
Solar wind-biomass	HOMER software	Australia 25.2744°S, 133.7751°E	Standalone	Gang Liu et al. described that solar wind biomass system has the advantage of cost and emission reduction [46] 2017
Solar wind-diesel-battery	HOMER software	South Africa 30.5595°S, 22.9375°E	Standalone	Temitope Adefarati et al. proposed system having reduced cost of energy and lower Net present cost [47] 2012
Solar wind-diesel	HOMER software	Saudi Arabia 29° 8.282′S, 44°19.817′E	Standalone	Shafiqur Rehman et al. proposed a system that has reduced the operating cost of system and CO_2 emissions. Wind energy contributed more efficiently than PV energy [48] 2018

(continued)

Table 5 (continued)

HES combinations	Method/technology	Site	System configuration	Observations
Solar wind-diesel-battery	HOMER software	Morocco 31° 03′ 43″N, 7° 54′ 58″E	Standalone	Hanane Benchran et al. analyzed the renewable energy potential of the Imlil village, Morocco. It is the most feasible and environmentally friendly solution as compared to grid extension [49] 2018
Solar wind-biodiesel	HOMER software, HAS and SSA implementation, MATLAB	Iran 32° 45′ 32″N, 58° 50′ 9″E	Standalone	Du Guangqian et al. developed an effective and reliable energy framework for size optimization [50] 2016
Solar wind-small hydro	HOMER software	Dejen district Ethiopia 10° 13′ 24″N, 38° 07′ 58″E	Standalone	Getachew Bekele and Getnet Tadesse proposed the most feasible system with lower cost of energy [51] 2016
Solar wind	iHOGA	Spain 40.4637°N, 3.7492°W	Standalone	Ms. Jyoti B. Fulzele and Dr. M. B. Daigavane discussed optimization of hybrid renewable energy system using variables such as global solar radiation, wind speed, and PV panel cost [52] 2012

3. Feasibility and reliability analysis for different sets of solar hybrid system.
4. Analysis of different sets of solar hybrid system combinations.
5. Comparison of different sets of solar hybrid system combinations.
6. Integration of optimal hybrid system using AI.

9 Conclusion

Sun power is turning into an option for the constrained non-renewable energy source assets. One of the least difficult and most direct uses of this energy is the transformation of sunlight radiations into electricity. The solar panel can be utilized either as an independent framework or as an enormous solar system that is associated with the power grid. The earth gets 84 terawatts of intensity, and world's consumption is around 12 terawatts of power for each day. So as to expand the conversion from sun power to electrical energy, the solar panels must be situated opposite to the sun. Therefore, the tracking of the sun's area and situation of the solar panel are significant. Likewise, the shape and size of solar collector is a significant thought.

A single source of renewable energy resources is not ready to give energy constantly to load; hence, hybrid energy systems turn into an essential solution. Hybrid systems are utilized for providing energy in various regions to beat the discontinuity and changeability of sun and wind resources.

Hybrid sustainable sources have been broadly valued all around the world as reliable source for the future energy needs. Remembering this, a broad literature survey on the independent and grid-associated PV/wind/battery hybrid energy framework has been introduced in this work. It is shown that the procedures such as GA, PSO, and ACO diminish the computational weight to accomplish the worldwide ideal arrangement. Different converter topologies for incorporating the HRES to the grid and furthermore extraordinary control systems have been secured according to the accessible literature. Alongside the previously mentioned regions, different computer tools for examining the HRES have been talked about, and their relative benefits and negative marks have been brought out. The perceptions drawn and the basic audit evaluation as exhibited in content make this work supportive for any specialist intrigued to investigate the research related to the different PV/wind hybrid energy frameworks.

References

1. Olatomiwa, L., Mekhilef, S., Ismail, M.S., Moghavvemi, M.: Energy management strategies in hybrid renewable energy systems: a review. Renew. Sustain. Energy Rev. **62**, 821–835 (2016)
2. Zahraee, S.M., Khalaji Assadi, M., Saidur, R.: Application of artificial intelligence methods for hybrid energy system optimization. Renew. Sustain. Energy Rev. **66**, 617–630 (2016)

3. Ma, T., Yang, H., Lu, L., Peng, J.: Technical feasibility study on a standalone hybrid solar-wind system with pumped hydro storage for a remote island in Hong Kong. Renew. Energy **69**, 7–15 (2014)

4. Ma, W., Xue, X., Liu, G.: Techno-economic evaluation for hybrid renewable energy system: application and merits. Energy **16**, 1–43 (2018)

5. Guo, S., Liua, Q., Sund, J., Jin, H.: A review on the utilization of hybrid renewable energy. Renew. Sustain. Energy Rev. **91**, 1121–1147 (2018)

6. Sharma, K.K., Thakur, G., Kaur, I., Singh, B.: Cost analysis of hybrid power system design using homer. Accepted at Lecture Notes in Electrical Engineering Springer. http://www.springer.com/series/7818 (in press) (Scopus Indexed)

7. Sharma, K.K., Thakur, G., Kaur, I.: Power management in hybrid micro grid system. Indian J. Sci. Technol. **10**(16), 1–5 (2017). ISSN: 0974-5645 (Scopus Indexed) (WoSc)

8. Sharma, K.K., Gupta, K., Kaur, I.: Simulation and optimization of MRES for remote power generation at Bharmour, India. Indian J. Sci. Technol. **10**(16), 6–14 (2017). ISSN: 0974-5645 (Scopus Indexed) (WoSc)

9. Sharma, K.K., Samia, Singh, B., Kaur, I.: Power system stability for the islanding operation of micro grids. Indian J. Sci. Technol. **9**(38), 1–5 (2016). ISSN: 0974-5645 (Scopus Indexed) (WoSc)

10. Inderpreet, K., Harpreet, K.: Study of hybrid renewable energy photovoltaic and fuel cell system: a review. Int. J. Pure Appl. Math. **118**(19), 615–633 (2018) (Scopus Indexed)

11. Harpreet, K., Manjeet, S., Yadwinder, B.S., Inderpreet, K.: Agri-waste assessment for optimal power generation in District Ludhiana, Punjab, India. In: 3rd-IEEE (conference record: 43722) ISBN NO: 978-1-5386-5130-8/18. ICEECCOT-2018 International Conference on Electrical, Electronics, Communication, Computer Technologies and Optimization Techniques on 14–15 Dec 2018. Organized by GSSSIETW Mysuru, p. 78 (SCOPUS)

12. Vipasha, S., Harpreet, K.: Design and implementation of multi junction PV cell for MPPT to improve the transformation ratio. In: International Conference on Green Technologies for Power Generation, Communication and Instrumentation (ICGPC, 2019) on 3–4 Apr 2019. ISBN NO: 978-93-5254-979-5. Organized by Department of ECE & EEE St. Peter's Institute of Higher Education and Research Avadi, Chennai, Tamilnadu, India (SCOPUS), p. 3

13. Obaidullah, L., Inderpreet, K., Harpreet, K.: Designing an effective and efficient solar tracking system to overcome the drawbacks of conventional P& O, International Conference on Green Technologies for Power Generation, Communication and Instrumentation (ICGPC, 2019) on 3–4 Apr 2019. ISBN NO: 978-93-5254-979-5. Organized by Department of ECE & EEE St. Peter's Institute of Higher Education and Research Avadi, Chennai, Tamil Nadu, India (SCOPUS), p. 15

14. Harpreet, K., Inderpreet, K.: Energy return on investment (EROI) analysis of 2KW solar photovoltaic system. In: International Conference on Green Technologies for Power Generation, Communication and Instrumentation (ICGPC, 2019) on 3–4 Apr 2019. ISBN NO: 978-93-5254-979-5. Organized by Department of ECE & EEE St. Peter's Institute of Higher Education and Research Avadi, Chennai, Tamil Nadu, India (SCI), p. 18

15. Sagar, P., Harpreet, K., Inderpreet, K.: Roof top solar installation: a case study. In: International Conference on Green Technologies for Power Generation, Communication and Instrumentation (ICGPC, 2019) on 3–4 Apr 2019. ISBN NO: 978-93-5254-979-5. Organized by Department of ECE & EEE St. Peter's Institute of Higher Education and Research Avadi, Chennai, Tamil Nadu, India, p. 24

16. Kim, J.S., Boardman, R.D., Bragg Sitton, S.M.: Dynamic performance analysis of a high-temperature steam electrolysis plant integrated within nuclear-renewable hybrid energy systems. Appl. Energy **228**, 2090–2110 (2018)

17. Dawouda, S.M., Lin, X., Okba, M.I.: Hybrid renewable microgrid optimization techniques: a review. Renew. Sustain. Energy Rev. **82**, 2039–2052 (2018)

18. Junga, J., Villaran, M.: Optimal planning and design of hybrid renewable energy systems for microgrids. Renew. Sustain. Energy Rev. 1–16 (2018)

19. Ghenai, C., Janajreh, I.: Design of solar-biomass hybrid microgrid system in Sharjah. In: Energy Proceedia 103th Conference on Renewable Energy Integration with Mini/Microgrid (REM 2016), Maldives, 19–21 Apr 2016, pp. 357–362
20. Venkatraman, R., Khaitan, S.K.: A survey of techniques for designing and managing microgrids. In: Proceedings of IEEE Conference, pp. 1–5 (2015). 978-1-4673-8040-9
21. Hina Fathima, A., Palanisamy, K.: Optimization in microgrids with hybrid energy systems—a review. Renew. Sustain. Energy Rev. 45, 431–446 (2015)
22. The International Energy Agency (IEA), World Energy Outlook, Paris, France (2018)
23. Mahesh, A., Sandhu, K.S.: Hybrid wind/photovoltaic energy system developments: critical review and findings. Renew. Sustain. Energy Rev. 52, 1135–1147 (2015)
24. Goel, S., Sharma, R.: Performance evaluation of stand alone, grid connected and hybrid renewable energy systems for rural application: a comparative review. Renew. Sustain. Energy Rev. 78, 1378–1389 (2017)
25. Mazzeo, D., Oliveti, G., Baglivo, C., Congedo, P.M.: Energy reliability-constrained method for the multi-objective optimization of a photovoltaic-wind hybrid system with battery storage. Energy 1–54 (2018)
26. Kaabeche, A., Ibtiouen, R.: Techno-economic optimization of hybrid photovoltaic/wind/diesel/battery generation in a stand-alone power system. Sol. Energy 103, 171–182 (2017)
27. Martin, S.S., Chebak, A.: Concept of educational renewable energy laboratory integrating wind, solar and biodiesel energies. Int. J. Hydrog. Energy 1–11 (2016)
28. Ghenai, C., Salameh, T., Merabet, A.: Technico-economic analysis of off grid solar PV/Fuel cell energy system for residential community in desert region. Int. J. Hydrog. Energy 1–11 (2018)
29. Dey, S., Dash, R., Swain, S.C.: Fuzzy based optimal load management in standalone hybrid solar PV/Wind/Fuel cell generation system. In: Proceedings of International Conference on Communication, Control and Intelligent Systems (CCIS), pp. 486–490. IEEE (2015)
30. http://mcensustainableenergy.pbworks.com. Accessed on 15/12/2018
31. https://www.energy.gov/. Accessed on 15/12/2018
32. https://www.nrel.gov/. Accessed on 15/12/2018
33. Hoseinzadeh, H., Kasaeian, A., Shafii, M.B.: Geometric optimization of parabolic trough solar collector based on the local concentration ratio using the Monte Carlo method. Energy Convers. Manag. 175, 278–287 (2018)
34. Abdulhamed, A.J., Adam, N.M., Ab-Kadir, M.Z.A., Hairuddin, A.A.: Review of solar parabolic-trough collector geometrical and thermal analyses, performance, and applications. Renew. Sustain. Energy Rev. 91, 822–831 (2018)
35. Cappelletti, A., Catelani, M., Kazimierczuk, M.K.: Practical issues and characterization of a photovoltaic/thermal linear focus 20 × solar concentrator. IEEE Trans. Instrum. Meas. 65(11), 2464–2475 (2016)
36. Khan, M.J., Yadav, A.K., Mathew, L.: Techno economic feasibility analysis of different combinations of PV-Wind-Diesel-Battery hybrid system for telecommunication applications indifferent cities of Punjab, India. Renew. Sustain. Energy Rev. 76, 577–607 (2017)
37. Chatterjee, A., Rayudu, R.: Techno-economic analysis of hybrid renewable energy system for rural electrification in India. In: Proceedings of IEEE Conference (2017)
38. Chauhan, A., Saini, R.P.: Techno-economic feasibility study on integrated renewable energy system for an isolated community of India. Renew. Sustain. Energy Rev. 59, 388–405 (2016)
39. Sinha, S., Chandel, S.S.: Prospects of solar photovoltaic–micro-wind based hybrid power systems in western Himalayan state of Himachal Pradesh in India. Energy Convers. Manag. 105, 1340–1351 (2015)
40. Krishan, O., Sathans: Design and techno-economic analysis of a HRES in a rural village. In: Proceedings of 6th International Conference on Smart Computing and Communications, 7–8 Dec 2017, vol. 125, pp. 321–328
41. Nag, A.K., Sarkar, S.: Modeling of hybrid energy system for futuristic energy demand of an Indian rural area and their optimal and sensitivity analysis. Renew. Energy 7, 1–31

42. Sawle, Y., Gupta, S.C., Bohre, A.K.: Review of hybrid renewable energy systems with comparative analysis of off-grid hybrid system. Renew. Sustain. Energy Rev. 1–19 (2017)
43. Singh, A., Baredar, P., Gupta, B.: Techno-economic feasibility analysis of hydrogen fuel cell and solar photovoltaic hybrid renewable energy system for academic research building. Energy Convers. Manag. **145**, 398–414 (2017)
44. Bhattacharjee, S., Acharya, S.: Performative analysis of an eccentric solar–wind combined system for steady power yield. Energy Convers. Manag. **108**, 219–232 (2016)
45. Tawfik, T.M., Badr, M.A., El-Kady, E.Y., Abdellatif, O.E.: Optimization and energy management of hybrid standalone energy system: a case study. Renew. Energy Focus **25**, 48–56 (2018)
46. Liu, G., Rasul, M.G., Amanullah, M.T.O., Khan, M.M.K.: Feasibility study of stand-alone PV-wind-biomass hybrid energy system in Australia. In: Proceedings of IEEE Conference (2011)
47. Adefarati, T., Bansal, R.C., Justo, J.J.: Techno-economic analysis of a PV–wind–battery–diesel standalone power system in a remote area. J. Eng. **2017**(13), 740–744 (2017)
48. Rehman, S., Mahbub Alam, M., Meyer, J.P., Al-Hadhrami, L.M.: Feasibility study of a wind-pv-diesel hybrid power system for a village. Renew. Energy **38**, 258–268 (2012)
49. Benchraa, H., Redouane, A., El Harraki, I., El Hasnaoui, A.: Techno-economic feasibility study of a hybrid Biomass/PV/Diesel/Battery system for powering the village of Imlil in High Atlas of Morocco. In: Proceedings of IEEE Sponsored 9th International Renewable Energy Congress (IREC-2018)
50. Guangqian, D., Bekhrad, K., Azarikhah, P., Maleki, A.: A hybrid algorithm based optimization on modeling of grid independent biodiesel-based hybrid solar/wind systems. Renew. Energy **122**, 551–560 (2018)
51. Bekele, G., Tadesse, G.: Feasibility study of small Hydro/PV/Wind hybrid system for off-grid rural electrification in Ethiopia. Appl. Energy **97**, 5–15 (2012)
52. Fulzele, J.B., Daigavane, M.B.: Design and optimization of hybrid PV-wind renewable energy system. Mater. Today Proc. **5**, 810–818 (2018)

Optimal Selection of Electric Motor for E-Rickshaw Application Using MCDM Tools

Abhinav Anand, Dipanjan Ghose, Sudeep Pradhan, Shabbiruddin and Akash Kumar Bhoi

Abstract Implementation and ultimate establishment of a sustainable environment require many transformations at minor levels. Vehicular emissions from traditional fuel cars have long been accounted for a major hindrance to such developments and hence to eliminate its intrusion, usage of electric vehicles (EV) is promoted. In a developing nation like India, an e-rickshaw can ensure an environment-friendly as well as economic support to the societal backbone. However, to ensure the efficient working of the e-rickshaw at optimum costs as well, the type of motor used should be placed under careful examination. In this study, brushless DC (BLDC) motors, series wound DC motors and three-phase AC induction motors were analyzed using seven multidimensional criteria comprising motor specifications, technical as well as economic factors, using the Technique for Order Preference by Similarity to Ideal Solution (TOPSIS) methodology to obtain a suitable motor applicable for usage in e-rickshaws. The weights for the criteria were obtained using Decision-Making Trial and Evaluation Laboratory (DEMATEL) method. The study stands quite novel in its area of application and the methodology used can be implemented in any study of similar nature.

Keywords Electrical machines · Sustainable development · Electric motors · Multi-criteria decision making (MCDM) · DEMATEL · TOPSIS method · EV

A. Anand · D. Ghose · S. Pradhan · Shabbiruddin (✉) · A. K. Bhoi
Department of Electrical and Electronics Engineering, Sikkim Manipal Institute of Technology, Sikkim Manipal University, East Sikkim, Sikkim, India
e-mail: shabbiruddin85@yahoo.com

A. Anand
e-mail: abhinav.raman13@gmail.com

D. Ghose
e-mail: ghosedipanjan1998@gmail.com

S. Pradhan
e-mail: sudeeppradhan76@gmail.com

A. K. Bhoi
e-mail: akash730@gmail.com

© Springer Nature Singapore Pte Ltd. 2020 501
P. K. Mallick et al. (eds.), *Cognitive Informatics and Soft Computing*,
Advances in Intelligent Systems and Computing 1040,
https://doi.org/10.1007/978-981-15-1451-7_52

1 Introduction

The basic purpose of an electric motor at its very fundamentals is to convert electrical energy into mechanical energy. Thus, its usage in terms of an EV is pretty evident—to convert the electrical energy stored in batteries of the EV into mechanical energy of the wheels, further making electric motors the heart of a propulsion system. Clearly, inappropriateness in its choice can adversely affect the overall efficiency and working of an EV, perhaps on many grounds, rendering it unusable for day-to-day purposes.

EVs, despite their long journey in complete adaption completely replacing conventional vehicles, seemingly have a long history. The first presence of an EV was noted as the small EV mode developed in the Dutch town of Groningen by Professor Stratingh in 1835 [1] though this option did not become viable until Frenchmen Camille Faure and Gaston Plante respectively added their innovations, including mending certain flaws associated with the storage batteries as late as 1881 [1]. Today, implementations of combustion engines have added up greatly to increase the speed of hybrid EVs. Recent data as of January 2018 shows the most sold plug-in car to be Nissan Leaf, with as many as 300,000 units sold worldwide globally [2]. Quite exorbitantly, due to the close inclination between the relation between motor used in an EV and its ultimate efficiency in terms of speed, it is essential that enough investment of thought is done before selection of the most suitable motor for any EV before manufacturing or purchasing.

The choice of motor can be attributed to a number of factors including cost, efficiency and reliability. Appropriate electric-propulsion system process selection is, however, difficult and should be carried out at the system level [3]. Clearly, taking into consideration of the variegated factors affecting the choice of motor, it is a choice involving multiple criteria regarding the characteristics of a motor and hence can be easily categorized as a decision-making problem. To solve a problem involving a multiple criteria analysis, various MCDM tools may be assessed for the evaluation process [4–9]. In this study, the weightage of the considered criteria toward their affection in the efficiency of an e-rickshaw [10] was found out using the DEMATEL's method and then the considered motor variations were assessed using the TOPSIS method. DC motors are used exclusively in EV sector at present but the change to AC brings advantage in weight and efficiency, besides providing improved range. The switched reluctance motor also has advantages for special applications [11]. The various motors considered for inclusion in this study are: AC induction motors, brushless DC motors (BLDC) and series wound DC motors.

1. **BLDC Motor**: Also known as synchronous DC motor, BLDC takes power by DC supply using a switching power supply or an inverter which produces AC current for driving each phase of the motor through electronic controller. The electronic controller provides pulses of current to the motor winding that controls the torque and speed of the motor, making it suitable for EV usage. In general, BLDC motors are used in maximum EVs mostly due to its small size, lightweight, simple construction, easy to maintain, simple rotor cooling, high efficiency and good reliability features. The pros of a brushless motor compared to brushed

motors are high speed, light in weight, and electronic control, further adding up to its EV-based suitability [12].

2. **Series Wound DC Motor**: A DC series motor takes its requisite power from DC supply where field and armature winding are in series. The motor flux is directly proportional to the armature current so it provides very high torque at starting phase whereas it takes low starting current; therefore, it is commonly used for starting high-inertia loads, such as elevators, trains. Due to this advantage of DC series motor, it can be considered for EV application [13].

3. **Three-Phase AC Induction Motor**: An induction motor is one which runs on alternating current needed to produce torque obtained by electromagnetic induction from the stator winding's magnetic field. Two kinds of rotor can be used for induction motor either squirrel-cage type or wound type. Single-phase induction motors are often used for smaller loads whereas three-phase induction motors are widely used for industrial drives because they are reliable, rugged and economical. Although earlier it was used as fixed-speed service, nowadays these motors are increasingly being used for variable-speed service. Induction motors are widely used in both variable frequency drive and fixed-speed applications as well as adding to its suitability for EV applications [14].

2 Methodology

For the methodology, seven parameters, namely, voltage rating (V), current rating (I), power (P), torque (τ), revolutions per minute (RPM), weight and cost, pertaining to three kinds electric motors $X1$, $X2$ and $X3$, where $X1$ is BLDC motor, $X2$ is a 3 phase motor and $X3$ is a DC series motor, were assessed on basis of a model as shown in Fig. 1.

The involved methodology can further be elaborated as:

DEMATEL method:

The DEMATEL is based on the idea that higher the dispersal in the values of a criteria involved in the selection process, more the precedence of it toward its suitability [15]. In case of the optimum selection of the electric motors for usage, the DEMATEL method of weighing has a very sensitive impact, considering that it specializes in formalizing all the criteria and alternatives. The principle of this method works by relating the preference of the criteria over each other in a scale of 0–9 with increasing dependency, of course with 0 meaning no dependency at all [15] and hence the diagonal elements in the comparison matrix are always zero.

TOPSIS method:

For the TOPSIS method, using the advice from experts, the initial decision matrix for each alternative for corresponding criteria is prepared. From the initial decision matrix, the normalized matrix is obtained using Eq. (1) as shown below:

Fig. 1 Proposed model for calculation

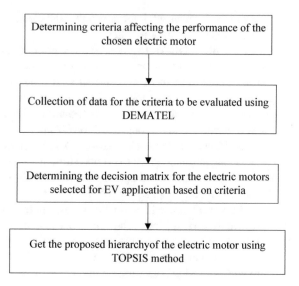

Determining criteria affecting the performance of the chosen electric motor

Collection of data for the criteria to be evaluated using DEMATEL

Determining the decision matrix for the electric motors selected for EV application based on criteria

Get the proposed hierarchy of the electric motor using TOPSIS method

$$a_{mn} = b_{mn} \sqrt{\sum_{m-1}^{y} b_{mn}^2} \quad m = 1, 2, \ldots, y \text{ and } n = 1, 2, \ldots, z \quad (1)$$

From the normalized matrix, the weighted normalized matrix is formed using the weights obtained after the DEMATEL's method [16] to finally get the positive ideal (E^+) and negative ideal (E^-) solutions as shown using Eqs. (2a) and (2b).

$$E^* = \left\{ \left(\max_i c_{mn} \middle| n \in F_b \right), \left(\min_i c_{mn} \middle| n \in F_c \right) \right\} = \left\{ c_n^* \middle| n = 1, 2, \ldots, y \right\} \quad (2a)$$

$$E^- = \left\{ \left(\min_i c_{mn} \middle| n \in F_b \right), \left(\max_i c_{mn} \middle| n \in F_c \right) \right\} = \left\{ c_n^- \middle| n = 1, 2, \ldots, y \right\} \quad (2b)$$

Now with the help of y-dimensional Euclidean distance, the separation measures are obtained [16]. The separation measures of the positive ideal answer and the negative ideal answer for the entire alternative, respectively, are shown below.

And finally the relative nearness is figured to get the priority vectors for each alternative [16].

3   Calculations

Assuming a person wants to buy parts of e-rickshaw and assemble it by own, in the beginning he/she has to select a motor according to the requirement. In India,

Table 1 Weights of criteria obtained after DEMATEL's method

Criteria considered	Weights
Volts (V)	0.154321
I (A)	0.161728
P (kW)	0.137058
Torque (N m)	0.142898
RPM	0.146017
Weight (kg)	0.123636
Cost (Rs.)	0.134342

Table 2 Decision matrix for TOPSIS method [19, 20]

Engine	Volts (V)	I (A)	P (kW)	Torque (N m)	RPM	Weight (kg)	Cost (Indian rupees)
X1	48	90	2	7	6000	7	8500
X2	48	83.3	4	70	6500	29	20,600
X3	48	41	1.5	9.5	1500	35	15,700

e-rickshaws mostly involve importing parts from China [17]. According to notifications, four passengers with 40 kg luggage have been allowed, net power of its motor not being more than 2000 W and maximum speed of the vehicle being up to 25 km per hour [18]. Taking care of the prescribed factors, the selection process to find the optimum motor type for an e-rickshaw was carried out. At first, using the DEMATEL's method [15], the weights for the considered criteria were obtained as shown in Table 1.

After having obtained the weights of the criteria, the TOPSIS method was carried out. For the TOPSIS method, using the advice of an expert from TATA Motors Limited and practical data [19, 20] wherever feasible, the decision matrix was made, as shown in Table 2.

From the decision matrix, the normalized and then the weighted normalized matrix were obtained [16] and following the steps of the TOPSIS method [16], the resultant priority vectors for each alternate motor were procured.

4   Results and Discussions

The graph shown in Fig. 2 clearly depicts the results obtained after the TOPSIS method. Thus, the order of preference of the motors for usage in an e-rickshaw would be:

3 phase AC Motor > BLDC Motor > DC Series Motor

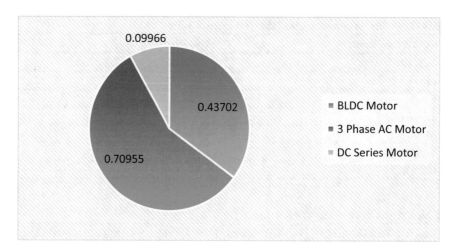

0.09966

0.43702

0.70955

- BLDC Motor
- 3 Phase AC Motor
- DC Series Motor

Fig. 2 Results obtained after TOPSIS method

Three-phase induction motors have long been known for their rugged, light and economical architecture. Initially, they were mostly considered useful for usage in fixed speed applications but over the years their usages have also been extended to variable speed and variable frequency areas by development of advanced control mechanisms. Clearly, induction motors provide a safe vent of optimism when speaking in terms of both technical excellence and economic factors, owing to their large preference when it comes to EV applications. BLDC motors, as the name suggests, are brushless, thus the disadvantages of maintenance and losses attributed to the usage of brushes are easily eliminated. Also, they are lightweight, with simplicity in construction and highly efficient as well, making it second in the list of preferentiality in using in an e-rickshaw. Finally, DC series motor, despite having its own advantages, is lagging in many aspects considering the other two alternatives. Although as discussed, due to their lower starting current and higher torque they can be used for loads with very high inertia, but when it comes to simpler applications as lightweight e-rickshaws, their advantages do not come much into play. Factors dealing with costs manage to notch up to higher places in terms of influences over a choice, hence arousing the necessity for checking the robustness of the result obtained sensitivity analysis is performed.

4.1 Sensitivity Analysis

The governing equations underlining the mathematical model for this analysis can be represented as shown in Eqs. (3) and (4) [21].

$$\mathrm{SI}_i = [(\beta M_i) + (1 - \beta)O_i] \tag{3}$$

$$O_i = \frac{1}{\left[F_i \sum_{i=1}^{v} F^{-1} \right]} \tag{4}$$

where the nomenclated terms can be represented as:

O Objective Factor Measure,
F Objective Factor Cost,
M Subjective Factor Measure,
SI   Selection Index,
v Number of alternate renewable energy systems, here $v = 7$, and
β Objective Factor Decision Weight.

Subjective Factor Measure values are obtained as depicted in Eq. (3) are basically the normalized priority vectors for the motors obtained after applying TOPSIS method. Table 2 represents the corresponding cost attribute values, represented by the Objective Factor Cost for each alternative. The Objective Factor Measure is derived from the Objective Factor Cost values using Eq. (4), to ultimately obtain the Sensitivity Index from Eq. (3).

Sensitivity analysis graph obtained is shown in Fig. 3. Value of the Decision Weight at any instance would give the preference of the cost related criteria over the other factors utilized for selection. Evidently, lower values of this weight denote a greater dominance of the cost-related criteria.

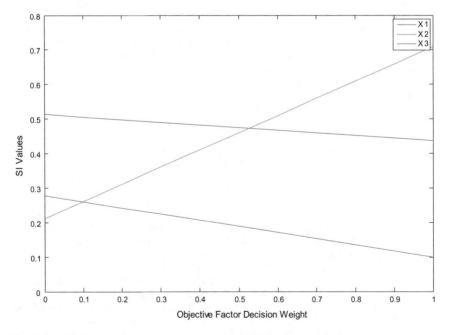

Fig. 3 Sensitivity analysis

5 Conclusion

In this work, three different kinds of electric motors generally used in industrial applications were taken up and assessed on the basis of seven criteria, ranging from multidimensional areas relating to their technical specifications, costs, weight, utility and so on. A gap of research on many aspects is one of the major causes why despite being in the market since a long time now, EVs still have a long way to completely overtake pollution causing conventional fuel cars. However, attributing to degrading conditions of the environment, it is up utmost necessity to accelerate the development of a sustainable development. The work projected hopes to vent a way for both manufacturers and consumers alike to invest further in efficient EVs in the form of e-rickshaws, thus accounting for little means to develop an environment-friendly society at large. Inclusion of a larger number of factors for calculation or greater alternatives of the motors further adds to a practical suitability of the results obtained.

References

1. Rajashekara, K.: History of electric vehicles in General Motors. IEEE Trans. Ind. Appl. **30**(4), 897–904 (1994)
2. Article: Electric Vehicle, https://en.wikipedia.org/wiki/Electric_vehicle#Plug-in_electric_vehicle
3. Zeraoulia, M., Benbouzid, M., Diallo, D.: Electric motor drive selection issues for HEV propulsion systems: a comparative study. IEEE Trans. Veh. Technol. **55**(6), 1756–1764 (2006)
4. Ghose, D., Naskar, S., Shabbiruddin: Multi-criteria analysis for industrial site selection in West Bengal (India) using Q-GIS. In: International Research Conference on Innovation, Technology and Sustainability, 24–25 Jan 2019, Manila, Philippines (2019)
5. Ghose, D., Naskar, S., Shabbiruddin: Q-GIS-MCDA based approach to identify suitable biomass facility location in Sikkim (India). In: International Conference on Advanced Computational and Communication Paradigms 2019. Proceedings will be published in IEEE (2019)
6. Badang, D.A.Q., Sarip, C.F., Tahud, A.P.: Geographic information system (GIS) and multicriteria decision making (MCDM) for optimal selection of hydropower location in Rogongon, Iligan City. In: International Conference on Humanoid, Nanotechnolgy, Information Technology, Communication and Control, Environment and Management, Philippines (2018)
7. Srihari, K., Raval, P., Shabbiruddin: Selection of an electric motor for an equivalent internal combustion engine by TOPSIS method. In: Springer International Conference on Emerging Trends and Advances in Electrical Engineering and Renewable Energy, ETAEERE 2016. Proceedings Advances in Power Systems and Energy Management, Lecture Notes in Electrical Engineering, vol. 436. Springer, Singapore (2016)
8. Das, A., Shabbiruddin: Renewable energy source selection using analytical hierarchy process and quality function deployment: a case study. In: Proceeding Published in Science Technology Engineering and Management (ICONSTEM), IEEE Xplore Digital library (2016)
9. Kaya, T., Kahraman, C.: Multi-criteria renewable energy planning using an integrated fuzzy VIKOR and AHP methodology: the case of Istanbul. Energy **35**(6), 2517–2527 (2010)
10. Article: E-Rickshaw, https://en.wikipedia.org/wiki/Electric_rickshaw
11. West, J.G.W.: DC, induction, reluctance and PM motors for electric vehicles. Power Eng. J. **8**(2), 77–88 (1994)

12. Article: Brushless DC Motor, http://gemsmotor.com/nema-42-brushless-dc-motor
13. Husain, A.: Electric Machines, pp. 497–548. Dhanpat Rai & Company Publications Limited
14. Husain, A.: Electric Machines, pp. 307–416. Dhanpat Rai & Company Publications Limited
15. Lin, H.H., Cheng, J.H.: Design process by integrating DEMATEL, and ANP methods. In: International Conference on Applied System Invention. IEEE, Japan (2018)
16. Niu, D.: Studies on TOPSIS arithmetic with DEMATEL weights scheme. In: Proceedings 2013 International Conference on Mechatronic Sciences, Electric Engineering and Computer (MEC), Shengyang, China (2013)
17. Article: A. Kalra: E-rickshaws legal but aren't regulated (2014). https://wap.business-standard.com/article-amp/current-affairs/r-rickshaws-legal-but-aren-t-regulated-114070200086_1.htm
18. Study: E-Rickshaws operating in Siliguri, West Bengal, India, https://smartnet.niua.org/sites/default/files/resources/assessment_of_e-rickshaw_operations_in_siliguri.pdf
19. A. C. Induction Motor Specifications, https://www.alibaba.com/product-detail/3600rpm-4kW-48V-AC-Motor-for_60857364882.html?spm=a2700.7724857.normalList.26.360e506blCYJoC
20. D. C. Series Motor Specifications, https://www.alibaba.com/product-detail/48V-1-5KW-Electric-DC-Motor_60274220632.html?spm=a2700.7724857.normalList.17.415034a9MmInnD&s=p
21. Bhattacharya, A., Sarkar, B., Mukherjee, S.K.: A new method for plant location selection: a holistic approach. Int. J. Ind. Eng. Theory Appl. Pract. **11**(4), 330–338 (2004)

Design and Analysis of THD with Single-Tuned Filter for Five-Phase DC to AC Converter Drive

Obaidullah Lodin, Inderpreet Kaur and Harpreet Kaur

Abstract The purpose of this article is to reduce the effect of unwanted frequency with single-tuned filter (STF) in five-phase DC to AC converters. This STF filter removes the majority of the unwanted frequencies and signals at the end of alternative voltage or current converters. STF enhanced the design feature of five-phase DC to AC converter by eliminating the multiple of fundamental frequency in power flow to the load. Five phases DC to AC converter design is more useful and provides superior performance than the common three phase in DC to AC converter. This technique is confirmed by making five-phase PWM five-phase DC to AC converter with STF and without STF in MATLAB simulation, and comparison of the obtained outputs is done and tabulated.

Keywords Single-tuned filter · Multiple of fundamental frequency (MFF) or harmonics · Induction motor (IM) · DC to AC converter

1 Introduction

Manufacturers make use of IM profusely because of its numerous benefits, as applied science is advancing the new motor plot are set upped per the user necessities. Modern industrial development conveys the new concepts within the production companies to settle on the most effective designs for them. Current situation witnesses redoubled use of three-phase IM within the production companies everywhere in globe. As scientists engaged on numerous drive designs in ASD exploitation of VSI and CSI, functioning is the main thing that convinces the client, MFF (losses) are foremost impact in PWM DC to AC converters, and this turns to a problem for the research

O. Lodin (✉) · I. Kaur · H. Kaur
Chandigarh University, Mohali, India
e-mail: obaidullah.lodin@gmail.com

I. Kaur
e-mail: hod.eee@cumail.in

H. Kaur
e-mail: harpreetchanni@yahoo.in

© Springer Nature Singapore Pte Ltd. 2020
P. K. Mallick et al. (eds.), *Cognitive Informatics and Soft Computing*,
Advances in Intelligent Systems and Computing 1040,
https://doi.org/10.1007/978-981-15-1451-7_53

scholars to minimize or repress the unwanted signals at the end of the electrical converter drive that starting point for several issues to the load also to the electric converters.

Development within the PE results in the establishment of multi-phase IMs in motorized countries. Since range of phases raised the benefits will increase step by step, three-phase IM is common device; however, five-phase IM is most well-liked than common three-phase IM because of its numerous benefits like decrementing the amplitude and incrementing the frequency of torque pulsation, lowering the rotor current losses, decreasing DC-link current losses and better liability and intense fault tolerance [1].

The approach-orientated analysis magnified nowadays, the most important application of multi-phase IMs is ship thrust, traction (electrical and integrated electrical cars) and also the conception of electrical craft, rail technology and aviation [1–7].

Altogether DC to AC converter the foremost drawback happens once the loss due to unwanted frequencies arises; undesired frequency contents are in current and voltages. DC to AC converter consists of nonlinear components like power electronics devices that produce odd MFF, and it is overlying on the main frequency. The MFF in any power systems causes losses within the components and wires, heating in ASD, LPF issue and torque pulsations in IMs. Many scientists worked on losses to minimize or repress the content of the MFF conjointly with help of filters they decreased the effect of MFF (losses) in electrical converters. In five-phase electrical converter drive, the main MFF (loss) is that the third order. The fifth and each fifth MFF (loss) order is not present within the five-phase DC to AC converter drive. Priority is to cut back these unwanted losses before sending to the consumers.

To eliminate the losses and to vary frequency with phase voltage and amplitude, filters are used, and filters have lots of usage in PE system to refuse and permit specific limits of frequencies. The passive components offer forceful reduction of interrupting signals and total harmonic distortion. A filter could be a device implemented to stop flowing of unwanted frequency from supply to consumers. The MFF (losses) can have an effect on to the less electrical resistance devices. Several of the filters construction is on the market these days, and it consists of parallel linked capacitance to form a less electrical resistance. These devices are going to be accepted as inactive filters. The inactive filters attract unwanted current from the supply linked to the load. Many researchers use inactive filters to suppress the disturbing signals, and inside this article, a passive shunt STF is considered. Benefit is easy to construct and fewer costs to design. A STF consists of both capacitance and inductance. The capacitance values are made in a way so that the circuit electrical resistance reached to the 0. The capacitance is used to compensate the reactive power. A resistance added to eliminate waveforms impurities keeps the range of frequencies [8–17].

2   Construction of STF

STF is often made by focusing on the Q.

$$Q = \sqrt{\frac{\frac{L}{C}}{R}} \qquad (1)$$

Q, L, C and R stand for quality factor, inductance, capacitance and resistance one by one.

Whereas in the construction of STF, the resonance condition must satisfy (2).

$$X_C = X_L \qquad (2)$$

where $X_C = \frac{1}{\omega C}$ and $X_L = \omega L$, $\omega = 2\pi f$, $f = $ fundamental frequency [18–24].

3   Five-Phase DC to AC Converter

Five-phase DC to AC converter made by using ten insulated-gate bipolar transistors. In Figure 1, P1–P10 are the pulse generators, switches (1–10) are in on condition for a period of 180° with 72° out of phase. Output of DC to AC converter can be achieved by accurately switching of the insulated-gate bipolar transistors.

Two IGBTs of the top and three IGBTs of the down parts are on, or two IGBTs of the down and three IGBTs of the top are on. The IGBT sequences of the five-phase DC to AC converter are in Fig. 2.

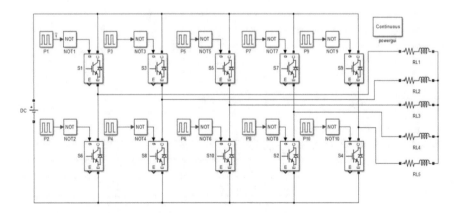

Fig. 1 CKT of five-phase DC to AC converter

Fig. 2 IGBT sequences

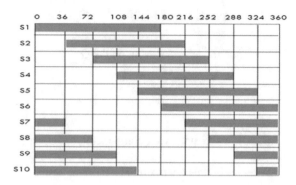

4 Five-Phase DC to AC Converter with STF

Five-phase DC to AC converter drive is connected with STF as shown in Fig. 3. The
STF construction capacitance = 21 µF, inductance = 0.87 and resistance = 86 Ω
(Table 1).

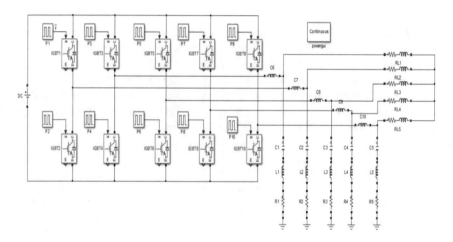

Fig. 3 CKT of five-phase DC to AC converter with STF

Table 1 For STF of 50 Hz

Apparatus	Range
Resistor: R1–R5	86 Ω
Capacitor: C1–C5	21 µF
Inductor: L1–L5	0.87

5 Software Implementation Output

Software implementation is done for two different frequencies of 50 and 9 Hz for rated speed and slow speed, respectively, with resistive and inductive load with resistance $= 1.7 \, \Omega \, \ddot{Y}$ and $L = 1.3 \times 10^{-3}$ H. The V_L is plotted and measured for both with and without STF. Total harmonic distortion is calculated for with and without STF

a. 50 Hz (Input):

The V_L waveforms of DC to AC converter are in Fig. 4.
The V_L waveforms of five-phase DC to AC converter with STF for 50 Hz are in Fig. 5.

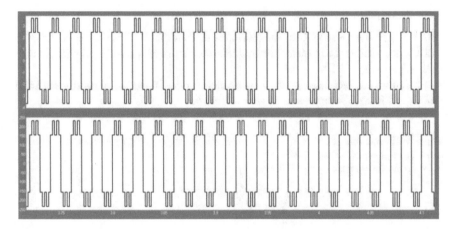

Fig. 4 V_L and I_L graph of PWM converter without STF for 50 Hz

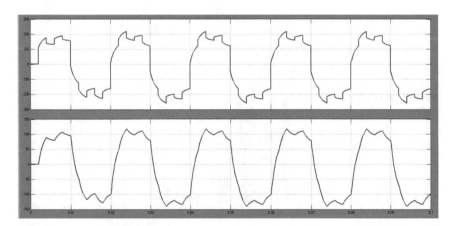

Fig. 5 V_L and I_L waveforms of a phase in five-phase PWM DC to AC converter with STF for 50 Hz

Table 2 Percentage of total harmonic distortion of different filter condition for 50Hz	Type	Total harmonics distortion (%)
	Five-phase DC to AC converter drive (without STF)	42.75
	Five-phase DC to AC converter drive with STF	26.20

The THD in DC to AC converter is presented in Table 2. The FFT studies are done for both with and without STF and shown in Figs. 6 and 7, respectively.

By software implementation, the single MFF order voltage could be evaluated for DC to AC converter with and without STF as shown in Table 3.

b. 9 Hz (Input):

FFT study of five-phase DC to AC converter drive for an I/p 9 Hz and input voltage = 26 V is considered to measure the total harmonic distortion with and without STF (Table 4 and Figs. 8, 9).

Percentage of each MFF order is calculated for the DC to AC converter with and without STF, and the same are shown in Table 5.

Fig. 6 FFT sequences of normal five-phase DC to AC converter for fin = 50 Hz

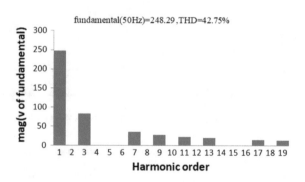

Fig. 7 FFT sequences of normal five-phase DC to AC converter with STF for fin = 50 Hz

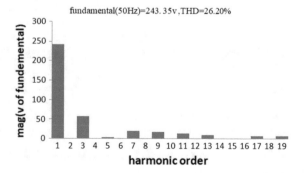

Table 3 Voltage for different harmonics order

Orders of MFF	Five-phase DC to AC converter drive (V)	Five-phase DC to AC converter drive with STF (V)
1st	247.29	244.35
2nd	Zero	Zero
3rd	83.77	57.5
4th	Zero	Zero
5th	Zero	2.14
6th	Zero	Zero
7th	36.46	21.32
8th	Zero	Zero
9th	28.58	18.10
10th	Zero	Zero
11th	23.56	14.67
12th	Zero	Zero
13th	18.0	9.14
14th	Zero	Zero
15th	Zero	0.83
16th	Zero	Zero
17th	13.7	8.62
18th	Zero	Zero
19th	14.07	8.61

Table 4 Percentage of total harmonic distortion of different filter condition for 9Hz

Type	Total harmonics distortion (%)
Five-phase DC to AC converter drive (without STF)	43.28
Five-phase DC to AC converter drive with STF	23.72

Fig. 8 FFT sequences of normal five-phase DC to AC converter for fin = 9 Hz

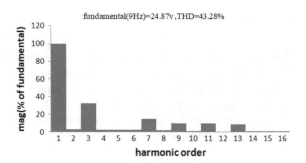

fundamental(9Hz)=24.87v,THD=43.28%

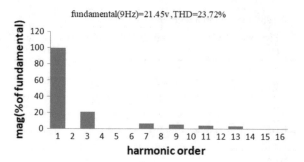

Fig. 9 FFT sequences of five-phase DC to AC converter drive with STF for fin = 9 Hz

Table 5 For fin = 9 Hz

Orders of harmonics	Five-phase DC to AC converter drive (%)	Five-phase DC to AC converter drive with STF (%)
1st	100	100
2nd	2.9	Zero
3rd	31.95	19.83
4th	1.99	Zero
5th	1.98	Zero
6th	1.49	Zero
7th	15.88	5.52
8th	2.86	Zero
9th	9.78	4.84
10th	1.04	Zero
11th	9.63	3.86
12th	0.47	Zero
13th	8.88	3.11
14th	0.78	Zero

6   Conclusion

The five-phase pulse-width modulation drive is made by MATLAB software. The STF is constructed and linked to the output of the AC to DC converter, to study THD of five-phase DC to AC converter drive. The outputs are matched up with five-phase DC to AC converter drive with and without STF, and the outputs show that THD is reduced by using STF. This software implementation practically used to evaluate the temperature of IM in the future.

References

1. Anandhi, T.S., Natarajan, S.P., Senthil Kumar, S., Vijayarajeswaran, R.: Circuit level comparison of single-phase five-level inverters. In: 2005 International Conference on Power Electronics and Drives Systems, Kuala Lumpur, pp. 799–804 (2005)
2. Jyothi, B., Rao, M.V.G.: Comparison of five leg inverter and five phase full bridge inverter for five phase supply. In: 2014 International Conference on Smart Electric Grid (ISEG), Guntur, pp. 1–4 (2014)
3. Wyszynski, A., Schaumann, R.: Frequency and phase tuning of continuous-time integrated filters using common-mode signals. In: Proceedings of IEEE International Symposium on Circuits and Systems—ISCAS'94, London, vol. 5, pp. 269–272 (1994)
4. Das, J.C.: Passive filters—potentialities and limitations. IEEE Trans. Ind. Appl. **40**(1), 232–241 (2004)
5. Ahmed, S.M., Member IEEE, Abu-Rub, H., Senior Member IEEE, Iqbal, A., Senior Member IEEE, Rizwan Khan, M., Payami, S.: A three-to-five-phase matrix converter based five-phase induction motor drive system. Int. J. Recent Trends Eng. Technol. **8**(2) (2013)
6. Kim, N., Kim, M.: Modified direct torque control system of five phase induction motor. J. Electr. Eng. Technol. **4**(2), 266–271 (2009)
7. Bhattacharya, S., Cheng, P.T., Divan, D.M.: Hybrid solutions for improving passive filter performance in high power applications. IEEE Trans. Ind. Appl. **33**(3), 732–747 (1997)
8. Kikovka, S., Tytiuk, V., Ilchenko, O.: Exploring the operational characteristics of a three-phase induction motor with multi-zone stator windings. In: 2017 International Conference on Modern Electrical and Energy Systems (MEES), Kremenchuk, pp. 120–123 (2017)
9. Manjesh, Ananda, A.S.: Harmonics and THD analysis of five phase inverter drive with single tuned filter using Simulink/MATLAB. In: 2016 International Conference on Emerging Technological Trends (ICETT) (2016)
10. Rao, A.M., Sivakumar, K.: Five level single phase inverter scheme with fault tolerance for islanded photovoltaic applications. In: 2015 7th International Conference on Information Technology and Electrical Engineering (ICITEE), Chiang Mai, pp. 194–199 (2015)
11. Tawfeeq, O.T.: THD reduction of a current source rectifier-DC motor drive using single tuned filters. Int. J. Inven. Eng. Sci. (IJIES) **1**(12) (2013). ISSN: 2319-9598
12. Lihitkar, S., Rangari, S.: Comparative study of IPD and POD for three level five phase inverter. In: 2018 International Conference on Smart Electric Drives and Power System (ICSEDPS), Nagpur, pp. 313–316 (2018)
13. Ananda, A.S., Manjesh: Analysis of harmonics in a five phase PWM inverter with LR load and mitigation of harmonics by π filter. In: IEEE 2016 Biennial International Conference on Power and Energy Systems: Towards Sustainable Energy (PESTSE) (2016)
14. Irfan, M.M., Prasad, P.H.K., Rao, P.V.: Simulation of five-level five-phase SVPWM voltage source inverter. In: 2010 International Conference on Power, Control and Embedded Systems, Allahabad, pp. 1–5 (2010)
15. Moinoddin, S., Iqbal, A.: Space vector model of a five-phase current source inverter. In: 2016 Biennial International Conference on Power and Energy Systems: Towards Sustainable Energy (PESTSE), Bangalore, pp. 1–6 (2016)
16. Rangari, S.C., Suryawanshi, H.M., Shah, B.: Harmonic content testing for different stator winding connections of five-phase induction motor. In: 2016 IEEE 6th International Conference on Power Systems (ICPS), New Delhi, pp. 1–5 (2016)
17. Kamel, T., Abdelkader, D., Said, B.: Vector control of five-phase permanent magnet synchronous motor drive. In: 2015 4th International Conference on Electrical Engineering (ICEE), Boumerdes, pp. 1–4 (2015)
18. Jones, M., Levi, E.: A literature survey of state-of-the-art in multiphase ac drives. In: Proceedings of 37th International UPEC, Stafford, U.K., pp. 505–510 (2002)
19. Sakthisudhursun, B., Pandit, J.K., Aware, M.V.: Simplified three-level five-phase SVPWM. IEEE Trans. Power Electron. **31**(3), 2429–2436 (2016)

20. Prasad Rao, K.P., Krishna Veni, B., Ravithej, D.: Five-leg inverter for five-phase supply. Int. J. Eng. Trends Technol. **3**(2) (2012)

21. Padmanaban, S., Blaabjerg, F., Wheeler, P., Lee, K.-B., Bhaskar, M.S., Dwivedi, S.: Five-phase five-level open-winding/star-winding inverter drive for low-voltage/high-current applications. In: 2016 IEEE Transportation Electrification Conference and Expo, Asia-Pacific (ITEC Asia-Pacific), Busan, pp. 066–071 (2016)

22. Chinmaya, K.A., Udaya Bhasker, M.: Analysis of different space vector pulse width modulation techniques for five-phase inverters. In: 2014 Annual IEEE India Conference (INDICON), Pune, pp. 1–6 (2014)

23. Purushothaman, S.K.: A quantitative approach to minimize harmonics elimination using filter design. Int. J. Adv. Res. Comput. Sci. Softw. Eng. **3**(12) (2013)

24. Barik, S.K., Jaladi, K.K.: Performance characteristic of five-phase induction motor with different conduction modes in VSI. In: 2015 IEEE Students Conference on Engineering and Systems (SCES), Allahabad, pp. 1–5 (2015)

Interference Mitigation Methods for D2D Communication in 5G Network

Subhra S. Sarma and Ranjay Hazra

Abstract Interference poses to be a grave threat to wireless communication systems. Quest for higher data rate and higher energy efficiency has led to a paradigm shift. Since the inception of cellular technology in 1980, researchers are in search of a communication system which can exhibit high data rate, low latency, higher capacity, energy efficiency and spectrum efficiency. This led the researchers to explore 5G, operating in mmWave band (30–300 GHz) which is expected to support uncompressed video streaming, 3D telepresence, holographic communication and other multimedia applications. But all these technological advances suffer from a foe, i.e., cross-channel interference and co-channel interference. Thus, this paper chronologically discusses the various models applied to minimize or eradicate the problem of interference in 5G communication systems. Future research directions have also been discussed.

Keywords Interference · D2D communication · 5G

1 Introduction

Mobile communication has become an indispensable part of human life. It affects the social lives in a way like no other things, the way we live, work and play. Thus, with the massive adoption of cellular technology, the requirement for data rate has increased exponentially. With the inception of cellular technology in 1980, it has evolved dramatically. Thus, researchers have been exploring new and novel ideas for enhancing data rate, energy efficiency, latency, spectrum efficiency, etc., which led them to 5G and 6G. 5G operates in mmWave band of 30–300 GHz with licensed bands for mmWave being 28, 38 and 72 GHz [1] for providing better quality of service. 5G

S. S. Sarma (✉) · R. Hazra
Department of Electronics & Instrumentation Engineering, National Institute of Technology Silchar, Silchar, India
e-mail: subhra3s@gmail.com

R. Hazra
e-mail: ranjayhazra87@gmail.com

© Springer Nature Singapore Pte Ltd. 2020
P. K. Mallick et al. (eds.), *Cognitive Informatics and Soft Computing*,
Advances in Intelligent Systems and Computing 1040,
https://doi.org/10.1007/978-981-15-1451-7_54

technology is garnering accolades for multi-gigabit data rate, ultra-low latency and multimedia functionality as spectacular improvement over the prior generations. But in spite of these feats, these technological advancements suffer from a major issue, i.e., cross-channel interference and co-channel interference. Device-to-device (D2D) communication is also no exception to this. So, it is imperative to develop a model to minimize or eradicate the issue of interference from D2D communication systems. Many such methods have been developed using various models and algorithms, such as game theory, machine learning, etc., to reduce interference. Interference can prove to be hazardous for the smooth running of communication systems.

This paper presents an extensive review on the models developed for mitigation of interference for D2D communication in 5G network. The interference considered for the purpose is cross-channel interference and co-channel interference. Various models have been considered for the purpose such as game theory and machine learning. All of these models have been discussed for mitigation of interference from the communication systems. Potential future prospects are also showcased which can prove to be a game changer.

2   D2D Communication: An Overview

This section presents an overview of D2D communication and related interference issue. To begin with, D2D communication may be defined as direct communication among two wireless devices which are in close proximity. For a conventional cellular network, all the information must process through a base station (BS). On the contrary, D2D users can communicate among themselves without traversing the BS as depicted in Fig. 1. Thus, it exhibits certain advantages such as higher spectral efficiency, higher data rate, low latency, resource fairness and energy efficiency [2]. D2D users are able to access the spectrum through underlay spectrum sharing and overlay spectrum sharing. Underlay spectrum sharing proposes to use cellular spectrum for both D2D and cellular communications, whereas overlay spectrum sharing dedicates a part of cellular resources only to D2D communication [3]. Overlay spectrum sharing provides a flexible way to access the spectrum, but it has its own disadvantages of low efficiency and the inability to utilize the spectrum properly. Thus, underlay approach provides a relatively better efficiency. However, signaling overhead becomes higher, and the need for more channel state information becomes complicated in underlay approach.

Interference in D2D communication can be classified into cross-channel interference and co-channel interference. Cross-channel interference may be defined as the interference among users present in different channels, for example, cellular user interfering with D2D user. Again, co-channel interference is the interference between two users of the same frequency. Overlay spectrum sharing can totally eradicate cross-channel interference. However, underutilization of the spectrum poses to be an acute problem. On the other hand, interference must be properly managed for underlay

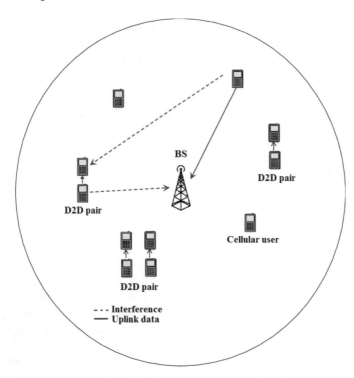

Fig. 1 D2D communication underlaying cellular network including cellular communication

spectrum sharing since it may jeopardize the functioning of D2D communication if not handled properly [4].

3   Issues of 5G

This section portrays the features of 5G and also the issues related to it. The emergence of 5G facilitates the demand for high data rates required for uncompressed video streaming, interactive gaming, etc. Potential interferences that pose to be a threat to communication system are among local D2D communication and global D2D communication. Local D2D communication may be defined as the path between two devices within same BS. Global D2D communication may be defined as the path between two devices associated with different BS. mmWave has relatively low multi-user interference due to directional antenna [5–8]. The advantages of mmWave communication are huge bandwidth (of up to GHz), higher throughput and enormous capacity which make it a suitable candidate for the upcoming generation of wireless communication. The key features of mmWave are higher propagation loss in comparison with microwave band, shorter wavelength and limited penetration capability. Shorter wavelength causes problem in diffraction, whereas limited penetration capability restricts the usage of D2D communication to indoor environments only.

4 Different Models for Interference Management

In this section, we discuss several models used for mitigation of interference for D2D communication.

4.1 Game Theory Model

4.1.1 Pricing-Based Model

Presence of intra-layer interference and inter-layer interference may prove to be hazardous for the better performance of the communication system. Thus, a pricing-based noncooperative game-theoretic method has been used for mitigation of interferences. The base station (BS) serves as a supervisor, and D2D pairs are treated as subordinates. It then sets a price for interference occupied by the cellular users for each subchannel based on the throughput requirements. The D2D pairs are then made to compete for the subchannels for various power allocation strategies as noncooperative Nash game. The prices on each subchannel get raised if there is any increase in the inter-layer interference incurred by D2D pairs until QoS of cellular users is satisfied. Maintaining the quality of service (QoS) of both D2D and cellular users is of paramount importance. Thus, pricing strategy should be designed with utmost care, following which, the existence and uniqueness of Nash equilibrium should be investigated so as to satisfy the data requirements of D2D pairs [9].

4.1.2 Mean Field Game for Interference and Energy Control

The idea of mean field characterizes the two-dimensional space-time dynamics of context. It ensures that the generic player makes an optimal decision based on the mean field in contrast to the strategies of all the other players. It provides an improved performance when simulated under interference dynamics. The generic player remains unaffected by the interactive impacts from other players but only the mass impact. Thus, for a particular time instant, mean field can be defined as the probability distribution of the states over the set of all players. The algorithm initializes the related parameters of state dynamics and also transmission power of D2D pairs. The mean field is calculated by predefined threshold parameters. Hamilton–Jacobi–Bellman (HJB) equation controls the interaction of players with mean field, whereas Fokker–Planck–Kolmogorov (FPK) equation defines the movement of mean field junction in accordance with the player's action. Finite difference method is applied to solve these equations. Through this updation procedure, power control mechanism has been achieved [10–12].

4.1.3 Stackelberg Game Model

Apart from the advantages of introducing ultra-dense network, it depicts new interference problems exhibiting performance deterioration. In [13], a Stackelberg game model has been applied to solve the uplink interference problem of macrocells. Macrocell user equipment (MUE) and remote user equipment (RUE) constitute the leader and the follower, respectively, in the game. So as to increase its own spectral efficiency, MUEs propose a power price which in turn is accepted by RUEs and thereby adjusts the transmit power by maximizing the energy efficiency. In accordance with power price from MUE, all RUEs constitute a noncooperative game achieving Nash equilibrium through iteration. Profit of MUEs is achieved by the combination of its own communication rate and power pricing.

4.1.4 Noncooperative Game Model

An important QoS requirement is interference threshold which ensures optimal operation of cellular user equipments. Spectrum reuse enhances intra-layer interferences. In [14], a noncooperative game is considered for maximizing each user equipment energy efficiency under stringent interference-limited condition. Each player in this game counts for its own payoff and thereby maximizes its own energy efficiency. With the increase in number of user equipments, computational complexity and signaling overhead also increase. Karush–Kuhn–Tucker (KKT) conditions are employed for finding optimal solutions, thereby arising two subproblems, i.e., maximizing power allocation problem and finding corresponding Lagrange multiplier. In the maximization problem, the only thing needed for each D2D pair is to calculate the interference on all available channels instead of knowing specific strategies of other user equipments for determining power optimization. Updation of information would eventually converge to Nash equilibrium which also gives the price of anarchy. It may be defined as ratio of sum energy efficiency of overall network to the energy efficiency achieved at its worst-case equilibrium.

4.1.5 Exact Potential Game Model

Game-theoretic approaches have found extensive usage in mitigation of interferences from communication systems. Most of the researches carried out have solved the power control problem. In [15], joint problem of resource block and power allocation in a D2D underlay cellular network has been considered.

Interference minimization approach is adopted for all the links in the network. To solve the issue of resource block allocation, noncooperative game has been formulated and then proved to be an exact potential game. A number of Nash equilibrium points may exist which can coincide with points maximizing potential function. The desired Nash equilibrium point is solved by polynomial local search (PLS).

4.2   Machine Learning Model

4.2.1   Cooperative Online Learning Model

Q-learning is extensively used nowadays in the field of communication for its own set of advantages. Online learning uses reinforcement Q-learning for determining optimal policy for decision making. It includes four parameters, viz. state, action, probabilistic transition function from one state to the other and reward function. The reward function monitors the feedback for the quality of the action taken which in turn helps the system in gaining experience. An agent may be either a femto base station (FBS) or a D2D transmitter (DUE), and only agents of the same type can coordinate among themselves. The system state is defined as information gathered by the respective agent. It enforces the learning using global reward. Since the model is considered cooperative, Q-table is sent to all neighbors, and their state-action information is collected. Thus, if a certain agent maximizes its Q-value, global Q-value would be maximized. For heterogeneous environment, which requires a plethora of interactions, online learning provides a convenient way to make decisions regarding allocation [16].

4.2.2   Hierarchical Extreme Learning Model

Due to multiple interactions, Q-learning algorithm is time consuming and thus in need of a more sophisticated algorithm which yields better output keeping time consumption limited. In [17], hierarchical extreme learning model (HELM) has been used to minimize the incurred interference. It is an extension of extreme learning machine (ELM), an upcoming algorithm beneficial for generalized single-hidden layer feed-forward neural networks. Helm is supervised learning which requires lot of labeled data in order to train itself. But the application time needed is very less compared to Q-learning. As it is a known fact that with due time, the number of D2D users is growing exponentially, so HELM provides a better alternative with respect to limited time consumption.

4.2.3 Decision Tree Classifier Model

In hybrid underlaying cellular and D2D users for uplink transmission [18], generally three kinds of interference can be found, and they are interference of D2D to cellular links, interference of cellular to D2D links and interference between D2D users. It has been observed that Q-learning does not need any prior information but it suffers from the drawback of time consumption. So, a data-driven tree classifier algorithm has been formulated so that D2D power control issue can be solved by maximizing either the system capacity or the energy efficiency keeping the QoS of cellular users intact. Distributed-Q algorithm has been adopted for achieving global optimal policy and uses the output to generate training sample for Decision Tree.

4.3 Miscellaneous Model

4.3.1 Cucker–Smale Flocking Model

According to this method, an agent adjusts its velocity relative to its own velocity and also the weighted mean of the relative velocities of the other members of the group [19]. This theory has been applied to vehicle-mounted mobile relays (VMR) for fair resource allocation. The algorithm enables each node to adjust transmission power to make its own desirable bandwidth equal to mean of its neighbor's desirable bandwidths. A fair bandwidth allocation can thus be obtained in a completely distributed manner. However, the higher count of users may affect the system performance.

5 Potential Future Prospects

Widespread research has been done to minimize or mitigate the effects of interference from the D2D communication operating in 5G and has achieved substantial progress in this area. But research can be pursued in many directions with different approaches. Thus, few future prospects are furnished below.

5.1 Noncooperative Model with Incomplete CSI

Most of the models consider the channel state information (CSI) to be known, and a cooperative model is developed. But in practical scenario, knowing the CSI for all the links is not possible, and each service providing vendors may not disclose all their parameters for monetary reasons. Thus, it is fruitful to work on a noncooperative model where incomplete information will be known to the BS.

5.2 Enhanced Learning Method on Interference Dynamics

The machine learning methods considered in the review do not take into account the interference dynamics of 5G mmWave network. Since a number of attenuation factors come into play in this specific band such as buildings, walls, environmental factors, etc., it is advisable that a robust model should be developed which can counter the various random interferences that may occur.

6 Conclusion

In recent years, game theory and machine learning are finding use in designing models for mitigation of interference in D2D-based 5G communication systems. Several models have been developed to control the interference so that the QoS of the system can be maintained above the threshold. Interference problem hampers the energy efficiency, transmission power and spectrum efficiency. In this article, we have provided a detailed discussion on the methods undertaken for minimizing the interference. These models have been categorized on the basis of game theory, machine learning and miscellaneous method. Table 1 summarizes all the models developed and their achieved performances. It is found that only few models consider random interference dynamics. We have also presented a few research directions that can be carried out in the future.

Table 1 Summary of various methods of interference mitigation

References	Analytical tool	Direction	Achieved performance	Published year
[9]	Pricing-based noncooperative game	Uplink	High convergence probability	2015
[10]	Mean field game	Uplink/downlink	Increased battery average lifetime	2018
[11]	Mean field game	Uplink/downlink	Higher energy efficiency	2017
[13]	Stackelberg game	Uplink	Controlled interference with higher energy efficiency	2016
[14]	Noncooperative game	Uplink	Significant energy efficiency subjected to little spectrum efficiency loss	2014

(continued)

Table 1 (continued)

References	Analytical tool	Direction	Achieved performance	Published year
[15]	Exact potential game	Uplink	Higher spectrum efficiency	2017
[16]	Cooperative online learning model	Uplink/downlink	Higher throughput and spectral efficiency	2016
[17]	Hierarchical extreme learning model	Uplink	Reduced time complexity	2018
[18]	Decision Tree classifier model	Uplink	Reduced time complexity with higher energy efficiency and system capacity	2017
[19]	Cucker–Smale flocking model	Uplink/downlink	Fair bandwidth allocation	2019

References

1. Swetha, G.D., Murthy, G.R.: Selective overlay mode operation for d2d communication in dense 5g cellular networks. In: Computers and Communications (ISCC) 2017 IEEE Symposium on IEEE, pp. 704–709 (2017)
2. Venugopal, K., Valenti, M.C., Heath, R.W.: Device-to-device millimeter wave communications: interference, coverage, rate, and finite topologies. IEEE Trans. Wirel. Commun. **15**(9), 6175–6188 (2016)
3. Yu, S., Ejaz, W., Guan, L., Anpalagan, A.: Resource allocation schemes in D2D communications: overview, classification, and challenges. Wirel. Pers. Commun. **96**(1), 303–322 (2017)
4. Zhang, R., Cheng, X., Yang, L., Jiao, B.: Interference graph-based resource allocation (InGRA) for D2D communications underlaying cellular networks. IEEE Trans. Veh. Technol. **64**(8), 3844–3850 (2014)
5. Qiao, J., Shen, X.S., Mark, J.W., Shen, Q., He, Y., Lei, L.: Enabling device-to-device communications in millimeter-wave 5G cellular networks. IEEE Commun. Mag. **53**(1), 209–215 (2015)
6. Wei, L., Hu, R.Q., Qian, Y., Wu, G.: Key elements to enable millimeter wave communications for 5G wireless systems. IEEE Wirel. Commun. **21**(6), 136–143 (2014)
7. Li, L., Niu, X., Chai, Y., Chen, L., Zhang, T., Cheng, D., Xia, H., Wang, J., Cui, T., You, X.: The path to 5G: mmWave aspects. J. Commun. Inf. Netw. **1**(2), 1–18 (2016)
8. Rappaport, T.S., Sun, S., Mayzus, R., Zhao, H., Azar, Y., Wang, K., Wong, G.N., Schulz, J.K., Samimi, M., Gutierrez, F.: Millimeter wave mobile communications for 5G cellular: it will work! IEEE Access **1**, 335–349 (2013)
9. Yin, R., Yu, G., Zhang, H., Zhang, Z., Li, G.Y.: Pricing-based interference coordination for D2D communications in cellular networks. IEEE Trans. Wirel. Commun. **14**(3), 1519–1532 (2015)
10. Yang, C., Li, J., Sheng, M., Anpalagan, A., Xiao, J.: Mean field game-theoretic framework for interference and energy-aware control in 5G ultra-dense networks. IEEE Wirel. Commun. **25**(1), 114–121 (2017)

11. Yang, C., Li, J., Semasinghe, P., Hossain, E., Perlaza, S.M., Han, Z.: Distributed interference and energy-aware power control for ultra-dense D2D networks: a mean field game. IEEE Trans. Wireless Commun. **16**(2), 1205–1217 (2016)
12. Mkiramweni, M.E., Yang, C., Li, J., Han, Z.: Game-theoretic approaches for wireless communications with unmanned aerial vehicles. IEEE Wirel. Communs. **25**(6), 104–112 (2018)
13. Gu, X., Zhang, X., Zhou, Z., Cheng, Y., Peng, J.: Game theory based interference control approach in 5g ultra-dense heterogeneous networks. In: Asia-Pacific Services Computing Conference, pp. 306–319. Springer, Cham (2016)
14. Zhou, Z., Dong, M., Ota, K., Shi, R., Liu, Z., Sato, T.: Game-theoretic approach to energy-efficient resource allocation in device-to-device underlay communications. IET Commun. **9**(3), 375–385 (2015)
15. Katsinis, G., Tsiropoulou, E.E., Papavassiliou, S.: Joint resource block and power allocation for interference management in device to device underlay cellular networks: a game theoretic approach. Mob. Netw. Appl. **22**(3), 539–551 (2017)
16. AlQerm, I., Shihada, B.: A cooperative online learning scheme for resource allocation in 5G systems. In: 2016 IEEE International Conference on Communications (ICC), pp. 1–7 (2016)
17. Xu, J., Gu, X., Fan, Z.: D2D power control based on hierarchical extreme learning machine. In: 2018 IEEE 29th Annual International Symposium on Personal, Indoor and Mobile Radio Communications (PIMRC), pp. 1–7 (2018)
18. Fan, Z., Gu, X., Nie, S., Chen, M.: D2D power control based on supervised and unsupervised learning. In: 2017 3rd IEEE International Conference on Computer and Communications (ICCC), pp. 558–563 (2017)
19. Park, J., Choi, H.H., Lee, J.R.: Flocking-inspired transmission power control for fair resource allocation in vehicle-mounted mobile relay networks. IEEE Trans. Veh. Technol. **68**(1), 754–764 (2019)

Design of PWM Triggered SEPIC Converter Using Zero-Voltage Zero-Current Switching

Sagar Pradhan, Dibyadeep Bhattacharya and Moumi Pandit

Abstract This paper presents the design, simulation and fabrication of single-ended primary-inductance converter (SEPIC) model and analyzes the results of implementation of zero-voltage and zero-current switching (ZV-ZCS). ZV-ZCS SEPIC converter has been simulated in MATLAB in buck as well as boost mode. A fabricated model of the SEPIC converter has been designed to be operated in ZV-ZCS mode. This converter has been designed with an aim of reducing power losses during switching of higher frequency devices, such that the overall efficiency of the system can be improved.

Keywords SEPIC · Zero-voltage · Zero-current · PWM

1 Introduction

With increasing demand in the power sector, converters play a very important role. But there is a major drawback of using converters as the switching losses involved decreases the efficiency of the system. There are various techniques available like hard switching, snubber, ZVS and ZCS separately, etc. This paper presents a combination of both ZV-ZCS technique with the aim of minimizing the switching loss in single-ended primary-inductor converter (SEPIC) converter.

SEPIC is a DC/DC converter that can increase or decrease the dc voltage without producing an inverted output. The SEPIC topology is integrated with ZV-ZCS technique which will increase efficiency of the converter due to very low switching losses and non-inverted output compared to conventional Buck-Boost converter. Buck-Boost converters are often less costlier compared to SEPIC converters as they consist of single capacitor and inductor, but on the other hand, Buck-Boost converter has a drawback of having high input current ripple which creates harmonics which will lead converters to add a LC filter which in turn makes converter inefficient and costly [1].

S. Pradhan · D. Bhattacharya · M. Pandit (✉)
Electrical and Electronics Department, Sikkim Manipal Institute of Technology, Rangpo, India
e-mail: moumi.p@smit.smu.edu.in

© Springer Nature Singapore Pte Ltd. 2020   531
P. K. Mallick et al. (eds.), *Cognitive Informatics and Soft Computing*,
Advances in Intelligent Systems and Computing 1040,
https://doi.org/10.1007/978-981-15-1451-7_55

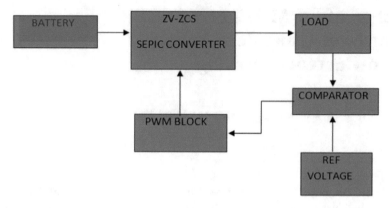

Fig. 1 Block diagram representing closed-loop system

SEPIC converters are also preferred because of not producing inverted output and electrical stress on the components of converter as in CUK converter and Buck-Boost converter which can result in overheating leading to complete damage of the components.

2 Block Diagram of a Model

Figure 1 shows the block diagram which represents a complete closed-loop model of a SEPIC converter. SEPIC converters are often used for regulation of power [2] as it can either increase or decrease the voltage according to the requirement of the load.

3 MATLAB Simulation of Conventional SEPIC Converter

3.1 SEPIC Converter Operating in Buck Mode

Figure 2 represents a MATLAB simulation of a SEPIC converter operating as Buck converter where the input voltage is of 12 V DC and the output observed is 4.8 V DC. Where,

Duty ratio (D) = 30%,
T_{on} period = 9.375×10^{-6} s,
T_{off} period = 2.1875×10^{-6} s,
T (total time) = 3.125×10^{-5} s,
F_s (Frequency) = 32 kHz.

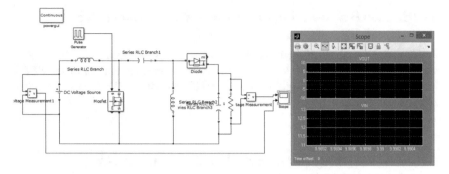

Fig. 2 Simulation of conventional SEPIC converter operating in buck mode

3.2 SEPIC Converter in Boost Mode

Figure 3 refers to the simulation design of SEPIC converter in boost mode.

Both of the simulation processes of SEPIC converter are done in MATLAB, and the only the duty ratio of the PWM block is maintained for both of the modes [3]. For operating the converter in the buck mode, the duty ratio is kept below 50% as of the requirement, and for the boost mode of operation, the duty ratio is kept above 50%. Where,

Duty ratio $(D) = 70\%$,
T_{on} period $= 2.1875 \times 10^{-5}$ s,
T_{off} period $= 9.38 \times 10^{-6}$ s,
T (total time) $= 3.125 \times 10^{-5}$ s,
F_s (Frequency) $= 32$ kHz.

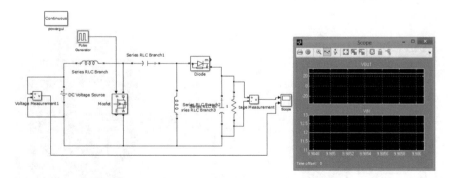

Fig. 3 MATLAB simulation of conventional SEPIC converter in boost mode

4   MATLAB Simulation of ZV-ZCS SEPIC Converter

Performance of the SEPIC converter can be enhanced by implementing ZV-ZCS switching mode. Therefore, to analyze the performance, simulation of the SEPIC has been done in MATLAB platform by switching the MOSFET in zero-voltage and zero-current mode.

4.1   ZV-ZCS Mode Implemented in SEPIC Converter Operating in Buck Mode

In zero-voltage and zero-current switching of any converter whether the operation is for a buck mode or boost mode, switching losses compared to other techniques in ZV-ZCS are comparatively less [4]. Zero voltage and zero current take place according to the pulse triggered to the switches.

In above circuit, SEPIC converter consists of two switches (MOSFETS), where MOSFET 1 is triggered ON/OFF with a duty cycle as required as for the application, and the other MOSFET which is MOSFET 2 is triggered with PWM signal and also a controlled duty cycle to realize a zero voltage and zero current in the main switch, which is MOSFET 1 in the above circuit.

Figure 4 shows a MATLAB simulation of ZV-ZCS SEPIC converter, where the input is of 24 V dc, and the output observed is of 20 V dc. Where,

Duty ratio $(D) = 12\%$,
T_{on} period $= 3.972 \times 10^{-6}$ s,
T_{off} period $= 2.9128 \times 10^{-5}$ s,
T (total time) $= 3.31 \times 10^{-5}$ s,
F_s (Frequency) $= 32$ kHz.

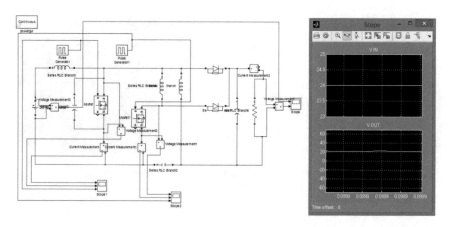

Fig. 4   MATLAB simulation of ZV-ZCS SEPIC converter in buck mode

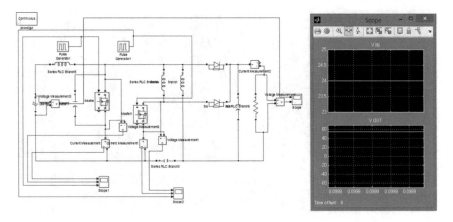

Fig. 5 MATLAB simulation of ZV-ZCS SEPIC converter in boost mode

4.2 ZV-ZCS SEPIC Converter in Boost Mode

Figure 5 represents a ZV-ZCS SEPIC converter in boost mode, where input voltage 24 V is boosted to 48 V dc. Operation of the ZV-ZCS SEPIC converter in both the boost and buck mode is kept same except the variation in duty cycle of both the MOSFETS [5].

As seen in the buck mode operation of the converter, MOSFET 1 will decide the output as required to that of the application and if the duty cycle is below 50%, then there will be a buck mode operation for the converter, but if the duty cycle is above 50%, then there will be a boost mode of operation. Hence, MOSFET 2 will be used for the realization of (ZV-ZCS) in for the main switch which is MOSFET 1. Applying a method to realize zero-voltage and zero-current switching (ZV-ZCS) in converters will increase the efficiency of the converter and the whole system.

Thus, to minimize switching losses and improve the overall efficiency of the system, a method to realize ZV-ZCS is a very efficient method, and PWM triggering of any switch in our case MOSFET will have an added benefit which will in turn improve the efficiency of the converter and also decrease the ripples in the output of the converter. Use of this method of switching and PWM triggering is not just advantageous in this project but beneficial in other converters also [6].

Where,

Duty ratio $(D) = 54\%$,
T_{on} period $= 1.7874 \times 10^{-5}$ s,
T_{off} period $= 1.5226 \times 10^{-5}$ s,
T (total time) $= 3.31 \times 10^{-5}$ s,
F_s (Frequency) $= 32$ kHz.

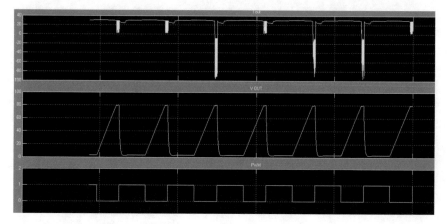

Fig. 6 Output of a MOSFET showing ZV-ZCS switching

5 MOSFET Output Showing a ZV-ZCS Switching

Implementation of ZV-ZCS in a main switch is observed by the duty cycle in the secondary switch which is MOSFET 2 in both of the modes (buck and boost) [7]. Figure 6 represents a PWM signal, a voltage signal and a current signal of the main switch which is MOSFET 1. Here, when the PWM signal is high, it can be seen that voltage signal and current signal are toward zero or close to zero. Switching at zero voltage and zero current or close to zero has increased the efficiency and decreased the switching losses of the converter.

6 Experimental Setup and Design Analysis of SEPIC Converter

An experimental setup for a SEPIC converter was designed and tested with all the values of parasitic elements similar to that of the MATLAB simulated SEPIC converter. Results obtained were exact to that of the simulation result for the conventional SEPIC converter, when the duty cycle of PWM signal given to converter was below 50%, buck mode was achieved, and the same converter can be operated in boost mode by increasing the duty cycle above 50% as that of the simulation result.

7 Experimental Output of the SEPIC Converter

A full analog PWM signal and a DC output are observed from the experimental setup shown in Fig. 7. As seen in Fig. 8, a non-inverted dc output is obtained from

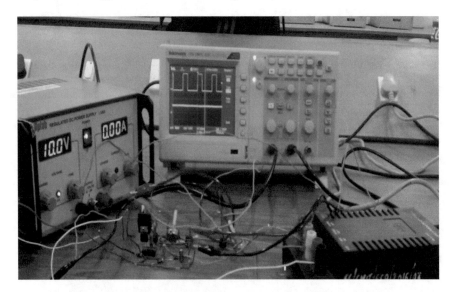

Fig. 7 Experimental setup of PWM triggered SEPIC converter

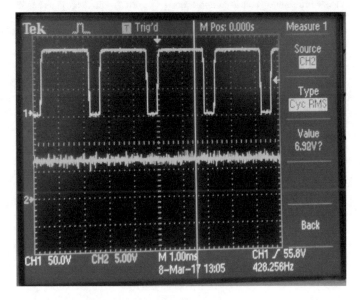

Fig. 8 1. PWM output, 2. DC output of SEPIC converter

the SEPIC converter [8], and by changing the duty cycle of the PWM signal, the output voltage of the converter is maintained as required by the application.

8 Conclusion

A ZV-ZCS SEPIC converter was simulated and designed using MATLAB, an experimental setup for the SEPIC converter was designed using a full analog model of PWM triggering to the MOSFET, and its outputs were observed and are evident from the hardware waveforms.

The theoretical analysis and the practical operation of this converter have been validated using simulation platform and fabricated prototype in hardware using PCB. As per the experimentation results, the two switches of the SEPIC converter, i.e., primary and the auxiliary switches were turned on at zero voltage and zero current. Thus, the output obtained was as expected and proved that by implementing a zero-voltage and zero-current switching improves the efficiency of the converter.

References

1. Arunkumar, G., Elangovan, D., Patra, J.K., Tania, H.M., James, C., Vats, S.: A solar based SEPIC converter for street lighting application. In: 2016 International Conference on Computation of Power, Energy Information and Communication (ICCPEIC), Apr 2016, pp. 482–486. IEEE
2. De Freitas, L.C., Gomes, P.C.: A high-power high-frequency ZCS-ZVS-PWM buck converter using a feedback resonant circuit. In: Proceedings of IEEE Power Electronics Specialist Conference-PESC'93, June 1993, pp. 330–336. IEEE
3. Lee, F.C.: High-frequency quasi-resonant converter technologies. Proc. IEEE **76**(4), 377–390 (1988)
4. Wang, K., Hua, G., Lee, F.C.: Analysis, design and experimental results of ZCS-PWM boost converters. In: Proceeding of the International Power Electronics Conference, Jan 1995
5. Panda, A.K., Aroul, K.: A novel technique to reduce the switching losses in a synchronous buck converter. In: 2006 International Conference on Power Electronic, Drives and Energy Systems, Dec 2006, pp. 1–5. IEEE
6. Cory, M.: Conventional and Zvt Synchronous Buck Converter Design, Analysis, and Measurement (2010)
7. Biswas, R.: Active power filter based on interleaved buck converter (Doctoral dissertation) (2013)
8. Lakshmi, D., Zabiullah, S., Venu Gopal, D.: Improved step down conversion in interleaved buck converter and low switching losses. Int. J. Eng. Sci. **4**, 15–24 (2014)

Array Radar Design and Development

**Subhankar Shome, Rabindranath Bera, Bansibadan Maji,
Akash Kumar Bhoi and Pradeep Kumar Mallick**

Abstract Array antenna technology is well established in communication which is mainly used for beam formation and beam sharpening. A narrow beam creation will increase the antenna gain towards a specific direction which is very helpful in target detection using modern Radar. Array antenna is consisting of several antenna elements, which help to create beam formation in space. Three types of beam formation are possible using array formation they are, Analog Beamforming, Digital Beamforming, and Hybrid Beamforming. Digital beamforming is little bit low-cost solution, where active beam rotation is not required, in case of analog and hybrid beam formation active phase shifter, amplifiers are required in behind of each antenna element to rotate the physical beam in the desired direction which is a little costly solution. In this article digital beamforming based array radar is simulated for moving target detection which may be very much helpful for automotive application. This kind of radar system may help the vehicular industry to produce the automated driverless car.

S. Shome (✉) · R. Bera
Department of Electronics & Communication Engineering, Sikkim Manipal Institute of
Technology, Sikkim Manipal University, Gangtok, Sikkim, India
e-mail: subho.ddj@gmail.com

R. Bera
e-mail: rbera50@gmail.com

B. Maji
National Institute of Technology Durgapur, Mahatma Gandhi Rd, A-Zone, Durgapur, West
Bengal 713209, India

A. K. Bhoi
Department of Electrical & Electronics Engineering, Sikkim Manipal Institute of Technology,
Sikkim Manipal University, Gangtok, Sikkim, India
e-mail: akash.b@smit.smu.edu.in

P. K. Mallick
School of Computer Engineering, Kalinga Institute of Industrial Technology (KIIT) University,
Bhubaneswar, India
e-mail: pradeepmallick84@gmail.com

© Springer Nature Singapore Pte Ltd. 2020
P. K. Mallick et al. (eds.), *Cognitive Informatics and Soft Computing*,
Advances in Intelligent Systems and Computing 1040,
https://doi.org/10.1007/978-981-15-1451-7_56

Keywords MIMO · Array · URA · Phased array · Radar

1 Introduction

The dream of an autonomous vehicle is facing a big challenge from the last few decades. All big car manufacturers are trying for it in collaboration with electronic superpowers. But still, engineering is not achieved at that level where a fully auto-mated car can be realized. Only a few levels of partial automation is tested in the world. Car automation is mostly dependent on the use of a huge number of advanced sensors, and the decision making ability by processing the sensors data. Radio sen-sors are one of the most effective sensors are getting used in an autonomous vehicle. In almost, every autonomous vehicle trail Radar is used. Long Range Radar, Medium Range Radar, Short Range Radar, every type is having some different application for car automation. It is almost impossible to automate a car without Radar. Considering this modern Radar development is in very high demand. Researchers are trying to upgrade every Radar section for better moving object detection. Radar baseband is upgraded to digital and several codes are used as a Radar source to get an advantage of pulse compressing technique. Also, continuous transmission upgraded into pulse transmission which required less power than the previous one. In this way latest antenna technologies like MIMO, Array technologies are applied to get the advan-tages like coherency gain, diversity gain, etc. [1–3]. Array technology adds another great advantage which is beamforming into Radar. Using beamform [4, 5], Radar can form a beam into space towards desired Target. This feature will help Radar to detect the desired Target using the main beam and reject all undesired one by placing it into the null. In this research, an end to end Array Radar simulation is done using System-Vue for moving object ct detection. Ranges, Velocity are calculated for the detected object. Not only had a single target rather a multi-target environment is sim-ulated which is more difficult to detect. System-Vue base simulation empowered us with object-oriented programming which is a little easy to implement. Any number of array elements can be simulated to get a better result. A number of array elements help to create more shape beam towards a target which will improve the detectability of the Radar.

2 Simulation

The complete Array Radar end to end model is simulated for Target detection, Veloc-ity and Range measurement in a multi-target environment. The end to end model mainly consists of Radar Transmitter, Radar RF part in which Array Antennas are simulated as Transmit antenna as well as Receive Antenna. In between Transmit and

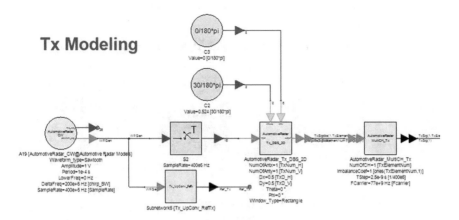

Fig. 1 Radar baseband design

Received antenna Target echo generator is also simulated, from which only transmitted echo signal will be received by the Received Array. After that Radar signal processing are used to find out the target Range and velocity for each target.

In Fig. 1, the Radar baseband simulation is done using a Frequency-Modulated Continuous-Wave (FMCW) signal as a baseband source. Then an Automotive Radar baseband transmitter is used which is having the capability to synthesize the main lobe along the direction determined by theta and phi [6]. This beam synthesis capability is helping the array antenna section to form a beam in the direction of given theta and phi [7]. Then the multi-channel transmitter helps us to simulate the ideal multichannel behaviors. In this block, the transmitted signal is tapped to use as a reference signal in the target detector block of the receiver side (Fig. 2).

Here transmitter signal is facing the simulated channel environment. The main advantage of the Radar is coming under this section, where Array Antenna is used for beamforming in specified Target direction. [4 × 4] antenna elements are used in uniform rectangular array formation for transmitter and received antenna [8]. Target Echo Generation block representing the reflected signal from the target which will be received by array antenna of the receiver side [9] (Fig. 2).

In Receiver side RF components chain are consist of Low Noise Amplifier (LNA), for better detection of low Signal to Noise Ration signal [10]. Then received signal is passed through the correlation based detector which helps for Target detection, in this block an original transmitted FMCW reference is used for correlation. After the Target is detected, a Range and Doppler FFT block is used to find out the Range and velocity of the target which is measured in next section (Fig. 3).

This processed data now passed through the Doppler Constant false alarm rate (CRAF) detection [11]. This is a very important block for Radar detection. In general, the detection is done based on a threshold value which is a function of both the probability of detection and the probability of false alarm. This block is very much

Antenna Models and Echo generation

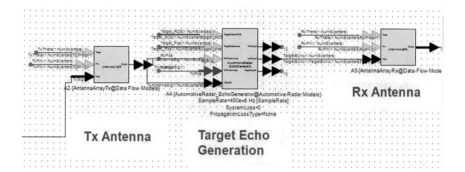

Fig. 2 Radar channel simulation inclusive radar RF and radar target

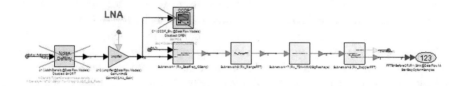

Rx Modeling

Fig. 3 Radar receiver baseband

important to detect the original target out of false alarm. Another Matlab script is implemented to find out the angel and range measurement [12, 13] for multi-target (Fig. 4).

In Fig. 5 the simulation parameters are set. TX and RX platform position, Target positions, are set in this section which has to be estimated in the receiver section. In this simulation radar transmitter and receiver, the platform is Static only the Target is moving. This platform is simulated using TX, RX platform block. Two targets Location block is used to simulate two different targets, which is a Multi-target environment. Target 1 range values in *X*, *Y*, *Z* coordinates are set to [10 0 150] and target 2 range value is set to [0 0 200]. Velocity for target 1 and 2 are the same, which is [0 0 3]. These values are estimated in the receiver and outputs are shown in the result section.

DSP-Range, Velocity Measurements

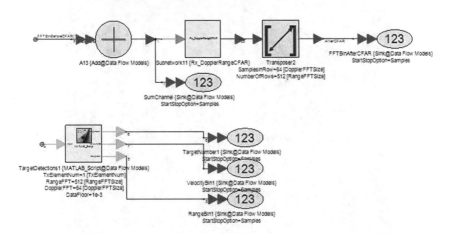

Fig. 4 Receiver signal processing for parameter estimation

Radar Tx Platform　　　　**Targtes' Location**　　　　**Radar Rx Platform**

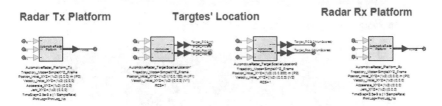

Fig. 5 Radar parameter settings

3   Results

Figure 6 showing the Multi-target detection without CRAF processing, after range FFT, target are placed visible in range but it is included with false alarm.

In Fig. 7, the effect of CFAR processing is clearly visible which increase the detection ability. Also it's suppressing the noise floor up to certain level.

In Fig. 8, the Target Position is placed in X, Y coordinate. The measure coordinates value is shown in Fig. 9. The given target range was [150] for target 1 and [100] for target 2, which is measure in shown in Fig. 9.

The measured range are showing 150.75 for target 1 and 200.25 for target 1 and the target velocity is measured in Fig. 10, which is 2.984 for both the target. The given target velocity was 3. So for the measurements of range and velocity for multi-target

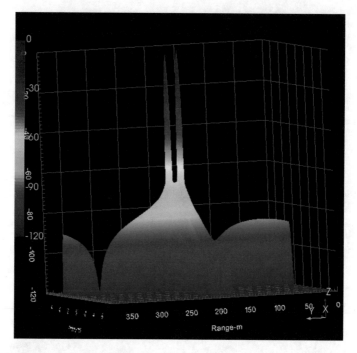

Fig. 6 Multi-target in range without CFAR

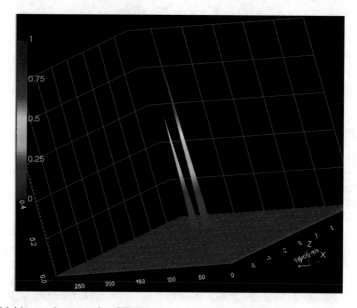

Fig. 7 Multi-target in range after CFAR

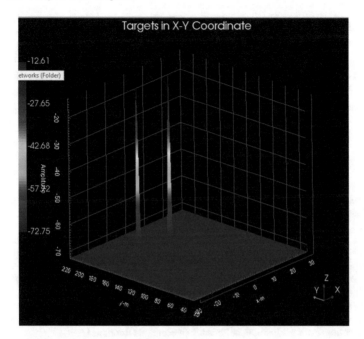

Fig. 8 Target detection and target position in *X, Y, Z* coordinate

Fig. 9 Measured target range for multi target

TargetRanges_Index	TargetRanges[1,1]	TargetRanges[1,2]
0	150.75	200.25

Fig. 10 Measured target velocity for multi target

TargetVelocties_Index	TargetVelocties[1,1]	TargetVelocties[1,2]
0	2.984	2.984

is very accurate using designed array radar, which may be very much effective for moving target like a car.

4 Conclusion

The simulation results are able to prove the performance of Array Antenna based radar system, which may be very good for Long range and medium range Radar in case of autonomous vehicular application. The accuracy level achieved in the measured parameter of the car in the road scenario will help the system to avoid collision and will increase the stability of the autonomous vehicle. Array antenna based beamforming is one of the key factors to achieve this accuracy, as it will provide a consistency main lobe towards the target. Even this beam formation is the 1st step to track the desired vehicle which may invoke another interesting which is Adaptive cruise control (ACC) for an autonomous vehicle.

References

1. Patole, S.M., Torlak, M., Wang, D., Ali, M.: Automotive radars: a review of signal processing techniques. IEEE Sig. Process. Mag. **34**(2), 22–35 (2017). https://doi.org/10.1109/MSP.2016.2628914
2. Fu, H., Fang, H., Cao, S., Lu, M.: Study on the comparison between MIMO and phased array antenna. In: IEEE Symposium on Electrical & Electronics Engineering (EEESYM). Kuala Lumpur (2012)
3. El-Din Ismail, N., Mahmoud, S.H., Hafez, A.S., Reda, T.: A new phased MIMO radar partitioning schemes. In: IEEE Aerospace Conference. Big Sky, MT (2014)
4. Van Trees, H.L.: Optimum Array Processing. Wiley-Interscience, New York (2002)
5. Pandey, N.: Beamforming in MIMO Radar. Department of Electronics and Communication Engineering, National Institute of Technology Rourkela Rourkela (2014)
6. Muralidhara, N., Rajesh, B., Biradar, R.C., Jayaramaiah, G.V.: Designing polyphase code for digital pulse compression for surveillance radar. 2nd International Conference on Computing and Communications Technologies (ICCCT), Chennai (2017)
7. Zalawadia, K.R., Doshi, T.V., Dalal, U.D.: Study and design narrow band phase shift beamformer. In: 2013 International Conference on Intelligent Systems and Signal Processing (ISSP), pp. 238–241. Gujarat (2013)
8. https://doi.org/10.1109/issp.2013.6526910
9. Mucci, R.: A comparison of efficient beamforming algorithms. In: IEEE Trans. Acoust. Speech Sig. Process. **32**(3) (1984)
10. Kiruthika, R., Shanmuganantham, T.: Design and measurement of novel shaped microstrip antenna with DGS for radar applications. Int. J. Wirel. Microw. Technol. (IJWMT) **8**(3), 33–41 (2018). https://doi.org/10.5815/ijwmt.2018.03.04
11. Vera-Dimas, J.G., Tecpoyotl-Torres, M., Vargas-Chable, P., Damián-Morales, J.A., Escobedo-Alatorre, J., Koshevaya, S.: Individual Patch Antenna and Antenna Patch Array for Wi-Fi Communication. Center for Research of Engineering and Applied Sciences (CIICAp), Autonomous University of Morelos State (UAEM), 62209, Av. Universidad No. 1001, Col Chamilpa, Cuernavaca, Morelos, México (2010)
12. Ghosh, C.K., Parui, S.K.: Design, analysis and optimization of a slotted microstrip patch antenna array at frequency 5.25 GHz for WLAN-SDMA system. Int. J. Electr. Eng. Inf. **2**(2) 2010
13. Wehner, D.R.: High-Resolution Radar, 2nd edn. ArtechHouse, Norwood, MA (1995)

Design of Fuzzy Controller for Patients in Operation Theater

Mohan Debarchan Mohanty and Mihir Narayan Mohanty

Abstract In recent days, technology develops for the application in versatile fields. Mostly, soft computing techniques enhance the technological aspects in the field of engineering and medicine. In this paper, fuzzy logic-based controller is designed to support the surgeons in operation theater. At the time of surgical operation, the anatomical parameters are to be well controlled especially blood sugar and blood pressure. It provides intelligent control applied in the field of medicine. A fuzzy logic controller for mean arterial pressure (MAP) control is considered as limit for the depth of anesthesia. The fuzzy membership functions and the linguistic rules using fuzzy logic are developed for the design. The adaptive neuro-fuzzy inference system (ANFIS) has been applied for implementation and simulation of the controller. The performance of the fuzzy logic based PID controller is found excellent in computer simulations with respect to noise tolerance, tracking ability. Simulation result is shown that the exceptional regulation of blood pressure around set-point targets is realized with application of fuzzy logic controller.

Keywords Fuzzy logic · PID controller · Fuzzy PID controller · ANFIS · MAP

1 Introduction

Automation in every field is widely spread. This application in the field of medicine is a challenging task. The anatomical behavior of the patient is reveled initially before operation. Further, anesthetists control significant control variables such as blood pressure, heart rate, temperature, blood oxygenation within the acceptable bounds.

M. D. Mohanty
Department of Electronics and Instrumentation Engineering, College of Engineering and Technology, Bhubaneswar, India
e-mail: mohan.debarchan97@gmail.com

M. N. Mohanty (✉)
Department of Electronics and Communication Engineering, ITER, Sikshsa 'O' Anusandhan (Deemed to be University), Bhubaneswar, India
e-mail: mihirmohanty@soa.c.in

© Springer Nature Singapore Pte Ltd. 2020
P. K. Mallick et al. (eds.), *Cognitive Informatics and Soft Computing*,
Advances in Intelligent Systems and Computing 1040,
https://doi.org/10.1007/978-981-15-1451-7_57

To ensure the patient safety, required anesthesia must be kept during the complete surgical process. The objective is to build up an automatic control systems to control the profundity of anesthesia.

For concerning the blood pressure, numerous researches have been done by different authors [1–12]. There are some standard PID controller like adaptive controller, LQ-type optimal controller, rule-based controllers which are the standard instruments for delay-plants. For taking the pure delay of the plant, there are some other types of PID controllers like smith controller, model predictive controller, and adaptive controller. Disturbances are not considered in case of smith controller because this controller does not have sufficient rejection ability of rejection. In noisy condition like any surgical operation, adaptive and self-tuning controllers are not able to perform well.

The anesthetists utilize uncertain, individual standards do not keep them from giving a sheltered and compelling analgesic; each specialist utilizes some sort of guidelines; however, in some cases, straightforward principles are clouded by an air of significance. In estimating the profundity of anesthesia, most anesthetists respect the mean blood vessel weight (MAP) as the most solid estimation. The dimension of MAP fills in as a guide for the conveyance of breathed in anesthesia. The control objective is to direct MAP to any ideal set point and kept up the recommended set point within the sight of undesirable unsettling influences (clamor), and the control variable is MAP. The parameters of the controller are resolved dependent on the portion reaction attributes of the individual patient, which is distinguished toward the start of the activity utilizing a rectangular test signal. The MAP of the patient is kept up at roughly 60–75 mm Hg, where the precise reference estimation of the MAP is controlled by the therapeutic specialist dependent on the state of the patient.

The original model was first introduced by Zwart et al. [13]. Later, this model is developed by Derighetti [14]. Generally, this model is based on two parts. In the first part, drug distribution and uptake can be done, and the second part can be used for blood flow circulation. The overall system is related with the anesthetic system that describes uptake and distribution of drug. Mean arterial pressure (MAP) can be derived as follows:

$$\text{MAP} = \text{DO}_0 \frac{1 + b_1 q_1 + b_2 q_2 + b_3 q_A}{\sum_{i=1}^{9} k_{i,0}(1 + c_i q_i)} \tag{1}$$

where $k_{j,0}$ and DO_0 are the terms can be derived from the partial blood pressure or may be constant that is either patient or drug dependant. A refers to the artery, and q is the anesthetic gas air. DO is the cardiac output prior to any type of anesthetic can be given. More amount of O_2 can cause problem for the patients having lung problem.

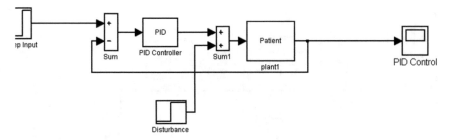

Fig. 1 PID control system

2 Control Using Fuzzy PID Controller

Some ongoing embedded systems settle on control choices. These choices are generally made by programming and dependent on input from the equipment under its control (named the plant). Such criticism usually appears as a analog sensor that can be examined by means of an A/D converter. An example from the sensor may speak to position, voltage, temperature, or some other proper parameter. Each example furnishes the product with extra data whereupon to base its control choices.

PID controllers are the combinations of many controllers, and the whole system is to be controlled by some complex, well-understood system. For designing such type of system, fuzzy PID controller is more suitable. In Fig. 1, a fuzzy PID controller system to control the anesthesia system is presented. This system can be useful for the patients in operation theaters (OT).

The patient's transfer functions $P(s)$ is considered as in [15] and can be derived as

$$P(s) = \frac{K\left(s + \frac{1}{10.6}\right)}{\left(s + \frac{1}{34}\right)\left(s + \frac{1}{3.08}\right)\left(s + \frac{1}{4.8}\right)}$$

While a disturbance is added to the system, Fig. 2 shows the effect.

3 Anesthesia Control: A Fuzzy Controller

For designing a real-world system, some problems may arise. These problems may increase the complexity, make the system agreeable, stochastic, nonlinear, and time varying in nature. These factors can make a challenge for designing a well-enabled PID controller that can be utilized in some real-world problem. In the literature, different types of PID controller have been proposed by the researchers. Most of those controllers are based on fuzzy IF-THEN rule. This rule can be derived by the following form:

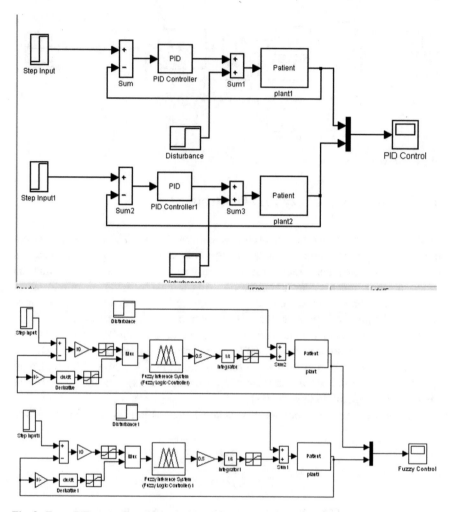

Fig. 2 Fuzzy PID controller with and without disturbances

IF Y_1 is A^1 and... and Y_n is A^N THEN X is B. where $Y = (Y_1, Y_2,... Y_n)^T$ and X is the input and output linguistic variable, respectively. A^1 and B are the linguistic values characterized by using the membership function. It can be considered that this fuzzy rule representation provides a suitable framework to integrate the human knowledge. Nine rules derived from the Mamdani fuzzy logic control system are used in the proposed system. MATLAB ANFIS model is considered for the simulation purpose.

4 Simulation and Results

Different rules adapted for the control are shown in Fig. 3. FIS editor used for the proposed system is displayed in Fig. 4. Figure 5 shows the membership function. Figures 6 and 7 display the rule viewer and surface viewer result of optimal drug at any point which depends on the error and error rate.

Rate of error or any change in error for the system response is shown in the above figure. Error rate is considered as −0.28 and error is 0.318 is selected in the proposed work. Optimal quantity of drug provided in the proposed system is −0.105. This PID controller is embedded with FLC ANFIS model, and the comparison between these two types of controller is presented in Fig. 8. Response of the PID controller is good in the first experiment, but the problem can be arisen if the disturbance will increase, it may create some changes in the behavior, and response may be late. It is presented in Fig. 9. These types of problems are improved in the proposed PID controller system.

Figure 9 shows different between two systems FLC and PID. In FLC, we can notice the system response time is faster than PID and has not as much of noise and additional stability.

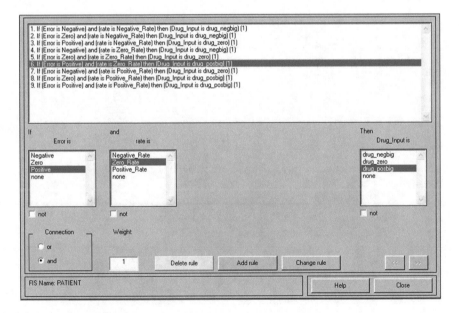

Fig. 3 Rules adapted for the control system

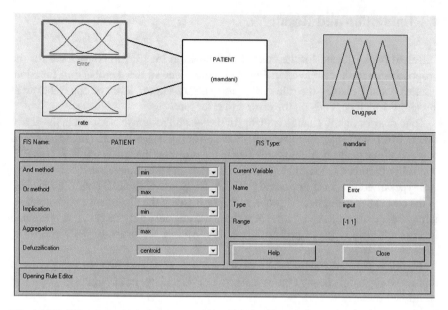

Fig. 4 FIS editor

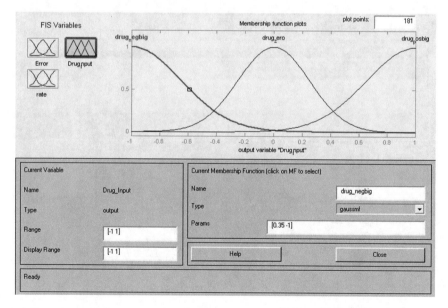

Fig. 5 Membership function

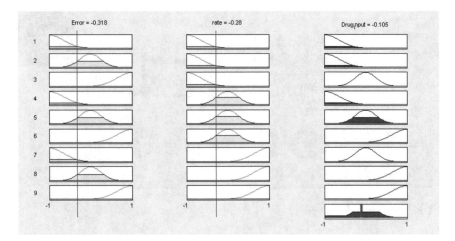

Fig. 6 Rule viewer

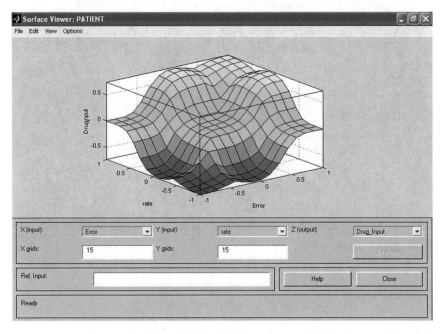

Fig. 7 Surface viewer

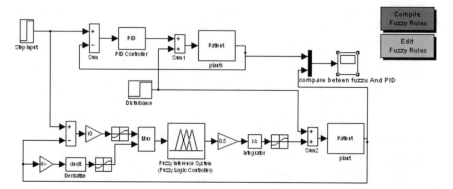

Fig. 8 Complete PID structure

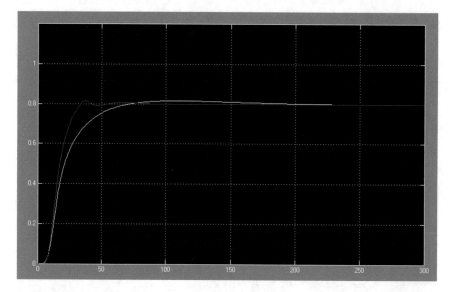

Fig. 9 PID and FLC

5 Conclusion

In this project, we applied a new algorithm using fuzzy logic controller to control the oblivion via measuring the blood pressure. A simulation platform was constructed approximately a nonlinear recirculatory physiological system, and it was customized to comprise a more well-organized method of bringing the anesthetic. It can be noticed from the simulation result that the proposed fuzzy logic-based PID controller is more efficient in terms of set-point tracking and consumption of drug. A comparison with the PID controller is carried out, and the performances of the FLC over the PID are shown.

References

1. Sheppard, L.C.: Computer control of the infusion of vasoactive drugs. Ann. Biomed. Eng. **8**, 431–444 (1980)
2. Widrow, B.: Adaptive model control applied to real-time blood-pressure regulation. In: Pattern Recognition and Machine Learning, pp. 310–324. Springer, Boston, MA (1971)
3. Koivo, A.J.: Automatic continuous-time blood pressure control in dogs by means of hypotensive drug injection. IEEE Trans. Biomed. Eng. **10**, 574–581 (1980)
4. Koivo, A.: Microprocessor-based controller for pharmacodynamical applications. IEEE Trans. Autom. Control **26**(5), 1208–1213 (1981)
5. Arnsparger, J.M., Mcinnis, B.C., Glover, J.R., Normann, N.A.: Adaptive control of blood pressure. IEEE Trans. Biomed. Eng. **3**, 168–176 (1983)
6. Woodruff, E.A., Northrop, R.B.: Closed-loop regulation of a physiological parameter by an IPFM/SDC (Integral Pulse Frequency Modulated/Smith Delay Compensator) controller. IEEE Trans. Biomed. Eng. **8**, 595–602 (1987)
7. Furutani, E., Araki, M., Kan, S., Aung, T., Onodera, H., Imamura, M., Maetani, S.: An automatic control system of the blood pressure of patients under surgical operation. Int. J. Control Autom. Syst. **2**(1), 39–54 (2004)
8. Fukui, Y., Masuzawa, T.: Development of fuzzy blood pressure control system. Iyo denshi to seitai kogaku. Jp. J. Med. Electron. Biol. Eng. 27(2), 79–85 (1989)
9. Pajunen, G.A., Steinmetz, M., Shankar, R.: Model reference adaptive control with constraints for postoperative blood pressure management. IEEE Trans. Biomed. Eng. **37**(7), 679–687 (1990)
10. Yu, C., Roy, R.J., Kaufman, H., Bequette, B.W.: Multiple-model adaptive predictive control of mean arterial pressure and cardiac output. IEEE Trans. Biomed. Eng. **39**(8), 765–778 (1992)
11. Delapasse, J.S., Behbehani, K., Tsui, K., Klein, K.W.: Accommodation of time delay variations in automatic infusion of sodium nitroprusside. IEEE Trans. Biomed. Eng. **41**(11), 1083–1091 (1994)
12. Isaka, S., Sebald, A.V.: Control strategies for arterial blood pressure regulation. IEEE Trans. Biomed. Eng. **40**, 4, 353–363 (1993)
13. Zwart, A., Smith, N.T., Beneken, J.E.: Multiple model approach to uptake and distribution of halothane: the use of an analog computer. Comput. Biomed. Res. **5**(3), 228–238 (1972)
14. Derighetti, M.: Feedback Control in Anaesthesia. Ph.D. thesis, Swiss Federal Institute of Technology, Zurich, Lawson (1999)
15. Mahfouf, M., Linkens, D.A., Asbury, A.J., Gray, W.E., Peacock, J.E.: Generalised predictive control (GPC) in the operating theatre. In IEE Proceedings D (Control Theory and Applications), vol. 139(4), pp. 404–420. IET Digital Library, July 1992

Classification of Faults in Power Systems Using Prediction Algorithms

C. Subramani, M. Vishnu Vardhan, J. Gowtham and S. Ishwaar

Abstract Power transformers are devices used in the transmission networks to step up and step down a voltage value. It primarily operates during large loads and has the highest efficiency at full load and near it. Power transformers generally have higher power ratings (>200 MVA) and are present near the generation sites. The power transformer becomes an integral part of the system and has an immediate and direct impact on the stability and reliability of the network. Hence, in order to ensure that the system remains stable, the protection of the power transformer becomes a task of utmost importance. In this paper, we describe a method to optimize the fault detection by differentiating between actual faults and other conditions that mimic a fault with the help of classification algorithms such as artificial neural networks (ANNs) and decision trees.

1 Introduction

Transformer is an electrical equipment that is used to exchange electrical energy by electromagnetic induction within two circuits [1, 2]. A power transformer is a category in transformers that is employed to transmit electrical energy to multiple parts of an electrical circuit, from the generator to the distribution lines. Such a transformer is employed in the electrical systems to facilitate step up or step down of voltages.

C. Subramani (✉) · M. Vishnu Vardhan · J. Gowtham · S. Ishwaar
Department of Electrical and Electronics Engineering, SRM Institute of Science and Technology, Kattankulathur, India
e-mail: csmsrm@gmail.com

M. Vishnu Vardhan
e-mail: vishnublasted@gmail.com

J. Gowtham
e-mail: gowthamsjp@gmail.com

S. Ishwaar
e-mail: ishwaar17797@gmail.com

© Springer Nature Singapore Pte Ltd. 2020
P. K. Mallick et al. (eds.), *Cognitive Informatics and Soft Computing*,
Advances in Intelligent Systems and Computing 1040,
https://doi.org/10.1007/978-981-15-1451-7_58

557

The protection system of electrical equipment has traditionally been set up with the help of electromechanical relays that use the physical properties of the flowing current to excite a coil and produce a trip signal if the pre-set parameters are violated. Though this is a simple and tested technique, advances in technology allow us to gather and leverage more data than ever before and the ease with which we can extract and store data along with the computing power offered by the new age processors allow us to develop more efficient and accurate methods to improve the operation of the protective devices [3].

The present protection methods used are based on detecting higher magnitude signals and tripping the circuits based on the rating of the relay or similar protective components. However, these methods do not account for signals (like magnetic inrush current) that are not actually faults, but still produce a higher magnitude of current [4, 5]. Though these conditions do not affect the machine, the breakers trip anyway due to the high magnitude.

This project proposes a method to improve the protection of power transformers by using classification algorithms such as artificial neural networks (ANNs) and decision trees in order to differentiate between actual fault conditions and the other system states that may cause unwanted tripping of the circuit. From the data, we obtain about the operation and different operating conditions of the equipment; we build a database with which we train the algorithms.

Artificial neural network (ANNs) are essentially algorithms that have been for-mulated, inspired by the functioning of the human brain. Structurally, ANN consists of massively parallel, interlinked processing elements. In the thesis, each processing element, or a neuron, is very simple to learn anything understandable on its own. Significant learning capacity and, therefore, processing power come only from the collection of many neurons within a neural network [6].

Decision tree learning involves the usage of a decision tree to imitate a predictive algorithm in order to begin from observations of an item, feature, or an attribute to decisions about the item's value. It is a type of predictive approach used in statistics, machine learning, and data mining as well [7, 8].

Every time a new transformer is introduced in the network or every time a fault is cleared, and the phenomenon of magnetic inrush current takes place and may affect the operation of the protection system. When the protection system gets activated and trips the breaker in such conditions, it further affects the stability and overall reliability of the system. This problem can be solved by implementing a method where the relay, or the equivalent protective device, learns to differentiate between an actual fault and a condition that mimics the characteristics of the fault [9]. We develop an algorithm using the parameters like voltage, current, the rate of rise/fall of the signal, the changes in time period between normal and fault conditions and other parameters which can accurately represent the signal to train an artificial neural network that can optimize the relay operation by efficiently differentiating actual fault occurrences and other system conditions that may incorrectly cause the circuit breaker to trip.

2 Faults in a Power Transformer

Internal faults are characterized by abnormal operation within the power transformer, and these may include conditions where the insulation of the conductors gets deteriorated and may cause internal short circuits. Other conditions that may cause an internal fault to occur may include winding failure, overheating, or oil contamination in the transformer [10]. Also, a core fault can damage the laminations of the machine and it is detrimental to its operation. The shorting of the internal windings can cause a very high current to flow which may cause overheating and permanent damage to the windings resulting in the shutdown of the transformer.

Due to the repeated cooling action that takes place in the transformer, the oil gets contaminated leading to inefficient cooling that may cause further abnormal operation of the transformer.

External faults are those faults that occur on the outside but have a direct impact in the operation of the transformer. These faults include external short circuits, overloading of the system, or other natural factors like lightning strikes that may cause surges that damage the equipment. These faults are capable of damaging the insulation of the device, thus causing permanent damage and subsequent shutdown of the transformer. Other types of external faults that may affect the operation of the transformer include overvoltage, overcurrent, or undervoltage conditions. Most of the faults that occur can be reduced by regular maintenance and testing of the equipment [11].

HV disturbances in the machine may be caused due to surge transient or power frequency over voltage. The high frequency surge may occur due to arcing, lightning impulses or the different switching operations of various electrical equipment [12]. That may impact the operation of the transformer machine.

3 Fault Modelling

This section delves into the different faults that we have used to build a predictive model, and the respective simulation has been described further. They were simulated using MATLAB with pre-set values of supply and load. The simulink model is shown in Fig. 1.

3.1 Magnetic Inrush Current

Inrush current or input surge current is the maximum instantaneous input current drawn by an electrical device when first turned on. AC electric motors and transformers would draw several times their normal full-load current when first energized, for a few cycles of the input waveform. When a transformer is first energized, a transient

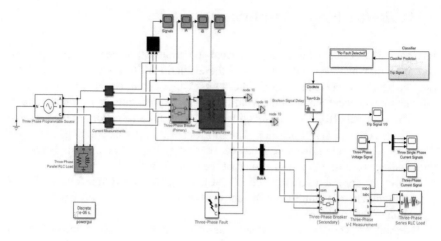

Fig. 1 Simulation model

current up to 10–15 times larger than the rated transformer current can flow for few cycles. The data obtained by simulating this fault is shown in Fig. 1.

3.2 Overcurrent Faults

Overcurrent is created due to device overloads, due to shorting or due to grounding problems as well. An overload takes place when equipment is met with currents much higher than their rated capacity; hence, more heat is let out. The hardware setup was connected to a lighting load with a demand higher than that of the rated value. The rated value was set at different levels for experimentation purposes, and the resulting data from the signals were acquired.

3.3 Line Faults

Typical cases of the line fault are the unsymmetrical LG, LL, LLG faults or symmetrical faults like the LLL or LLLG faults. When these faults occur, it produces one low impedance route for the current to flow. As a result, a very large current is taken from the input supply, creating opening of relays, damaging of insulation, and parts of the equipment. Due to the operational difficulties of simulating such faults in the hardware setup, they were analysed with the help of software simulations.

3.4   Over/Undervoltage Faults

The overvoltage faults, also known as voltage transients or voltage surges, can occur due to several reasons like system faults, disconnection of high-power industrial loads, sudden load deductions, and energization of capacitor banks. The undervoltage faults are more common and may be caused due to underloaded or overloaded equipment and transformers as well. During the peak power demand hours or when a service is experiencing some problem, the need for energy surpasses the ability of the transformer to meet it; hence, the voltage drops. This was simulated with the hardware setup by connecting a motor load with rated voltage higher than the supply voltage.

4   Prediction Algorithms

Once data has been collected for the various types of faults, we train a couple of different architectures of neural networks and decision trees using the acquired data to differentiate between a normal signal and the different fault signals.

4.1   ANN Training and Architecture

We use an iterative learning algorithm called backpropagation that involves the following:

1. Random initialization of weights
2. Feed forward using initial weights
3. Determination of error at the output nodes
4. Backpropagation of the errors
5. Updating the weights
6. Feed forward using updated weights.

The stages 3–6 are iterated several times in order to increase the prediction/classification accuracy of the neural network.

This backpropagation algorithm to train a 3L FFN shown in Fig. 2.

Signals which are fed into the input layer pass via the hidden layer, thus reaching the output layer. A learning way mainly consists of finding the linking weights and schema of the connections. The NN calculates the nonlinear connection between the input given and output values obtained by adjustment of the weights internally. Furthermore, the neural network can be fed with inputs that is not included during the training.

A backpropagation algorithm searches for the minimum cost in the function $J(\theta)$ in accordance with weight space with the help of the conjugate gradient method.

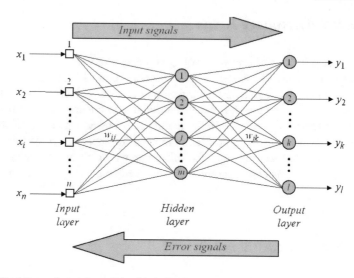

Fig. 2 Feed-forward neural network with the backpropagation algorithm

$$J(\theta) = \frac{1}{m}\left[\sum_{i=1}^{m}\sum_{k=1}^{K}(y_k^i\log\log\left(h_\theta(x^i)\right)_k + \left(1 - y_k^i\right)\log\left(1 - \left(h_\theta(x^i)\right)_k\right)\right]$$

$$+ \frac{\lambda}{2m}\left[\sum_{l=1}^{L-1}\sum_{i=1}^{s_l}\sum_{j=1}^{s_{l+1}}(\theta_{ji}^l)^2\right]$$

Cost function is to be minimized where θ refers to the weights.

The combination of weights that helps to minimize the cost function is considered as the solution to the learning problem, and it can further be used for prediction with new data.

4.2 Decision Tree Classification

Decision tree classification involves using a decision tree model in order to move from observations about an item, feature, or an attribute to final decisions about the product's final value. It is a type of predictive approach used for statistics, machine learning, and data mining.

During decision analysing, a decision tree is used blatantly to show a decision or decision-making. A tree can be trained using the source data that is split into subsets on the basis of values in different attributes. This process is done consecutively for every obtained set in an iterative way named recursive partitioning. An iteration is termed over when the values of the target variable is the same as the values of a subset at a node, or whenever a split doesn't add value to the predictions further.

5 Simulation Result and Discussion

The results of various faults were observed using MATLAB simulation and the results are shown in Figs. 3, 4, 5, 6, 7, and 8.

5.1 ANN Implementation

When the training was performed with the help of ANN, the accuracy of prediction of the faults was low since a large number of faults were to be predicted. ANN gives high accuracy, only when the outcomes of the prediction is less is quantity.

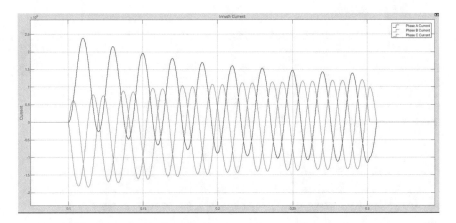

Fig. 3 Inrush condition

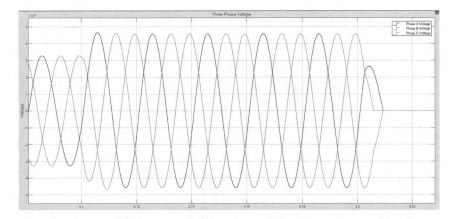

Fig. 4 Overvoltage fault

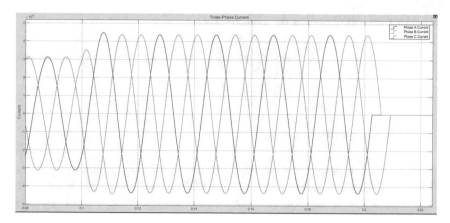

Fig. 5 Overcurrent fault

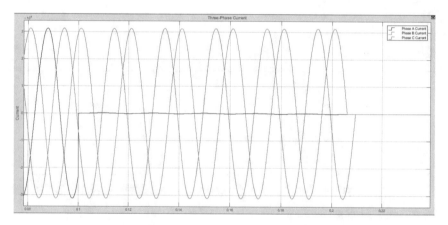

Fig. 6 LG fault

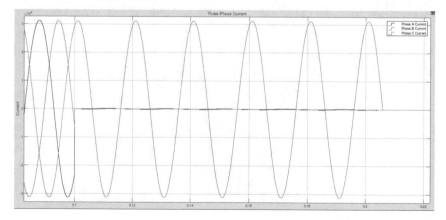

Fig. 7 LLG fault

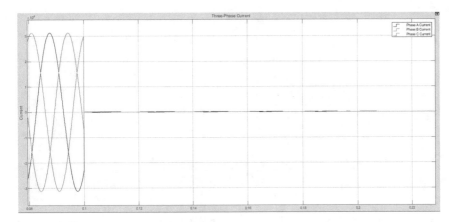

Fig. 8 LLLG fault

When the prediction was carried out with only two faults, it had high accuracy, but when the number of faults was increased the weights that are present in the algorithm did not work well. This is the reason as to why ANN was not used.

5.2 Decision Tree Classification Training

The data taken from the fault waveforms is used to train the prediction algorithm. Every fault along with the data for no fault condition is taken from the simulation waveforms. The data is saved separately in a file named FAULTDATA (Table 1).

1. Normal condition
2. Double line to ground fault
3. Line to ground fault
4. Overcurrent fault
5. Overvoltage fault
6. Inrush fault condition which shown in Fig. 9.

Table 1 Fault data from Simulink

S. No.	Peak-to-peak magnitude	Magnitude/time period
1.	62,400	6.1800e−06
2.	37,800	3.7400e−06
3.	142	0.0140
4.	87,400	8.7400e−06
5.	91,600	9.1300e−06
6.	254,000	2.5700e−05

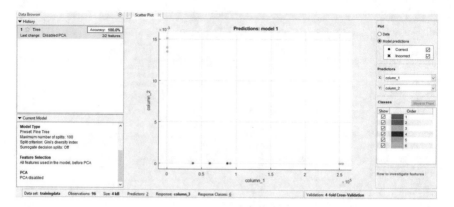

Fig. 9 Visualization of predicted classes for all fault data produced through simulation in the MATLAB classification learner application

The training for the prediction algorithm is shown above using the data taken from the waveforms in the Simulink. Each fault is represented in a separate colour corresponding to the numbers above.

The Fig. 10 shows how the fine tree classification is performed using the parameters that was given to train the algorithm. Since the effect of variation is more for the parameter of amplitude of the wave only that parameter is used to classify the faults even though two parameters are given to train the network.

The classification is done with the training that is given to the algorithm, and the type of fault is displayed in the MATLAB–Simulink diagram as shown in Fig. 10.

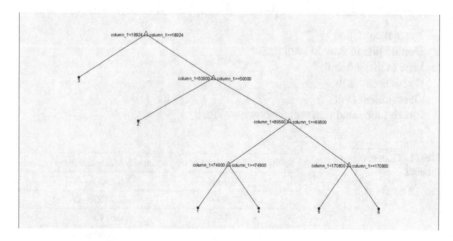

Fig. 10 Decision tree for Simulink data

6 The Hardware Implementation

A three-phase step-down transformer was fed through an autotransformer with a primary voltage of 365 V and a stepped-down secondary voltage of 200 V. This 200 V supply was fed to a contactor through the three 6A circuit breakers. Uncharacteristic to normal relays, contactors are made to be connected directly to load devices that produce high currents. The contactor was controlled externally, and the output was connected to the load. Lighting loads as well as motor loads were used to simulate the different faults. The load currents were measured with the help of a differential current probe connected directly to the digital storage oscilloscope (DSO). The data obtained in the DSO with the help of the current probe was stored and used for training the model which shown in Fig. 11.

6.1 Hardware Results and Discussion

The hardware setup for all the faults was performed with the same source voltage which was kept at 365 V at the input of the transformer which gave output of 200 V.

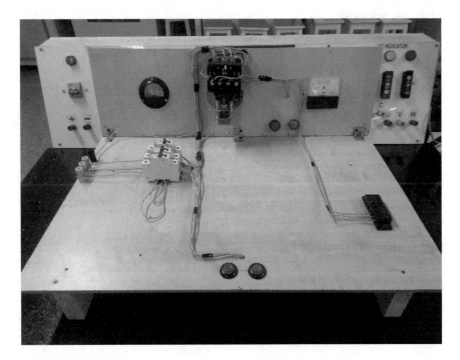

Fig. 11 Hardware setup

Table 2 Inrush fault values

S. No.	Electrical parameter	Values
1.	Inrush current (A)	2
2.	Fuse rating (A)	1.5
3.	Fault identification	0.1 s
4.	Power consumed	None

Table 3 Overcurrent fault values

S. No.	Electrical parameter	Values		
1.	Load current (A)	4.8	5.7	7.2
2.	Fuse rating (A)	2.25	2.25	2.25
3.	Breaker operation time (S)	21	17	11
4.	Power consumed (W)	1600	2000	2600

The data obtained during the hardware testing of the various system conditions is reported in the Tables 2, 3, 4, and 5.

6.1.1   Inrush Fault

See Table 2.

6.1.2   Overcurrent Fault

See Table 3.

6.1.3   Undervoltage Fault

See Tables 4 and 5.

Table 4 Undervoltage fault values

S. No.	Electrical parameter	Values	
1.	Load current (A)	4.4 (R)	7.2 (e)
2.	Fuse rating (A)	2.25	
3.	Breaker operation time (S)	0.1	
4.	Power consumed (W)	2.2	

Table 5 Short-circuit fault values

S. No.	Electrical parameter	Values
1.	Short-circuit current (A)	47
2.	Fuse rating (A)	6
3.	Breaker operation time (S)	0.1
4.	Power consumed (W)	1600

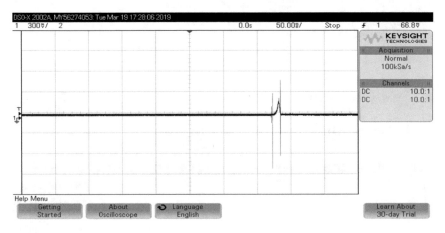

Fig. 12 Inrush current condition observed in hardware

6.1.4 Line-to-Line Fault

The waveforms taken during fault conditions via the DSO to train the decision tree algorithm are shown in Figs. 12, 13 and 14.

The data from the above fault waveforms is taken via the DSO. In this case, the data points of current at various places in the waveforms were taken out from the DSO via a digital storage device to train the algorithm as shown below.

Figures 15 and 16 shows how the data obtained from the hardware setup is used to train the classifier and arrive at a decision based on the features chosen and the conditions used for splitting. The figures represent the decision-making process of the tree based on the parameters on which it is programmed and shown in Fig 17.

7 Results

As evident from the results shown in the previous sections, this method of identifying and classifying the faults to improve the protection system of the power transformer is an improvement over the method currently in use. This method can be implemented without any major investment required with respect to storing and processing the data.

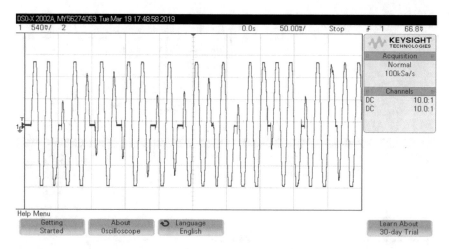

Fig. 13 Undervoltage signal obtained from the hardware setup

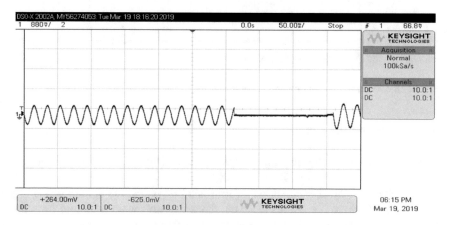

Fig. 14 Line fault condition observed from the hardware setup

Using this method to identify the fault and subsequently controlling the relay or the circuit breaker improves the reliability of the system by optimizing the operation of the protection system. The accuracy that we were able to attain was 100% for the MATLAB simulation and 95.1% for the hardware setup.

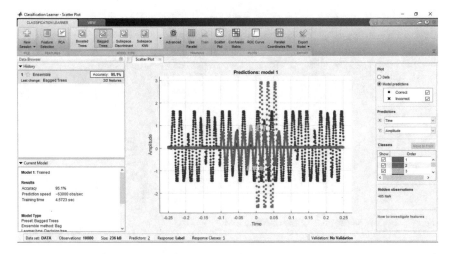

Fig. 15 Visualization of predicted classes for all fault data produced using hardware setup in the MATLAB classifier learner application

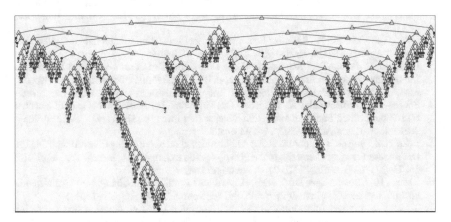

Fig. 16 Decision tree trained using hardware data

8   Conclusions

As visible from the work shown above, a new setup for the fault classification of a power transformer has been developed and listed. It allows to design and add more faults into the scheme as required. The works depict that a data-based knowledge system reduces the workload of the user and helps reduce cost needed in setting up different relays for different faults. This system provides a very high accuracy and can be developed as per the needs of the user.

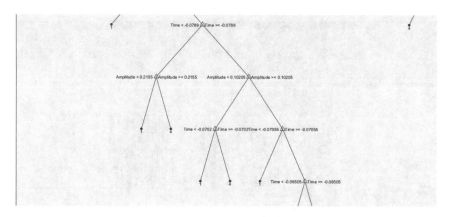

Fig. 17 Zoom-in at one node of the decision tree

References

1. Jan, S.T., Afzal, R., Khan, A.Z.: Transformer failures, causes & impact. In: International Conference Data Mining, Civil and Mechanical Engineering (ICDMCME '2015), Bali (Indonesia), 1–2 Feb 2015
2. Bashi, S.M., Mariun, N., Rafa, A.: Power transformer protection using microcontroller-based relay. J. Appl. Sci. **7**(12), 1602–1607 (2007). ISSN 1812-5654
3. Pehardaa, D., Ivanković, I., Jaman, N.: Using Data from SCADA for Centralized Transformer Monitoring. www.elsevier.com/locate/procedia 1877–7058 © 2017 The Authors. Published by Elsevier Ltd, 4th International Colloquium "Transformer Research and Asset Management"
4. Shenoy, U.J., Parthasarathy, K., Khincha, H.P., Thukaram, D., Sheshadri, K.G.: Artificial Neural Network Based Improved Protection Scheme for Power Transformers. Indian Institute of Technology, Kharagpur, 721302, 27–29 Dec 2002
5. Loko, A.Z., Bugaje, A.I., Bature, A.A.: Automatic method of protecting transformer using PIC microcontroller as an alternative to the fuse protection technique. Int. J. Tech. Res. Appl. **3**(2), pp. 23–27 (2015). e-ISSN 2320-8163. www.ijtra.com
6. Balaga, H., Vishwakarma, D.N., Nath, H.: Artificial neural network based backup differential protection of generator-transformer unit. Int. J. Electron. Electr. Eng. **3**(6) (2015)
7. Tripathy, M.: Power transformer differential protection based on neural network principal component analysis, Harmonic Restraint and Park's Plots. Advances in Artificial Intelligence, vol. 2012, Article ID 930740, 9 p. Hindawi Publishing Corporation. https://doi.org/10.1155/2012/930740
8. Khorashadi-zadeh, H., Li, Z.: A sensitive ANN based differential relay for transformer protection with security against CT saturation and tap changer operation. Turk J. Electr. Eng. **15**(3) (2007)
9. Sonwani, L., Singh, D.K., Sharma, D.: Simulation of transformer for fault discrimination using wavelet transform &neural network. Int. J. Sci. Eng. Technol. Res. (IJSETR) **4**(7) (2015)
10. Tripathy, M., Maheshwari, R.P., Nirala, N.: Transformer differential protection based on wavelet and neural network. Int. J. Electron. Electr. Eng. **7**(7), 685–695 (2014). ISSN 0974-2174
11. Sahu, V.K., Vaidya, A.P.: Power transformer protection using ANN, fuzzy system and Clarke's transform. Int. J. Adv. Electron. Comput. Sci. **2**(8) (2015). ISSN 2393-2835
12. Khandait, A.P., Kadaskar, S., Thakare, G.: Real time monitoring of transformer using IOT. Int. J. Eng. Res. Technol. (IJERT)

A Systematic Literature Review of Machine Learning Estimation Approaches in Scrum Projects

Mohit Arora, Sahil Verma, Kavita and Shivali Chopra

Abstract It is inevitable for any successful IT industry not to estimate the effort, cost, and duration of their projects. As evident by Standish group chaos manifesto that approx. 43% of the projects are often delivered late and entered crises because of overbudget and less required functions. Improper and inaccurate estimation of software projects leads to a failure, and therefore it must be considered in true letter and spirit. When Agile principle-based process models (e.g., Scrum) came into the market, a significant change can be seen. This change in culture proves to be a boon for strengthening the collaboration between developer and customer. Estimation has always been challenging in Agile as requirements are volatile. This encourages researchers to work on effort estimation. There are many reasons for the gap between estimated and actual effort, viz., project, people, and resistance factors, wrong use of cost drivers, ignorance of regression testing effort, understandability of user story size and its associated complexity, etc. This paper reviewed the work of numerous authors and potential researchers working on bridging the gap of actual and estimated effort. Through intensive and literature review, it can be inferred that machine learning models clearly outperformed non-machine learning and traditional techniques of estimation.

Keywords Effort estimation · Scrum · Machine learning · Agile software development

M. Arora · S. Verma (✉) · Kavita · S. Chopra
Lovely Professional University, Phagwara, Punjab, India
e-mail: sahilkv4010@yahoo.co.in

M. Arora
e-mail: mohit.15980@lpu.co.in

Kavita
e-mail: kavita.21914@lpu.co.in

S. Chopra
e-mail: shivali.19259@lpu.co.in

© Springer Nature Singapore Pte Ltd. 2020
P. K. Mallick et al. (eds.), *Cognitive Informatics and Soft Computing*,
Advances in Intelligent Systems and Computing 1040,
https://doi.org/10.1007/978-981-15-1451-7_59

1 Introduction

Agile estimation has always been challenging for IT experts across the globe, and this issue has been constantly put on by various researchers in their literature. A typical estimation framework opted by most of the IT industries is given in Fig. 1, wherein requirements *aka* the desired user stories are being stacked in the product backlog and further tagged with their respective sizes. Story point is most used unit to size a user story, i.e., 61.67% of industries employing it.

As per ISPA [1], two-thirds of software projects neglect to be conveyed on time and inside budget. There are two principle reasons for software project disappointments: One is improper estimation as far as task size, cost, and staff required, and second being the uncertainty of system and software requirements. The major challenges for estimating of Scrum-based projects are change and sprint-wise estimation. Most of the IT industries have adopted hybrid process models which are mostly driven by Agile umbrella methodologies.

As per [2] the transition of process models from heavyweight like iterative waterfall to lightweight like Agile, a change can also be seen in effort estimation approaches. All the tradition estimation approaches [3] like expert judgement, top-down estimation, and Delphi cost estimation are well suited in one or other form for heavyweight process models but lack in bridging the estimated and actual effort gap of Agile methodologies. Thus, due to volatile nature of Agile-based project requirements, researchers started exploring alternatives and end up at soft computing techniques [4]. A standout among the most widely recognized uses of neuro-fuzzy frameworks [5] is delivering rules for unpredictable issues. On alternate hands, software projects are characteristically uncertain and complex with the goal that the accessible data is not sufficient at the beginning time of task and the issue of effort estimation is totally unclear. In this circumstance, fuzzy and neuro-fuzzy models can deal with the vulnerability and increment the estimation exactness. Also, encouraging outcomes has been accounted from fuzzy-based models connected to the field of software effort estimation.

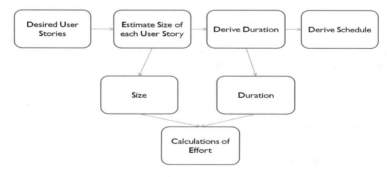

Fig. 1 Typical estimation process in Scrum

Because of unpredictability of effort estimation issue and trouble of project and people attribute relational analysis, the optimization procedure assumes an essential job here. The optimization [6] can specifically be connected to effort estimation process like quality weighting in analogy-based estimation or in a roundabout way connected to machine learning strategies, for example, ANN and ANFIS. It can be further extended to attribute weighting, tuning ANN adjustment (weight and bias), ANFIS adjustment, structure configuration, variable positioning.

To the best of our information, there is no current review that centers around ML models of Scrum-based projects, which rouses our work in this paper. This paper contains technical abbreviated terms which can be viewed in Table 1. The upcoming section, viz., Sect. 2, highlights a collaborated context of ML impact in ASD, Sect. 3 explains the review method, Sect. 4 discusses the review results, and Sect. 5 provides conclusion and future research directions.

Table 1 List of abbreviations

SDEE	Software development effort estimation	ANFIS	Adaptive neuro-fuzzy inference system
ML	Machine learning	ANN	Artificial neural network
ASD	Agile software development	CBR	Case-based reasoning
DT	Decision tree	BN	Bayesian network
SVR	Support vector regression	GA	Genetic algorithm
GP	Genetic programming	AR	Association rule
ISBSG	International Software Benchmarking Standards Group	MMRE	Mean magnitude of relative error
PRED	Percentage relative error deviation	LR	Linear regression
RF	Random forest	MLP	Multilayer perceptron
SGB	Stochastic gradient boosting	RBF	Radial basis function
ABC	Artificial bee colony	PSO	Particle swarm optimization
CART	Classification and regression tree	TLBO	Teaching–learning-based optimization
TLBABC	Teaching–learning-based artificial bee colony	DABC	Directed artificial bee colony
LM	Levenberg–Marquardt	ISPA	International Society of Parametric Analysis
NB	Naïve Bayes	KNN	K-nearest neighbor
EJ	Expert judgement	FPA	Function point analysis
MRE	Magnitude of relative error	NF	Neuro-fuzzy

2 Related Literature

Jørgensen and Shepperd presented a core review in [7] which recognizes more than 10 estimation methods in 80 s used for effort estimation, wherein regression-based techniques are better as compared to empirical techniques of estimation. In spite of the expansive number of exact investigations on machine learning models in the estimation of software projects irrespective of the process model approach, conflicting outcomes have been accounted for with respect to the estimation exactness of these models. For instance, it was accounted that estimation exactness shifts under a similar machine learning model when it is developed with various datasets [3, 8] or scenarios [9]. With respect to the correlation between ML model and regression model, thinks about in [3] announced that ML model is better than regression model, while examines in [10] reasoned that regression model beats ML model. ANN and case based reasoning techniques outperformed each other when applied on different datasets in [8] and [18]. The difference in the current empirical examinations on ML models has not yet completely comprehended and may keep experts from embracing ML models by comparing with different areas in which ML systems have been connected effectively. Besides, the hypothesis of ML systems is greatly entangled than that of traditional estimation procedures. To encourage the uses of ML procedures in SDEE area, it is pivotal to deliberately condense the empirical proof on ML models in ongoing research and practice. Industry experts use expert judgement and Delphi cost estimation techniques more as compared to ML. Some of the ML strategies that have been utilized for SDEE are [11–14] CBR, ANN, DT, BN, SVR, GA, GP, AR, etc., and most of them are not yet applied in Agile estimation. The above ML systems are utilized either alone or in blend with other ML or non-ML methods. For example, GA has been utilized with CBR, ANN, and SVR for highlight weighting and choice. Fuzzy logic [15] is utilized with CBR, ANN, and DT for execution. Different datasets have been utilized for estimation, viz., ISBSG, JIRA and Atlassian repositories, PROMISE data repository, and so on. W.R.T approval techniques, holdout, n times overlay cross-validation ($n > 1$), and leave-1-out cross-validation [3, 16] are the predominant ones. MMRE, PRED (25) (percentage of forecasts that are inside 25% of the real estimate), and MdMRE [17] are the three most well-known precision measurements.

Out of all ML techniques, BN [3, 18, 19] found to have most exceedingly bad MMRE in contrast to CBR (51%), ANN (37%), DT (55%), SVR (34%), and GP and AR (49%) separately for estimating projects includes both traditional and lightweight methodologies, but in some cases it did not. Research demonstrates ANN and SVR [9] beat other ML models, yet it does not mean that we can utilize them without confinement as to expand precision, expanding the number of concealed layers will build the preparation time and may create over-fitting issues [3]. Examination of ML models with regression models, COCOMO estimation, EJ, and FPA [20] has also been carried out. Studies indicate CBR and ANN are more exact than regression models. GP is less precise than regression. So, based on the stats we have concluded generally that ML models outflank non-ML strategies. Distinctive estimation settings are made with reference to in writing, for example, little informational collection, anomalies, absolute highlights, and missing qualities.

Analysts recommend [21, 22] that it is more productive to decide the best model in a specific setting instead of deciding the best single model, since estimate models carry on uniquely in contrast to one dataset to other, which makes them precarious. Studies directed on information mining report that group strategies furnish exact outcomes in examination with single strategies as every strategy has quality and shortcoming so joining will moderate the shortcoming. Outfit effort estimation systems might be gathered into two noteworthy classes [22–24, 16]: homogeneous (e.g., bagging and SVR, RF, MLP, LR, RBF, ANFIS, CBR, RF, SGB, CART, and so forth) and heterogeneous perceived by their base models and blend rules. ANN was utilized most with outfits. Studies demonstrate that solitary ML procedures are the predominant methodology used to develop ensembles. It has been discovered that homogeneous troupes dependent on DT are the most exact, trailed by homogeneous one's dependent on CBR, and from there on came SVR homogeneous development.

ANN, DT, CBR, SVR, regression, and neuro-fuzzy [5] are most utilized for group, wherein request of best outcomes pursues DT, regression, CBR, SVR, and afterward NF. Mix rules have additionally been extricated for consolidating endeavors of base models and are partitioned into two sections such as linear and nonlinear. Mean, mean weighted, and middle are most utilized straight mix rules. MLP, SVM, CART, and FIS utilizing c imply subtractive grouping are most utilized non-straight principles. All the techniques mentioned and discussed in this section are derived from general estimation approaches to demonstrate a trail of estimation trends. The next section will include some research questions which will be revolve around Scrum-based project estimation only.

3   Method

In this section, we have discussed the various research questions, review inclusion and exclusion criterion, data source description, and study select process.

3.1   Research Questions

This review paper aims to summarize the present status of implication of machine learning models in Scrum-based projects. The following research questions have been framed in this context and are given as follows:

RQ1: Which ML models have been used for Scrum estimation?
RQ2: Do ML models distinctively outperform other ML models for Scrum estimation?

RQ3: What is the overall estimation accuracy of ML techniques used in Scrum-based projects?

RQ4: Does estimation accuracy of Scrum-based projects increase by using meta-heuristic algorithms?

RQ5: What are the various Scrum project datasets available on Web?

RQ6: Are ensemble estimation methods better than single estimation for Scrum projects?

RQ7: What are the various significant factors affecting effort of Scrum projects?

3.2 Include and Exclude Criterion

This study incorporates the papers which have connected the diverse soft computing techniques for estimation in Agile software development. Papers are incorporated from different online sources, journals, conferences, and so forth distributed till date. A few papers which are not explicitly based on ASD are also likewise included because of some essential data. Papers and data which are not important to the exploration subject are excluded from the examination.

3.3 Data and Literature Sources Description

This data source used in the study includes papers from TOSEM (ACM), IEEE transactions, ScienceDirect, Google Scholar, Springer, etc. Some search strings have been used to search papers from aforementioned online databases, viz. Software AND (effort OR cost) AND (estimate) AND (learning OR "machine learning") OR "machine" OR "case-based reasoning" OR "decision tree" OR "regression analysis" OR "neural net" OR "Bayesian network" OR "Bayesian net" OR "support vector machine" OR "support vector regression" OR "deep" OR "learning" OR "fuzzy" OR "neuro-fuzzy" OR "ANFIS" OR "meta-heuristic" OR "Scrum" OR "Agile" AND "software" AND "development" OR "genetic algorithm" OR analogy OR "expert judgement" OR "planning poker."

3.4 Study Selection Process

After applying the include and exclude criterion, the selection has been primarily carried out in two steps:

- Choosing abstract and title: The review procedure is brought through a few research papers where some of them were chosen by looking on to their titles and modified works.

- Choosing complete article. A good number of papers and articles are reviewed and thoroughly analyzed, and the same has been discussed in Sect. 4 research questions.

4 Results and Discussion

The various research questions mentioned in Sect. 3.1 will be answered here.

4.1 Which ML Models Have Been Used for Scrum Estimation (RQ1)?

A wide variety of ML models has been extensively used in Agile software development and its associated methodologies under its umbrella. Table 2 contains ML techniques used in Scrum estimation with their frequency and year of publication.

Table 2 ML techniques used in Scrum estimation

ML techniques	Use in paper	YOP
Fireworks algorithm optimized neural network	Thanh Tung Khuat and My Hanh Le in [25]	2018
Multiagent techniques	Muhammad D Adnan et al. in [26]	2017
Mamdani fuzzy inference systems	Jasem M. Alostad et al. in [15]	2017
General regression neural networks	Aditi Panda et al. in [27]	2015
Probabilistic neural networks	Aditi Panda et al. in [27]	2015
GMDH polynomial neural network	Aditi Panda et al. in [27]	2015
Cascade correlation neural network	Aditi Panda et al. in [27]	2015
Stochastic gradient boosting	Shashank Mouli Satapathy et al. in [28]	2017
Random forest	Shashank Mouli Satapathy et al. in [28]	2017
Decision tree	Shashank Mouli Satapathy et al. in [28]	2017
Bayesian networks	Dragicevic Srdjana et al. in [19]	2017
Hybrid ABC–PSO algorithm	Thanh Tung Khuat and My Hanh Le in [29]	2017
SVM, NB, KNN, DT	Simone Porru et al., in [30]	2016
Naïve Bayes (NB)	K Moharreri et al. in [31]	2016
Deep learning—long short-term memory (recurrent neural networks)	M. Choetkiertikul et al. in [32]	2015
Particle swarm optimization (PSO)	Manga I, et al. in [33]	2014
SVR kernel methods	Shashank Mouli Satapathy et al. in [9]	2014

From the above table, it can be seen that most of the authors have used different ML techniques and as per their respective year of publication a trend can be inferred that researchers are now shifted to ML techniques to create an auto-estimate environment. In the subsequent section, a comparative analysis has been carried out

4.2 Do ML Models Distinctively Outperform Other ML Models for Scrum Estimation? (RQ2)

It has been given in Sect. 2 that ML techniques outperform non-ML techniques. Moreover, expert-based estimations are suffered from individual bias. In this research question, a comparative analysis of all the ML techniques applied for Scrum-based project estimation is mentioned in Table 3.

As an accuracy parameter, various metrics like MRE and PRED have been mentioned as per the availability of data in the literature. Various ML techniques outperform other ML techniques by applying on either same dataset or different datasets.

It can be inferred from Table 3 that best current ML technique as per the accuracy metric MMRE for 21 project data is fireworks algorithm optimized neural network with 2.93% MMRE. It cannot be deduced exactly as the projects/datasets used by other authors are different and may have less or more predication. An improved predication can also be seen in others with different datasets. The compiled MMRE results can be seen in Fig. 2.

4.3 What Is the Overall Estimation Accuracy of ML Techniques Used in Scrum-Based Projects? (RQ3)

To the best of our knowledge, 16 ML techniques have been used for Scrum-based project estimation till date. From Table 3, the average mean magnitude of relative error for the same dataset ML techniques comes out to be 0.2822.

4.4 Does Estimation Accuracy of Scrum-Based Projects Increase by Using Meta-Heuristic Algorithms? (RQ4)

In the literature, very less, empirical evidence can be seen in the context of inclusion of meta-heuristic algorithms in Scrum-based projects. As mentioned in Table 3, only two such papers have been refereed, i.e., fireworks algorithm [25] and ABC–particle swarm optimization (PSO) [29]. Among these two, fireworks algorithm has good estimation accuracy as compared to all ML techniques used for Scrum. In this

Table 3 Comparative accuracies of different ML estimation techniques

Estimation techniques	Use in paper	Accuracy parameters	Dataset used	Outperformed
Fireworks algorithm optimized Neural network	[25]	MMRE-0.0293	21 projects developed by six software companies presented in Zia's work	TLBO, TLBABC, DABC, LM
Multiagent techniques	[24]	MMRE—0.1	12 Web projects	Delphi and planning poker
Mamdani fuzzy inference systems	[14]	MMRE (sprint1)—0.28 MMRE (sprint2)—0.15 MMRE (sprint3)—0.09	Three sprints of real software projects	Comparison with actual est.
General regression neural network (GRNN)	[25]	MMRE—0.3581	21 projects developed	Regression (Zia's work) and PNN
Probabilistic neural network (PNN)	[25]	MMRE—1.5776	21 projects developed	Zia's work
GMDH polynomial neural network. (GMDHPNN)	[25]	MMRE—0.1563	21 projects developed	GRNN and PNN
Cascade correlation neural network (CCNN)	[25]	MMRE—0.1486	21 projects developed	GRNN, PNN, GMDHPNN
Stochastic gradient boosting (SGB)	[26]	MMRE—0.1632	21 projects developed	RF and DT
Random forest (RF)	[26]	MMRE—0.2516	21 projects developed	DT
Decision tree (DT)	[26]	MMRE—0.3820	21 projects developed	Zia's work
Bayesian networks	[18]	Accuracy—above 90% for six datasets	160 tasks in real Agile projects	Comparison with actual estimate
Hybrid ABC–PSO algorithm	[27]	MMRE—0.0569	21 projects developed	ABC, PSO, GRNN, PNN, GMDHPNN, CCNN

(continued)

Table 3 (continued)

Estimation techniques	Use in paper	Accuracy parameters	Dataset used	Outperformed
SVM, NB, KNN, DT	[28]	SVM MMRE—0.50 NB MMRE—0.85 KNN MMRE—0.70 DT MMRE—0.98	699 issues of inventive s/w designers 5607 issues from 8 open-source projects	Comparison with actual estimates
Naïve Bayes (NB)	[29]	MMRE—2.044	10 teams in IBM rational team concert	None
Deep learning	[30]	Improved MMRE	23,313 issues from 16 projects	Empirical estimation technique like educated guess
Particle swarm optimization (PSO)	[31]	MMRE—0.1988	21 projects	Zia's work
SVR kernel methods SVR linear kernel SVR polynomial kernel SVR RBF kernel SVR sigmoid kernel	[11]	MMRE—0.1492 MMRE—0.4350 MMRE—0.0747 MMRE—0.1929	21 projects	SVR linear, polynomial, and sigmoid kernel

context, we can deduce that estimation accuracy does increase with the inclusion of meta-heuristic algorithms.

4.5 What Are the Various Scrum Project Datasets Available on the Web? (RQ5)

Datasets for Scrum projects can be found on various online repositories and are shown in the table either as a Web link or as a paper link. Some of the repositories like ISBSG also contain Agile data that may be present as such, and an appropriate data cleaning and filtering techniques need to be applied to get the same (Table 4).

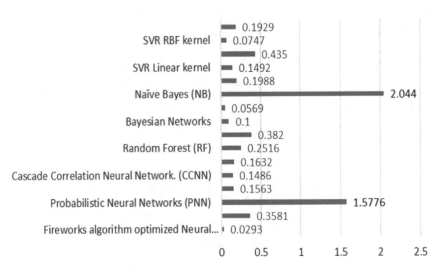

Fig. 2 Comparative accuracies of ML techniques used in Scrum

Table 4 Agile projects dataset links

Dataset name	Dataset links
ISBSG datasets	[6]
Three sprints of real software projects	[14]
Twelve Web projects data for an e-commerce site	[26]
699 issues from industrial projects and 8 open-source projects	[28]
Story point dataset	[32]
Twenty-One projects developed by six software companies presented in Zia's work	[34]

4.6 Are Ensemble Estimation Methods Better Than Single Estimation in Scrum Projects? (RQ6)

Yes, it can be inferred from the review that ensemble estimation techniques yield better results that is just single estimation method. The estimation accuracy of particle swarm optimization alone is less than artificial bee colony–PSO. On the similar grounds, when we backtrack our literature for estimation techniques used for heavyweight process models, ensemble wins in majority.

Table 5 Factors affecting Scrum-based project effort [35, 36]

Project-related factors	People-related factors	Resistance factors
Project domain	Communication skills	Perfect team composition and defects in third-party tools
Quality requirement	Familiarity in team	Working place un-comfort and stakeholder response
Hardware and software requirements	Managerial skills	Drifting to Agile, lack of clarity in requirements, volatility of requirements
Operational ease	Security	Team dynamics and change in working environment
Complexity	Working time	Expected team changes and other project responsibilities
Data transaction	Past project experience	Introduction to new technology and prerequisite availability of resources
Multiple sites	Technical ability	Usability

4.7　What Are the Various Significant Factors Affecting Effort of Scrum Projects (RQ7)?

Effort in Scrum projects has been widely affected by various people, resistance, and project factors. Many authors have proposed various factors in this context, and the same can be viewed in Table 5.

5　Conclusion and Future Research Directions

In this review paper, the following research gaps have been identified that open up an opportunity for all the potential researchers across the globe.

- Missing estimation factors may result in poor estimation as there are potential accelerating and decelerating factors to affect the estimate of Agile-based projects.
- Total effort is a result of effort of all elements of a sprint and reiterates again after the potential shippable release so there is a need of adding regression test effort to make it more accurate.
- There are so many machine learning and optimization approaches missing in the literature, and they have not yet been applied for estimating effort of Scrum-based projects.
- No standard/generic scale for story size and story complexity found in the literature.
- No generic or single estimation model made for Scrum estimation.

- Non-functional requirement effort missing in calculating the total effort of the sprint or project.

References

1. Nassif, A.B., Azzeh, M., Capretz, L.F., Ho, D.: Neural network models for software development effort estimation: a comparative study. Neural Comput. Appl. **27**(8), 2369–2381 (2016)
2. Popli, R., Chauhan, N.: Cost and effort estimation in agile software development. In: 2014 International Conference on Reliability Optimization and Information Technology (ICROIT), pp. 57–61 (2014)
3. Wen, J., Li, S., Lin, Z., Hu, Y., Huang, C.: Systematic literature review of machine learning based software development effort estimation models. Inf. Softw. Technol. **54**(1), 41–59 (2012)
4. Bilgaiyan, S., Mishra, S., Das, M.: A review of software cost estimation in Agile software development using soft computing techniques. In: 2016 2nd International Conference on Computational Intelligence and Networks (CINE), pp. 112–117 (2016)
5. Sharma, A., Ranjan, R.: Software effort estimation using neuro fuzzy inference system: past and present. Int. J. Recent Innov. Trends Comput. Commun. **5**(8), 78–83 (2017)
6. Samareh Moosavi, S.H., Khatibi Bardsiri, V.: Satin bowerbird optimizer: a new optimization algorithm to optimize ANFIS for software development effort estimation. Eng. Appl. Artif. Intell. **60**, 1–15 (2017)
7. Jorgensen, M., Shepperd, M.: A systematic review of software development cost estimation studies. IEEE Trans. Softw. Eng. **33**(1), 33–53 (2007)
8. Pospieszny, P., Czarnacka-Chrobot, B., Kobylinski, A.: An effective approach for software project effort and duration estimation with machine learning algorithms. J. Syst. Softw. **137**, 184–196 (2018)
9. Satapathy, S.M., Panda, A., Rath, S.K.: Story point approach based Agile software effort estimation using various SVR Kernel methods. In: International Conference on Software Engineering and Knowledge Engineering, pp. 304–307 (2014)
10. Mendes, E., Watson, I., Triggs, C., Mosley, N., Counsell, S.: A comparative study of cost estimation models for web hypermedia applications. Empir. Softw. Eng. **8**(2), 163–196 (2003)
11. Usman, M., Mendes, E., Börstler, J.: Effort estimation in Agile software development: a survey on the state of the practice. In ACM International Conference Proceedings Series, pp. 1–10 (2015)
12. Azzeh, M., Nassif, A.B., Banitaan, S.: Comparative analysis of soft computing techniques for predicting software effort based use case points. IET Softw. **12**(1), 19–29 (2018)
13. Yousef, Q.M., Alshaer, Y.A.: Dragonfly estimator: a hybrid software projects' efforts estimation model using artificial neural network and Dragonfly algorithm. Int. J. Comput. Sci. Netw. Secur. **17**(9), 108–120 (2017)
14. Menzies, T., Yang, Y., Mathew, G., Boehm, B., Hihn, J.: Negative results for software effort estimation. Empir. Softw. Eng. **22**(5), 2658–2683 (2017)
15. Alostad, J.M., Abdullah, L.R.A., Aali, L.S.: A fuzzy based model for effort estimation in Scrum projects. Int. J. Adv. Comput. Sci. Appl. (IJACSA) **8**(9), 270–277 (2017)
16. Idri, A., Hosni, M., Abran, A.: Systematic literature review of ensemble effort estimation. J. Syst. Softw. **1**, 1–35 (2016)
17. Bilgaiyan, S., Sagnika, S., Mishra, S., Das, M.: A systematic review on software cost estimation in Agile software development. J. Eng. Sci. Technol. Rev. **10**(4), 51–64 (2017)
18. Radlinski, L.: A survey of bayesian net models for software development effort prediction. Int. J. Softw. Eng. Comput. **2**(2), 95–109 (2010)

19. Dragicevic, S., Celar, S., Turic, M.: Bayesian network model for task effort estimation in agile software development. J. Syst. Softw. **127**, 109–119 (2017)
20. Salmanoglu, O.D.M., Hacaloglu, T.: Effort estimation for Agile software development : comparative case studies using COSMIC functional size measurement and story points. In: ACM Mensura, pp. 1–9 (2017)
21. Padmaja, M., Haritha, D.: Software effort estimation using meta heuristic algorithm. Int. J. Adv. Res. Comput. Sci. **8**(5), 196–201 (2017)
22. Murillo-Morera, J., Quesada-López, C., Castro-Herrera, C., Jenkins, M.: A genetic algorithm based framework for software effort prediction. J. Softw. Eng. Res. Dev. **5**(1), 1–33 (2017)
23. de Araújo, R.A., Oliveira, A.L.I., Meira, S.: A class of hybrid multilayer perceptrons for software development effort estimation problems. Expert Syst. Appl. **90**, 1–12 (2017)
24. Dave, V.S., Dutta, K.: Neural network based models for software effort estimation: a review. Artif. Intell. Rev. **42**(2), 295–307 (2014)
25. Khuat, T., Le, H.: An effort estimation approach for Agile software development using fireworks algorithm optimized neural network. Int. J. Comput. Sci. Inf. Secur. **14**(7), 122–130 (2018)
26. Adnan, M., Afzal, M.: Ontology based multiagent effort estimation system for Scrum Agile method. IEEE Access, 25993–26005 (2017)
27. Panda, A., Satapathy, S.M., Rath, S.K.: Empirical validation of neural network models for Agile software effort estimation based on story points. Procedia Comput. Sci. **57**, 772–781 (2015)
28. Satapathy, S.M., Rath, S.K.: Empirical assessment of machine learning models for agile software development effort estimation using story points. Innov. Syst. Softw. Eng. **13**(2–3), 191–200 (2017)
29. Khuat, T.T., Le, M.H.: A novel hybrid ABC-PSO algorithm for effort estimation of software projects using Agile methodologies. J. Intell. Syst. **27**(3), 489–506 (2018)
30. Porru, S., Murgia, A., Demeyer, S., Marchesi, M., Tonelli, R.: Estimating story points from issue reports. In: Proceedings of the 12th International Conference on Predictive Models and Data Analytics in Software Engineering—PROMISE 2016, pp. 1–10 (2016)
31. Moharreri, K., Sapre, A.V., Ramanathan, J., Ramnath, R.: Cost-effective supervised learning models for software effort estimation in Agile environments. In: 2016 IEEE 40th Annual Computer Software and Applications Conference (COMPSAC), pp. 135–140 (2016)
32. Choetkiertikul, M., Dam, H.K., Tran, T., Pham, T.T.M., Ghose, A., Menzies, T.: A deep learning model for estimating story points. IEEE Trans. Softw. Eng. **14**(8), 1–12 (2016)
33. Manga, I., Blamah, N.V.: A particle swarm optimization-based framework for Agile software effort estimation. Int. J. Eng. Sci. **3**(6), 30–36 (2014)
34. Tipu, S.K., Zia, S.: An effort estimation model for agile software development. Adv. Comput. Sci. Appl. **2**(1), 314–324 (2012)
35. Popli, R., Chauhan, N.: Agile estimation using people and project related factors. In: 2014 International Conference on Computing for Sustainable Global Development (INDIACom), pp. 564–569 (2014)
36. Arora, M., Chopra, S., Gupta, P.: Estimation of regression test effort in Agile projects. Far East J. Electron. Commun. **3**(II), 741–753 (2016)

HCI Using Gestural Recognition for Symbol-Based Communication Methodologies

Prerna Sharma, Nitigya Sharma, Pranav Khandelwal and Tariq Hussain Sheikh

Abstract Sign language in itself is the only tool of communication for the society which is not able to hear voices and speak words. Using sign language, they can express their emotions and thoughts and can convey what they want to say. But not everyone understands sign language, only the people who require it do. So people with such kinds of handicaps need a translator with them in order to convert their language to a common tongue and that is the main reason of sign language recognition becoming such a crucial task. Since sign language consists of different movements and positions of the hand, therefore, the accuracy of sign language depends on how accurately the machine could recognize the gesture. We are trying to develop such a system what we call translating HCI for sign language. In this system, the user has to place their hand in front of the webcam performing sign gestures and in real time, the system will read your hand gesture and will return the respective character/alphabet on the screen. Utilizing the proposed system normal people can understand sign language and can easily communicate with hearing-impaired people.

Keywords Deep learning · Image processing · HCI · Machine learning · Convolutional neural network

P. Sharma · N. Sharma (✉) · P. Khandelwal
Maharaja Agrasen Institute of Technology, Delhi, India
e-mail: nitigya.sharma12@gmail.com

P. Sharma
e-mail: prernasharma@mait.ac.in

P. Khandelwal
e-mail: khandelwalpranav1996@gmail.com

T. H. Sheikh
Government Degree College, Poonch, India
e-mail: Tariqsheakh2000@gmail.com

© Springer Nature Singapore Pte Ltd. 2020
P. K. Mallick et al. (eds.), *Cognitive Informatics and Soft Computing*,
Advances in Intelligent Systems and Computing 1040,
https://doi.org/10.1007/978-981-15-1451-7_60

1   Introduction

Sign language (SL) [1] is a visual–gestural language used by deaf and hard-hearing people for communication purposes. Three-dimensional spaces and the hand movements are used (and other parts of the body) to convey meanings. It has its own vocabulary and syntax which is purely different from spoken languages/written language. Spoken languages use the oratory faculties to produce sounds mapped against specific words and grammatical combinations to convey meaningful information. Then the oratory elements are received by the auditory faculties and processed accordingly. Sign language uses the visual faculties which are different from spoken language. Spoken language makes use of rules to produce comprehensive messages; similarly, sign language is also governed by a complex grammar words and grammatical combinations to convey me deaf and hard-hearing people use visual and gestural language for communication purposes which is known as sign language. Meaning is conveyed using three-dimensional spaces and hand movements. The language spoken/written is purely different from sign language in terms of vocabulary and syntax. Oratory facilities are utilized to convey meaningful information by making combinations of sounds. Then the sound produced is received by the auditory organs and processed accordingly. On the contrary, sign language uses the visual faculties. Every language in the world utilizes complex grammatical rules to make it understandable and unified among the masses so is done in sign language. It has its own rules and grammar [2]. Sign language has also got grammar and vocabulary but unlike most of the available languages, visual modality is applied for exchanging information. Like for any other language, it becomes difficult for a foreigner to understand language of other unknown masses similarly problem arises when sign language utilizing community has to deal with other communities. As a result, a deaf person gets confined to his/her family or the deaf community for communication. As they start feeling hesitant to talk to others. This hesitation is evolved due to the difficulties they have to face while making an attempt to make others understand what they are trying to say [3]. Sign language is mainly gestural language, and hence this study aims to develop an HCI for understanding hand gestures and develop alphabets, words and sentences. Rather than going for expensive electronic technology like Kinect or gloves, this work aims to resolve this in a cost-effective way utilizing computer vision and machine learning algorithms.

According to our study, all the user need to do is to place their hand in front of the webcam and with help of OpenCV, that frame will be captured by the machine and get pass on to the deep learning model for further prediction of the character according to the hand gesture specified by the user.

In this work, for reducing the chances of error while recognizing the respective hand gesture we have defined a specific rectangle on the webcam for the user. So, the model will only predict the respective character if the hand placed inside that rectangular area.

Before passing the captured frame to the deep learning model it is converted from RGB to HSV, as the HSV format will intensify the light intensity at the hand gesture and decrease the light intensity in the background. Main reason behind this is because our model is trained on that dataset which was already in HSV format.

2 Literature Review

Different ideas and methods have been implemented by the researchers for the recognition of sign language. Few approaches were based on different ideas like sensory gloves, artificial neural network, support vector machine and other classification algorithms [4].

In this work, we will be using convolutional neural network for the recognition of hand gestures and orientations. Till 2013, the field of computer vision or image processing was not that efficient because that time image processing was done by feature extraction, and machine learning which was relatively slow and slightly less accurate for image recognition. Because earlier the best algorithm used for image recognition was histogram of oriented gradient (HOG) and support vector machine (SVM) [5] which was very good for any classification problem but the training time as well as the prediction time both were relatively slow as compared to the new deep learning algorithm that are now available for application.

But after that when convolutional neural network was introduced to developers for image processing it changed the field of computer vision because CNN was much better from machine learning in extracting the specific feature from the image and was more accurate in recognizing them.

3 Methods and Material

3.1 Proposed Idea

The user/deaf people will place their hand in front of the webcam. Now the frames per second would be captured from the camera. The captured image will be cropped to 64 × 64 pixel that will be containing only the hand gesture of the user. Now the cropped image is in RGB format it will be get converted in HSV format (black and white) because only the orientation of the fingers and palm are required for recognizing the character and not the skin color of the hand. The converted image is now passed through the convolutional neural network which is the most appropriate neural network used for computer vision. Data augmentation is also performed within the convolutional neural network in order to make more robust. After the neural network model is trained, OpenCV would be used for capturing the images for the user. The result of the image will be displayed on the screen, that is, the character assigned to that hand gesture (Fig. 1).

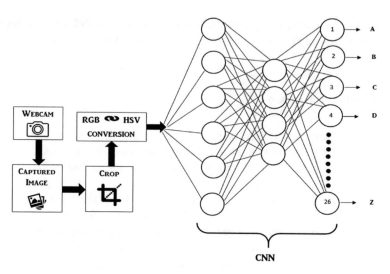

Fig. 1 Process architecture

3.2 Data Acquisition

Camera is the crucial device that will be used in this model. As the input is in the form of captured image of hand gesture that was made by the user. Some researchers depend on simple camera to get the image. Sensory gloves are also used by few researchers previously but they require more investment and a different type of machine training algorithms. But at our level, we will be using a webcam for real-time hand gesture detection for the user. Before passing the image to the convolutional neural network for training the image is converted from RGB format to HSV format. But fortunately, the dataset for the training the neural network model was available on GitHub. The dataset available was the images of 26 different hand gestures refer-ring to the 26 characters on the English alphabet and for each alphabet there were 2000 images, respectively. The images are of 64×64 pixel and were already in HSV format (Fig. 2).

3.3 Processing Method

Now, the most appropriate choice for image recognition used in deep learning is 3D convolutional neural network. The idea behind 3D CNN is nothing but the 2D CNN itself. In CNN, the feature extraction is done by using feature map or filter map or also known as kernel. Feature map is matrix which is slide on to the whole image and calculating its matrix multiplication with the image to enhance the feature present in the image. As the kernel slide through the whole image for extracting both upper as

Fig. 2 Dataset images for
some characters

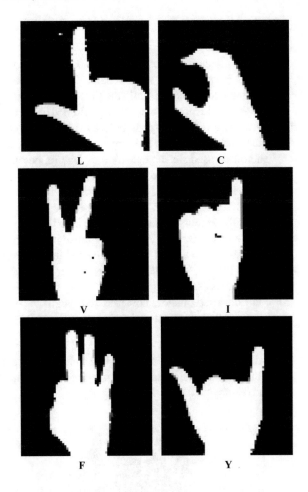

well the lower features. Each CNN is three layers deep. Local contrast normalization
is also applied to the first to layer of CNN and then the output of neurons is passed
through rectifier linear unit (ReLU). Further, the output of ReLU is passed on to the
pooling layer (in our model we have used max pooling), in order to decrease the size
of matrix as well as enhancing the feature value existing in the feature matrix. ReLU
provides nonlinearity to the model which thereby making it better for classification
because the image pixels are already complex. Training gets faster, even in the case
of bigger datasets because of ReLU and pooling as compared to the standard CNN
model. While ReLU does not require the idea of normalization but still local contrast
normalization is used. The normalization aids in generalization and it decreases the
test error rates. To make CNN better for prediction data augmentation is used and for
reducing errors and also to reduce overfitting, dropouts are used. In dropouts, few
neurons are closed so the model has to learn a new path for training [6].

3.4 Implementation

(1) The implementation of our work is basic and shown below in Fig. 3. Now, the image is captured by the webcam. The green box on the screen is where the hand would be placed while the time of recognition. The green box area would get cropped from the whole image frame (Fig. 4).

(2) Now the cropped image is in RGB format. Basically, RGB refers to the combinations of the color red, green and blue. As it is important to decide which should be used for the detection of feature, colors or black and white. For the needs of this study, the physical appearance of the hand is not utilized but only its orientation is used, hence we will be using the intensity (that is, black and white) for recognition and analyzing the image and because there only two color; the amount of data to analyze is reduced and the load on the processor is also reduced. But black and white make it difficult to differentiate between the feature to be extracted (that is, the hand in our case) and the background of the image [4]. So, we will be converting our cropped image into HSV format for further processing (Fig. 5).

(3) The converted image is now passed through the convolutional neural network to match with the character/alphabet referring to the gesture in the cropped image.

The last layer of our convolutional neural network contains 26 neurons because of the 26 different characters in the English alphabet. Once the image is passed on to the CNN, that is, the orientation of the image pixels, it matches the character referring to it and displays it on the screen (Fig. 6).

Fig. 3 HCI recognizing letter V

Fig. 4 Input for HCI

(4) Adding the fact that communication is not done just by using single charac-
ter/alphabets but through words and sentences. So, to make possible we added
a key feature in this process. If the user wants to make word from the character,
they just need to press the '*p*' key on their keyboard when the matching character
display on the screen as it will stack up the character to its previous character.
This key is sufficient to make a complete word but for making a complete sentence we need to add space between the words. For that '*w*' key will be pressed
after stacking up the last character of the previous and before the first character
for next word is added.

Further adding an extra feature, when the sentence/word is complete and read by
the second person '*n*' key will be used to clear the text are for the next word/sentence
(Fig. 7).

4   Results and Discussion

While using image processing, deep learning is widely considered a better approach
than machine learning because feature extraction is an inbuilt process in deep learning
and more efficient than used in machine learning. And in the terms of accuracy, deep
learning also performs better than machine learning but the data required is more in
deep learning but fortunately in 2018 we have sufficient data to work on (Fig. 8).

Fig. 5 RGB to HSV
conversion

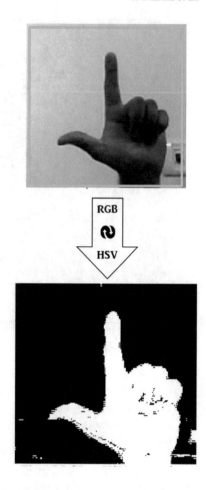

Convolutional neural network worked very well in terms of performance and
accuracy. After training the CNN model with 15 epochs and with the batch size of
800, the accuracy of the model observed to be 97.8%.

5   Limitations

The prototype we build successfully detects several basic words to form a complete a sentence but still there are some limitations which are understated with easy
expandable solutions.

Problem: We have to place our hands in certain orientation and position.
Solution: It is done for attaining a higher accuracy but a rather flexible hand
positioning system could be achieved by rising the density of the training data.

Fig. 6 Output for symbol L

Fig. 7 Sentence formation

Problem: The work currently does not recognize words formed through hand gestures. It detects the data character to character according to the dataset.

Solution: This problem can be resolved only through further research which will be done later. The dataset for words related to certain gestures is vast and will need to be updated frequently after feeding it into the system.

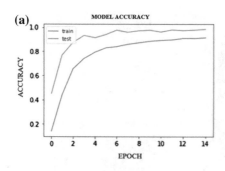 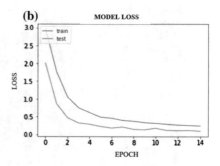

Fig. 8 **a** Model accuracy, **b** model loss

Problem: The image detection is light sensitive because of the conversion of the image from RGB to HSV (black–white) before feeding it to the neural network.
Solution: The HSV levels need to be adjusted by the user according to the background lighting so that the hand gestures can be detected correctly and clearly.

6 Conclusion

The outcome of our research and hard work is that we have developed a prototype of a system utilizing computer vision and machine learning which can actually very cost-effectively fulfill the need of an external costly human translator.

While working on the project we went through various studies and came across a project which helped us a lot to get the desired dataset for our study. Rupesh, "https://github.com/rrupeshh" narrowed down our search and helped our study with the dataset. The dataset can be found here "https://github.com/rrupeshh/Simple-Sign-Language-Detector/tree/master/mydata".

Some previous work that has been done on sign language recognition has more accuracy because the dataset was formed from single person [7]. But if add images of more than five people in the dataset the resulting accuracy would decrease but it would be more reasonable because different people will have a slightly different way of representing the same character.

6.1 Future Scope

There were many factors and idea which we took in our sight while started working on this topic and this work have the ability to be extended for all those other usage like:

The prototype utilizes human inputs as callbacks to develop sentences. In the future, this can be converted to automatic system by making the system learn the basic time a person requires to perfectly show a symbol. By understanding this we can calculate resolution time and trigger automatic callbacks accordingly.

Reverse communication channel can be built easily to convert natural language to sign language.

The real use of this research is in integration of this technology with currently working technologies and empowers ease of access. It can be utilized with application like video chats, speech recognition, etc., to give differently abled citizen a chance to get mixed with the mainstream.

References

1. Konstantinidis, D., Dimitropoulos, K., Daras, P.: Sign language recognition based on hand and body skeletal data. (Research Gate)
2. Sahoo, A.K., Mishra, G.S., Ravulakollu, K.K.: Sign Language Recognition: State of the Art (2014)
3. Kagalkar, R.M., Gumaste, S.V.: ANFIS Based Methodology for Sign Language Recognition and Translating to Number in Kannada Language (2018)
4. Jadhav, A., Tatkar, G., Hanwate, G., Patwardhan, R.: Sign Language recognition. Int. J. Adv. Res. Comput. Sci. Softw. Eng. 7(3) (2017)
5. Jain, S., Sameer Raja, K.V., Mukerjee, A.: Indian Sign Language Character Recognition. Indian Institute of Technology, Kanpur Course Project-CS365A (2016)
6. Pramada, S., Saylee, D., Pranita, N., Samiksha, N., Vaidya, A.S.: Intelligent sign language recognition using image processing. IOSR J. Eng. (IOSRJEN) 3(2), pp. 45–51 (2013). e-ISSN: 2250-3021, p-ISSN: 2278-8719, ‖V2‖
7. Suharjitoa, Andersonb, R., Wiryanab, F., Ariestab, MC, Kusumaa, G.P.: Sign language recognition application systems for Deaf-Mute people: a review based on input-process-output. In: 2nd International Conference on Computer Science and Computational Intelligence, ICCSCI (2017)

Modified Ant Lion Optimization Algorithm for Improved Diagnosis of Thyroid Disease

Naman Gupta, Rishabh Jain, Deepak Gupta, Ashish Khanna and Aditya Khamparia

Abstract Thyroid is one of the most common diseases affecting millions of individuals across the world. According to the findings from numerous studies and surveys on thyroid disease, it is estimated that about 42 million people in India and around 20 million people in America are suffering from some form of thyroid diseases, and women make up the majority of thyroid patients among them. It is caused due to the under (Hypothyroidism) or over (Hyperthyroidism) functionality of thyroid gland, which is responsible for maintaining the metabolism of the body, and it is imperative to diagnose its effects as early as possible so that a possible cure or treatment can be performed at the earliest. This paper aims to propose a modified ant lion optimization algorithm (MALO) for improving the diagnostic accuracy of thyroid disease. The proposed MALO is employed as a feature selection method to identify the most significant set of attributes from a large pool of available attributes to improve the classification accuracy and to reduce the computational time. Feature selection is one of the most significant aspects of machine learning which is used to remove the insignificant features from a given dataset to improve the accuracy of machine learning classifiers. Three different classifiers, namely Random Forest, k-Nearest Neighbor (kNN) and Decision Tree, are used for diagnosing the thyroid disease. The experimental results indicate that MALO eliminates 71.5% insignificant features out of the total number of features. The best accuracy achieved on the reduced

N. Gupta (✉) · R. Jain · D. Gupta · A. Khanna
Maharaja Agrasen Institute of Technology, Guru Gobind Singh Indraprastha University, Delhi, India
e-mail: namangupta0227@gmail.com

R. Jain
e-mail: rishabh.jain1379@gmail.com

D. Gupta
e-mail: deepakgupta@mait.ac.in

A. Khanna
e-mail: ashishkhanna@mait.ac.in

A. Khamparia
School of Computer Science and Engineering, Lovely Professional University, Punjab, India
e-mail: aditya.khamparia88@gmail.com

© Springer Nature Singapore Pte Ltd. 2020
P. K. Mallick et al. (eds.), *Cognitive Informatics and Soft Computing*,
Advances in Intelligent Systems and Computing 1040,
https://doi.org/10.1007/978-981-15-1451-7_61

set of features is 95.94% with Random Forest Classifier. Also, a notable accuracy of 95.66% and 92.51% has been achieved by Decision Tree classifier and k-Nearest Neighbor classifier, respectively. Additionally, MALO has been compared with other optimized variants of evolutionary algorithms to show the effectiveness and superiority of the proposed algorithm. Hence, the experimental results indicate that the MALO significantly outperforms the other algorithms present in the literature.

Keywords Thyroid disease · Ant Lion optimization algorithm · Modified Ant Lion optimization algorithm · Evolutionary algorithm · Feature selection · Machine learning

1 Introduction

Thyroid is an endocrine gland which secrets thyroid hormones to regulate the metabolism and protein synthesis inside the body. It consists of two lobes connected by an isthmus and is located below the Adam's apple, in front of the neck. Other than the two well-known hormones, triiodothyronine (T3) and thyroxine (T4), calcitonin is another hormone which is produced by thyroid gland and is responsible for the regulation of calcium ions by a process called calcium homeostasis [1]. Thyroid hormones are secreted by thyroid gland which is signaled by thyroid-stimulating hormone (TSH). Pituitary gland releases TSH, signaled by thyroid-releasing hormone (TRH). TRH is produced by hypothalamus which is a small organ present in the brain. Finally, T4 and T3 (the active hormone) are released into the blood. Later, the hormone T4 is converted to T3 in the liver and other parts throughout the body. The basic types of thyroid which have been clinically identified are hyperthyroidism and hypothyroidism.

There are several diseases which may affect thyroid, for example, one of the most common causes for occurring hyperthyroidism is Graves' disease. Several genetic combinations and environmental factors causing Graves' disease are explained in [2]. Likewise, iodine deficiency is the most common cause of hypothyroidism and can be preventable intellectual disability [3] at further stages.

An evolutionary algorithm is considered to be one most important component of evolutionary computation in artificial intelligence. The working of an evolutionary algorithm is performed by an evaluating process in which each member of the search space, running for being an optimal solution, is evaluated. The evaluation process is done with the help of a fitness function which is either data dependent or data independent. The least fit member is eliminated in each iteration, and rest of the members are allowed to compete in successive iterations, till the optimal solution is obtained. The mathematical model behind each algorithm is inspired by biological processes like mutation, reproduction and hunting.

In recent years, the computational data in various sectors has drastically increased and is stored in a tabular format containing a huge number of instances and attributes (features), which gradually increases noise in the datasets. Computational processing like model training, knowledge discovery is not feasible on such datasets due to the increase in computational time and cost. Due to these factors, feature selection is performed primarily on these datasets before processing them in further stages. It aims to improve the efficiency and accuracy of classifiers by obtaining the most significant features from all the features present in the dataset. Unlike traditional algorithms for feature selection, evolutionary algorithms have surpassed them in terms of both time and cost. Several evolutionary algorithms have been widely used as feature selection methods, namely Binary Gray Wolf Optimization [4], Binary Bat Algorithm [5], Modified Whale optimization algorithm [6], etc.

In 2015, Mirjalili [7] proposed a metaheuristic bio-inspired evolutionary algorithm called Ant Lion optimizer (ALO), inspired from the hunting behavior of an ant lion. Later in 2015, Zawbaa et al. [8] presented a feature selection model using ALO to improve the feature selection process. Further advancement in ALO was done by Mirjalili–Jangir–Saremi in [9], proposing a multi-objective version of Ant Lion optimization algorithm (MOALO) to solve complex constrained engineering problems. In [10], Emary–Zawbaa–Hassanien suggested a binary variant of ALO and employed it as a wrapper-based feature selection method. Ali et al. [11] gave a multi-objective variant of ALO to find the optimal locations of distributed generation (DG) systems to improve the overall performance of DG.

In the presented work, a modified Ant Lion optimization (MALO) algorithm has been introduced for improved classification of thyroid disease. MALO is a modified variant of traditional ALO algorithm with better optimization. All the equations of the standard algorithm have been quantized. The proposed model generates a set of most significant features leaving behind the irrelevant features.

The presented exposition is organized in the following manner: Section 2 discusses the preliminary, algorithm and techniques of traditional ALO. The proposed MALO algorithm is presented in Sect. 3. Section 4 presents the dataset description, experimental setup details and input parameters. The results obtained are discussed in Sect. 5. Finally, conclusion along with the potential future aspects of the proposed work is discussed in Sect. 6.

2 Preliminaries

Ant lion optimization is a metaheuristic algorithm which inspired from nature. It was proposed by Mirjalili [7] to imitate the hunting pattern of Ant Lions.

Ant Lions—They are also known as doodlebugs which hunt their prey in larvae stage and reproduce in their adulthood stage. They pursue their hunting process by digging a cone-shaped hole or trap by moving around in a circular path. After the trap gets dug, Ant Lion waits for their prey at the bottom of the trap. Once the prey is caught in the trap, sand is thrown at it to gradually bury it under the sand. Finally, after catching it in their jaw, it is consumed by them, and the remains are thrown out of the trap to make it ready for another prey.

Based on the hunting process of doodlebugs, Mirjalili [7] put down conditions for the optimization algorithm.

- Prey—Ants. They use different random walks to move around the search space. The traps of the Ant Lion affect their random walks.
- The size and the depth of the hole it can dig depend on its fitness (high fitness corresponds to large hole).
- The probability of catching prey for an Ant Lion will be higher if the hole dug is large.
- The elite (fittest Ant Lion) and an Ant Lion in each iteration can catch the ant.
- To simulate the sliding of the ants, their random walk is decreased rapidly, once they enter the hole.

The pseudocode of ALO is described in Algorithm 1.

Algorithm 1: Ant Lion Optimization Algorithm (ALO)

Input: Search space, fitness function, Number of ants and Antlions, number of iterations (T)
Output: The Elitist Antlion and the its fitness

1. Initialize a population of n ant's positions and n antlion's positions randomly.
2. Calculate the fitness of all ants and antlions.
3. Find the fittest antlion; Elite.
4. t=0
5. while $(t \leq T)$
    for each ant_i **do**
        Select an antlion using Roulette wheel (building trap).
        Slide ants towards the antlion; see equations (2), (3).
        Create a random walk for this ant_i and normalize it; see equations (5) and (6) for modeling trapping, equation (7) for random walk, and equation (9) for normalization of walk.
    end for
6. Calculate the fitness of all ants.
7. Replace an antlion with its corresponding ant it if becomes fitter (Catching Prey); see equation (1).
8. Update elite if an antlion becomes fitter than the elite.

    end while

When an ant reaches to a fitness value greater than the Ant Lion, it gets pulled into the sand and is caught by the Ant Lion. After every catch, the ant lion amends the trap and replaces its position with the last caught prey and gets ready to catch another prey. The Ant Lion optimizer [7] works on the following steps for each Ant Lion:

1. **Building Trap**: To increase the probability of selecting Ant Lions with higher fitness, Roulette Wheel is used.
2. **Catching Prey and re-building the hole**: In this stage, the Ant Lion replaces its position with the position of its prey after consuming it. This happens because the hunting of prey can only be possible if the prey (ant) becomes fitter than the Ant Lion. The following equation shows how the position of Ant Lion gets updated after a hunting round:

$$\text{Antlion}_j^t = \text{Ant}_i^t, \text{ if } f\left(\text{Ant}_i^t\right) > f\left(\text{Antlion}_j^t\right) \tag{1}$$

where current iteration is represented by t. Here, Antlion_j^t and Ant_i^t represent the position of jth antlion and ith ant at tth iteration, respectively.

The Ant Lion optimizer works on the following steps for each ant:

1. **Sliding ants toward Ant Lions**: Once the Ant Lion realizes that their prey has entered the trap, they shoot sand outward to catch them. This activity slides the ants toward the center of the hole. To simulate this sliding of ants, we have mathematical Eqs. 2, 3 and 4 which decrease the radius of their random walk hyper-sphere.

$$c^t = \frac{c^t}{I} \tag{2}$$

$$d^t = \frac{d^t}{I} \tag{3}$$

$$I = 10^w \frac{t}{T} \tag{4}$$

Here, c^t and d^t denote the minimum and maximum of all variables at tth iteration in Eqs. 2 and 3, respectively. In Eq. 4, current iteration is represented by t, and the maximum number of iterations is represented by T. W is a constant which is defined based on the current iteration as shown in Table 1.

2. **Trapping in Ant Lion holes**: The selected ant lion gets the ant trapped in their hole. The walk of the ants gets affected and confined by the position of the selected Ant Lion. The following mathematical Eqs. 5 and 6 represent their random walk in search space:

$$c_i^t = c^t + \text{Antlion}_j^t \tag{5}$$

$$d_i^t = d^t + \text{Antlion}_j^t \tag{6}$$

3. **Random Walk of ants**: The random walk of ants around the Ant Lions is simulated mathematically as shown in Eq. 7.

Table 1 Values of W at different values of current iteration t

t	W
$>0.1T$	2
$>0.5T$	3
$>0.75T$	4
$>0.9T$	5
$>0.95T$	6

$$X(t) = [0, \text{cumsum}(2r(t_1) - 1); \text{cumsum}(2r(t_2) - 1); \ldots;$$
$$\text{cumsum}(2r(t_T) - 1)] \tag{7}$$

where cumsum denotes the cumulative sum. The maximum number of iterations and current iteration are represented by T and t, respectively. Here, t also denotes the step of random walk, and $r(t)$ is the stochastic function from Eq. 8.

$$r(t) = \{1 \text{ifrand} > 0.5 \land 0 \text{ifrand} \leq 0.5\} \tag{8}$$

To ensure that the random walks are always within the search space, normalization is performed using Eq. 9.

$$X_i^t = \frac{\left[\left(X_i^t - a_i\right) * \left(d_i - c_i^t\right)\right]}{b_i^t - a_i} + c_i \tag{9}$$

Here, the minimum and maximum of the random walk of the ith variable are represented by a_i and b_i, respectively.

3 Modified MALO Algorithm for Feature Selection

Elitism: To make sure that the best possible solution is obtained across iterations, elitism is used, which means both the selected Ant Lions and elite Ant Lion obtained so far are saved, and it is assumed that they both affect the random walk of ants in the search space. Hence, the crossover of both the random walks is used to change the position of ants, stated in Eq. 10:

$$X_i^{t+1} = \text{Crossover}(\text{RW}_1, \text{RW}_2) \tag{10}$$

where the function Crossover $(\text{RW}_1, \text{RW}_2)$ is the appropriate crossover between solutions RW_1, RW_2. Here, RW_1 and RW_2 are the random walks. The used crossover is the simple stochastic as given in Eq. 11:

$$x^d = \{x_1^d \text{ifrand} \geq 0.5, x_2^d \text{otherwise}\} \tag{11}$$

where x_1 and x_2 are vectors at dimension d. After the crossover, the sigmoid function is applied along with a threshold function given below in Eq. 12 to convert the continuous values into binary values.

$$\text{Ant}_d^{t+1} = \{1, \text{if } y^d \geq \text{rand}, 0 \text{ otherwise}\} \tag{12}$$

where the output of the sigmoid function is y^d.

The flow graph of traditional ALO and proposed MALO is shown in Fig. 1.

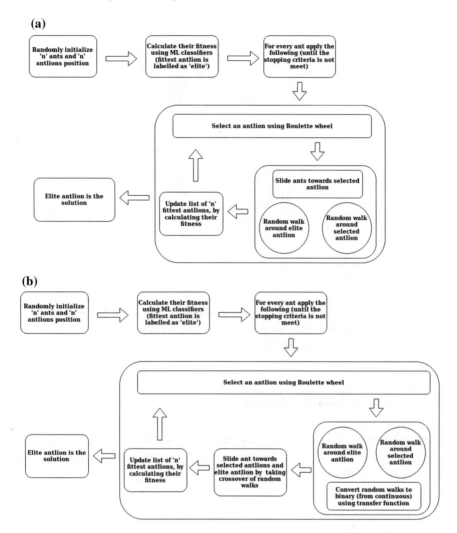

Fig. 1 Flowchart: **a** Traditional ALO and **b** proposed MALO

4 Implementation

4.1 Experimental Setup

The entire experiment was performed on a mobile computing setup, a laptop with Core i5 (5th Generation) clocked at 2.2 GHz coupled with 12 Gigabytes of RAM and 1 TB Hard disk, running Ubuntu OS (ver. 18.04). The software setup included Anaconda with Python 3.2 with libraries, namely numpy, scikit-learn, plotly and pandas.

Table 2 Input variables

Variables	Value	Description
SearchAgents_no	21	Total number of search agents
dim	50	Total number of dimensions of each search agent
Max_iter	10	Total number of iterations

Table 3 Classifiers with their tuning parameters

Classifier	Parameters
Random Forest	n_estimators = 8, criterion = 'entropy', random_state = 0
K-Nearest Neighbor	n_neighbors = 3
Decision tree	criterion = "entropy", max_depth = 3, max_leaf_nodes = 2

4.2 Input Variables

The input variables to be initialized at the beginning of the algorithm are number of search agents (population size), dimension and maximum number of iterations. The values of these variables are presented in Table 2.

4.3 Machine Learning Models

To classify thyroid disease, three different machine learning classifiers were used, namely Random Forest, Decision Tree and k-Nearest Neighbor. The tuned parameters used in each classifier are shown in Table 3.

4.4 Thyroid Dataset

The thyroid dataset was obtained from University of California at Irvine (UCI) storehouse. It consists of data for both hypothyroid and hyperthyroid conditions. It contains 21 attributes and 7200 records along with a prediction column consisting three different types of classes, i.e., normal, hyper and hypo. The names of the attributes and their domain are provided in Table 4.

Table 4 Attribute information

Attribute	Domain	Attribute	Domain	Attribute	Domain
Age	[0.01, 0.97]	Thyroid_surgey	[0, 1]	Hypopituitary	[0, 1]
Sex	[0, 1]	131_treatment	[0, 1]	Psych	[0, 1]
On_throxine	[0, 1]	Query_hypothyroid	[0, 1]	TSH	[0.0, 0.53]
Query_on_thyroxine	[0, 1]	Query_hyerothyroid	[0, 1]	T3	[0.0005, 0.18]
On_antithyroid_medication	[0, 1]	Lithium	[0, 1]	TT4	[0.0020, 0.6]
Sick	[0, 1]	Goiter	[0, 1]	T4U	[0.017, 0.233]
Pregnant	[0, 1]	Tumor	[0, 1]	FTI	[0.0020, 0.6421]

5 Results and Discussion

This section discusses the results obtained when the proposed MALO and optimized cuttlefish algorithm (OCFA) were applied to the thyroid dataset. The dataset consists of 7200 records combining normal, hyper and hypothyroid patients. The total number of significant thyroid features selected by each algorithm and accuracy achieved with them using different classifiers is presented in Tables 5 and 6, respectively.

The accuracy plot for modified ant lion optimization algorithm (MALO) and optimized cuttlefish algorithm (OCFA) is shown in Figs. 2 and 3, respectively.

Table 5 Total no. of thyroid features selected by MALO and OCFA

Bio-inspired feature selection algorithms	Number of features selected
Modified Ant Lion algorithm (MALO)	6
Optimized cuttlefish algorithm (OCFA)	9

Table 6 Accuracy for each algorithm

Algorithm	ML classifiers	Accuracy (%)
Modified Ant Lion algorithm	Random Forest	95.94
	K-Nearest Neighbor	92.51
	Decision Tree	95.66
Optimized cuttlefish algorithm	Random Forest	86.76
	K-Nearest Neighbor	92.58
	Decision Tree	92.55

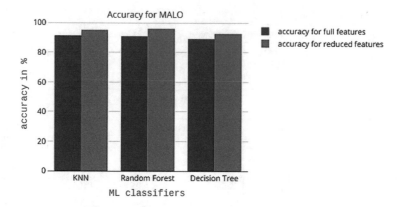

Fig. 2 Accuracy obtained by MALO

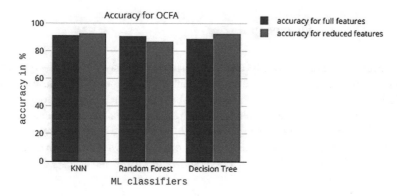

Fig. 3 Accuracy obtained by OCFA

The comparison of accuracy and the number optimal features selected by MALO and OCFA is shown in Figs. 4 and 5, respectively:

6 Conclusion and Future Scope

In the presented work, a novel modified version of Ant Lion optimization algorithm called as MALO has been proposed for early and accurate diagnosis of thyroid disease. The proposed MALO has been implemented as a feature selection method that follows a wrapper-based approach to eliminate insignificant features, thus increasing the classification accuracy and decreasing the computational time and costs. The experimental results demonstrate a very high accuracy of 95.94% with MALO, and it outperforms the other algorithms present in the literature for thyroid detection. Having demonstrated the potential of proposed algorithm, it can be combined with other

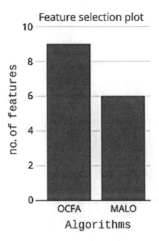

Fig. 4 Feature selection plot

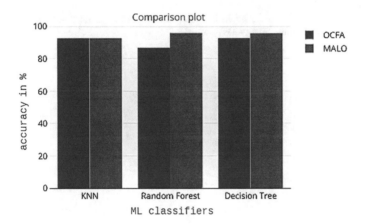

Fig. 5 Accuracy for algorithms

bio-inspired algorithms by following a hybrid approach to improve the diagnosis of thyroid disease. Many other bio-inspired can also be applied as feature selection methods combined with other machine learning classification models and deep learning approaches to diagnose thyroid disease. Apart from the thyroid disease, the proposed method can also be used to diagnose other diseases like Parkinson's, Lung cancer and many more.

Acknowledgements This article acknowledges the dataset used in this study is obtained from University of California, Irvine Machine Learning Repository. The authors declare that they have no conflicts of interest. This article does not contain any studies involving animals performed by any of the authors. This article does not contain any studies involving human participants performed by any of the authors.

References

1. Hall, J.E.: Guyton and Hall Textbook of Medical Physiology e-Book. Elsevier Health Sciences (2010)
2. Weetman, A.P.: Graves' disease. N. Engl. J. Med. **343**(17), 1236–1248 (2000)
3. Fauci, A.S. (ed.): Harrison's Principles of Internal Medicine, vol. 2. Mcgraw-hill, New York (1998)
4. Emary, E., Zawbaa, H.M., Hassanien, A.E.: Binary grey wolf optimization approaches for feature selection. Neurocomputing **172**, 371–381 (2016)
5. Mirjalili, S., Mirjalili, S.M., Yang, X.S.: Binary bat algorithm. Neural Comput. Appl. **25**(3–4), 663–681 (2014)
6. Jain, R., Gupta, D., Khanna, A.: Usability feature optimization using MWOA. In: International Conference on Innovative Computing and Communications. Springer, Singapore (2019)
7. Mirjalili, S.: The ant lion optimizer. Adv. Eng. Softw. **83**, 80–98 (2015)
8. Zawbaa, H.M., Emary, E., Parv, B.: Feature selection based on antlion optimization algorithm. In: 2015 Third World Conference on Complex Systems (WCCS). IEEE (2015)
9. Mirjalili, S., Jangir, P., Saremi, S.: Multi-objective ant lion optimizer: a multi-objective optimization algorithm for solving engineering problems. Appl. Intell. **46**(1), 79–95 (2017)
10. Emary, E., Zawbaa, H.M., Hassanien, A.E.: Binary ant lion approaches for feature selection. Neurocomputing **213**, 54–65 (2016)
11. Ali, E.S., Elazim, S.M.A., Abdelaziz, A.Y.: Ant lion optimization algorithm for renewable distributed generations. Energy **116**, 445–458 (2016)

A New Single-Phase Symmetrical Multilevel Inverter Topology with Pulse Width Modulation Techniques

Shubham Kumar Gupta, Anurag Saxena, Nikhil Agrawal, Rishabh Kumar Verma, Kuldeep Arya, Anand Rai and Ankit Singh

Abstract MLI is an emerging technology for many industrial purposes. It is widely used to solve power quality issues. Presented topology required reduced switches with less gate driver circuits compared to traditional topologies, so the complexity, area of installation, and overall cost of the proposed structure are reduced. Proposed symmetrical topology is simple and optimal, which can be efficiently continued for higher levels in the output of MLI. As the levels in MLI increases, THD decreases, losses decrease, and efficiency increases. This paper shows the comparison of THD, with different PWM techniques at various modulation indexes, and analyzes the simulating results of presented seven-level symmetrical multilevel inverter, which is executed by MATLAB/Simulation R2013a version software.

Keywords Multilevel inverter · Pulse width modulation · Total harmonic distortion

1 Introduction

Inverter is the most recognizable converter which transforms DC power into desired AC power at desired frequency. Sources for inverter circuit are DC which can be DC battery source, fuel cell, etc., and output of inverter is sinusoidal AC waveform but in practice the output of inverter is not sinusoidal; it contains some distortions which create power quality issue. Multilevel inverter is an upgraded version of inverter, which starts with three levels in output voltage waveform [1–3]. Multilevel inverters have more attention and importance in the power industry because of its more power capability with lower number of harmonics and good power quality [4]. In multilevel inverter, THD is decreased by increasing the levels in output voltage. In the present generation, the power quality of electrical energy is a big issue. These problems overcome by MLI. MLI have advantages over a two-level inverters such as less electromagnetic interference, draw input current with less harmonic contents, and lower commutation losses because of low switching frequency, so the efficiency

S. K. Gupta (✉) · A. Saxena · N. Agrawal · R. K. Verma · K. Arya · A. Rai · A. Singh
S.R. Group of Institutions, Jhansi, India
e-mail: shubhamgsti@gmail.com

© Springer Nature Singapore Pte Ltd. 2020
P. K. Mallick et al. (eds.), *Cognitive Informatics and Soft Computing*,
Advances in Intelligent Systems and Computing 1040,
https://doi.org/10.1007/978-981-15-1451-7_62

of multilevel inverter is improved [5]. In comparison with the two-level inverter, multilevel inverter required more switching components and gate driver circuits, which increases the complexity and overall cost of MLI, count as the disadvantage of multilevel inverters. There are mainly three traditional topologies: diode-clamped (DCMLI), cascaded H-bridge (CHBMLI), and flying-capacitor (FCMLI). Generally, multilevel inverters are classified into two types: symmetrical configuration and asymmetrical configuration. In symmetrical multilevel inverters, all DC voltage sources have equal magnitude. In asymmetrical multilevel inverters, all DC voltage sources have different magnitude [6]. This paper presents a symmetrical MLI topology, required to reduce switches with gate driver compared to traditional topologies. In this proposed symmetrical topology, the levels in output waveform can be efficiently improved by adding less switches. In this proposed model, used switch consists of IGBT with anti-parallel diode [7, 8].

2   Proposed Multilevel Inverter Topology

General structure of proposed symmetrical MLI is displayed in Fig. 1, required N symmetrical DC voltage sources (V_1, V_2, V_3,..., V_N) and ($2N + 4$) unidirectional switches (S_1, S_2,..., S_{2N}, and T_1,..., T_4), for generating ($2N + 1$) number of levels in output. Proposed topology has two portions: One is the polarity part which is fixed, used to make the polarity of voltage level, and second is the level generation part, used to increase the voltage levels of MLI. In this proposed topology, all implemented DC link voltages are symmetrical in nature ($V_1 = V_2 = V_3 = ... = V_N = V$).

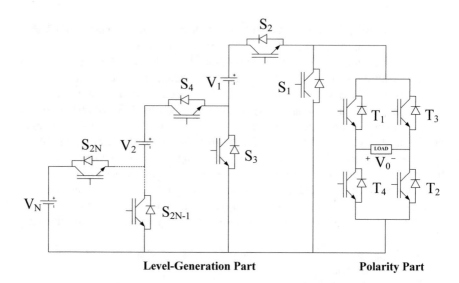

Fig. 1 General structure of anticipated symmetrical multilevel inverter topology

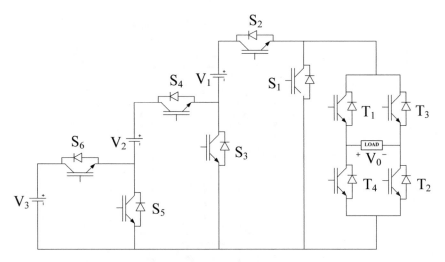

Fig. 2 Proposed model for seven-level symmetrical MLI

Maximum voltage generated by this proposed topology is the sum of all source voltages implemented in multilevel inverter. For seven-level MLI, this topology required three voltage sources and ten switches, shown in Fig. 2 and Table 1.

3 Operating Modes of Proposed 7-Level MLI

See Fig. 3.

4 PWM Techniques

PWM techniques provide switching pulses to the gate terminal of the switching component. In multi-carrier pulse width modulation, carrier signals are compared with the reference signal. When reference signal magnitude is more than carrier signal magnitude, then the result in the output is one otherwise zero. $(N - 1)$ carriers and 1 reference signal are required for N-Level MLI. In this paper, PD, POD, APOD, ISC, and VFISC pulse width modulation techniques are implemented [9, 10].

Table 1 Switching configuration for seven-level MLI

Output voltages level	Switching configurations										
	S_1	S_2	S_3	S_4	S_5	S_6	T_1	T_2	T_3	T_4
$(V_1 + V_2 + V_3)$	0	1	0	1	0	1	1	1	0	0
$(V_1 + V_2)$	0	1	0	1	1	0	1	1	0	0
(V_1)	0	1	1	0	0	0	1	1	0	0
0	1	0	0	0	0	0	1	1	0	0
$-(V_1)$	0	1	1	0	0	0	0	0	1	1
$-(V_1 + V_2)$	0	1	0	1	1	0	0	0	1	1
$-(V_1 + V_2 + V_3)$	0	1	0	1	0	1	0	0	1	1

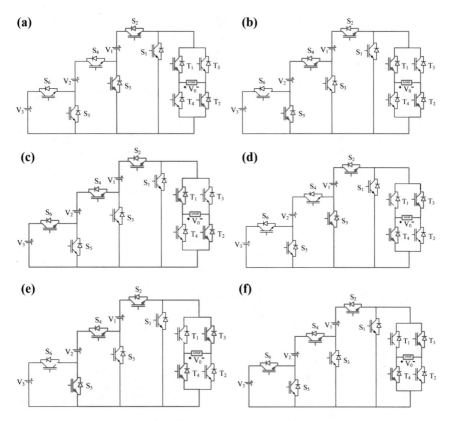

Fig. 3 **a** (V_1) level. **b** ($V_1 + V_2$) level. **c** ($V_1 + V_2 + V_3$) level. **d** $-(V_1)$ level. **e** $-(V_1 + V_2)$ level. **f** $-(V_1 + V_2 + V_3)$ level. Proposed topology operating modes for seven-level MLI

4.1 PD-PWM Technique

İn phase disposition technique, all implemented ($N - 1$) carrier signals are in the same phase, as shown in Fig. 4.

4.2 POD-PWM Technique

İn phase opposition disposition technique, implemented carrier signals above zero are 180° phaseout with implemented carrier signals below zero, as shown in Fig. 5.

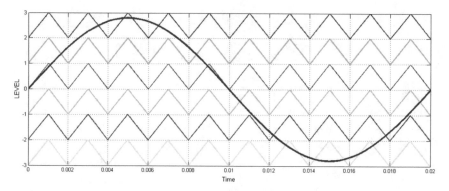

Fig. 4 PD-PWM techniques

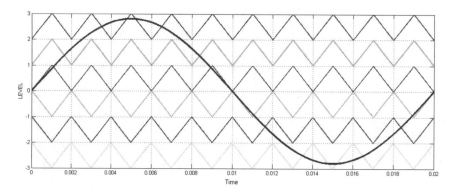

Fig. 5 POD-PWM techniques

4.3 APOD-PWM Technique

İn alternate phase opposition disposition technique, all implemented carrier signals are 180° phaseout with its neighboring carrier signal, as shown in Fig. 6.

4.4 ISC-PWM Technique

İn inverted sine carrier technique, implemented $(N - 1)$ carrier signals are inverted sinusoidal, as shown in Fig. 7.

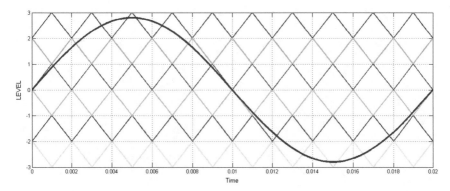

Fig. 6 APOD-PWM techniques

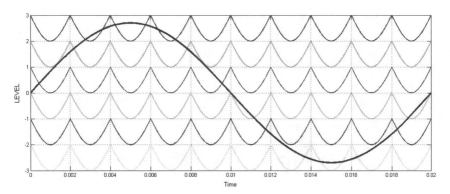

Fig. 7 ISC-PWM techniques

4.5 VFISC-PWM Technique

İn variable frequency inverted sine carrier technique, all implemented carrier signals are sinusoidal with different carrier frequency, as shown in Fig. 8.

5 Simulation Results

This proposed seven-level MLI, is shown in Fig. 2, and is analyzed in MAT-LAB/Simulink R2013a software. Parameters for seven-level symmetrical MLI are as follows: DC voltage source $V_1 = V_2 = V_3 = 80$ V and resistance $R = 10\ \Omega$. Output result for the proposed seven-level MLI is shown in Fig. 9. Table 2 displays the comparison between % THD for PD, POD, APOD, ISC, and VFISC-PWM techniques with 2000 Hz carrier frequency at various modulation indexes. THD is executed by FFT analyzer in MATLAB/Simulation R2013a software, as displayed in Fig. 10.

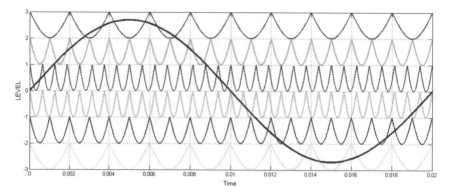

Fig. 8 VFISC-PWM techniques

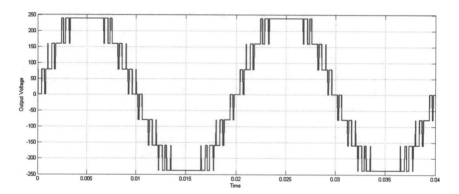

Fig. 9 Output voltage waveform for proposed seven-level MLI

Table 2 % THD result for proposed seven-level symmetrical MLI

Level	Modulation index	PWM techniques				
		PD	POD	APOD	ISC	VFISC
7 level	1.1	15.71	15.52	15.77	16.58	16.63
	1.0	18.23	18.18	18.39	19.67	19.97
	0.9	22.40	22.11	21.98	22.77	22.65
	0.8	24.26	24.06	24.18	24.64	24.54

6 Conclusion

A novel symmetrical MLI topology with greater performance over traditional MLI topologies has been studied. This proposed MLI topology uses less switches with respective gate driver circuits compared to traditional MLI topologies, so the complexity and overall cost of inverter reduced. Proposed topology presents the simple

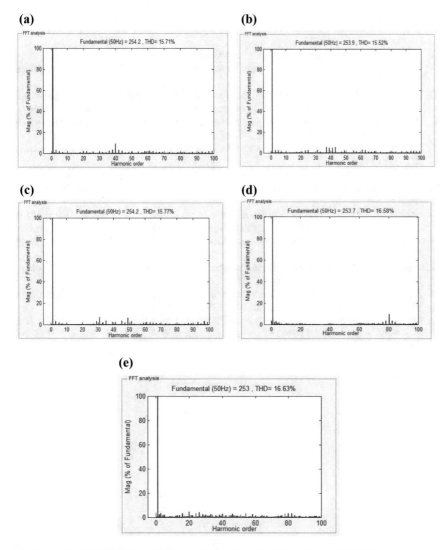

Fig. 10 **a**. PD-PWM. **b**. POD-PWM. **c**. APOD-PWM. **d**. ISC-PWM. **e**. VFISC-PWM. %THD of proposed seven-level symmetrical MLI with modulation index ($M_a = 1.1$, & $M_f = 40$)

and optimal method of increasing the levels in output. For proposed seven-level symmetrical MLI, %THD is 15.52%, with phase opposition disposition (POD-PWM) technique at modulation index ($M_a = 1.1$).

References

1. Baker, R.H.: High-voltage converter circuits. US Patent 4,203,151, May 1980
2. Nabe, I.T., Akagi, H.: A new neutral point clamped PWM inverter. IEEE Trans. Ind. Applicant **1A-17**, 518–523 (1981)
3. Samsami, H., Taheri, A., Samanbakhsh, R.: New bidirectional multilevel inverter topology with staircase cascading for symmetric and asymmetric structures. IET Power Electronics (2017)
4. Agrawal, N., Bansal, P.: Analysis of new 7-level an asymmetrical multilevel inverter topology with reduced switching devices. Inter. J. Sci. Eng. Technol. Res. (IJSETR) **6**(6) (2017)
5. Babaei, E.: Optimal topologies for cascaded sub-multilevel converters. Power Electron. **2**(3), 251–261 (2010)
6. Holmes, D.G., McGrath, B.P.: Opportunities for harmonic cancellation with carrier-based PWM for two-level and multilevel cascaded inverters. IEEE Trans. Ind. Appl. **37**, 574–582 (2001)
7. Sahoo, S.K., Ramulu, A., Prakash, J., Deeksha.: Performance analysis and simulation of five level and seven level single phase multilevel inverters. In: Third International Conference on Sustainable Energy and Intelligent System (SEISCON 2012), VCTW, Tamilnadu, India, 27–29 Dec 2012
8. Najafi, E., Yatim, A.H.M.: Design and implementation of a new multilevel inverter topology. IEEE Trans. Ind. Electron. **59**(11), (2012)
9. Agrawal, N., Bansal, P.: A new 21-level asymmetrical multilevel inverter topology with different PWM techniques. In: IEEE Conference RDCAPE 2017, pp 224–229. Noida (2018)
10. Jeevananthan, S., Nandha, K.R., Dananjayan, P.: Inverted sine carrier for fundamental fortification in PWM inverters and FPGA based implementation. Sebian J. Electr. Eng. **4**(2), 171–187 (2007)

Single Band Notched UWB-BPF for Short-Range Communication Systems

Abu Nasar Ghazali, Jabir Hussain and Wriddhi Bhowmik

Abstract Presented here is a two-layered ultra-wideband (UWB) bandpass filter with a notched passband and suppressed stopband. The fundamental structure of the filter is aligned on layer-coupled microstrip-to-coplanar waveguide (CPW) technology. The bottom layer has a multiple-mode resonator (MMR) inside a CPW, and the upper layer has two microstrip lines separated by some distance. The MMR is optimized in dimension to place its initial resonant modes within the UWB spectrum and the tight coupling of the transition develops the UWB passband. Later, modified complimentary split ring resonator (MCSRR)-shaped defected ground structure (DGS) etched in the CPW develops the passband notch, whereas double U (DU) DGS units under the feeding lines provide suppressed stopband. The prototype proposed is essentially small in size capturing only 25.26 mm × 11.01 mm.

Keywords Bandpass filter · Broadside coupled · Coplanar waveguide · Microstrip-to-CPW transition · Ultra-wideband (UWB)

1 Introduction

Research on UWB filter design gathered speed after the spectrum of 3.1–10.6 GHz was released by the Federal Communication Commission (FCC) for use in UWB systems [1].

A. N. Ghazali (✉) · J. Hussain · W. Bhowmik
School of Electronics Engineering, Kalinga Institute of Industrial Technology (KIIT) Deemed University, Bhubaneswar, India
e-mail: abu.ghazalifet@kiit.ac.in

J. Hussain
e-mail: jabir.hussainfet@kiit.ac.in

W. Bhowmik
e-mail: wriddhi.bhowmikfet@kiit.ac.in

© Springer Nature Singapore Pte Ltd. 2020
P. K. Mallick et al. (eds.), *Cognitive Informatics and Soft Computing*,
Advances in Intelligent Systems and Computing 1040,
https://doi.org/10.1007/978-981-15-1451-7_63

A UWB-BPF is one which satisfies the criteria of flat passband, which are enhanced insertion/return loss in the passband, linear group delay, and 110% fractional bandwidth. The last decade saw innumerable UWB filters being conceptualized and developed in this regard [2–5]. However, the problem of interference caused researchers to develop another set of UWB filters which had notched band [6–9]. Of these structures proposed, most had narrow stopband problem.

In this letter, we overcome the above problem by proposing an BPF with stopband at 5.8 GHz (to suppress WLAN band) and extended stopband. Our UWB-BPF is based on broadband capacitive coupling mechanism of microstrip-to-CPW transition, wherein both are present on either sides of the substrate. The CPW (bottom) has an open-circuited MMR which when optimized in dimensions places its resonant modes inside the UWB spread and the capacitive coupling of the transition generates a smooth passband. Later, modified complimentary split ring resonator (MCSRR) etched on the MMR and two U-shaped DGS (DUDGS) [10] etched under the feeding lines develop passband notch and suppressed stopband, respectively. The optimized design of the prototype structure has been done using the EM software IE3D on the laminate with $\varepsilon_r = 10.8$ and thickness 0.635 mm.

2 Conceptualized UWB-BPF

Figure 1 shows the conceptualized UWB filter wherein the MMR inside the CPW is adjusted in dimensions to position its resonant modes inside the UWB spread. The central section of MMR is made approximately half of the guided wavelength (λ_g), whereas the narrow end sections are approximately quarter of λ_g. This MMR alone when excited develops a weak coupling response as shown in Fig. 2a. In order to generate tight coupling, the top-layer microstrips are designed corresponding to the MMR. The optimized coupling of the microstrip/CPW transition causes the enhanced coupling peak around the mid UWB frequency, thereby converting the weak coupling

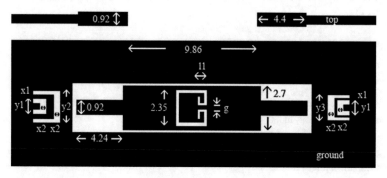

Fig. 1 Proposed UWB structure based on layered transition of microstrip and CPW. Dark and light shades are conductor and etched part, respectively. Dimensions in mm

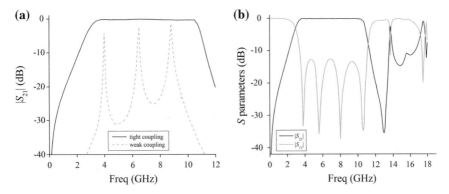

Fig. 2 **a** Comparative weak/tight coupling responses. **b** Optimized UWB spectrum of the microstrip/CPW transition

to tight coupling response as shown in Fig. 2a. The complete UWB spectrum is plotted and presented in Fig. 2b. The passband extends from 3 to 10.77 GHz with insertion/return loss improved at 0.8/12.6 dB. The stopband, however, is narrow with harmonics first at 13.7 and second at 17.6 GHz.

In order that the passband notch can be developed, we etch a MCSRR in the MMR. The dimensions of MCSRR are as depicted in Fig. 1. The length and width have been fixed at 2.35 mm so as to attain maximum notch tunability, whereas the length of folded line $l1$ is kept variable. Figure 3 depicts the variable passband notches for g

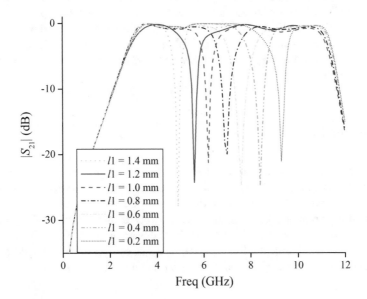

Fig. 3 Controllable notches for variable length of the folded line $l1$

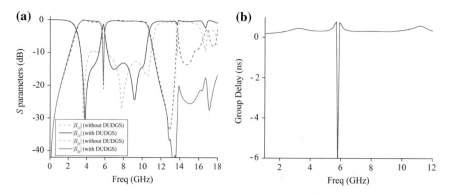

Fig. 4 Simulated frequency characteristics. **a** Comparative S parameters for and without DUDGS. **b** Group delay variation

$= 0.2$ mm and variable $l1$, from which the value of $l1 = 1.2$ mm is selected for notch at 5.8 GHz.

The notched band UWB filters have higher-order harmonics in the stopband which deteriorate its isolation. In order to improve the same, we need to extend the stopband. Here, we do it by inculcating DUDGS units under the I/O lines. These units develop transmission zeros at these harmonics points, thereby suppressing them and extending the stopband. For our proposed structure, the dimensions of DUDGS considered are $x1 = 0.2$ mm, $y1 = 0.96$ mm, $x2 = 0.5$ mm, $y2 = 2$ mm and $y3 = 1.6$ mm. For these dimensions, the TZs are placed at 13.9 and 17.3 GHz which causes the stopband to increase with attenuation higher than 20 dB and extend up to 18 GHz as presented in Fig. 4a. Also, it can be observed that the presence of DUDGS has no ill effect on the frequency characteristics of the proposed structure. The group delay observed is depicted in Fig. 4b, which shows flat response with variation less than 0.34 ns, except at the notch.

3 Conclusion

The manuscript presents a broadband BPF with band-suppressed WLAN band and attenuated stopband up to 18 GHz and linear and flat (linear) group delay (less than 0.34 ns within passband). It is shown that the location of the passband notch can be controlled by tuning the MCSRR dimensions, whereas the optimized design of the DUDGS units can suppress and attenuate the stopband up to 18 GHz. The prototype measures only 25.26 mm × 11.01 mm, depicting its compactness and making itself an excellent accessory for present/future UWB communication devices.

References

1. Federal Communications Commission: Revision of part 15 of the commission's rules regarding ultra-wideband transmission systems (2002)
2. Li, K., Kurita, D., Matsui, T.: An ultra-wideband bandpass filter using broadside-coupled microstrip-coplanar waveguide structure. IEEE MTT-S Int. Dig. 675–678 (2005)
3. Hsu, C., Hsu, F., Kuo, J.: Microstrip bandpass filters for ultra-wideband (UWB) wireless communications. In: International Microwave Symposium, Long Beach, CA (2005)
4. Zhu, L., Sun, S., Menzel, W.: Ultra-wideband (UWB) bandpass filters using multiple-mode resonator. IEEE Microw. Wirel. Compon. Lett. 5(11), 796–798 (2005)
5. Wang, H., Zhu, L.: Ultra-wideband bandpass filter using back-to-back microstrip-to-CPW transition structure. Electron. Lett. 41(24) (2005)
6. Ting, S.W., Tam, K.W., Martins, R.P.: Miniaturized microstrip lowpass filter with wide stopband using double equilateral U-shaped defected ground structure. IEEE Microw. Wirel. Compon. Lett. 6(5), 240–242 (2006)
7. Yang, G.M., Jin, R., Victoria, C., Harris, V.G., Sun, N.X.: Small UWB bandpass filter with notched band. IEEE Microw. Wirel. Compon. Lett. 18(3) (2008)
8. Li, Q., Li, Z.J., Liang, C.H., Wu, B.: UWB bandpass filter with notched band using DSRR. Electron. Lett. 46(10) (2010)
9. Chen, J.Z., Wu, G.C., Liang, C.H.: A novel compact UWB bandpass filter with simultaneous narrow notched band and out-of-band performance improvement. PIERS 24, 35–42 (2011)
10. Yan, T., Lu, D., Tang, X.H., Xiang, J.: High-selectivity UWB bandpass filter with a notched band using stub-loaded multi-mode resonator. Int. J. Electron. Commun. (AEÜ) 70, 1617–1621 (2016)

Network Intrusion Detection System Using Soft Computing Technique—Fuzzy Logic Versus Neural Network: A Comparative Study

Srinivas Mishra, Sateesh Kumar Pradhan and Subhendu Kumar Rath

Abstract Security of a data framework is its significant property, particularly today, when PCs are interconnected by means of web. Since no framework can be completely secure, the opportune and precise recognition of interruptions is essential. For this reason, intrusion detection systems (IDS) were planned. An interruption identification framework's principle aspiration is to sort exercises of a framework into two key classes: standard and suspicious exercises. Most IDS business devices are abuse frameworks with principle-based master framework structure. Be that as it may, these strategies are less effective when assault qualities shift from inherent marks. In such manner, a scope of pre-preparing practices, for example, information mining, neural systems, Petri nets, state change outlines, hereditary calculations, choice trees and fluffy-based rationales is worked out. In this paper, we have considered distinctive fluffy methodologies for interruption identification framework utilizing fuzzy set hypothesis, and we dissect fuzzy standard, expressly for peculiarity-based assault finding. For structure inconsistency framework, neural systems likewise can be utilized, on the grounds that they can figure out how to segregate the ordinary and anomalous conduct of a framework from models. Along these lines, they offer a promising strategy for structure oddity frameworks. Fake neural systems give the possibility to recognize and arrange organized action dependent on constrained, fragmented and nonlinear information sources. We have completed a similar report between two delicate registering procedures utilization of fluffy in the event that principles versus use of the feed forward neural network prepared by back-propagation calculation for interruption location. Reenactment result demonstrates that the feed

S. Mishra (✉)
Department of Computer Science & Engineering, Biju Patnaik University of Technology, Rourkela, Odisha, India
e-mail: srinivas_mishra@yahoo.com

S. K. Pradhan
Department of Computer Science, Utkal University, Bhubaneswar, Odisha, India
e-mail: sateeshind@yahoo.com

S. K. Rath
Biju Patnaik University of Technology, Rourkela, Odisha, India
e-mail: deputyregistrar@bput.ac.in

© Springer Nature Singapore Pte Ltd. 2020
P. K. Mallick et al. (eds.), *Cognitive Informatics and Soft Computing*,
Advances in Intelligent Systems and Computing 1040,
https://doi.org/10.1007/978-981-15-1451-7_64

forward neural network prepared by back-propagation calculation for interruption location yields better outcome which distinguishes the interruptions precisely and is well reasonable for constant applications as contrasted and fluffy on the off chance that rules.

Keywords Intrusion detection system · Fuzzy logic · Fuzzy if–then rule · Back-propagation neural network · Feed forward neural network · Soft computing

1 Introduction

Anything that can be developed utilizing regular structure procedures can likewise be worked with fluffy rationale, and the other way around [1]. Fluffy rationale is utilized in interruption recognition since 1997 since it is proficient to manage vagueness and multifaceted nature which is gotten from human perspective. By the assistance of fluffy factors or phonetic terms, interruption discovery qualities can be seen effectively and judgment of ordinary and strange movement in the system depends on its fluffiness nature that can perceive the level of disagreeableness of a hub as a substitute of yes or no conditions [2]. The capacity of fuzzy rationale in IDS's has the accompanying outline.

If Condition

then Consequence

where Condition is a fuzzy variable and Consequence is the fuzzy set.

Fluffy rationale is a kind of many-esteemed rationale or probabilistic rationale; it manages thinking that is rough as opposed to fixed and exact [3, 4]. In contrast and regular rationale, they can have changing qualities, where double sets have two-esteemed rationale, genuine (spoke to as 1) or false (spoke to as 0), fluffy rationale factors may have a reality esteem that extents in scale somewhere in the range of 0–1. Fluffy rationale has been reached out to manage the idea of halfway truth, where reality worth may shift between totally obvious and completely false. In addition, when semantic factors are utilized, these degrees might be constrained by specific capacities.

In fake neural systems, neurons were spoken to as models of natural systems into calculated parts for circuits that could perform computational assignments [4, 5]. The fundamental model of the counterfeit neuron is established upon the usefulness of the natural neuron [6]. By definition, "Neurons are essential flagging units of the sensory system of a living being in which every neuron is a discrete cell whose few procedures are from its cell body. One can separate between two fundamental kinds of systems, systems with input and those without it. In systems with criticism, the yield esteems can be followed back to the information esteems. Anyway there are systems wherein for each info vector laid on the system, a yield vector is determined and this can be perused from the yield neurons, there is no input. Thus, just, a forward progression of data is available. System having this structure is called as feed forward systems. This system has one information layer, one concealed layer and one yield layer. There can be any number of concealed layers. The information layer is associated with the concealed layer, and the shrouded layer is associated with the yield layer by methods for interconnection loads. The predisposition is accommodated both the covered up and the yield layer, to follow up on the net contribution to be determined. Neural system is made out of a few preparing units (hubs) and coordinated connections between them. These associations are weighted speaking to connection among information and yield neurons [7].

2 Experiments and Results of IDS Based on Fuzzy Logic

Following experiments and results were brought from previous work of the same author which is already published [1].

The whole district is isolated into five sets like fluffy set VERY LOW (VL), fluffy set LOW (L), fluffy set MEDIUM (M), fluffy set HIGH (H) and fluffy set VERY HIGH (VH) as appeared in Fig. 1 [7]. The x-hub demonstrates the qualities in the fluffy set, and the y-pivot demonstrates the enrollment work. The quantity of bundles is the fluffy variable while the LOW, MEDIUM and HIGH speak to the estimations of the fluffy variable. The usage of this framework was done in Snort, and the results were plotted utilizing MATLAB[8]. The outcomes are appeared (Figs. 2 and 3).

Fuzzification and Membership Functions: To get the enrollment esteems for every one of the characteristics, we watched the scope of qualities each trait obtain and

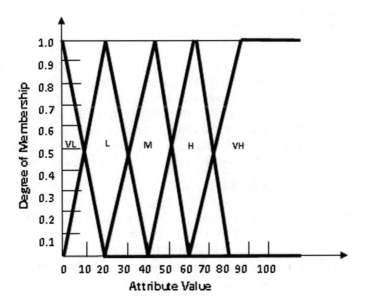

Fig. 1 Fuzzy break used for intrusion detection

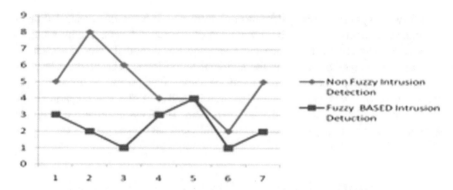

Fig. 2 False alarm generation

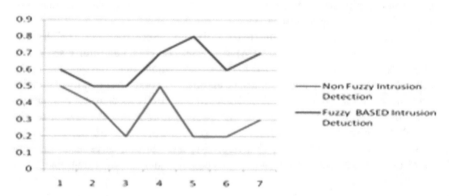

Fig. 3 Accuracy in intrusion detection

registered the normal incentive for each property in the ordinary, assault and blended which contains both typical and assault information of the preparation information. In light of the base, normal and most extreme estimations of every property, three participation esteems low (L), medium (M) and high (H) were characterized. The triangular participation capacity was utilized to recognize enrollment record related to every enrollment estimation of the three traits. The yield is in the state of level of interruption in the information. Scientifically, it is spoken to as underneath,

Rule:	if x is A then y is B
Relation:	$R = (A \times B) \, U \, (\bar{A} \times y)$
Membership Function:	$\mu(x, y) = \text{Max} \, [\text{Min}\{\mu A(x), \mu B(y)\}, \, 1 - \mu A(x)]$
Rule:	if x is A then y is B else y is C
Relation:	$R = (A \times B) \, U \, (\bar{A} \times C)$
Membership Function:	$\mu(x, y) = \text{Max} \, [\text{Min}\{\mu A(x), \mu B(y)\}, \, \text{Min}\{1 - \mu A(x), \mu C(y)\}]$

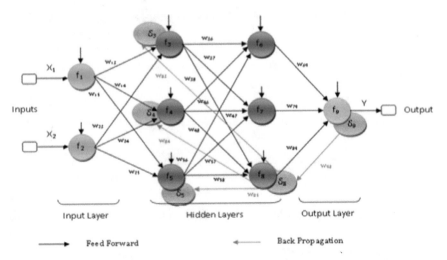

Fig. 4 Feed forward neural networks with back-propagation of errors

3 Experiments and Results Based on Feed Forward Neural Networks with Back-Propagation of Errors

Following experiments and results were brought from previous work of the same author which is already published [2] (Fig. 4).

The preparation model was performed by methods for root mean square (RMS) mistake examination [9, 10] utilizing learning rate of 0.80, two info layers, six shrouded layers and one yield layer. There were three classifications of off base yields: false positive, false negative and unimportant neural system yield [11]. The unimportant yields were those that did not speak to any of the yield classes [12–14] in the informational index. The mean square mistake accomplished by the system during preparing is 9.9979e-004. With six shrouded hubs, the system took 249.7030 s to achieve the blunder objective. The presentation of system during preparing is appeared in Figs. 5, 6 and 7.

4 Results and Discussion

The above graph shows that the % of intrusions detected using neural network technique is promisingly high as compared with % of intrusions detected using fuzzy logic technique. Although fuzzy reasoning and fuzzy if–then rules are best suited for specific situations like uncertainty [15–17], in this scenario neural network technique reaches 70%; whereas, the highest number of intrusions detected using fuzzy logic technique is 66% (Fig. 8).

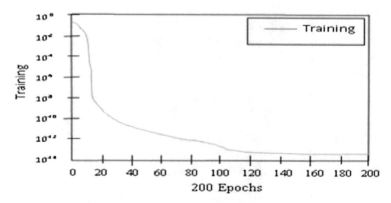

Fig. 5 Mean squared error (MSE) of the back-propagation training procedure versus training epochs

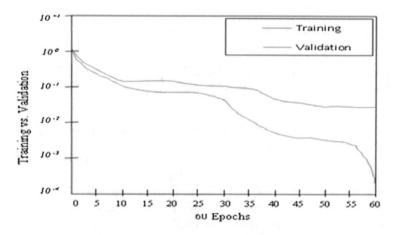

Fig. 6 Training process error when the early stopping validation method is applied

5 Summary

The paper examines on how fuzzy rationale is one of the incredible assets for thinking under vulnerability, and since vulnerability is a basic normal for interruption investigation, fuzzy rationale is in this manner a reasonable device to use to break down interruptions in a network. Moreover, the utilization of fluffy rationale can help in identifying irregularities which cannot be attentively regarded as ordinary or peculiar. In the structure of an abnormality location framework, one can exploit the neural system, capacity to learn and of its ability to sum up. Neural system can figure out how to segregate among typical and irregular conduct of the framework from precedents. No unequivocal meaning of strange conduct of the framework is essential, and consequently, the fundamental obstruction in structure irregularity framework

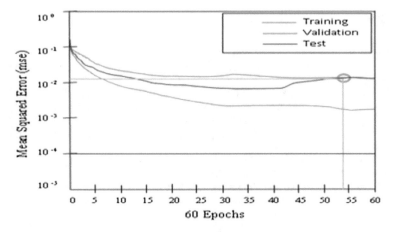

Fig. 7 Mean squared error (MSE)

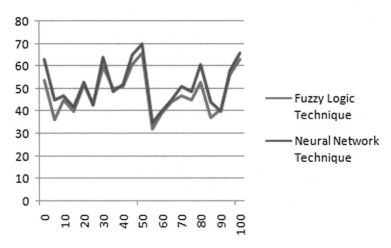

Fig. 8 % of intrusions detected using fuzzy logic versus neural network technique

could be survived. Number of the ages required to prepare the system is high as contrast with the other ANN strategy at the same time, and discovery rate is extremely high. BPNN can be utilized when one needs to distinguish the assault as well as to group the assault into explicit classification with the goal that preventive move can be made. The general framework demonstrates the characterization of 94.82%, with 0.6% both, false positive and false negative rate, and mean square mistake of 0.004. The whole similar examination demonstrates that the neural network technique yields preferred outcomes over fuzzy logic system for interruption location.

References

1. Mishra, S., Pradhan, S.K., Rath, S.K.: Network intrusion detection system using fuzzy if-then rules and fuzzy reasoning: a soft computing technique. Int. J. Comput. Eng. Appl. **XII**(IV), 301–310 (2018)
2. Mishra, S., Pradhan, S.K., Rath, S.K.: Performance analysis of network intrusion detection system using back propagation for feed forward neural network in MATLAB/SIMULINK. Int. J. Comput. Eng. Res. **8**(5), 58–65 (2018)
3. Borgohain, R.: FuGeIDS: fuzzy genetic paradigms in intrusion detection systems. Int. J. Adv. Netw. Appl. **3**(6), 1409–1415 (2012)
4. Tajbakhsh, A., Rahmati, M., Mirzaei, A.: Intrusion detection using fuzzy association rules. Appl. Soft Comput. **9**(2), 462–469 (2017)
5. Gomez, J., Dasgupta, D.: Evolving fuzzy classifiers for intrusion detection. In: Proceeding of the 2002 IEEE, Workshop on Information Assurance, United States Military Academy, West Point, NY (2002)
6. Bhattacharjee, P.S., Begum, S.A.: Fuzzy approach for intrusion detection system: a survey. Int. J. Adv. Res. Comput. Sci. **4**(2), 101–107 (2013)
7. Toosi, A.N., Kahani, M.: A new approach to intrusion detection based on an evolutionary soft computing model using neuro-fuzzy classifiers. Comput. Commun., 2201–2212 (2017). (Elsevier, Science Direct)
8. Ishibuchi, H., Nakashima, T., Murata, T.: A fuzzy classifier system that generates fuzzy if–then rules for pattern classification problems. In: Proceedings of 2nd IEEE International Conference on Evolutionary Computation, vol. 2, pp. 759–764. IEEE, Perth, Australia, 29 Nov to 1 Dec 1995 (1995)
9. Suthishni, D.N.P., Kumar, G.P.R.: Intrusion detection analysis by implementing fuzzy logic. Int. J. Appl. Eng. Res. **11**(5), 3216–3220 (2016)
10. Shanmugavadivu, R., Nagarajan, N.: Learning of intrusion detector in conceptual approach of fuzzy towards intrusion methodology. Int. J. Adv. Res. Comput. Sci. Softw. Eng. **2**(5), 246–250 (2012)
11. Haddadi, F., khanchi, S., Shetabi, M., Derhami, V.: Intrusion detection and attack classification using feedforward neural network. In: Second International Conference on Computer and Network Technology, 978-0-7695-4042-9/10 © 2010 IEEE, IEEE Computer Society, pp. 262–266. https://doi.org/10.1109/iccnt.2010.28 (2010)
12. Shum, J., Malki, H.A.: Network intrusion detection system using neural network. In: Proceeding of IEEE Fourth International Conference on Natural Computation, pp. 242–246 (2008)
13. Wang, G., Hao, J., Ma, J., Huang, L.: A new approach to intrusion detection using artificial neural networks and fuzzy clustering. Expert Sys. Appl. **37**(9), 6225–6232 (2010)
14. Pandit, T., Dudy, A.: A feed forward artificial neural network based system to minimize Dos attack in wireless network. Int. J. Adv. Eng. Technol. **7**(3), 938–947 (2014)
15. Nazir, A.: A comparative study of different artificial neural networks based intrusion detection systems. Int. J. Sci. Res. Publ. **3**(7) (2013)
16. Pradhan, M., Pradhan, S.K., Sahu, S.K.: Anomaly detection using artificial neural network. Int. J. Eng. Sci. Emerg. Technol. (2012)
17. Patel, P.J., Shah, J.S., Patel, J.: Performance analysis of neural networks for intrusion detection system. Int. J. Comput. Technol. Appl. **8**(2), 88–93 (2017)

Statistical Analysis of Target Tracking Algorithms in Thermal Imagery

Umesh Gupta and Preetisudha Meher

Abstract In the current scenario as we all know that target tracking is important aspect in almost every area of real life such as medical, ATM surveillance, border security, quality checking, weather forecasting, defence security, sea shore security and monitoring moving objects. For this reason, there is a great need to develop effective and efficient algorithm for target tracking which will be used in many research areas. To fulfil the need of current time, in this paper, a statistical analysis is performed using different detection algorithm with tracking algorithm like Kalman filter with single targets in infrared imagery. This will give a great help to researcher for developing an efficient algorithm in the context of target tracking.

Keywords Infrared imagery · Target tracking algorithms · Kalman filter · Target detection algorithms · Single target · Multiple targets

1 Introduction

1.1 Infrared Imagery

Infrared imagery follows the spectrum of infrared electromagnetic radiation which has range of wavelengths (0.74–$1000\,\mu\text{m}$) and includes thermal radiation discharged by objects near room temperature. To detect the objects as target in night or foggy environment, thermal imagery is one of the suitable technologies. By using this thermal imagery, one can create thermal images or series of thermal images that develop a video sequence of emitted light from the target at their boundary temperatures

U. Gupta (✉)
Department of Computer Science and Engineering, National Institute of Technology, Yupia, Arunachal Pradesh, India
e-mail: er.umeshgupta@gmail.com

P. Meher
Department of Electronics and Engineering, National Institute of Technology, Yupia, Arunachal Pradesh, India
e-mail: preetisudha1@gmail.com

© Springer Nature Singapore Pte Ltd. 2020
P. K. Mallick et al. (eds.), *Cognitive Informatics and Soft Computing*,
Advances in Intelligent Systems and Computing 1040,
https://doi.org/10.1007/978-981-15-1451-7_65

[1, 2]. Infrared energy released from the blackbody is understood by three laws such as follows [1]

(1) *Planck's law*: It gives the basic understanding of the electromagnetic emission which is released from the black target in thermal equilibrium at a specific temperature.

$$W_\lambda = C_1/\lambda^5 \left(e_2^{C/\lambda T} - 1\right) \tag{1}$$

where W_λ represents spectral radiant emittance/wavelength; λ is the wavelength; T stands for temperature; C_2 represents radiation constant which is equal to 1.4388×10^4; and h is the Planck's constant which is 6.6261×10^{-34}.

(2) *Stefan–Boltzmann's law*: According to this law, the total radiant energy of the black target is directly proportional to the target temperature for all wavelengths lies within zero to infinite. It is calculated by applying integration to Eq. (1) as follows:

$$W = e\sigma T^4 \tag{2}$$

where Stefan–Boltzmann constant $(s) = 5.6705 \times 10^{-12}$ and Boltzmann constant $(K) = 1.3807 \times 10^{-23}$.

(3) *Wien's displacement law*: According to this law, the spectral radiance of black target peak at the wavelength λ_m which is inversely proportional to the temperature is written in such a way

$$\lambda_m T = \Theta \tag{3}$$

where $\Theta = 2897.8$; Light velocity$(c) = 2.9979 \times 10^{10}$ and C_1 is the radiation constant $= 3.7418 \times 10^4$.

1.2 Target Tracking

Target tracking is the activity to monitor and follow the target by using static or dynamic camera at definite time [3–6]. The main goal of target tracking in video is related to track the object in video frames. In target tracking, all the sequences of static video frames are processed and focused on the movement of target with respect to all surroundings [7–10]. There are number of application where target tracking is playing a crucial role such as border security, line of control surveillance and traffic control. For more, study follows [11–16].

1.3 Target Detection and Tracking Algorithms

(1) *Single reference frame (SRF)*: It is one of the basic target detection algorithms in which background subtraction is considered and follows the difference with the next consecutive frame [17] (SRF)

$$T = u \pm 3\sigma \tag{4}$$

where the average of the difference image is represented by u; σ shows the standard deviation.

(2) *Running average (RA)*: Running average can be represented as [17]:

$$V(t + 1) = (1 - \epsilon)V(t) + \alpha G(t) \tag{5}$$

where updating rate is ϵ; the background image is $V(t)$; and the current image $G(t)$ is at time t.

(3) *Temporal median filter (TMF)*: The temporal median of the nth frame, the median operation of image $I_{\mathrm{Med}}(p, q, n)$ is calculated [18] by

$$I_{\mathrm{Med}}(p, q, n) = \mathrm{Med}[I(p, q, n - 1), I(p, q, n - 2)..I(p, q, n - N)] \tag{6}$$

where $I(p, q, n - 1),\ldots,I(p, q, n - N)$ shows the pixel values over the previous N frames of the nth frame at located (p, q).

(4) *Mean shift filter (MSF)*: The algorithm of mean shift filter is defined as [19] as

Step 1: Firstly, create a rectangular window that covers each data points.
Step 2: In this step, calculate the average of data points that are lying under the defined window.
Step 3: At last, shifting the consecutive window to the average of data points until convergence.

Typically, kernel K is defined as $||p||^2$ and $K(x) = k(||p||^2)$, where k is called the profile of kernel. If k is non-negative and non- increasing: $k(p) \geq k(q)$ if $p < q$ else k is piecewise continuous and $\int k(p)\mathrm{d}p < \infty$, where limit $(0, \infty)$.

Suppose the set M^d has p_i data points for $i = 1,\ldots, n$ in d-D Euclidean space and $K(p)$ represents that how much p relates to the approximation of the average. Then, the sample mean ms for p along with K is obtained as

$$\mathrm{ms}(p) = \frac{\sum K(p - p(i)p(i))}{\sum K(p - p(i))}, \text{ for } i = 1 \text{ to } n. \tag{7}$$

Here, mean shift is $(\mathrm{ms}(p) - p)$. For each iteration, p tends to $\mathrm{ms}(p)$ until $\mathrm{ms}(p) = p$. The sequence p, $\mathrm{ms}(p)$, $\mathrm{ms}(\mathrm{ms}(p))\ldots$is called the trajectory of (p).

(5) *Blob analysis (BA) with Gaussian mixture model (GMM)*: In this Gaussian mixture model, conditional density is considered for each pixel. There are multi-colour objects which have different densities. It is defined as [20]

$$P(\lambda|\beta) = \sum_{k=1}^{M} \backslash P(\lambda|\beta)P(k) \tag{8}$$

where pixels are represented as λ; $P(k)$ as prior probability that pixel λ was obtained by component k
    and where

$$\sum_{k=1}^{M} P(k) = 1$$

where GMM with mean μ and covariance matrix Σ in the consideration of a two-dimensional space:

$$P(\lambda|\beta) = \frac{1}{2\Pi|\frac{\sum_k |P(k)|}{2}}e^{-1/2(\lambda-\mu(k))}\sum_{j}^{-1}(\lambda - \mu(k)) \tag{9}$$

(6) Kalman Filter: The Kalman filter is very crucial estimator which is used recursively. It means that the next step is calculated on the basis of the previous step and the current step [21]. For representation of the algorithm, one notation is used such as Xn|m which signifies the estimation at time n on the basis of observation at time m. Kalman filter is including two variables:

   (1) The posteriori state estimator as $X_{w|w}$.
   (2) The posteriori error covariance matrix as $P_{w|w}$.

The Kalman filter has been processed in two steps named predict and update. In the first step, i.e. priori step estimator that finds and constructs the current state time step by using previous time step. In the update step, the output of first step, i.e. the current time step will be joined to the current observation that clarifies the actual state estimate. It is also known as posteriori state estimation.

Algorithms of Kalman filter

   (1) Priori step estimator: $X_{w|w-1} = F_k x_{w-1|w-1} + B_w u_w$
        Priori estimate covariance: $P_{w|w-1} = F_w P_{w|w-1} F_w T + Q_w$
   (2) Residual step estimation: $y_w = z_w - H_w x_{w-1|w-1}$
        Residual estimate covariance: $S_w = H_w P_{w|w-1} H_w T + R_w$
   (3) Optimal Kalman gain: $K_w = P_{w|w-1} H_w T S_{w-1}$

(4) Posteriori state estimate: $X_{w|w} = X_{w|w-1} + K_w y_w$
　　　Posteriori estimate covariance: $P_{w|w} = (I - K_w H_w) P_{w|w-1}$,
where Q_w is the covariance of the process noise and the observation model is P_w; F_w represents the state-transition model and H_w the covariance of the observation noise as R_w; Kalman gain is K_w; z_w shows an measurement of the true state x_w; B_w presents the control-input model; and u_w is the control vector for each time step w.

2 Numerical Analysis

In this numerical analysis section, we are computing and calculating the exact location of the target and tracking the object in all video frames by following four steps.

Step 1: *Data acquisition*—In this step, thermal videos have been acquired and divided into image frames.
Step 2: *Pre-processing*—All video frames are processed by removing noise and clutters while added in transmission time or acquisition time.
Step 3: *Background subtraction*—Different detection algorithms [22] for background or foreground have been applied [23–26].
Step 4: *Object tracking*—For this, Kalman filter has been applied with labelling-based connected component on detected target in the step 3 or directly to the step 2.

Here, five algorithms such as single reference frame (SRF) with Kalman filter, running average (RA) with Kalman filter, temporal median filter (TMF) along with Kalman filter, mean shift filter (MSF) and blob analysis with Gaussian mixture model (GMM) for target detection and tracking have been compiled on real-world data sets named IEEE OTCBVS WS Series Bench, Infrared data sets [27]. In this database, we have considered three data sets. The descriptions of data sets are as follows: In the first data set, one target or person is moving from west direction to east direction of the field of view with one gun named AK-47 where target occupied the 10th part of the field of view. Second concerned data set consists of a target that is moving from the west direction to east direction, but some occlusion either full or partial was present. In the third data set, target is involving in translator motion but makes 90° angle to the camera with one gun named AK-47. He is moving from front to far and then returns back to his position. In this data set, the size of target is also varying from high to low or low to high. Data acquisition sensors are Raytheon L-3 and Thermal-Eye 2000AS. The eight-bit greyscale image is considered which is stored in joint picture expert group format having 320 × 240 pixels. For more study one can follows [28–31].

The accuracy performance of all algorithms is computed and tabulated in Tables 1 and 2 for real-world data set. There are several parameters [32] which have been used for calculating the performance of algorithms as follows:

- *True Positive* (*TP*): It will be true when actual target is detected, but target is present.

Table 1 Result analysis of SRF with KF, RA with KF, TMF with KF, MSF and BA with GMM using TAR for IR data sequences

S. No.	IR data sequences	TF	Tracking accuracy rate (TAR)				
			SRF with KF	RA with KF	TMF with KF	MSF	BA with GMM
1	Single men with Gun	190	0.945	0.975	1	0.955	1
2	Single men with Partial and full occlusion	440	0.969	0.994	0.987	0.661	1
3	Single men move from front to last	1210	0.95	0.983	0.993	0.990	0.631
Total		1840	0.954	0.984	0.993	0.868	0.877

Table 2 Result analysis of SRF with KF, RA with KF, TMF with KF, MSF and BA with GMM using TTAR for IR data sequences

S. No.	IR data sequences	TF	Target tracking accuracy rate (TTAR)				
			SRF with KF	RA with KF	TMF with KF	MSF	BA with GMM
1	Single men with Gun	190	0.932	0.950	1	0.913	1
2	Single men with partial and full occlusion	440	0.983	0.988	0.980	0.472	1
3	Single men move from front to last	1210	0.94	0.995	0.987	0.981	0.368
Total		1840	0.95	0.977	0.989	0.788	0.789

- *False Negatives (FN)*: It will be true when it will not detect any target, but target is present.
- *True Negative (TN)*: It will be true when it will not detect any target but target is not present.
- *False Positives (FP)*: It will be true when it detects the target, but target is absent.
- *Sensitivity (S)*: It is number of actual detected target by the algorithm. The value of high and low sensitivity signifies the high and low detection rate but absence of false positive.
- *Positive Predictive Value (PPV)*: It gives those numbers of real targets that were present actually in video frames. A high value of PPV signifies low false alarm rate.
- *Tracking Accuracy Rate (TAR)*: By using F-measure, one can calculate the tracking accuracy rate through taking the harmonic mean of S and PPV.
- *Target Tracking Accuracy rate (TTAR)*: It is calculated by the target tracker in terms of TP, FP, TN and FN in video frames.
- *Mean (μ)*: It is arithmetic mean of the all detected and tracking targets.
- *Standard Deviation (σ)*: It is calculated by considering the power of ½ of the average of squared differences of the values for any definite set.
- *Standard Error (S.E.)*: It is derived by the ratio of standard deviation and sample size.

The tracking results of all the data sets with five algorithms are shown in Fig. 1. In this Fig. 1, some of the frames are placed such as for data set 1—60, 90 and 120; for data set 2—40,190 and 240; for data set 3—100, 900 and 1100. For detection and tracking, two boxes are used in green and blue colour that signify the current

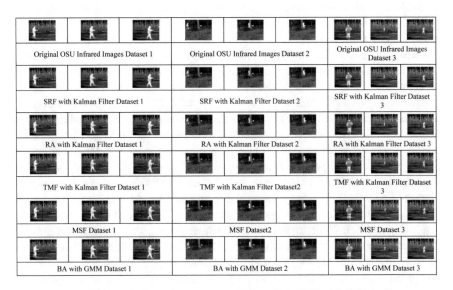

Fig. 1 Result of OSU infrared image sequence on SRF and RA, TMF with Kalman filter

estimation, and one box is shown as red colour that predicts the upcoming stage. According to Tables 1 and 2, first algorithm SRF with KF is not well suitable for target detection and tracking in infrared imagery. It is very sensitive in terms of noise and great susceptible for any minor changes in surrounding conditions. SRF with KF is working well with visible imagery. Second approach RA with KF has given comparable accuracy only with partial occlusion in infrared imagery rather than full occluded video frame. It is well not suited to nonuniform targets in thermal imagery. One can say that it will work for visible imagery [21].

In this Mean Shift algorithm method, black rectangle is moving around the target that represents the estimation of target location in these all frames. Mean Shift tracking is more suitable to visible imagery besides of infrared imagery. The initial value of mean shift iteration is unreasonable, so it is unable to track the rapid movement target. The target pixel information cannot be updated adaptively, and kernel window width cannot change adaptively, making it easy to generate the accumulated error. It is easy to lose target under complex background in thermal. Mean Shift tracking is also not working properly in partial occlusion and full occlusion in data sequence 2 due to high false negative. The Mean Shift algorithm almost lost target after 100th frame in data sequence 2, so it has poor tracking accuracy.

Blob analysis (BA) with Gaussian mixture model (GMM) is not tracked in target size variation of thermal imagery. Two things are not detected and tracked by GMM. Firstly, it is less sensitive of variation in target size, and second thing, it is not working when target stops for a while. The Gaussian mixture model as it is presented here has the disadvantage that all eigenvectors of the local covariance matrix need to be extracted. However, its computational problem greatly limits its application. It also has difficulty in segmenting objects that stop for a while during moving.

In the consideration of single target tracking, temporal median filter (TMF) with Kalman filter is performing better among SRF with KF, RA with KF, MSF and BA with GMM in infrared imagery as shown in Tables 1 and 2. In single target detection and tracking, this method gives best results in thermal data sequence 1 and 2. TMF with KF is working properly in partial occlusion and full occlusion. There is a pose variation and change in size of the target in sequence 3, but overall, it detects and tracks the single target efficiently in comparison with SRF with KF and RA with KF. It is more sensitive (near to unity) to detect and track the target in thermal imagery as shown in table. TMF with KF is more sensitive towards target detection and tracking, and its positive predictive value is also high in comparison with other methods as shown in Table 1. So positive predictive value will be high means low FAR. Here, the parameters such as TAR is near to target and TTAR is also good in infrared imagery due to less all false positive and false negative. All the results are shown in Fig. 2. Figure 2 shows the statistical analysis of TAR, TTAR, respectively, for single target OTCBVS thermal data sets in which graph shows standard deviation of different methods. With this graph, we can say TMF with KF is less deviate from the standard and at mean ($\mu = 0$) gives high occurrence of probability. In Tables 1 and 2, comparative analysis of TAR, TTAR for single target tracking is given, respectively. According to Tables 1, 2, target tracking accuracy rate (TAR, TTAR) is higher of

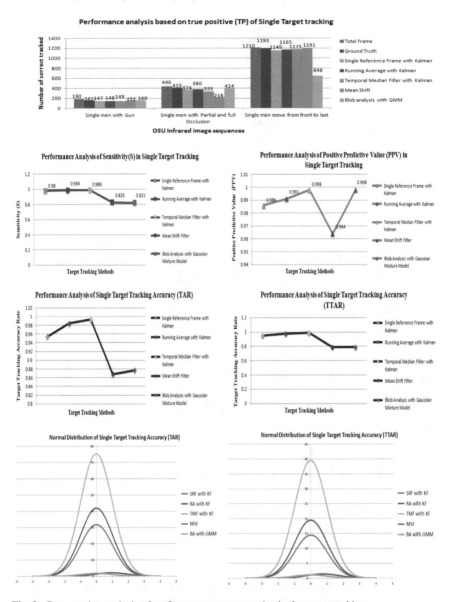

Fig. 2 Comparative analysis of performance parameters in single target tracking

TMF with KF where 0.993 and 0.983 are tracking accuracy rate out of 1890 frames, respectively.

According to Tables 1, 2, target tracking accuracy rate (TAR, TTAR) is higher of TMF with KF where 0.993 and 0.983 are tracking accuracy rate out of 1890 frames, respectively. TMF with KF tracking accuracy is near to unity that means it is a good

Table 3 Statistical analysis of single target OTCBVS IR data sets for SRF with KF, RA with KF, TMF with KF, MSF and BA with GMM using TAR

Statistical analysis of single target OTCBVS IR data sets (TAR)					
Methods	SRF with KF	RA with KF	TMF with KF	MSF	BA with GMM
Number of data set	3	3	3	3	3
Maximum	0.969	0.994	1	0.991	1
Minimum	0.945	0.975	0.993	0.661	0.631
Mean	0.957	0.984	0.996	0.826	0.815
Std. deviation	0.0126	0.0095	0.0053	0.1806	0.2130
Std. error	0.0072	0.0054	0.0030	0.1042	0.1229

Table 4 Statistical analysis of single target OTCBVS IR data sets for SRF with KF, RA with KF, TMF with KF, MSF and BA with GMM using TAR

Statistical Analysis of Single Target OTCBVS IR Data sets (TTAR)					
Methods	SRF with KF	RA with KF	TMF with KF	MSF	BA with GMM
Number of data set	3	3	3	3	3
Maximum	0.983	0.995	1	0.981	1
Minimum	0.932	0.950	0.980	0.472	0.368
Mean	0.95	0.977	0.989	0.788	0.789
Std. deviation	0.0275	0.0205	0.0101	0.2763	0.364
Std. error	0.0158	0.011	0.0058	0.1595	0.2101

tracking accuracy according to TAR and TTAR. In Tables 3 and 4, standard error of TMF with KF is also very less in comparison with other methods.

3 Conclusion

In the consideration of thermal imagery, different algorithms of target tracking have been discussed in this paper. Temporal median filter with Kalman filter gives good result in tracking the target either single target in thermal imagery in comparison with these four tracking methods (SRF with KF, RA with KF, MS, BA with GMM) with few limitation of TMF with KF. Temporal median filter with Kalman filter has less false positive and false negative in comparison with these four methods, but an effective and efficient tracking method needs more precise. Temporal median filter (TMF) with Kalman filter (KF) is best method to detect and track the targets in infrared imagery. This comparative analysis could be helpful to enhance the security in many areas such as biological systems, econometrics, robotics and sensor network, intelligent transportation system, remote sensing, perimeter security system, intelligent transportation system and surveillance where it may contribute to save, sustain and protect human life. There can be some extension for this thesis work. So recommendations are such as target classification may be done, and analysis of target tracking in hybrid imagery (visible + infrared spectrum) may be done and may be applied on more target such as birds, land and air vehicles. The limitations of this thesis work are such as, in this video sequences, camera is stationary. All the video sequences are thermal or infrared. There is no sudden change in the illumination of scene. This analysis is tested on person only as target only.

Acknowledgements We are highly grateful to IEEE OTCBVS WS Series Bench Real-World Infrared Database.

References

1. Bellis, B.M.: Origins of thermal imaging or infrared (IR) imaging. Available http://inventors.about.com/od/militaryhistoryinventions/p/thermal_image.htm (2019)
2. Sun, H., Wang, C., Wang, B., El-Sheimy, N.: Pyramid binary pattern features for real-time pedestrian detection from infrared videos. Neurocomputing **74**, 797–804 (2011)
3. Comaniciu, D., Ramesh, V., Meer, P.: Kernel-based object tracking. IEEE Trans. Pattern Anal. Mach. Intell. **25**, 564–577 (2003)
4. Yin, Y., Man, H.: Adaptive mean shift for target-tracking in FLIR imagery. In: Wireless and Optical Communications Conference, pp. 1–3 (2009)
5. Wang, Z., Hou, Q., Hao, L.: Improved infrared target-tracking algorithm based on mean shift. Appl. Opt. **51**, 5051–5059 (2012)
6. Li, C., Jiang, N., Si, J., Abousleman, G.P.: Robust target detection and tracking in outdoor infrared video. IEEE Int. Conf. Acous. Speech Signal Process., pp. 1489–1492 (2008)
7. Wang, J.-T., Chen, D.-B., Chen, H.-Y., Yang, J.-Y.: On pedestrian detection and tracking in infrared videos. Pattern Recogn. Lett. **33**, 775–785 (2012)
8. Dollar, P., Wojek, C., Schiele, B., Perona, P.: Pedestrian detection: an evaluation of the state of the art. IEEE Trans. Pattern Anal. Mach. Intell. **34**, 743–761 (2012)
9. Su, G., Ma, H., Hou, Y.: A robust approach for anti-jamming target tracking in forward looking infrared imagery. In: Sixth International Conference on Image and Graphics (ICIG), pp. 636–641 (2011)
10. Mirzaei, G., Majid, M.W., Ross, J., Jamali, M.M., Gorsevski, P.V., Frizado, J.P., et al.: Avian detection and tracking algorithm using infrared imaging. In: IEEE International Conference on Electro/Information Technology (EIT), pp. 1–4 (2012)
11. Heo, Duyoung, Lee, Eunju, Ko, ByoungChul: Pedestrian detection at night using deep neural networks and saliency maps. Electron. Imaging **2018**(17), 1–9 (2018)
12. De Oliveira, D., Wehrmeister, M.: Using deep learning and low-cost RGB and thermal cameras to detect pedestrians in aerial images captured by multirotor UAV. Sensors **18**(7), 2244 (2018)
13. Liu, Q., He, Z.: PTB-TIR: a thermal infrared pedestrian tracking benchmark. arXiv:1801.05944 (2018)
14. Song, E., et al.: AHD: thermal image-based adaptive hand detection for enhanced tracking system. IEEE Access **6**, 12156–12166 (2018)
15. Haelterman, R., et al.: Pedestrian detection and tracking in thermal images from aerial MPEG videos (2018)

16. Lahouli, I., et al.: Pedestrian tracking in the compressed domain using thermal images. In: International Workshop on Representations, Analysis and Recognition of Shape and Motion From Imaging Data. Springer, Cham (2017)
17. Zheng, Y., Fan, L.: Moving object detection based on running average background and temporal difference. In: International Conference on Intelligent Systems and Knowledge Engineering (ISKE), pp. 270–272 (2010)
18. Van Droogenbroeck, M., Paquot, O.: Background subtraction: experiments and improvements for vibe. In: IEEE Computer Society Conference on Computer Vision and Pattern Recognition Workshops (CVPRW), pp. 32–37 (2012)
19. Hung, M.-H., Pan, J.-S., Hsieh, C.-H.: Speed up temporal median filter for background subtraction. In: First International Conference on Pervasive Computing Signal Processing and Applications (PCSPA), pp. 297–300 (2010)
20. Dar-Shyang, L.: Effective Gaussian mixture learning for video background subtraction. IEEE Trans. Pattern Anal. Mach. Intell. **27**, 827–832 (2005)
21. Venkataraman, V., Fan, G., Havlicek, J., Fan, X., Zhai, Y., Yeary, M.: Adaptive Kalman filtering for histogram-based appearance learning in infrared imagery (2012)
22. Piccardi, M.: Background subtraction techniques: a review. IEEE Int. Conf. Syst. Man Cybern. **4**, 3099–3104 (2004)
23. Genin, L., Champagnat, F., Le, B.G.: Single frame IR point target detection based on a Gaussian mixture model classification, pp. 854111–854111 (2012)
24. Gupta, U., Dutta, M., Vadhavaniya, M.: Analysis of target tracking algorithm in thermal imagery. Int. J. Comput. Appl. **71**, 12443–19140 (2013)
25. Ma, Y., et al.: Pedestrian detection and tracking from low-resolution unmanned aerial vehicle thermal imagery. Sensors **16**(4), 446 (2016)
26. Goswami, M.S.: Unusual event detection for low resolution video using kalman filtering. M.E. Thesis, Computer Science and Engineering. Panjab University, India (2018)
27. Miezianko, R.: IEEE OTCBVS WS series bench: terravic motion infrared database. Available http://www.cse.ohio-state.edu/otcbvs-bench/ (2019)
28. Gonzalez, R.C.: Digital image processing, 3rd edn. Wiley (2019)
29. Perkins, S., Fisher, R., Walker, A., Wolfart, E.: Connected components labeling. Available http://homepages.inf.ed.ac.uk/rbf/HIPR2/label.htm (2000)
30. Schölkopf, B., Smola, A., Müller, K.R.: Nonlinear component analysis as a kernel eigenvalue problem. Neural Comput. **10**, 1299–1319 (1998)
31. Ling, H., Bai, L., Blasch, E., Mei, X.: Robust infrared vehicle tracking across target pose change using l1 regularization. In: International Conference on Info Fusion (2010)
32. Performance Evaluation Parameters. Available http://100dialysis.wordpress.com/tag/research/ (2019)

Fuzzy Approach to Determine Optimum Economic Life of Equipment with Change in Money Value

M. Balaganesan and K. Ganesan

Abstract One of the most pragmatic and topical areas of engineering economics is replacement analysis. Fuzzy set theory is the main tool which has been applied in many real-life applications to tackle the dubiousness situation. The aim of this paper is to determine the optimal replacement period of equipment with imprecise costs. In this problem, all these imprecise costs are taken as trapezoidal fuzzy numbers. Also, the proposed method provides fuzzy optimal solution of replacement model without converting to a classical model. A numerical example is illustrated to validate the proposed method.

Keywords Fuzzy set · Trapezoidal fuzzy number · Ranking · Fuzzy arithmetic · Fuzzy replacement model

1 Introduction

Replacement theory plays a major role in economic decision analysis for our real-world system. Due to new developments, the current equipments may become technologically obsolete or the current equipments have deteriorated on account of its long use over time and as such do not function efficiently or fail suddenly or it requires expensive maintenance which in turn affects the firm's performance such as the revenue and customer service. In other words, it requires expensive maintenance which affects the firm's performance such as the revenue and customer service. In order to avoid such negative effects of machine failures, it is necessary to find the appropriate machine maintenance and replacement policies. The classical mathematical techniques are successfully applied to model and solve a replacement problem when the decision parameters of the models are fixed at crisp values.

M. Balaganesan · K. Ganesan (✉)
Department of Mathematics, Faculty of Engineering and Technology, SRM Institute of Science and Technology, Kattankulathur, Chennai 603203, India
e-mail: ganesank@srmist.edu.in

M. Balaganesan
e-mail: balaganm@srmist.edu.in

© Springer Nature Singapore Pte Ltd. 2020
P. K. Mallick et al. (eds.), *Cognitive Informatics and Soft Computing*,
Advances in Intelligent Systems and Computing 1040,
https://doi.org/10.1007/978-981-15-1451-7_66

In our real-life situations, most of the information is imprecise in nature. Hence, the classical mathematical techniques may not be used to solve the real-world problems. To tackle such kind of situations, Zadeh [14] developed fuzzy set theory concept and it has been applied in most of the real-life application areas such as control theory and decision-making problems. Bellman [1] studied equipment replacement policy in classical nature. Many authors such as Buckley [2, 3], Biswas and Pramanik [4, 5], Choobineh et.al [6], Dreyfus et.al [7], Dimitrakoset.al [8], Mahdavi et.al [11], Nezhad et.al [12], Zhao et.al [13], etc have studied replacement problems under crisp and fuzzy environments. Biswas and Pramanik [5] determined the replacement time for fuzzy replacement problem using Yeger's ranking method. El-Kholy and Abdelalim [9] presented a comparative study for fuzzy ranking methods in determining economic life of equipment in crisp nature. In this paper, we discussed the replacement of an equipment under fuzzy nature without converting to classical version.

The rest of this paper is organized as follows. In Sect. 2, the basic concepts of fuzzy set, the trapezoidal fuzzy numbers and its arithmetic operations are stated. In Sect. 3, the replacement policy and the algorithm to determine fuzzy replacement time are discussed. Section 4 is dedicated to find the optimal replacement time of equipment with an example. Section 5 concludes this paper.

2 Preliminaries

Definition 2.1 A fuzzy set $\tilde{A}$ defined on the set of real numbers R is said to be a fuzzy number if its membership function $\tilde{A}: R \rightarrow [0, 1]$ has the following characteristics.

(i) $\tilde{A}$ is convex, i.e., $\tilde{A}(\lambda x_1 + (1 - \lambda)x_2) \geq \min\{\tilde{A}(x_1), \tilde{A}(x_2)\}$, $\lambda \in [0, 1]$, for all $x_1, x_2 \in R$
(ii) $\tilde{A}$ is normal; i.e., there exists an $x \in R$ such that $\tilde{A}(x) = 1$
(iii) $\tilde{A}$ is piecewise continuous.

Definition 2.2 A fuzzy number $\tilde{a}$ is a trapezoidal fuzzy number denoted by $\tilde{a} = (a_1, a_2, a_3, a_4)$ where a_1, a_2, a_3, a_4 are real numbers and its membership function is given by

$$\mu_{\tilde{a}}(x) = \begin{cases} \frac{x-a_1}{a_2-a_1} & \text{if } a_1 \leq x \leq a_2 \\ 1 & \text{if } a_2 \leq x \leq a_3 \\ \frac{a_4-x}{a_4-a_3} & \text{if } a_3 \leq x \leq a_4 \\ 0 & \text{otherwise} \end{cases}$$

Without loss of generality, we represent the trapezoidal fuzzy number $\tilde{a} = (a_1, a_2, a_3, a_4) = ([a_2, a_3], \alpha, \beta) = (m, w, \alpha, \beta)$, where $m = \left(\frac{a_2+a_3}{2}\right)$ and $w = \left(\frac{a_3-a_2}{2}\right)$ are the midpoint and width of the core $[a_2, a_3]$, respectively. Also, $\alpha = (a_2 - a_1)$ denotes the left spread and $\beta = (a_4 - a_3)$ denotes the right spread of the trapezoidal fuzzy number.

2.1 Trapezoidal Fuzzy Numbers and Their Arithmetic Operations

For any two arbitrary trapezoidal fuzzy numbers $\tilde{a} = (m(\tilde{a}), w(\tilde{a}), \alpha_1, \beta_1)$ and $\tilde{b} = (m(\tilde{b}), w(\tilde{b}), \alpha_2, \beta_2)$, we define the arithmetic operations as follows.

(i) Addition:

$$\tilde{a} + \tilde{b} = (m(\tilde{a}), w(\tilde{a}), \alpha_1, \beta_1) + (m(\tilde{b}), w(\tilde{b}), \alpha_2, \beta_2)$$
$$= (m(\tilde{a}) + m(\tilde{b}), \max\{w(\tilde{a}), w(\tilde{b})\},$$
$$\max\{\alpha_1, \alpha_2\}, \max\{\beta_1, \beta_2\})$$

(ii) Subtraction:

$$\tilde{a} - \tilde{b} = (m(\tilde{a}), w(\tilde{a}), \alpha_1, \beta_1) - (m(\tilde{b}), w(\tilde{b}), \alpha_2, \beta_2)$$
$$= (m(\tilde{a}) - m(\tilde{b}), \min\{w(\tilde{a}), w(\tilde{b})\},$$
$$\min\{\alpha_1, \alpha_2\}, \min\{\beta_1, \beta_2\})$$

(iii) Multiplication:

$$\tilde{a} \times \tilde{b} = (m(\tilde{a}), w(\tilde{a}), \alpha_1, \beta_1) \times (m(\tilde{b}), w(\tilde{b}), \alpha_2, \beta_2)$$
$$= (m(\tilde{a}) \times m(\tilde{b}), \max\{w(\tilde{a}), w(\tilde{b})\},$$
$$\max\{\alpha_1, \alpha_2\}, \max\{\beta_1, \beta_2\})$$

(iv) Division:

$$\tilde{a} \div \tilde{b} = (m(\tilde{a}), w(\tilde{a}), \alpha_1, \beta_1) \div (m(\tilde{b}), w(\tilde{b}), \alpha_2, \beta_2)$$
$$= (m(\tilde{a}) \div m(\tilde{b}), \min\{w(\tilde{a}), w(\tilde{b})\},$$
$$\min\{\alpha_1, \alpha_2\}, \min\{\beta_1, \beta_2\})$$

Here, the midpoint is taken in the ordinary arithmetic, whereas the width, and left and right spreads are considered to follow the lattice rule. That is for $a, b \in L$, define $a \vee b = \max\{a, b\}$ and $a \wedge b = \min\{a, b\}$.

2.2 Ranking of Trapezoidal Fuzzy Number

We recall the ranking function based on the graded mean discussed by Krishnaveni and Ganesan [10] $R(\tilde{a}) = \left[\left(\frac{a_2 + a_3}{2} \right) + \left(\frac{\beta - \alpha}{4} \right) \right]$.

For any two arbitrary trapezoidal fuzzy numbers $\tilde{a} = (\tilde{a}_1, \tilde{a}_2, \tilde{a}_3, \tilde{a}_4)$ and $\tilde{b} = (\tilde{b}_1, \tilde{b}_2, \tilde{b}_3, \tilde{b}_4)$, we have the following comparison:

(i) $\tilde{a} \prec \tilde{b}$ if and only if $R(\tilde{a}) \prec R(\tilde{b})$
(ii) $\tilde{a} \succ \tilde{b}$ if and only if $R(\tilde{a}) \succ R(\tilde{b})$
(iii) $\tilde{a} \approx \tilde{b}$ if and only if $R(\tilde{a}) \approx R(\tilde{b})$.

3 Algorithm to Find the Fuzzy Replacement Time of Equipment

To determine the fuzzy replacement time period, we proceed as follows:

Step 1. Represent each fuzzy trapezoidal cost into parametric form.
Step 2. Using present worth factor and maintenance cost, find $\sum \tilde{R}_n \tilde{v}^{n-1}$ for $n = 1, 2, \ldots$
Step 3. Using fuzzy machine cost and $\sum \tilde{R}_n \tilde{v}^{n-1}$, find $\tilde{C} + \sum \tilde{R}_n \tilde{v}^{n-1}$.
Step 4. By subtracting the fuzzy scrap value of equipment, determine $\tilde{C} - \tilde{S}_n \tilde{v}^n + \sum \tilde{R}_n \tilde{v}^{n-1}$.
Step 5. Compute $\sum \tilde{v}^{n-1}$ (i.e., cumulative present worth value up to each of the years).
Step 6. Using steps 4 and 5, compute the average total fuzzy cost $\tilde{W}(n) = \frac{\tilde{C} - \tilde{S}_n \tilde{v}^n + \sum \tilde{R}_n \tilde{v}^{n-1}}{\sum \tilde{v}^{n-1}}$.
Step 7. Determine the minimum weighted average fuzzy cost by the method of graded mean ranking index, which gives the optimal time for replacement of the equipment.

3.1 Optimal Replacement Policy of the Equipment

(i) Do not replace the equipment if the next period's cost is less than the weighted average of previous costs.
(ii) Replace the equipment if the next period's cost is greater than the weighted average of previous costs.

The present worth of all discounted costs at the end of every n years is

$$\tilde{P}(n) = (\tilde{C} + \tilde{R}_1)[1 + V^n + V^{2n} + \ldots] + \tilde{R}_2 V[1 + V^n + V^{2n} + \ldots] + \cdots$$
$$+ \tilde{R}_n V^{n-1}[1 + V^n + V^{2n} + \ldots]$$
$$= \left(\tilde{C} + \tilde{R}_1 + \tilde{R}_2 V + \cdots + \tilde{R}_n V^{n-1}\right)(1 + V^n + V^{2n} + \ldots)$$
$$= \left(\tilde{C} + \tilde{R}_1 + \tilde{R}_2 V + \cdots + \tilde{R}_n V^{n-1}\right)\left(\frac{1}{1 - V^n}\right)[V < 1]$$

$$= \frac{\tilde{F}(n)}{1 - V^n}, \text{ where } \tilde{F}(n) = \begin{cases} \tilde{C} + \sum\limits_{n=1} \tilde{R}_n V^{n-1}, & \text{if there is no solvage value} \\ \tilde{C} - S_n V^n + \sum\limits_{n=1} \tilde{R}_n V^{n-1}, & \text{if there is a solvage value} \end{cases}$$

Similarly, $\tilde{P}(n+1) = \frac{\tilde{F}(n+1)}{1 - V^{n+1}}$

$\tilde{P}(n)$ is minimum if $\tilde{P}(n+1) - \tilde{P}(n) > 0 > \tilde{P}(n) - \tilde{P}(n-1)$

That is, if $\dfrac{\tilde{F}(n+1)}{1 - V^{n+1}} - \dfrac{\tilde{F}(n)}{1 - V^n} > 0 > \dfrac{\tilde{F}(n)}{1 - V^n} - \dfrac{\tilde{F}(n-1)}{1 - V^{n-1}}$

Now $\dfrac{\tilde{F}(n+1)}{1 - V^{n+1}} - \dfrac{\tilde{F}(n)}{1 - V^n} = \dfrac{(1 - V^n)\tilde{F}(n+1) - (1 - V^{n+1})\tilde{F}(n)}{(1 - V^{n+1})(1 - V^n)}$

$$= \dfrac{\tilde{F}(n+1) - \tilde{F}(n) + \tilde{F}(n)V^{n+1} - \tilde{F}(n+1)V^n}{(1 - V^{n+1})(1 - V^n)}$$

But $\tilde{F}(n+1) = \tilde{F}(n) + \tilde{R}_{n+1} V^n$

$$= \dfrac{\tilde{R}_{n+1} V^n + \tilde{F}(n)V^{n+1} - V^n[\tilde{F}(n) + \tilde{R}_{n+1} V^n]}{(1 - V^{n+1})(1 - V^n)}$$

$$= \dfrac{\tilde{R}_{n+1} V^n[1 - V^n] + \tilde{F}(n)V^n[V - 1]}{(1 - V^{n+1})(1 - V^n)}$$

$$= \dfrac{V^n[1 - V][\tilde{R}_{n+1}(\frac{1 - V^n}{1 - V}) - \tilde{F}(n)]}{(1 - V^{n+1})(1 - V^n)}$$

That is, $\tilde{P}(n)$ is minimum if

$$\dfrac{V^n[1 - V][\tilde{R}_{n+1}(\frac{1 - V^n}{1 - V}) - \tilde{F}(n)]}{(1 - V^{n+1})(1 - V^n)} > 0 > \dfrac{V^{n-1}[1 - V][\tilde{R}_n(\frac{1 - V^{n-1}}{1 - V}) - \tilde{F}(n-1)]}{(1 - V^n)(1 - V^{n-1})}$$

$$\dfrac{V^n[\tilde{R}_{n+1}(\frac{1 - V^n}{1 - V}) - \tilde{F}(n)]}{(1 - V^{n+1})} > 0 > \dfrac{V^{n-1}[\tilde{R}_n(\frac{1 - V^{n-1}}{1 - V}) - \tilde{F}(n-1)]}{(1 - V^{n-1})}$$

$$\dfrac{V[\tilde{R}_{n+1}(\frac{1 - V^n}{1 - V}) - \tilde{F}(n)]}{(1 - V^{n+1})} > 0 > \dfrac{[\tilde{R}_n(\frac{1 - V^{n-1}}{1 - V}) - \tilde{F}(n-1)]}{(1 - V^{n-1})}$$

$$\tilde{R}_{n+1}\left(\dfrac{1 - V^n}{1 - V}\right) - \tilde{F}(n) > 0 > \tilde{R}_n\left(\dfrac{1 - V^{n-1}}{1 - V}\right) - \tilde{F}(n) + \tilde{R}_n V^{n-1}$$

$$\tilde{R}_{n+1}\left(\dfrac{1 - V^n}{1 - V}\right) > \tilde{F}(n) > \tilde{R}_n\left[\left(\dfrac{1 - V^{n-1}}{1 - V}\right) + V^{n-1}\right]$$

$$\tilde{R}_{n+1} > \tilde{F}(n)\left(\dfrac{1 - V}{1 - V^n}\right) > \tilde{R}_n$$

$$\tilde{R}_{n+1} > \frac{\tilde{F}(n)}{1 + V + V^2 + \cdots + V^{n-1}} > \tilde{R}_n$$

$$\tilde{R}_{n+1} > \frac{\tilde{F}(n)}{\sum_{r=0}^{n-1} V^r} > \tilde{R}_n$$

$\frac{\tilde{F}(n)}{\sum_{r=0}^{n-1} V^r}$ is the weighted average cost of previous n years with weights $0, 1, V, V^2, \ldots, V^{n-1}$, respectively.

The value of n satisfying the above relation will be the best replacement time.

4 Example

Consider a problem discussed by Biswas and Pramanik [5], and the fuzzy cost of a machine is US$ (5900, 5950, 6050, 6100). Table 1 gives the running cost and the salvage value at the end of the year in fuzzy nature.

Due to market fluctuations, the fuzzy rate of interest is $(0.08, 0.09, 0.10, 0.12)$ per year. Determine the optimal period of replacement.

The following procedure gives the optimum replacement period of the equipment. We first find $\tilde{v} = (1 + \tilde{r})^{-1}$, which is present worth factor of the money to be spent in a year, and it is equivalent to $(0.91, 0.005, 0.01, 0.02)$ (Table 2).

Table 3 gives the weighted average cost which is minimum in the 6th year. Therefore, the fuzzy optimum replacement period of the equipment is at the end of the 6th year. The weighted average cost obtained by our proposed method is (2850.7,

Table 1 Fuzzy maintenance cost and scrap value

Year (n)	Running cost	Salvage value
1	(1100, 1150, 1200, 1270)	(3800, 3900, 3950, 4000)
2	(1300, 1360, 1400, 1450)	(2600, 2650, 2700, 2760)
3	(1500, 1580, 1600, 1650)	(1900, 1950, 2000, 2060)
4	(1800, 1840, 1850, 1870)	(1450, 1470, 1480, 1500)
5	(2000, 2030, 2050, 2070)	(950, 980, 1000, 1050)
6	(2350, 2400, 2460, 2500)	(550, 570, 590, 600)
7	(2900, 2920, 2950, 3000)	(500, 530, 540, 550)

Table 2 Parametric form of fuzzy maintenance cost, scrap value and present worth factor in 100\$

Year (n)	$\tilde{R}_n$	$\tilde{S}_n$	$\tilde{v}^{n-1}$
1	(11.75, 0.25, 0.5, 0.7)	(39.25, 0.25, 1, 0.5)	(1, 0.005, 0.01, 0.02)
2	(13.8, 0.2, 0.6, 0.5)	(26.75, 0.25, 0.5, 0.6)	(0.91, 0.005, 0.01, 0.02)
3	(15.9, 0.1, 0.8, 0.5)	(19.75, 0.25, 0.5, 0.6)	(0.828, 0.005, 0.01, 0.02)
4	(18.45, 0.05, 0.4, 0.2)	(14.75, 0.05, 0.2, 0.2)	(0.754, 0.005, 0.01, 0.02)
5	(20.4, 0.1, 0.3, 0.2)	(9.9, 0.1, 0.3, 0.5)	(0.686, 0.005, 0.01, 0.02)
6	(24.3, 0.3, 0.5, 0.4)	(5.8, 0.1, 0.2, 0.1)	(0.624, 0.005, 0.01, 0.02)
7	(29.35, 0.15, 0.2, 0.5)	(5.38, 0.05, 0.3, 0.1)	(0.568, 0.005, 0.01, 0.02)

Table 3 Table determining the optimal replacement time

Year (n)	$\tilde{R}_n$	$\tilde{S}_n$	$\tilde{v}^{n-1}$	$\sum \tilde{R}_n \tilde{v}^{n-1}$	$\tilde{S}_n \tilde{v}^n$	$\sum \tilde{v}^{n-1}$	$\tilde{C} - \tilde{S}_n \tilde{v}^n + \sum \tilde{R}_n \tilde{v}^{n-1}$	$\tilde{W}(n)$	$R[\tilde{W}(n)]$
1	(11.75, 0.25, 0.5, 0.7)	(39.25, 0.25, 1, 0.5)	(1, 0.005, 0.01, 0.02)	(11.75, 0.25, 0.5, 0.7)	(35.718, 0.25, 1, 0.5)	(1, 0.005, 0.01, 0.02)	(36.032, 0.25, 1, 0.5)	(36.032, 0.005, 0.01, 0.02)	36.035
2	(13.8, 0.2, 0.6, 0.5)	(26.75, 0.25, 0.5, 0.6)	(0.91, 0.005, 0.01, 0.02)	(24.308, 0.25, 0.6, 0.7)	(22.149, 0.25, 0.5, 0.6)	(1.91, 0.005, 0.01, 0.02)	(62.159, 0.25, 0.5, 0.6)	(32.544, 0.005, 0.01, 0.02)	32.547
3	(15.9, 0.1, 0.8, 0.5)	(19.75, 0.25, 0.5, 0.6)	(0.828, 0.005, 0.01, 0.02)	(37.473, 0.25, 0.8, 0.7)	(14.892, 0.25, 0.5, 0.6)	(2.738, 0.005, 0.01, 0.02)	(82.581, 0.25, 0.5, 0.6)	(30.161, 0.005, 0.01, 0.02)	30.164
4	(18.45, 0.05, 0.4, 0.2)	(14.75, 0.05, 0.2, 0.2)	(0.754, 0.005, 0.01, 0.02)	(51.384, 0.25, 0.8, 0.7)	(10.119, 0.05, 0.2, 0.2)	(3.492, 0.005, 0.01, 0.02)	(101.265, 0.05, 0.2, 0.2)	(28.999, 0.005, 0.01, 0.02)	29.002
5	(20.4, 0.1, 0.3, 0.2)	(9.9, 0.1, 0.3, 0.5)	(0.686, 0.005, 0.01, 0.02)	(65.378, 0.25, 0.8, 0.7)	(6.178, 0.1, 0.3, 0.5)	(4.178, 0.005, 0.01, 0.02)	(119.200, 0.1, 0.3, 0.5)	(28.530, 0.005, 0.01, 0.02)	28.533
6	(24.3, 0.3, 0.5, 0.4)	(5.8, 0.1, 0.2, 0.1)	(0.624, 0.005, 0.01, 0.02)	(80.541, 0.3, 0.8, 0.7)	(3.294, 0.1, 0.2, 0.1)	(4.812, 0.005, 0.01, 0.02)	(137.247, 0.1, 0.2, 0.1)	**(28.522, 0.005, 0.01, 0.02)**	**28.525**
7	(29.35, 0.15, 0.2, 0.5)	(5.38, 0.05, 0.3, 0.1)	(0.568, 0.005, 0.01, 0.02)	(97.212, 0.3, 0.8, 0.7)	(2.781, 0.05, 0.3, 0.1)	(5.380, 0.005, 0.01, 0.02)	(154.431, 0.05, 0.3, 0.1)	(28.705, 0.005, 0.01, 0.02)	28.708

2851.7, 2852.7, 2854.7) which is sharper than (2598.10, 2752.26, 2936.25, 3127.77) obtained by Biswas and Pramanik [5] (Fig. 1).

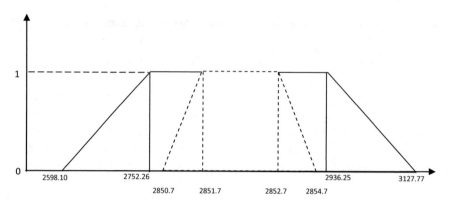

Fig. 1 Graphical representation for the comparison of weighted average costs

5 Conclusion

Wageness is present in all replacement decisions due to some unknown factors, such as revenue streams, maintenance costs and inflation. Fuzzy sets provide a mathematical technique for explicitly incorporating wageness into the decision-making model, especially when the system involves human subjectivity. The use of fuzzy sets to explicitly model uncertainty in replacement decisions via fuzzy numbers particularly trapezoidal fuzzy numbers is illustrated in this paper. We have proposed a fuzzy method for the fuzzy optimal solution of replacement model without converting to its classical version. A numerical example is solved to show that the wageness in the solution obtained by our proposed method is less than by the method discussed by Biswas and Pramanik [5]. This work can be extended to replacement models involving intuitionistic trapezoidal fuzzy numbers in which money fluctuates with time period.

References

1. Bellman, R.E.: Equipment replacement policy. SIAM J. Appl. Math. **3**, 133–136 (1955)
2. Buckley, J.: The fuzzy mathematics of finance. Fuzzy Sets Syst. **21**, 257–273 (1987)
3. Buckley, J.: Solving fuzzy equations in economics and finance. Fuzzy Sets Syst. **48**, 289–296 (1992)
4. Biswas, P., Pramanik, S.: Application of fuzzy ranking method to determine the replacement time for fuzzy replacement problem. Int. J. Comput. Appl. **25**(11), 41–47 (2011)
5. Biswas, P., Pramanik, S.: Fuzzy approach to replacement problem with value of money changes with time. Int. J. Comput. Appl. **30**(10), 28–33 (2011)
6. Choobineh, F., Behrens, A.: Use of intervals and possibility distributions in economic analysis. J. Oper. Res. Soc. **43**(9), 907–918 (1992)
7. Dreyfus, S.E., Law, A.M.: The Art and Theory of Dynamic Programming. Academic Press, New York (1977)

8. Dimitrakos, T.D., Kyriakidis, E.G.: An improved algorithm for the computation of the optimal repair/replacement policy under general repairs. Eur. J. Oper. Res. **182**, 775–782 (2007)

9. El-Kholy, A.M., Abdelalim, A.M.: A comparative study for fuzzy ranking methods in determining economic life of equipment. Int. J. Constr. Eng. Manage. **5**(2), 42–54 (2016)

10. Krishnaveni, G., Ganesan, K.: A new approach for the solution of fuzzy games. J. Phys. Conf. Ser. (JPCS) **1000**, 012017 (2018)

11. Mahdavi, M., Mahdavi, M.: Optimization of age replacement policy using reliability based heuristic model. J. Sci. Ind. Res. **68**, 668–673 (2009)

12. Nezhad, M.S.F., Niaki, S.T.A., Jahromi, A.E.: A one-stage two-machine replacement strategy based on the Bayesian inference method. J. Ind. Syst. Eng. **1**(3), 235–250 (2007)

13. Zhao, X.F., Chen, M., Nakagawa, T.: Three kinds of replacement models combined with additive and independent damages. In: Proceedings of the Ninth International Symposium on Operations Research and Its Applications (ISORA '10), pp. 31–38 (2010)

14. Zadeh, L.A.: Fuzzy sets. Inf. Control **8**, 338–353 (1965)

High-Voltage Gain DC–DC Converter for Renewable Energy Applications

M. Kavitha and V. Sivachidambaranathan

Abstract Eco-friendly sources of energy like solar are available in abundance and operate at low voltage. The output photovoltaic voltage can be increased by passive clamp-based DC–DC converter which employs coupled inductor. The energy from the source is preserved in inductor and in intermediate condenser. The passive clamp supports in regaining the inductor leakage energy and thus improves the converter voltage gain. Passive clamp network also reducess the conduction losses in power switchs. The photovoltaic voltage of 34 V is boosted to 340 V by the converter. A prototype is modeled to verify the results of simulation. The simulation is carried out in MATLAB/Simulink, and the prototype is implemented.

Keywords Photovoltaic · Passive clamp · DC–DC converter · Coupled inductor · Bidirectional converter

1 Introduction

High-voltage gain converter is used for application such as fuel cell and photovoltaic. The converter may be isolated or non-isolated. The isolated converter using transformer of high frequency gives high-voltage gain but due to leakage inductance the switch suffers voltage spikes. This type of converters has high current ripples [1–3]. Though the input and load are isolated using transformer in a DC–DC converter, the cost of transformer becomes high, and the size is bulky [4]. To overcome the problems associated with isolated converter, the coupled coil-based converter is employed. The converter with coupled inductor reduces the current stress, power switch components with low rating that can be preferred [5]. Interleaved converter with coupled inductor was also proposed in [6–8]. However, the converter of coupled inductor may have high leakage inductance. It also produces voltage spikes when switches are turned

M. Kavitha (✉) · V. Sivachidambaranathan
Department of Electrical and Electronics Engineering, Sathyabama Institute of Science and Technology, Chennai, India
e-mail: kaveem@gmail.com

V. Sivachidambaranathan
e-mail: sivachidambaram_eee@yahoo.com

© Springer Nature Singapore Pte Ltd. 2020
P. K. Mallick et al. (eds.), *Cognitive Informatics and Soft Computing*,
Advances in Intelligent Systems and Computing 1040,
https://doi.org/10.1007/978-981-15-1451-7_67

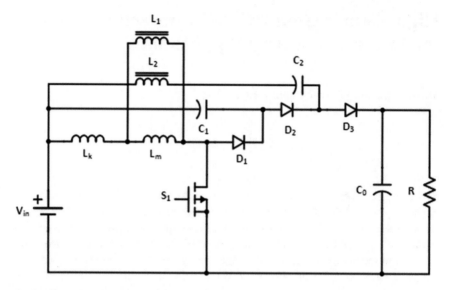

Fig. 1 High gain DC–DC converter

off and current spikes when switches are turned on, and therefore the efficiency of the converter is reduced. To overcome the above drawback, active clamp-based converter is used in which the leakage energy is recovered [1, 8]. But the active clamp network introduces switching conduction losses. Therefore, passive clamp-based converter is used which consists of resistor, capacitor and diode. The voltage stress of power switches is reduced [9].

For achieving a converter of high-voltage gain, coupled inductor with an intermediate capacitor is used [10–14]. The leakage energy is recovered by converter with coupled inductor which is connected in an interleaved manner along with passive clamp network and intermediate condenser [15].

In this paper, the passive clamp-based converter is implemented for photovoltaic application. This converter reduces the conduction loss of the switch. Improved voltage gain is obtained. The description of the converter and its parameter analysis and modes of operation is explained in Sect. 2. This parameter analysis of the converter helps the engineering students to design the passive clamp converter for any desired application, and they also can analyze the dynamic performance of the converter. Section 3 introduces the description of simulation circuit. Section 4 includes the simulation results of passive clamp-based converter. Section 5 includes the hardware implantation. Conclusion is given in Sect. 6.

2 High Gain DC–DC Converter

Figure 1 displays the high gain DC–DC converter. The converter consists of coupled inductor, output capacitor C_0 and intermediate capacitor C_2. L_1 and L_2 are the inductance of the coupled coil primary and secondary, respectively. The feedback diode D_2 and capacitor C_2 is connected to the secondary of the coupled inductor. The passive clamp circuit comprises diode D_1 and condenser C_1. The bidirectional converter is used when power flow in either direction is required [16–19]. The isolated bidirectional converter charges a low-voltage battery during buck mode and discharges a high voltage to load during boost mode. Active Power filter [20] and AC link converter are given in [21], and series resonant converter with half bridge circuit is given [22]. Buck–boost converter using fuzzy-based technique is given [23]. Parallel quasi resonant converter topology is explained in [24]. High-frequency isolated series–parallel resonant converter is given in [25]. Transformerless two-phase interleaved high gain DC converter is given in [26]. Bidirectional series–parallel resonant converter for power factor correction is given in [27]. Half bridge series resonant PFC DC–DC converter is given in [28]. In the present work, passive clamp-based DC–DC converter is given.

The five different operational modes are explained below:

A. *Mode 1*

When S_1 the power switch is excited, the inductor L_1 and L_m receive current through switch. Figure 2a represents operating mode 1. Passive clamp diode D_1 and feedback diode D_2 are reverse biased. The condenser C_1 and C_2 and coil L_2 receive current when diode D_2 is forward biased. The condenser C_2 voltage is given by $V_{C2} = V_{L2} + V_{C1}$. The feedback diode is reverse biased.

B. *Mode 2*

The magnetizing current circulates through L_1. The switch S_1 is turned ON. The diode D_2 is forward biased. Therefore, the current flows through L_2, C_1 and C_2. Figure 2b shows the mode 2 operation.

C. *Mode 3*

The clamp diode D_1 and output diode D_3 are forward biased. The diode D_2 is in reverse bias, and the current of diode D_2 falls zero. When the diode D_1 is forward biased, the leakage energy is stored in C_1 via D_1. Thus, the leakage energy is retrieved. The input energy is discharged to the load through diode D_3. Figure 2c represents the operating mode 3.

D. *Mode 4*

This mode commences when the inductor energy is recovered. The diodes D_1 and D_3 are in forward bias. The current flows from supply to load. Figure 2d gives the mode 4 operation.

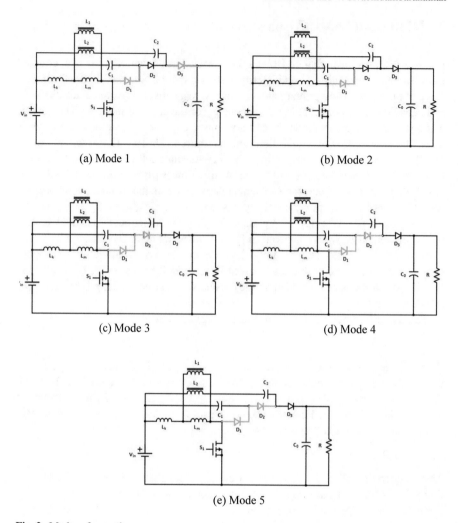

(a) Mode 1   (b) Mode 2

(c) Mode 3   (d) Mode 4

(e) Mode 5

Fig. 2 Modes of operation

E. *Mode 5*

The switch S_1 is excited. The current I_m is used for magnetizing the leakage inductor. Figure 2e shows the mode 5 operation. The clamp diode D_1 and feedback D_2 diode are in forward biased. This mode gets terminated when diode D_3 is in reverse bias. The direction of L_2 current is varied. The current flows through secondary of the inductor. Since the changes in current is not permitted in inductor, I_{L2} rises slowly. The switching voltage decreases slowly. Due to rise of current and fall of voltage, the switching losses are less.

F. *Analysis of parameters*

The various parameters such as voltage gain, switching stress and capacitor voltage [15] are discussed in this section.

1. *Voltage Gain*

When the power switch S_1 is excited, the voltage across the inductive coil is

$$V_{L1(on)} = V_{in} \tag{1}$$

$$V_{L2} = V_{C2} - V_{C1} \tag{2}$$

$$V_{L2} = n V_{in} \tag{3}$$

When S_1 switch is unexcited, then

$$V_{L1(off)} = -V_{c1} \tag{4}$$

$$V_{L2} = V_{in} + V_{C2} - V_O \tag{5}$$

From Eq. (2),

$$V_{C2} = V_{C1} + V_{L2} \tag{6}$$

Substitute Eqs. (3) and (4) in (6)

$$V_{C2} = n V_{in} - V_{L1(off)} \tag{7}$$

Substitute V_{C2} in (5)

$$V_{L2} = V_{in} + n V_{in} - V_{L1(off)} - V_O \tag{8}$$

We know that

$$n = \frac{V_{L2}}{V_{L1}}$$

Therefore, $V_{L1} = \frac{V_{L2}}{n}$

$$V_{L1(off)} = \frac{V_{in} + n V_{in} - V_{L1(off)} - V_O}{n} \tag{9}$$

$$V_{L1(off)} = \frac{V_{in} + n V_{in} - V_O}{n + 1} \tag{10}$$

Taking voltage-sec equation across the L_1 inductor is

$$V_{L1(on)} D + V_{L1(off)} (1 - D) = 0 \tag{11}$$

$$V_{in} D + \frac{V_{in} + n V_{in} - V_O}{n + 1} (1 - D) = 0 \tag{12}$$

$$\frac{V_o}{V_{in}} = \frac{n + 1}{1 - D} \tag{13}$$

2. *Power Switch Voltage*

The switch voltage is represented as

$$V_{SW} = -V_{L1(off)} + V_{in}$$

$$V_{SW} = \frac{V_{in}}{1 - D} \tag{14}$$

3. *Capacitor Voltages*

The clamp capacitor C_1 voltage is specified as

$$V_{C1} = \frac{D V_{in}}{1 - D} \tag{15}$$

The voltage across the intermediate capacitor C_2 is given by

$$V_{C2} = V_{L2} + V_{C1}$$

$$V_{C2} = n V_{in} + \frac{D V_{in}}{1 - D} \tag{16}$$

3 Description of Simulation Circuit

Figure 3 shows the simulation circuit of high gain DC–DC converter which uses photovoltaic as an input source. The production power of photovoltaic module is intermittent. Since the PV voltage is varying in nature, algorithm such as Perturb and Observe is applied for tracking the maximum PV output power. The bidirectional converter is used when power flow in either direction is required. The isolated bidirectional converter charges a low-voltage battery during buck mode and discharges a high voltage to load during boost mode. The load can either be supplied by photovoltaic module or by battery through bidirectional converter.

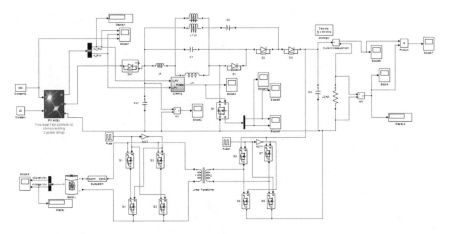

Fig. 3 Simulation circuit for high gain DC–DC converter

4 Simulation Results

The PV module is connected to the converter with the irradiance and temperature of 1000 W/m^2 and 20 °C, respectively. The output voltage across PV is 34 V which is fed to the passive clamp-based DC–DC converter. The converter steps up the voltage to 340 V with the output current of 0.7 A. The power delivered to the load is 250 V. Due to voltage fall and current rise in the inductor L_2, the stress across the switch is found to be 120 V.

Figure 4 shows the output voltage of the photolvoltaic. This voltage is given to the converter. Figure 5 displays the switching pulse with the switching frequency of 50 kHz. Figure 6 shows the switching stress. Figure 7 and Fig. 8 explain the waveform of current and voltage of the storage element, respectively. Figures 9, 10 and 11 represent the voltage waveform, current waveform and power delivered to the load.

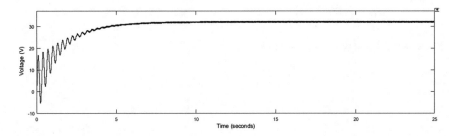

Fig. 4 PV output voltage

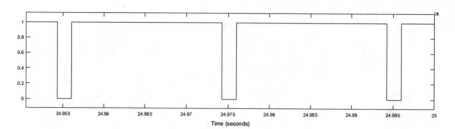

Fig. 5 Switching pulse

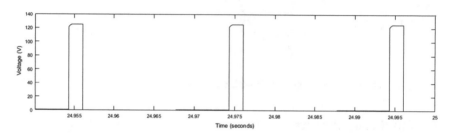

Fig. 6 Voltage stress of switch S_1

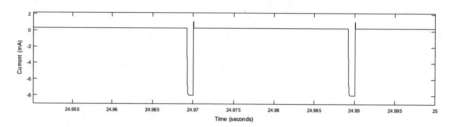

Fig. 7 Waveform of storage element current

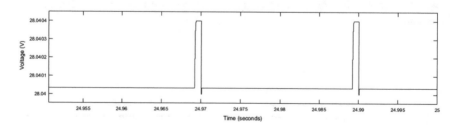

Fig. 8 Waveform of storage element voltage

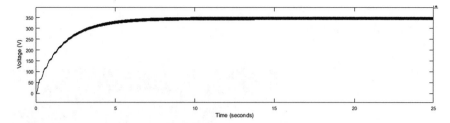

Fig. 9 Load voltage waveform

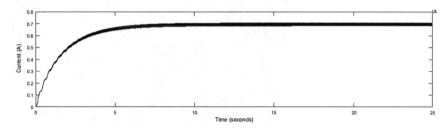

Fig. 10 Load current waveform

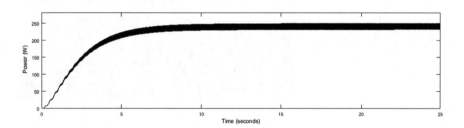

Fig. 11 Power delivered to the load

5 Hardware Description

A prototype is constructed to validate the simulation results. The experimental implementation is shown in Fig. 12. The experimental setup consists of a power supply unit, PIC microcontroller, gate driver circuit, battery and a resistive load. A step down transformer of 230 V/12 V is connected to bridge rectifier, which provides 12 V DC. The rectified DC 12 V supply is given to IC 7805 voltage regulator, in order to attain 5 V DC constant voltage. This voltage is in turn supplied to PIC16F887 microcontroller. The control signal is created by microcontroller which is not sufficient to activate the switches; therefore, driving switching pulses were produced by IR2112 optocoupler. The switches present in converter and bidirectional converter receive the pulses through respective gate driver unit. Figure 13 shows the switching pulse given to passive clamped converter. Based on the duty ratio, the converter output voltage is obtained. The input 12 V DC supply is boosted to 88 V using the converter. The converter voltage output is measured using the multimeter and shown in Fig. 14.

Fig. 12 Experimental setup

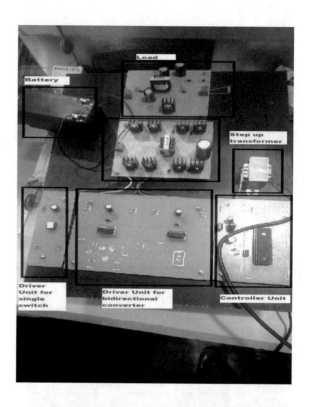

Fig. 13 Gate pulse given to converter switch

Fig. 14 Voltage delivered to the load

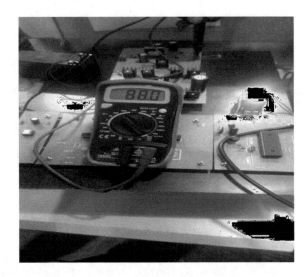

6 Conclusion

The output voltage of photovoltaic module is increased with high gain DC–DC converter. The passive clamp supports in regaining the inductor leakage energy, thus improves the converter voltage gain. The voltage fall of the switch and inductor current I_{L2} rise causes less switching stress. The isolated bidirectional converter charges a low-voltage (12 V) battery during buck mode and discharges a high voltage to load during boost mode. The PV voltage of 34 V is boosted to 340 V. The switching stress was found to be 120 V. A prototype is fabricated to validate the simulation results.

References

1. Lee, J.-H., Liang, T.-J., Chen, J.-F.: Isolated coupled-inductor-integrated DC–DC converter with nondissipative snubber for solar energy applications. IEEE Trans. Ind. Electron. **61**(7), 3337–3348 (2014)
2. Choi, C.T., Li, C.K., Kok, S.K.: Modeling of an active clamp discontinuous conduction mode flyback converter under variation of operating conditions. In: Proceedings of the IEEE 1999 International Conference on Power Electronics and Drive Systems, 1999. PEDS'99, vol. 2, pp. 730–733. IEEE (1999)
3. Xuewei, P., Rathore, A.K.: Novel bidirectional snubberless naturally commutated soft-switching current-fed full-bridge isolated DC/DC converter for fuel cell vehicles. IEEE Trans. Ind. Electron. **61**(5), 2307–2315 (2014)
4. Prudente, M., Pfitscher, L.L., Emmendoerfer, G., Romaneli, E.F., Gules, R.: Voltage multiplier cells applied to non-isolated DC–DC converters. IEEE Trans. Power Electron. **23**(2), 871–887 (2008)

5. Kavitha, M., Sivachidambaranathan, V.: Transformer less inverter using unipolar sinusoidal pulse width modulation technique for grid connected photovoltaic power system. Int. J. Appl. Eng. Res. **10**(2), 3089–3100 (2015). ISSN 0973-4562

6. Ramesh Babu, A., Raghavendran, T.A., Sivachidambaranathan, V., Barnabas Paul Glady, J.: Novel cascaded H-bridge sub-multilevel inverter with reduced switches towards low total harmonic distortion for photovoltaic application. Int. J. Ambient Energy 1–5 (2017). ISSN 0143-0750 (Print) 2162-8246 (Online)

7. Saravanan, M., Ramesh Babu, A.: High power density multi-mosfet-based series resonant inverter for induction heating applications. Int. J. Power Electron. Drive Sys. (IJPEDS) **7**(1), 107–113 (2016). ISSN 2088-8694

8. Ramesh, B.A., Raghavendiran, T.A.: Analysis of non-isolated two phase interleaved high voltage gain boost converter for PV application. In: Control, Instrumentation, Communication and Computational Technologies (ICCICCT), 2014 International Conference, pp. 491–496. IEEE (2014)

9. Alcazar, Y.J.A., de Souza Oliveira, D., Tofoli, F.L., Torrico-Bascopé, R.P.: DC–DC nonisolated boost converter based on the three-state switching cell and voltage multiplier cells. IEEE Trans. Ind. Electron. **60**(10), 4438–4449 (2013)

10. Li, W., Zhao, Y., Deng, Y., He, X.: Interleaved converter with voltage multiplier cell for high step-up and high-efficiency conversion. IEEE Trans. Power Electron. **25**(9), 2397–2408 (2010)

11. Zimny, J., Bielik, S., Michalak, P., Bojko, M.: The laboratory stand for measurements and analysis of photovoltaic modules. Int. J. Electr. Eng. Educ. **55**(2), 142–154 (2018)

12. Das, M., Agarwal, V.: A novel, high efficiency, high gain, front end DC–DC converter for low input voltage solar photovoltaic applications. In: Proceedings of IEEE IECON, pp. 5744–5749, 25–28 Oct 2012

13. Kavitha, M., Sivachidambaranathan, V.: Comparison of different control techniques for interleaved DC–DC converter. Int. J. Power Electron. Drive Syst. (IJPEDS) **9**(2), 641–647 (2018). ISSN 2088-8694

14. Balasubramanian, V., Senthil Nayagam, V., Pradeep, J.: Alleviate the voltage gain of high step-up DC–DC converter using quasi active switched inductor structure for renewable energy. In: 2017 International Conference on Computation of Power, Energy Information and Communication (ICCPEIC), pp. 835–841. IEEE (2017)

15. Das, M., Agarwal, V.: Design and analysis of a high-efficiency DC–DC converter with soft switching capability for renewable energy applications requiring high voltage gain. IEEE Trans. Ind. Electron. **63**(5), 2936–2944 (2016)

16. Indira, D,. Sivachidambaranathan, V., Dash, S.S.: Closed loop control of hybrid switching scheme for LLC series-resonant half-bridge DC–DC converter. In: Proceedings of the "Second International Conference on Sustainable Energy and Intelligent System" (SEISCON 2011), 20–22 July, pp. 295–298. IET Chennai and Dr. MGR University (2011)

17. Kavitha, M., Sivachidambaranathan, V.: PV based high voltage gain quadratic DC–DC converter integrated with coupled inductor. In: IEEE International Conference on Computation of Power, Energy Information and Communication (ICCPEIC), 20–21 Apr 2016, pp. 607–612. IEEE (2016). ISBN 978-1-5090-0901-5

18. Karthikeyan, P., Sivachidambaranathan, V.: Bidirectional buck–boost converter-fed DC drive. In: Artificial Intelligence and Evolutionary Computations in Engineering Systems (Proceedings of International Conference ICAIECES 2015), vol. 394, pp. 1195–1203. Springer, New Delhi (2016). ISBN: 978-81-322-2656-7

19. Averbukh, M.: Improved dimensionless nomograms approach in the electric drives and power electronics courses. Int. J. Electr. Eng. Educ. 0020720918776459 (2018)

20. Preeti Pauline Mary, M., Sivachidambaranathan, V.: Enhancement of active power filter operational performance using SRF theory for renewable source. Indian J. Sci. Technol. **8**(21), 71562, 1–7 (2015). ISSN 0974-6846

21. Preeti Pauline Mary, M., Sivachidambaranathan, V.: Design of new bi-directional three phase parallel resonant high frequency AC link converter. Int. J. Appl. Eng. Res. **10**(4), 8453–8468 (2015). ISSN 0973-4562

22. Sivachidambaranathan, V., Dash, S.S., Santhosh Rani, M.: Implementation of half bridge DC–DC converter using series resonant topology. Eur. J. Sci. Res. **74**(3), 381–388. ISSN: 1450–216X (2012)
23. Kavitha, M., Sivachidambaranathan, V.: Power factor correction in fuzzy based brushless DC motor fed by bridgeless buck boost converter. In: IEEE International Conference on Computation of Power Energy Information and Communication (ICCPEIC), 22–23 Mar 2017, pp. 549–553 (2017). ISSN: 978-1-5090-4324-8/17/$31.00 ©2017 IEEE
24. Geetha, V., Sivachidambaranathan, V.: A single switch parallel quasi resonant converter topology for induction heating application. Int. J. Power Electron. Drive Syst. (IJPEDS) **9**(4), 1718–1724 (2018)
25. Sivachidambaranathan, V.: High frequency isolated series parallel resonant converter. Indian J. Sci. Technol. **8**(15), 52311, 1–6 (2015). ISSN 0974-6846
26. Kavitha, M., Sivachidambaranathan, V.: Performance analysis of transformer-less two phase interleaved high gain DC converter using MPPT algorithm. Indian J. Sci. Technol. **8**(15), 61428, 1–8 (2015). ISSN 0974-6846
27. Sivachidambaranathan, V.: Bi-directional series parallel resonant converter for power factor correction. Int. J. Appl. Eng. Res. **9**(21), 10953–10961 (2014). ISSN 0973-4562
28. Sivachidambaranathan, V., Dash, S.S.: Simulation of half bridge series resonant PFC DC to DC converter. In: IEEE International Conference on "Recent Advances in Space Technology Services & Climate Change—2010" (RSTS&CC-2010), Nov 13–15, pp. 146–148. Sathyabama University in association with Indian Space Research Organisation (ISRO), Bangalore and IEEE, IEEE Explore (2010). ISBN 978-1-4244-9184-1

Integrated Voltage Equalizer Enhanced with Quasi-Z-Source Inverter for PV Panel Under Partial Shading

V. Sivachidambaranathan and A. Rameshbabu

Abstract The single-switch voltage equalizer with Quasi-Z-Source inverter for partial shading is proposed. The occurrence of multiple maximum power points in the PV system is settled, and the partial shading issue is focussed. Various parameters like output voltage of the converter, voltage gain and efficiency are analysed for the proposed system. The efficiency of the proposed system is 94% with a proper voltage balancing. The performance comparison between the ZSI and Q-ZSI is made validating the Q-ZSI for the PV application. The implementation of a 20 W system is made, and its results are tabulated.

Keywords Quasi-Z-Source inverter · Voltage equalizer · Partial shading

1 Introduction

The renewable energy plays a major role in all the fields like industries and domestic purposes. The power generation and balancing of the power for the usage for the system are to be taken care. The partial shading condition occurs when we have less amount of solar irradiance; this partial shading is taken care with a SEPIC converter for voltage balancing. The power loss between the long string and the short string based on the connection and on the photovoltaic modules due to partial shading is discussed in [1]. The different scheme for tracking the maximum power operating under different shading condition and also the voltage balancing of the designed system is taken care [2, 3]. The converters like buck/boost and multi-stacked buck/boost SEPIC converter are the circuits used for equalizing the voltage [4]. The inverter plays a major role in the industries and also in the domestic areas showing the proper coupling effect. There are different converters employed in the field of renewable energy conversion, but ZSI is the specific inverter used for its higher performance

V. Sivachidambaranathan (✉) · A. Rameshbabu
Department of EEE, Sathyabama Institute of Science and Technology, Chennai 600119, India
e-mail: sivachidambaram_eee@yahoo.com

A. Rameshbabu
e-mail: rameshbabuaa@gmail.com

P. K. Mallick et al. (eds.), *Cognitive Informatics and Soft Computing*,
Advances in Intelligent Systems and Computing 1040,
https://doi.org/10.1007/978-981-15-1451-7_68

671

in the system developed for distributed generated application, photovoltaic power system applications and electric vehicle application [5–11]. The controlling of the converter or the inverter can be done with the help of proper pulse-width modulation technique [12]. The current control method can make the system more stable and ripple free [13, 14]. DC-DC resonant converter for PFC using half-bridge technique is used for medium and high power applications [15]. High-frequency parallel resonant using AC link converter [16]. SRF theory using active power filter for renewable energy [17] resonant converter using hybrid switching scheme for DC-DC converter [18] buck/boost converter for DC motor application [19]. Quasi-resonant converter with single-switch [20] two-stage Quasi-Z-Source DC-DC converter technique [21]. Coupled inductor-based DC-DC converter [22]. Comparison of different control techniques for interleaved DC-DC converter [23]. High-frequency isolated series/parallel resonant converter [24]. Half-bridge DC-DC converter using series resonant topology [25]. The partial shading condition is depicted in Fig. 1.

The section of the paper is elaborated as in sections, description of the proposed system followed by simulation result of ZSI and proposed Q-ZSI system and the Q-ZSI experimental set-up for validating the simulated proposed system.

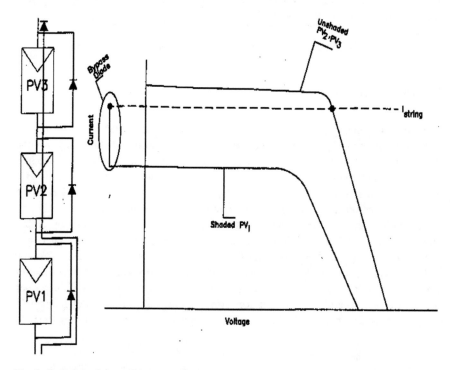

Fig. 1 Partial shaded condition

2 Description of the Proposed Quasi-ZSI

The block diagram depicts the overall operation of the proposed application for the specified loaded condition. Here there are three PV panels are given. The parameters such us voltage and the current are sent to the voltage equalizer to balance the voltage and to reduce or eliminate the unbalanced condition of the system. The Quasi-ZSI has introduced to have a reduced partial shading to drive the load in a proper and efficient manner. Figure 2 shows the brief note on the operation of the proposed system. The schematic diagram of the proposed converter with the load is shown as in Fig. 3.

The circuit diagram of the voltage equalizer circuit is developed, and the circuit of multi-stacked boost converter is replaced with the single-switch SEPIC for proper and efficient balancing of the system which is shown in Fig. 4. The added advantage of the system is that the converter has single switch, and the number modules can be connected to have a most controlled system. Here, the voltage-based control is preferred compared to the current-based one because the modules are mostly connected in series. This converter operates in both continuous and discontinuous modes considering the shading degree. When the shading degree is high, the continuous mode is used, and when it is low, discontinuous mode is used to have a voltage balancing in a perfect way to improve the performance.

The operation of the equalizer moves to T_{OFF} period as switch Q is turned off, and all the inductor currents start linearly decreasing. Applied voltages of inductors are equal to

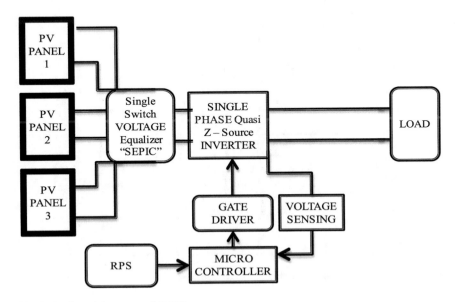

Fig. 2 Outline of the proposed Q-ZSI system

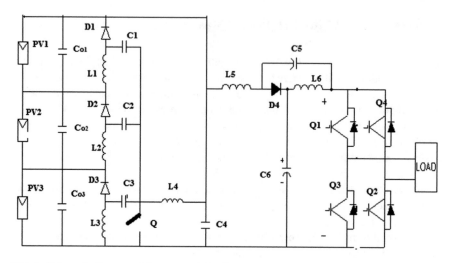

Fig. 3 Schematic diagram of the proposed Q-ZSI system

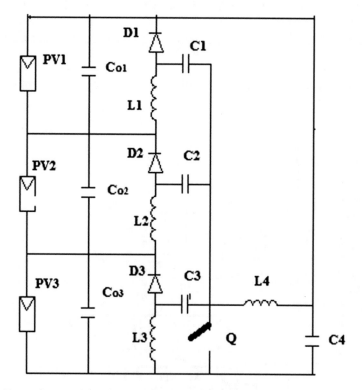

Fig. 4 Circuit diagram of the single-switch voltage equalizer

$$V_L = V_{PV}^* - V_D \tag{1}$$

where V_{PV}^* is the lowest voltage of shaded modules in the string and V_D is the forward voltage of diodes.

$$V_{PV}^* = \frac{D}{1-D}V_{string} - V_D \tag{2}$$

where D is the duty cycle of the switch Q. D is adjusted so that the voltage difference (ΔV) between the measured highest and lowest module voltages (V_H and V_L, respectively) in the string is to be certain fixed value.

The duty cycle 'D' can be yielded as

$$D = \frac{V_{PV}^* - +V_D}{V_{string} + V_{PV} + V_D} \tag{3}$$

The voltage equalizer is followed by Quasi-ZSI; it requires the proper switching so the carrier base PWM technique is chosen. Here, the saw tooth acts as the carrier wave, and the DC voltage acts as the modulating signal. Modification is made in the conventional PWM technique to include the additional shoot-through state. There are three ways of providing shoot-through states to the pulses. They are simple boost control (SBC), maximum boost control (MBC), and constant boost control (CBC). These strategies differ by the magnitude of the shoot-through line. In this paper, constant boost technique is implemented compared to the other two techniques

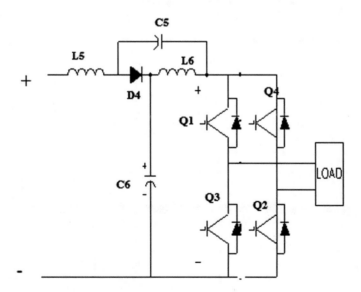

Fig. 5 Circuit description of the Q-ZSI

because of constant shoot-through duty ratio, reduced voltage stress, and better boost factor. Figure 5 shows the Quasi-ZSI employed with the applied load.

3 Simulation Results of the ZSI Inverter System

The simulation of the ZSI-based system developed is carried out with the help of the software MATLAB/Simulink; here, the PV panel is found to produce a voltage of 12 V each which is shown in Fig. 6. The PV voltage is sent to the voltage equalizer. Figure 7 shows the input voltage and current of the voltage equalizer. The equalizer

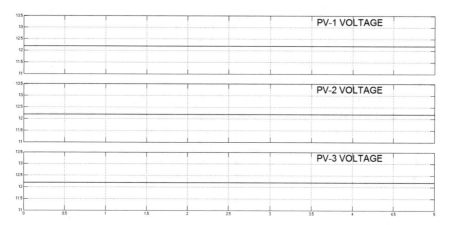

Fig. 6 Output voltage of the PV panels

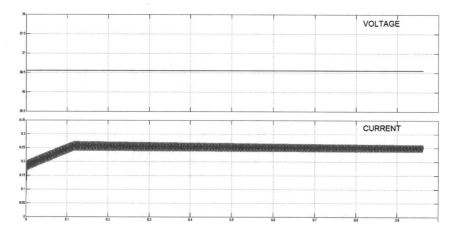

Fig. 7 Input voltage and current of voltage equalizer

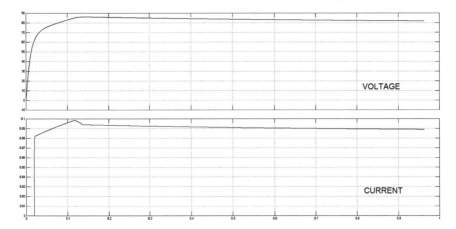

Fig. 8 Output voltage and current of voltage equalizer

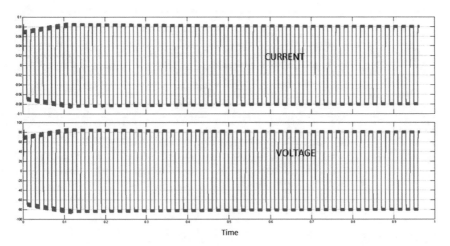

Fig. 9 Output voltage and current of ZSI of the system

produces an output voltage of 36.5 V and 0.25 A, which is revealed in Fig. 8. The measured output voltage of the ZSI is 82 V as depicted in Fig. 9.

4 Simulation Results of Proposed Quasi-ZSI System

The simulation of the proposed Q-ZSI-based system developed is carried out with the help of the software MATLAB/Simulink; here, the PV panel is found to produce a voltage of 12 V each which is shown in Fig. 10. The PV voltage is sent to the voltage equalizer. Figure 11 shows the input voltage and current of the voltage

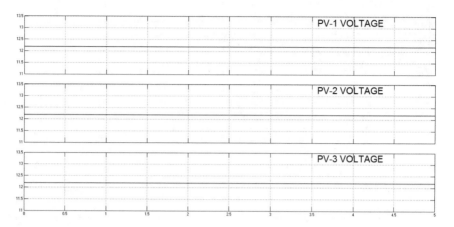

Fig. 10 Output voltage of the PV panel

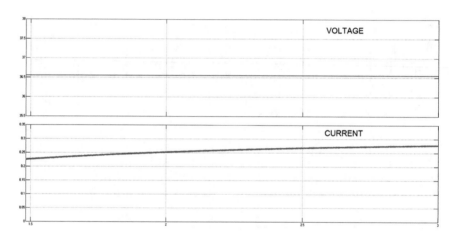

Fig. 11 Input voltage and current of equalizer for proposed Q-ZSI

equalizer. The equalizer produces the output voltage of 36.5 V and 0.25 A, which is revealed in Fig. 12. The measured output voltage of the Q-ZSI is 92 V as depicted in Fig. 13. The efficiency of the proposed Q-ZSI system is increased to be 94%. The measured quantities of the proposed system prove that the Quasi-ZSI shows the better performance, and the comparison of the results is shown and summarized in Table 1.

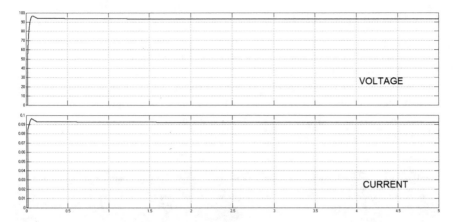

Fig. 12 Output voltage and current of equalizer for proposed Q-ZSI

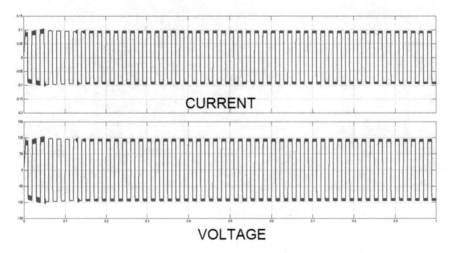

Fig. 13 Output voltage and current of proposed Q-ZSI

Table 1 Performance comparison

Parameter	ZSI	Q-ZSI
Equalizer output voltage (V)	36	36
Converter output voltage (V)	82	92
Voltage gain	2.2	2.6
Efficiency (%)	82	94

5 Experimental Set-up

The developed prototype model of the proposed Q-ZSI system is shown in Fig. 14. The result obtained from developed hardware is shown in Table 2. It is described in four cases, i.e., Case (i) no shading, Case (ii) PV panel no. 1 is shading, Case (iii) PV panel no. 2 is shading and Case (iv) PV panel no. 3 is shading. For all the cases, the output voltages of the proposed Q-ZSI are same. The measured value is 94 V. The output voltage measured is shown in Fig. 15.

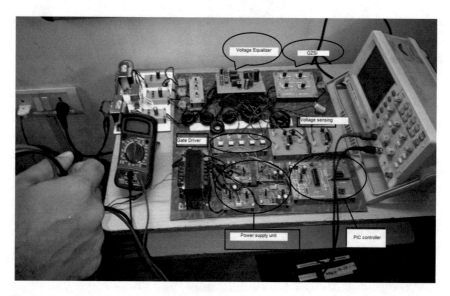

Fig. 14 Experimental set-up of the proposed system

Table 2 Validation of experimental set-up

Parameter	Measured value no shading (V)	Measured value PV panel no. 1 is shaded (V)	Measured value PV panel no. 2 is shaded (V)	Measured value PV panel no. 3 is shaded (V)
PV panel no. 1 voltage	12.5	9	12.5	12.5
PV panel no. 2 voltage	12.5	12.5	9.0	12.5
PV panel no. 3 voltage	12.5	12.5	12.5	9.0
Equalizer O/P voltage	34	33	33.5	33.5
Q-ZSI O/P voltage	94	94	94	94

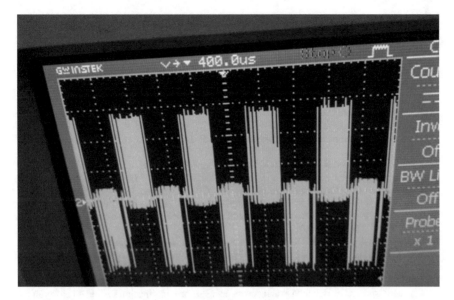

Fig. 15 Proposed Q-ZSI output voltage measured in CRO

6 Conclusion

Single-switch voltage equalizers with Q-ZSI for partially shaded PV modules have been analysed and implemented. The voltage equalizers can be derived by stacking CLD filters on SEPIC, and Q-ZSI is obtained by modifying the circuit diagram of Z-Source inverter. Both the SEPIC-based voltage equalizer with Z-Source inverter and SEPIC-based voltage equalizer with Q-ZSI simulation were carried out, and the simulation results were compared in terms of equalizer input voltage, voltage gain and efficiency. The results show that the SEPIC voltage equalizer with Q-ZSI shows the better performance than Z-Source inverter. According to the simulation result, SEPIC voltage equalizer with Q-ZSI prototype hardware kit was made for 20 W power output. The inverter output voltage for different values of input voltage to the equalizer which results from PV partial shading condition was verified successfully. Hence, the proposed PV-shaded SEPIC voltage equalizer with Q-ZSI is suitable for PV application.

References

1. Maki, A., Valkealahti, S.: Power losses in long string and parallel-connected short strings of series-connected silicon-based photovoltaic modules due to partial shading conditions. IEEE Trans. Energy Conv. **27**(1), 173–183 (2012)

2. Babu, A.R., Raghavendiran, T.A.: Performance enhancement of high voltage gain two phase interleaved boost converter using MPPT algorithm. J. Theoret. Appl. Inf. Technol. **68**(2), 360–368 (2014). ISSN 1992-8645
3. Inoue, T., Koizumi, H.: A voltage equalizer applying a charge pump for energy storage systems. In: European Conference on Circuit Theory and Design, Aug 2009, pp. 169–172
4. Uno, M., Kukita, A.: Single-switch voltage equalizer using multistacked buck-boost converters for partially shaded photovoltaic modules. IEEE Trans. Power Electron. **30**(6) (2015)
5. Babu, A.R., Raghavendiran, T.A.: Analysis of non-isolated two phase interleaved high voltage gain boost converter for PV application. In: 2014 International Conference on Control, Instrumentation, Communication and Computational Technologies (ICCICCT), pp. 491–496. IEEE (2014)
6. Babu, A.R., Raghavendiran, T.A.: Performance analysis of novel three phase high step-up dc-dc interleaved boost converter using coupled inductor. In: 2015 International Conference on Circuit, Power and Computing Technologies (ICCPCT), pp. 1–8. IEEE (2015)
7. Peng, F.Z.: Z-source inverter. IEEE Trans. Ind. Appl. **39**(2), 504–510 (2003)
8. Loh, P.C., Blaabjerg, F.: Magnetically coupled impedance-source inverters. IEEE Trans. Ind. Appl. **49**(5), 2177–2187 (2013)
9. Guo, F., Fu, L.X., Lin, C.H., Li, C., Choi, W., Wang, J.: Development of an 85 kW bidirectional quasi-Z-source inverter with DC-link feedforward compensation for electric vehicle applications. IEEE Trans. Power Electron. **28**(12), 5477–5488 (2013)
10. Battiston, A., Miliani, E.-H., Pierfederici, S., Meibody-Tabar, F.: Efficiency improvement of a quasi-Z-source inverter-fed permanent-magnet synchronous machine-based electric vehicle. IEEE Trans. Transport. Electrif. **2**(1), 14–23 (2016)
11. Li, Y., Jiang, S., Cintron-Rivera, J.G., Peng, F.Z.: Modeling and control of quasi-Z-source inverter for distributed generation applications. IEEE Trans. Ind. Electron. **60**(4), 1532–1541 (2013)
12. Ge, B., Abu-Rub, H., Peng, F.Z., Lei, Q., de Almeida, A.T., Ferreira, F.J.T.E., Sun, D., Liu, Y.: An energy-stored quasi-Z-source inverter for application to photovoltaic power system. IEEE Trans. Ind. Electron. **60**(10), 4468–4481 (2013)
13. Ayad, A., Hanafiah, S., Kennel, R.: A comparison of quasi-Z-source inverter and traditional two-stage inverter for photovoltaic application. In: Proceedings in International Exhibition and Conference for Power Electronics, Intelligent Motion, Renewable Energy and Energy Management, May 2015, pp. 1–8. Nuremberg, Germany
14. Liu, Y., Abu-Rub, H., Ge, B.: Z-source/quasi-Z-source inverters: derived networks, modulations, controls, and emerging applications to photovoltaic conversion. IEEE Ind. Electron. Mag. **8**(4), 32–44 (2014)
15. Sivachidambaranathan, V., Dash, S.S.: Simulation of half bridge series resonant PFC DC to DC converter. In: IEEE International Conference on "Recent Advances in Space Technology Services & Climate Change—2010" (RSTS&CC-2010), Nov 13–15, pp. 146–148. Sathyabama University in association with Indian Space Research Organisation (ISRO), Bangalore and IEEE, IEEE Explore (2010). ISBN 978-1-4244-9184-1
16. Preeti Pauline Mary, M., Sivachidambaranathan, V.: Design of new bi-directional three phase parallel resonant high frequency AC link converter. Int. J. Appl. Eng. Res. **10**(4), 8453–8468 (2015). ISSN 0973-4562
17. Preeti Pauline Mary, M., Sivachidambaranathan, V.: Enhancement of active power filter operational performance using SRF theory for renewable source. Indian J. Sci. Technol. **8**(21), 71562, 1–7 (2015). ISSN 0974-6846
18. Indira, D., Sivachidambaranathan, V., Dash, S.S.: Closed loop control of hybrid switching scheme for LLC series-resonant half-bridge DC-DC converter. In: Proceedings of the "Second International Conference on Sustainable Energy and Intelligent System" (SEISCON 2011), July 20–22, pp. 295–298. IET Chennai and Dr. MGR University (2011)
19. Kavitha, M., Sivachidambaranathan, V.: Power factor correction in fuzzy based brushless DC motor fed by bridgeless buck boost converter. In: IEEE International Conference on Computation of Power Energy Information and Communication (ICCPEIC), IEEE, 22–23 Mar 2017, pp. 549–553 (2017). ISSN: 978-1-5090-4324-8/17/$31.00 ©2017

20. Geetha, V., Sivachidambaranathan, V.: A single switch parallel quasi resonant converter topology for induction heating application. Int. J. Power Electron. Drive Syst. (IJPEDS) **9**(4), 1718–1724 (2018). ISSN 2088-8694
21. Revathi, N., Sivachidambaranathan, V.: Load resonant for step up DC-DC converter by two stages quasi Z source network. J. Chem. Pharm. Sci. (JCHPS), Special Issue 10, 114–121 (2015). ISSN 0974-2115
22. Kavitha, M., Sivachidambaranathan, V.: PV based high voltage gain quadratic DC-DC converter integrated with coupled inductor. In: IEEE International Conference on Computation of Power, Energy Information and Communication (ICCPEIC), 20–21 Apr 2016, pp. 607–612 (2016). ISBN 978-1-5090-0901-5
23. Kavitha, M., Sivachidambaranathan, V.: Comparison of different control techniques for interleaved DC-DC converter. Int. J. Power Electron. Drive Syst. (IJPEDS) **9**(2), 641–647 (2018). ISSN 2088-8694
24. Sivachidambaranathan, V.: High frequency isolated series parallel resonant converter. Indian J. Sci. Technol. **8**(15), 52311, 1–6 (2015). ISSN 0974-6846
25. Sivachidambaranathan, V., Dash, S.S., Santhosh Rani, M.: Implementation of half bridge DC to DC converter using series resonant topology. Eur. J. Sci. Res. **74**(3), 381–388 (2012). ISSN: 1450–216X

PV-Based Multiple-Input Single-Output DC–DC Luo Converter for Critical Load Application

A. Rameshbabu and V. Sivachidambaranathan

Abstract This paper proposes PV-based multiple-input single-output DC–DC Luo converter for critical load application. In this proposed work, the converter is simulated with PID controller and fuzzy logic controller, and the simulation results are analyzed. From the simulation results, the best performance is obtained from the fuzzy controller design. The two types of fuzzy logic controller are used for the proposed research, one for mode selection based on source strength and another for control the output voltage irrespective of changes in input and output load variation. The voltage-lift technique employed in this circuit converts the low DC input voltage to a high DC voltage with low ripple and high efficiency based on pulse-width modulated (PWM) control technique by controlling duty cycle. An integration of PV, grid, and backup battery storage device ensures the continuity and reliability of the power supply to the critical loads.

Keywords Luo converter · PID controller · Fuzzy controller · Voltage lift · Critical loads

1 Introduction

Among the available renewable energy resources, the photovoltaic energy is being widely utilized because of their abundance, low running cost, ease of installation, less maintenance, and sustainability to generate electricity, and also in present scenario, solar energy is one of the quite attractive pollution-free, essentially inexhaustible, and broadly available renewable energy source as a future energy supply [1, 2]. However, PV system has low output panel voltage and low power conversion efficiency [3]. The output of the PV depends on the variation of climatic condition. The output

A. Rameshbabu (✉) · V. Sivachidambaranathan
Department of Electrical and Electronics Engineering, Sathyabama Institute of Science and Technology, Chennai 600119, India
e-mail: rameshbabuaa@gmail.com

V. Sivachidambaranathan
e-mail: sivachidambaram_eee@yahoo.com

© Springer Nature Singapore Pte Ltd. 2020
P. K. Mallick et al. (eds.), *Cognitive Informatics and Soft Computing*,
Advances in Intelligent Systems and Computing 1040,
https://doi.org/10.1007/978-981-15-1451-7_69

of the PV is regulated using maximum power point tracking (MPPT). It is used to extract the maximum power from the PV system [4]. This paper presents the analysis of PV-based voltage-lift (VL) Luo converter hybrid with grid source and controlled with PID and fuzzy controller. In the modern technology of Luo DC–DC converter integrating the grid source, PV and battery source to meet our daily demand effectively and to get an uninterrupted power supply maintaining reliable power to critical loads [5–8] like radiation monitoring instruments (alpha, beta and gamma monitors), Fire alarm system, chemical process pumps, CCTV system, LED emergency exit lamps, telephone exchange load, Public address system relays in the actuators etc., and this method enables the stand alone operation. By using Luo converter, power density obtained is high with less ripple content in voltage and current profile [5–9]. Half-bridge series resonant PFC DC–DC converter is discussed in [10], closed loop control for LLC series resonant half-bridge DC–DC converter in [11], power factor correction in fuzzy-based brushless DC motor fed by bridgeless buck/boost converter in [12], single-switch parallel quasi resonant converter topology for induction heating application in [13], PV-based high voltage gain quadratic DC–DC converter integrated with coupled inductor in [14], and comparison of different control techniques for interleaved DC–DC converter in [15]. The output voltage increases in geometric progression, thereby efficiency of the converter increases. In voltage-lift Luo converter, its function and voltage output are better in compared with conventional boost step-up DC/DC converter. The output voltage maintains constant and steady value with fuzzy controller. The main features of the proposed system are given below,

- The operating cost of the system is low due to solar power.
- Reliability of the system is improved by the use of two sources and battery.
- Luo converter has high efficiency.
- Voltage ripple is less.
- This system enables standalone operation.

2 Existing System

Figure 1 shows the block diagram of the existing system. The critical load required continuous power supply from the source in order to maintain safety and security of the system. In general, the DC supply of the loads is fed from automatic constant voltage rectifier (ACVR). It consists of float charger (FC) and float-cum-boost charger (FCBC) along with battery bank. The FC and FCBC are fully controlled rectifiers in which they produce DC voltage in controlled manner. The DC supply maintained to the load depends upon the available source condition.

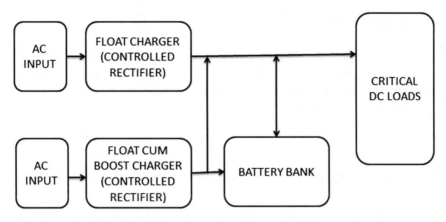

Fig. 1 Existing block diagram

The float charger is nothing but fully controlled rectifier, and it produces DC voltage. Under normal operating conditions, the FC fed the power to load and charges the battery bank with trickle charge near about milliampere. And its operating voltage is also restricted with minimum value based on the battery manufacture recommendations. The FCBC is also smaller to FC, but it can deliver more voltage compared with FC to fast charging of battery banks under fully discharged condition. In case of maintenance or fault on FC, the FCBC will feed power supply to load and act as float charger. However, during under boost charging mode, the FCBC and FC are isolated from each other to safeguard the critical load. Numbers of cells connected in series or parallel are known as battery banks. The voltage of the battery bank must be equal to the load voltage; hence, more numbers of cell are to be connected in series to meet the load. The cell may flood-type lead acid battery or sealed maintenance-free (SMF) battery depends upon the cost, space, and maintenance point of view by the customer and design requirement. In the event of both FC and FCBC supply not available, the energy stored in the battery bank is supplied to the critical loads. The disadvantages of the existing system listed below:

- High initial cost.
- FC and FCBC are capacity to be more.
- More number of batteries is required to get more voltage.
- Hence, more floor space also required.
- Maintenance is more when more batteries are connected as the bank.
- No voltage boost technique is employed under battery alone operation.
- Voltage regulation is poor.
- Standalone operation is not possible.

3 Proposed System

Figure 2 shows the block diagram of the proposed system with Luo converter. This Luo converter system comprises PV source, grid source, and a battery. Based on the availability and strength of the source, the operation of the Luo converter is classified into normal mode (PV only), alternate mode (grid only), and emergency mode (battery only).

3.1 Normal Mode of Operation

In this mode, PV source is high, grid is low or high, and battery is high or low; the output is obtained with PV source only. In case the battery is low, the predefined SOC (state of charge) on the controller then the PV charges the battery as well as delivers the power to the load irrespective of grid status.

3.2 Alternate Mode of Operation

When absence of PV power, based on the grid source and battery SOC, this alternate mode operates and provides boost voltage to the load and charges the battery.

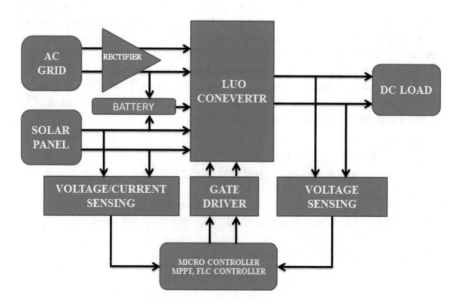

Fig. 2 Block diagram for proposed system with PV-based Luo converter

3.3 Emergency Mode of Operation

Under this mode, non-availability of both PV and grid source the energy stored in the battery is fed to the load till the any one source availability or complete discharging of battery. This mode is designed for short duration because higher the AH capacity, bulky and costly is the battery.

4 Negative-Lift Luo Converter

The circuit diagram of the lift Luo converter is shown in Fig. 3. The negative-lift Luo converter provides output voltage which is negative from positive source voltage. It consists of single switch (MOSFET), two diodes (D_1, D_2), an inductor (L), and two capacitors (C, C_o) with a DC source. Here C_o is the output capacitor, and R is the load resistance. When a pulse given to gate of the MOSFET and its turned ON, the diode D_1 is forward bias and D_2 is reverse bias, and the inductor (L) and current (I_L) increase with the slope of (V_i/L) for the period of KT, and capacitor (C) gets charge and load resistor gets the voltage from the output capacitor (C_o).

When a pulse removed from gate of the MOSFET and it is turned OFF, the diode D_2 is forward bias and D_1 is reverse bias, and the inductor (L) and current (I_L) decrease with the slope of (($V_0 - V_i$)/L) for the period of $1 - KT$, and capacitor (C)

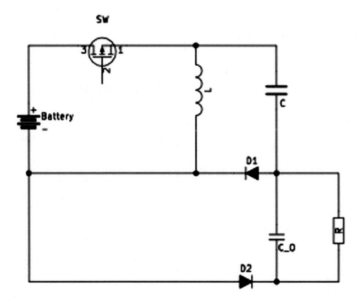

Fig. 3 Negative-lift Luo converter

also gets discharge. The energy from the inductor and capacitor charges the output capacitor and provides output voltage to the load.

Total time period

$$T = T_{\text{on}} + T_{\text{off}} \tag{1}$$

Duty cycle

$$K = \frac{T_{\text{on}}}{T} \tag{2}$$

With respect

$$T_{\text{on}} = KT \text{ and } T_{\text{off}} = (1 - K)T \tag{3}$$

The increasing current (ΔI_L) is equal to decreasing current.

At T_{on} period

$$\Delta I_L = \frac{V_i}{L} KT \tag{4}$$

At T_{off} period

$$\Delta I_L = \frac{V_o - V_i}{L}(1 - K)T \tag{5}$$

By equating the Eqs. (1) and (2), we get

Then

$$V_o = V_i \left(\frac{1}{1 - k} \right) \tag{6}$$

$$V_o = V_i \left(\frac{2 - K}{1 - K} - 1 \right) \tag{7}$$

$$\text{Gain} = \frac{V_o}{V_i} = \left(\frac{2 - K}{1 - K} - 1 \right) \tag{8}$$

5 Integrated Luo Converter with PV Grid and Battery

Figure 4 shows an integrated negative output super-lift Luo converter with PV source, grid source, and backup battery. As mentioned earlier, there are three mode operations obtained, and operation of the circuit is explained as follows with different conditions of the source strength.

5.1 Normal Mode of Operation

5.1.1 Mode I—(PV-High, Grid-High/Low, Battery-High)

T_{ON} period: In this mode, PWM signal is given to switch SW_1 which is ON and SW_4 which is ON, and all other switches are OFF. During time Ton period, diode D_6, D_1 is forward bias and L_1, L_2 and C charge with PV voltage, C_o provides output to load.

T_{OFF} period: SW_1 is OFF, and SW_4 is ON, the diode D_3 is forward bias, and diode D_1 is reverse bias. The stored energy on L_1, L_2, and C delivers the boost voltage to load and charges the output capacitor C_o.

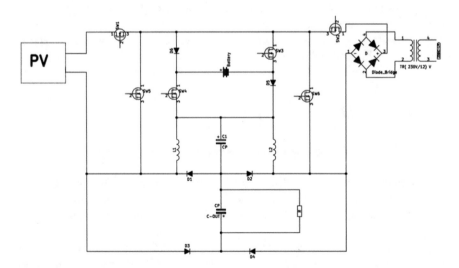

Fig. 4 Integrated negative output Luo converter

5.1.2   Mode II—Condition (PV-Medium, Grid-High/Low, Battery-High)

T_{ON} **period**: In this mode, PWM signal is given to switch SW_1 which is ON and SW_3, SW_4 which is ON, and all other switches are OFF. During time Ton period, diodes D_5 and D_6 are reverse bias with battery voltage since the PV voltage is medium level, D_1 is forward bias and L_1, L_2, and C charges with PV voltage, and C_o provides output to load.

T_{OFF} **period**: SW_1 is OFF, and SW_4 is ON SW_4, the diode D_3 is forward bias, and diode D_1 is reverse bias. The stored energy on L_1, L_2, and C delivers the boost voltage to load and charges the output capacitor C_o.

5.2   Functional Operation of Emergency Mode

5.2.1   Mode III—Condition (PV-Low, Grid-Low, Battery-High)

In this mode, the switches SW_5 and SW_6 act as operating switches turned ON and OFF by the PWM signal, and SW_3 and SW_4 act as control switches.

T_{ON} **period**: In this mode, PWM signal is given to switches SW_5, SW_6, and SW_4, SW_5 are simultaneously ON. Inductor L and capacitor C charge C_o supplies power to load, and the battery drains.

T_{OFF} **period**: SW_5, SW_6 are switched OFF. The stored energy on L_1, L_2, and C delivers the boost voltage to load and charges the output capacitor C_o.

5.3   Functional Operation of Alternate Mode

5.3.1   Mode IV—Condition (PV-Low, Grid-High, Battery-High)

T_{ON} **period**: In this mode, PWM signal is given to switch SW_2 which is ON and SW_3 is ON, and all other switches are OFF. During time Ton period, diodes D_5, D_2 are forward bias, L_1, L_2, and C charge with PV voltage, and C_o provides output to load.

T_{OFF} **period**: SW_2 is OFF, SW_3 is ON, the diode D_4 is forward bias, and diode D_2 is reverse bias. The stored energy on L_1, L_2, and C delivers the boost voltage to load and charges the output capacitor C_o.

Whenever the battery becomes low, it is disconnected from the actual circuit and charges separately with the grid source.

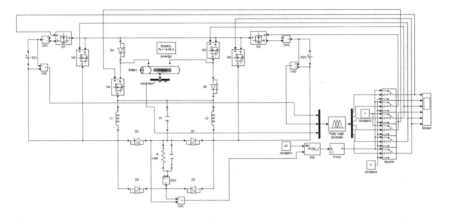

Fig. 5 Simulation diagram for PID and fuzzy controller

6 Simulation Result

6.1 PID and Fuzzy Controller

Figure 5 shows the simulation diagram of the converter with PID controller. There are two inputs given to the PID controller, one is actual voltage across the load in the Luo converter and other is the set voltage value. The controller compares the both values and gives error output. Depending on the error value, the switching frequency of the converter is varied. Figures 6 and 7 show the simulation results of PID controller; the graph shows the variation of voltage and current with respect to time.

6.2 Fuzzy Controller

Figure 8 shows the simulation diagram of the converter with fuzzy controller. Here the actual voltage and the set voltage are compared in a comparator, and the difference is stored in the memory block. The present error and the error already stored in the memory block are given to the fuzzy controller as two inputs. These inputs are converted into membership functions, and rules are formed by using the membership functions. Depending on the rules, the fuzzy controller gives output to the gate driver of the converter. Figures 9 and 10 show the simulation results of fuzzy controller; the graph shows the variation of voltage and current with respect to time.

It is shown from the graph some improved performance getting from the fuzzy controller, compared with the PID controller.

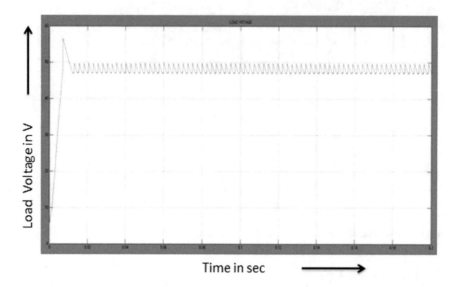

Fig. 6 Output voltage for PID and fuzzy controller

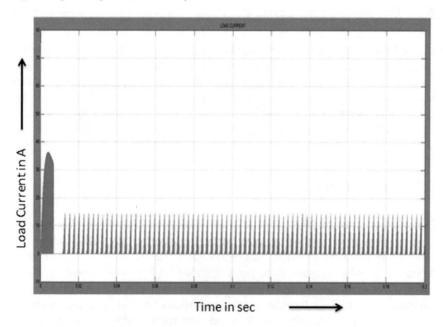

Fig. 7 Output current for PID and fuzzy controller

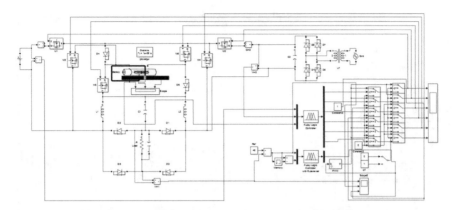

Fig. 8 Simulation diagram for fuzzy controller

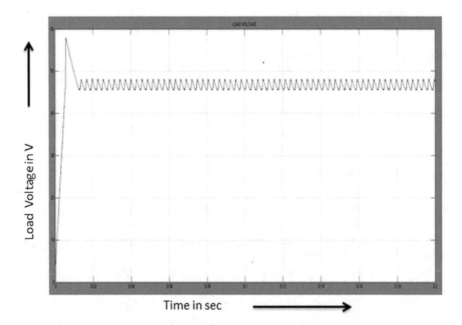

Fig. 9 Simulation output of fuzzy controller

6.3 Comparison Result of PID and Fuzzy Controller

From the Table 1, simulation results of the two controller are compared, and it clearly indicates that the performance of the fuzzy controller is better than other controller. Hence, the fuzzy-based controller is considered for hardware implementation.

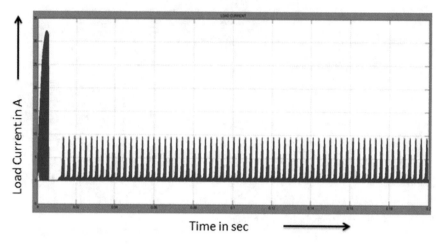

Fig. 10 Simulation output of fuzzy controller

Table 1 Comparison result of PID and fuzzy controller

Name of controller	Set value (V)	Peak over shoot (%)	Rise time (ms)	Settling time (ms)	% Error
PID	48	18.3	6.90	16	0.625
Fuzzy	48	22.0	5.65	14	2.290

7 Hardware Description and Result

Figure 11 shows the block diagram of the hardware components. The block diagram consists of transformers, PV input module, step-down transformer with rectifier for grid input, input and output parameter sensing unit, regulated power supply, gate driver, microcontroller, rechargeable battery, and Luo converter.

Source-I input is obtained from PV panel, and Source-II input obtained from the grid as AC source of 12 V 50 Hz power supply is achieved by step-down transformer. It is rectified as DC supply with bridge rectifier. Third 12 V, 3 Ah battery is also connected active part of the circuit. Power MOSFET devices, inductor and capacitor filter are connected to the power supply. These MOSFETs and filter circuits constitute Luo converter. Here the input power source is modified to a desired voltage of the load.

The regulated power supply block constitutes of IC7812 and 7912 and provides regulated power supply. The DC 12 V power supply uses to gate driver circuits and the DC 5 V power supply uses to microcontroller auxiliary supply. Power MOSFETs are triggered by the optical gate driver IC TLP250 through the control signal received from the microcontroller. The microcontroller follows controller instructions which are already programmed and generates 20 kHz PWM signal with different duty cycle based on the reference voltage set in the controller. This signal is sent to

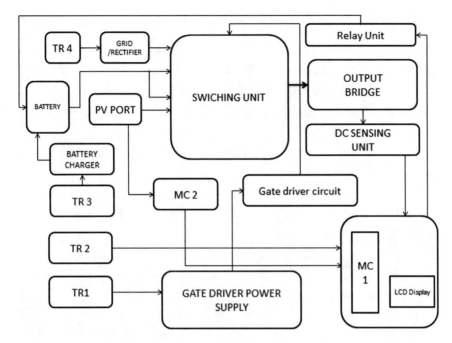

Fig. 11 Block diagram of hardware description

the optical gate driver and triggers the MOSFETs. The control logic used in the project is composed of fuzzy logic controller. This FLC coding is programmed in the microcontroller.

Figure 12 shows the actual hardware snapshot of the MISO Luo converter. This shows various printed circuit boards used for power and control circuits. Figure 13 shows the complete circuit diagram of the hardware. This shows the power MOSFETs are connected to the gate drivers and Luo converter. This also shows the microcontroller configured to the LCD and gate driver. The PV module used in this project is shown in Fig. 13, and its specification is given in Table 2. By changing the duty cycle of the Luo converter appropriately, the source and load impedance are matched together.

The Fig. 14 shows the hardware results were taken from the digital oscilloscope image, which shows the output voltage with respect to time, and it is, moreover, identical to the simulation results.

8 Conclusion and Future Scope

This research work deals with 'PV-based multiple-input single-output DC–DC Luo converter for critical load application.' A new multiple-input single-output

Fig. 12 Snapshot of developed hardware

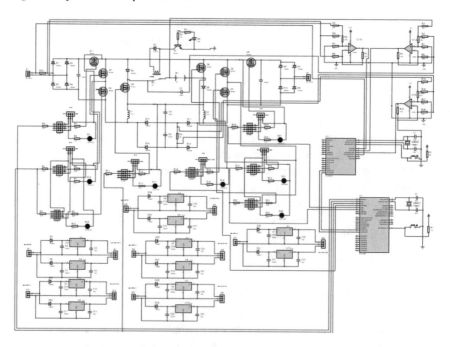

Fig. 13 Overall circuit diagram for developed hardware

Table 2 PV specification

Description	Value
Make	Microtek
Cell type	Polycrystalline
Module area	0.36 m^2
Power	50 W
Operating voltage	12 V
P_{max}	53.8 W
V_{mp}	18.9 V
I_{mp}	2.85 A
V_{oc}	22.2 V
I_{sc}	2.98 A

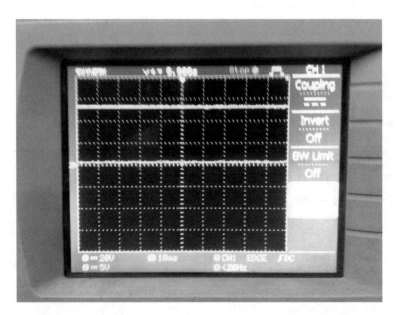

Fig. 14 Output voltage across load

Luo converter-based topology is proposed and simulated, and hardware results are obtained. It is also integrated with bidirectional battery port. It can be extended to interface number of sources having different characteristics. This converter has high-power handling capability, reduced number of passive elements, and ability to extract maximum power from each source makes it a competitive choice for standalone grid integration of renewable energy sources. FLC controller provides better controllability on the output voltage range under various modes of operation and ensures that reliable power feeding to any DC critical loads. The present work can be extended by increasing the more sources to improve more reliability, output voltage of the Luo converter is increased with the help of super voltage-lift technique, and switching losses of Luo converter are reduced by specially designed super-junction MOSFET.

References

1. Babu, A.R., Raghavendiran, T.A.: Performance analysis of novel three phase high step-up DC–DC interleaved boost converter using coupled inductor. In: 2015 International Conference on Circuit, Power and Computing Technologies (ICCPCT), pp. 1–8. IEEE (2015)
2. Babu, A.R., Raghavendiran, T.A.: Analysis of non-isolated two phase interleaved high voltage gain boost converter for PV application. In: 2014 International Conference on Control, Instrumentation, Communication and Computational Technologies (ICCICCT), pp. 491–496. IEEE (2014)
3. Babu, A.R., Raghavendiran, T.A.: High voltage gain multiphase interleaved DC–DC converter for DC micro grid application using intelligent control. Comput. Electr. Eng. **74**, 451–465 (2019). ISSN: 0045-7906
4. Babu, A.R., Raghavendiran, T.A.: Performance enhancement of high voltage gain two phase interleaved boost converter using MPPT algorithm. J. Theor. Appl. Inf. Technol. **68**(2), 360–368 (2014). ISSN 1992-8645
5. Luo, F.L., Ye, H., Rashid, M.H.: DC/DC conversion techniques and nine series Luo-converters. In: Power Electronics Handbook, pp. 335–406 (2001)
6. Luo, F.L., Ye, H.: Negative output super-lift converters. IEEE Trans. Power Electron. **18**(5), 1113–1121 (2003)
7. Shan, Z., Liu, S., Luo, F.L.: Investigation of a super-lift Luo-converter used in solar panel system. In: 2012 China International Conference on Electricity Distribution, pp. 1–4. IEEE (2012)
8. Kayalvizhi, R., Natarajan, S.P., Kavitharajan, V., Vijayarajeswaran, R.: TMS320F2407 DSP based fuzzy logic controller for negative output Luo re-lift converter: design, simulation and experimental evaluation. In: 2005 International Conference on Power Electronics and Drives Systems, vol. 2, pp. 1228–1233. IEEE (2005)
9. Behjati, H., Davoudi, A.: A multiple-input multiple-output DC–DC converter. IEEE Trans. Ind. Appl. **49**(3), 1464–1479 (2013)
10. Sivachidambaranathan, V., Dash, S.S.: Simulation of half bridge series resonant PFC DC to DC converter. In: IEEE International Conference on "Recent Advances in Space Technology Services & Climate Change—2010" (RSTS&CC-2010), Nov 13–15, pp. 146–148. Sathyabama University in association with Indian Space Research Organisation (ISRO), Bangalore and IEEE, IEEE Explore (2010). ISBN 978-1-4244-9184-1
11. Indira, D., Sivachidambaranathan, V., Dash, S.S.: Closed loop control of hybrid switching scheme for LLC series-resonant half-bridge DC–DC converter. In: Proceedings of the "Second International Conference on Sustainable Energy and Intelligent System" (SEISCON 2011), July 20–22, pp. 295–298. IET Chennai and Dr. MGR University (2011)
12. Kavitha, M., Sivachidambaranathan, V.: Power factor correction in fuzzy based brushless DC motor fed by bridgeless buck boost converter. In: IEEE International Conference on Computation of Power Energy Information and Communication (ICCPEIC), IEEE, 22–23 Mar 2017, pp. 549–553 (2017). ISSN: 978-1-5090-4324-8/17/$31.00 ©2017
13. Geetha, V., Sivachidambaranathan, V.: A single switch parallel quasi resonant converter topology for induction heating application. Int. J. Power Electron. Drive Syst. (IJPEDS) **9**(4), 1718–1724 (2018). ISSN 2088-8694

14. Selvamuthukumar, K., Satheeswaran, M., Babu, A.R.: Single phase thirteen level inverter with reduced number of switches using different modulation techniques. ARPN J. Eng. Appl. Sci. **10**(22), 10455–10462 (2015). ISSN 1819-6608
15. Saravanan, M., Babu, A.R.: High power density multi-mosfet-based series resonant inverter for induction heating applications. Int. J. Power Electron. Drive Syst. (IJPEDS) **7**(1), 107–113 (2016). ISSN: 2088-8694

Author Index

© Springer Nature Singapore Pte Ltd. 2020
P. K. Mallick et al. (eds.), *Cognitive Informatics and Soft Computing*,
Advances in Intelligent Systems and Computing 1040,
https://doi.org/10.1007/978-981-15-1451-7

Printed in the United States
By Bookmasters